T0292504

CAMBRIDGE LIBRARY COLLECTION

Books of enduring scholarly value

Physical Sciences

From ancient times, humans have tried to understand the workings of the world around them. The roots of modern physical science go back to the very earliest mechanical devices such as levers and rollers, the mixing of paints and dyes, and the importance of the heavenly bodies in early religious observance and navigation. The physical sciences as we know them today began to emerge as independent academic subjects during the early modern period, in the work of Newton and other 'natural philosophers', and numerous sub-disciplines developed during the centuries that followed. This part of the Cambridge Library Collection is devoted to landmark publications in this area which will be of interest to historians of science concerned with individual scientists, particular discoveries, and advances in scientific method, or with the establishment and development of scientific institutions around the world.

Baltimore Lectures on Molecular Dynamics and the Wave Theory of Light

The mathematical physicist and engineer William Thomson, 1st Baron Kelvin (1824–1904) is best known for devising the Kelvin scale of absolute temperature and for his work on the first and second laws of thermodynamics. The lectures in this collection demonstrate an attempt by Baron Kelvin to formulate a physical model for the existence of ether. This concept of a medium for light propagation became prominent in the late nineteenth century, arising from the combination of Maxwell's equations stating that light is an electromagnetic wave with the demands of Newtonian physics that light must move in a unique reference frame. First published in 1904, Kelvin's lectures describe the difficulties inherent in this model. These problems with the concept of ether are credited for inspiring Einstein to devise the theory of special relativity and the photoelectric effect, both of which are central to modern physics.

Cambridge University Press has long been a pioneer in the reissuing of out-of-print titles from its own backlist, producing digital reprints of books that are still sought after by scholars and students but could not be reprinted economically using traditional technology. The Cambridge Library Collection extends this activity to a wider range of books which are still of importance to researchers and professionals, either for the source material they contain, or as landmarks in the history of their academic discipline.

Drawing from the world-renowned collections in the Cambridge University Library, and guided by the advice of experts in each subject area, Cambridge University Press is using state-of-the-art scanning machines in its own Printing House to capture the content of each book selected for inclusion. The files are processed to give a consistently clear, crisp image, and the books finished to the high quality standard for which the Press is recognised around the world. The latest print-on-demand technology ensures that the books will remain available indefinitely, and that orders for single or multiple copies can quickly be supplied.

The Cambridge Library Collection will bring back to life books of enduring scholarly value (including out-of-copyright works originally issued by other publishers) across a wide range of disciplines in the humanities and social sciences and in science and technology.

Baltimore Lectures on Molecular Dynamics and the Wave Theory of Light

LORD KELVIN

CAMBRIDGE UNIVERSITY PRESS

Cambridge, New York, Melbourne, Madrid, Cape Town, Singapore,
São Paolo, Delhi, Dubai, Tokyo

Published in the United States of America by Cambridge University Press, New York

www.cambridge.org
Information on this title: www.cambridge.org/9781108007665

© in this compilation Cambridge University Press 2009

This edition first published 1904
This digitally printed version 2009

ISBN 978-1-108-00766-5 Paperback

BALTIMORE LECTURES

ON

MOLECULAR DYNAMICS

AND

THE WAVE THEORY OF LIGHT

London: C. J. CLAY AND SONS,
CAMBRIDGE UNIVERSITY PRESS WAREHOUSE,
AVE MARIA LANE.
Glasgow: 50, WELLINGTON STREET.

Leipzig: F. A. BROCKHAUS.
Bombay and Calcutta: MACMILLAN AND CO., Ltd.

BALTIMORE LECTURES

ON

MOLECULAR DYNAMICS

AND

THE WAVE THEORY OF LIGHT

FOUNDED ON MR A. S. HATHAWAY'S STENOGRAPHIC REPORT OF
TWENTY LECTURES DELIVERED IN JOHNS HOPKINS
UNIVERSITY, BALTIMORE, IN OCTOBER 1884:
FOLLOWED BY TWELVE APPENDICES ON ALLIED SUBJECTS

BY

LORD KELVIN, O.M., G.C.V.O., P.C., F.R.S., &c.

PRESIDENT OF THE ROYAL SOCIETY OF EDINBURGH,
FELLOW OF ST PETER'S COLLEGE, CAMBRIDGE,
AND EMERITUS PROFESSOR OF NATURAL PHILOSOPHY IN THE UNIVERSITY OF GLASGOW.

LONDON:
C. J. CLAY AND SONS,
CAMBRIDGE UNIVERSITY PRESS WAREHOUSE,
AVE MARIA LANE.

1904

𝕮𝖆𝖒𝖇𝖗𝖎𝖉𝖌𝖊:

PRINTED BY J. AND C. F. CLAY,

AT THE UNIVERSITY PRESS.

PREFACE.

HAVING been invited by President Gilman to deliver a course of lectures in the Johns Hopkins University after the meeting of the British Association in Montreal in 1884, on a subject in Physical Science to be chosen by myself, I gladly accepted the invitation. I chose as subject the Wave Theory of Light with the intention of accentuating its failures; rather than of setting forth to junior students the admirable success with which this beautiful theory had explained all that was known of light before the time of Fresnel and Thomas Young, and had produced floods of new knowledge splendidly enriching the whole domain of physical science. My audience was to consist of Professorial fellow-students in physical science; and from the beginning I felt that our meetings were to be conferences of coefficients, in endeavours to advance science, rather than teachings of my comrades by myself. I spoke with absolute freedom, and had never the slightest fear of undermining their perfect faith in ether and its light-giving waves: by anything I could tell them of the imperfection of our mathematics; of the insufficiency or faultiness of our views regarding the dynamical qualities of ether; and of the overwhelmingly great difficulty of finding a field of action for ether among the atoms of ponderable matter. We all felt that difficulties were to be faced and not to be evaded; were to be taken to heart *with the hope of solving them if possible*; but at all events with the certain assurance that there is an explanation of every difficulty though we may never succeed in finding it.

It is in some measure satisfactory to me, and I hope it will be satisfactory to all my Baltimore coefficients still alive in our world of science, when this volume reaches their hands; to find in it dynamical explanations of every one of the difficulties with which we were concerned from the first to the last of our twenty lectures of 1884. All of us will, I am sure, feel sympathetically interested in knowing that two of ourselves, Michelson and Morley, have by their great experimental work on the motion of ether relatively to the earth, raised the one and only serious objection* against our dynamical explanations; because they involve the assumption that ether, in the space traversed by the earth and other bodies of the solar system, is at rest absolutely except in so far as it is moved by waves of light or radiant heat or variations of magnetic force. It is to be hoped that farther experiments will be made; to answer decisively the great question:—is, or is not, ether at rest absolutely throughout the universe, except in so far as it is moved by waves generated by motions of ponderable matter? I cannot but feel that the true answer to this question is in the affirmative, in all probability: and provisionally, I assume that it is so, but always bear in mind that experimental proof or disproof is waited for. As far as we can be contented with this position, we may feel satisfied that all the difficulties of 1884, set forth in Lectures I, X, and XV, are thoroughly explained in Lectures XVIII, XIX, and XX, as written afresh in 1902 and 1903.

It seems to me that the next real advances to be looked for in the dynamics of ether are:—

(I) An explanation of its condition in the neighbourhood of a steel magnet or of an electromagnet; in virtue of which mutual static force acts between two magnets whether in void ether or in space occupied also by gaseous, liquid, or solid, ponderable matter.

(II) An investigation of the mutual force between two moving electrions, modified from purely Boscovichian repulsion; as it must be by the composition, with that force, of a force due to the inertia

* See Appendix A § 18 and Appendix B § 10.

of the ether set in motion by the motion of each of the electrions. It seems to me that, of these, (II) may be at present fairly within our reach ; but that (I) needs a property of ether not included in the mere elastic-solid-theory worked out in the present volume. My object in undertaking the Baltimore Lectures was to find how much of the phenomena of light can be explained without going beyond the elastic-solid-theory. We have now our answer : *every thing non-magnetic; nothing magnetic.* The so-called "electromagnetic theory of light" has not helped us hitherto : but the grand object is fully before us of finding a comprehensive dynamics of ether, electricity, and ponderable matter, which shall include electrostatic force, magnetostatic force, electromagnetism, electrochemistry, and the wave theory of light.

I take this opportunity of expressing the gratitude with which I remember the hearty and genial cooperation of my coefficients in our meetings of 19 years ago in Baltimore, and particularly the active help given me by the late Prof. Rowland, from day to day all through our work.

I desire also to specially thank one of our number, Mr A. S. Hathaway, for the care and fidelity with which he stenographically recorded my lectures, and gave his report to the Johns Hopkins University in the papyrograph volume published in December 1884. The first eleven lectures, as they appear in the present volume, have been printed from the papyrograph, with but little of even verbal correction; and with a few short additions duly dated.

Thirteen and a half years after the delivery of the lectures, some large additions were inserted in Lecture XII. In Lectures XIII, XIV, XV, freshly written additions supersede larger and larger portions of the papyrograph report, which still formed the foundation of each Lecture. Lectures XVI—XX have been written afresh during 1901, 1902, 1903.

In my work of the last five years for the present volume I have received valuable assistance successively from Mr W. Craig Henderson, Mr W. Anderson, and Mr G. A. Witherington; not

a 5

only in secretarial affairs, but frequently also in severe mathematical calculation and drawing; and I feel very grateful to them for all they have done for me.

The printing of the present volume began in August, 1885; and it has gone on at irregular intervals during the 19 years since that time; in a manner which I am afraid must have been exceedingly inconvenient to the printers.

I desire to thank Messrs J. and C. F. Clay and the Cambridge University Press for their never-failing obligingness and efficiency in working for me in such trying circumstances, and for the admirable care with which they have done everything that could be done to secure accuracy and typographical perfection.

KELVIN.

NETHERHALL,
January, 1904.

CONTENTS.

LECTURE III.

PART I.

LECTURE IV.

LECTURE V.

LECTURE VI.

LECTURE VII.

LECTURE VIII.

LECTURE IX.

LECTURE X.

LECTURE XI.

LECTURE XII.

LECTURE XIX.

RECONCILIATION BETWEEN FRESNEL AND GREEN.

LECTURE XX.

APPENDIX A.

ON THE MOTION PRODUCED IN AN INFINITE ELASTIC SOLID BY THE MOTION THROUGH THE SPACE OCCUPIED BY IT OF A BODY ACTING ON IT ONLY BY ATTRACTION OR REPULSION.

APPENDIX B.

Nineteenth Century Clouds over the Dynamical Theory of Heat and Light.

APPENDIX C.

On the Disturbance produced by two particular forms of Initial Displacement in an infinitely long Material System for which the velocity of Periodic Waves depends on the Wave-length.

APPENDIX D.

ON THE CLUSTERING OF GRAVITATIONAL MATTER IN ANY PART OF THE UNIVERSE.

APPENDIX E.

AEPINUS ATOMIZED.

APPENDIX F.

APPENDIX G.

HYDROKINETIC SOLUTIONS AND OBSERVATIONS.

APPENDIX H.

ON THE MOLECULAR TACTICS OF A CRYSTAL.

APPENDIX I.

ON THE ELASTICITY OF A CRYSTAL ACCORDING TO BOSCOVICH.

APPENDIX J.

MOLECULAR DYNAMICS OF A CRYSTAL.

APPENDIX K.

ON VARIATIONAL ELECTRIC AND MAGNETIC SCREENING.

APPENDIX L.

ELECTRIC WAVES AND VIBRATIONS IN A SUBMARINE TELEGRAPH WIRE.

INDEX

ERRATA.

Page 14, line 19, for "refracted" read "reflected."

,,   15, delete footnote.

,,   39, line 7 from foot, for "vibrations" read "vibration."

,,   46, line 7 from foot, for "V" read "v."

,,   48, in denominator of last term of equation (18), for "x^2" read "r^2."

,,   70, in heading, for "Part I." read "Part II."

,,   73, line 5, before "the" insert "the reciprocals of the squares of."

,,   74, lines 2 and 6, for "ϵ" read "t."

,,   75, in footnote, for "Appendix C" insert "Lect. XIX., pp. 408, 409," and delete "with illustrative curves."

,,   86, line 10 from foot, for "dx^2" read "dt^2."

,,   87, line 10, delete "that."

,,   87, line 3 from foot, for "$\nabla^2\phi - \dfrac{4\pi^2}{\lambda^2}\phi$" read "$\nabla^2\phi = -\dfrac{4\pi^2}{\lambda^2}\phi$."

,,   88, line 15, for "words painting" read "word-painting."

,,   90 and 92, in heading, for "Part I." read "Part II."

,,   103, last line, for "$\dfrac{1}{r^2}$" read "$\dfrac{1}{r^2}$."

,,   106, line 14, for "X" read "$\times$."

,,   239, for marginal "molecular" read "molar."

NOTES OF LECTURES

ON

MOLECULAR DYNAMICS

AND

THE WAVE THEORY OF LIGHT.

DELIVERED AT THE JOHNS HOPKINS UNIVERSITY, BALTIMORE,

BY

SIR WILLIAM THOMSON,

PROFESSOR IN THE UNIVERSITY OF GLASGOW.

STENOGRAPHICALLY REPORTED BY

A. S. HATHAWAY,

LATELY FELLOW IN MATHEMATICS OF THE JOHNS HOPKINS UNIVERSITY.

1884.

ADVERTISEMENT.

IN the month of October, 1884, Sir William Thomson of Glasgow, at the request of the Trustees of the Johns Hopkins University in Baltimore, delivered a course of twenty lectures before a company of physicists, many of whom were teachers of this subject in other institutions. As the lectures were not written out in advance and as there was no immediate prospect that they would be published in the ordinary form of a book, arrangements were made, with the concurrence of the lecturer, for taking down what he said by short-hand.

Sir William Thomson returned to Glasgow as soon as these lectures were concluded, and has since sent from time to time additional notes which have been added to those which were taken when he spoke. It is to be regretted that under these circumstances he has had no opportunity to revise the reports. In fact, he will see for the first time simultaneously with the public this repetition of thoughts and opinions which were freely expressed in familiar conference with his class. The "papyrograph" process which for the sake of economy has been employed in the reproduction of the lectures does not readily admit of corrections, and some obvious slips, such as Canchy for Cauchy, have been allowed to pass without emendation; but the stenographer has given particular attention to mathematical formulas, and he believes that the work now submitted to the public may be accepted, on the whole, as an accurate report of what the lecturer said.

A. S. HATHAWAY.

Dec., 1884.

LECTURE I.

5 P.M. WEDNESDAY, *Oct.* 1, 1884.

THE most important branch of physics which at present makes demands upon molecular dynamics seems to me to be the wave theory of light. When I say this, I do not forget the one great branch of physics which at present is reduced to molecular dynamics, the kinetic theory of gases. In saying that the wave theory of light seems to be that branch of physics which is most in want, which most imperatively demands, applications, of molecular dynamics just now, I mean that, while the kinetic theory of gases is a part of molecular dynamics, is founded upon molecular dynamics, works wholly within molecular dynamics, to it molecular dynamics is everything, and it can be advanced solely by molecular dynamics; the wave theory of light is only beginning to demand imperatively applications of that kind of dynamical science.

The dynamics of the wave theory of light began very molecularly in the hands of Fresnel, was continued so by Cauchy, and to some degree, though much less so, in the hands of Green. It was wholly molecular dynamics, but of an imperfect kind in the hands of Fresnel. Cauchy attempted to found his mathematical investigations on a rigorous molecular treatment of the subject. Green almost wholly shook off the molecular treatment, and worked out all that was to be worked out for the wave theory of light, by the dynamics of continuous matter. Indeed, I do not know that it is possible to add substantially to what Green has done in this subject. Substantial additions are scarcely to be made to a thing perfect and circumscribed as Green's work is, on the explanation of the propagation of light, of the refraction and the reflection of light at the bounding surface of two different mediums, and of the propagation of light through crystals, by a strict mathematical treatment, founded on the consideration of homogeneous elastic matter. Green's treatment is really complete in this respect, and there is nothing essential to be added to it. But there is a great deal of exposition wanting to let us make it our own. We must study it; we must try to see what there is in the very concise and sharp treatment, with some very long formulas, which we find in Green's papers.

The wave theory of light, treated on the assumption that the medium through which the light is propagated is continuous and homogeneous, except where distinctly separated by a bounding inter-face between two different mediums, is really completed by Green. But there is a great deal to be learned from that kind of treatment that perhaps scarcely has yet been learned, because the subject has not been much studied nor reduced to a very popular form hitherto.

Cauchy seemed unable to help beginning with the consideration of discreet particles mutually acting upon one another. But, except in his theory of dispersion, he virtually came to the same thing somewhat soon in his treatment every time he began it afresh, as if he had commenced right away with the consideration of a homogeneous, elastic solid. Green preceded him, I believe, in this subject. I have read a statement of Lord Rayleigh that there seems to have been an attributing to Cauchy of that which Green had actually done before. Green had exhausted the subject; but there is I believe no doubt that Cauchy worked in a wholly independent way.

What I propose in this first Lecture—we must have a little mathematics, and I therefore must not be too long with any kind of preliminary remarks—is to call your attention to the outstanding difficulties. The first difficulty that meets us in the dynamics of light is the explanation of dispersion; that is to say, of the fact that the velocity of propagation of light is different for different wave lengths or for light of different periods, in one and the same medium. Treat it as we will, vary the fundamental suppositions as much as we can, as much as the very fundamental idea allows us to vary them, and we cannot force from the dynamics of a homogeneous elastic solid a difference of velocity of wave propagation for different periods.

Cauchy pointed out that if the sphere of action of individual molecules be comparable with the wave lengths, the fact of the difference of velocities for different periods or for different wave lengths in the same medium is explained. The best way, perhaps, of putting Cauchy's fundamental explanation is to say that there is heterogeneousness through space comparable with the wave length in the medium,—that is, if we are to explain dispersion by Cauchy's unmodified supposition. We shall consider that, a little later. I have no doubt the truth is perfectly familiar already to many of you that it is essentially insufficient to explain the facts.

Another idea for explaining dispersion has come forward more recently, and that is the assumption of molecules loading the luminiferous ether and somehow or other elastically connected with it. The first distinct statement that I have seen of this view is in Helmholtz's little paper on anomalous dispersion. I shall have occasion to speak of that a good deal and to mention other names whom Helmholtz quotes in this respect, so that I shall say nothing about it historically, except that there we have in Helmholtz's paper, and by some German mathematicians who preceded him, quite another departure in respect to the explanation of dispersion. The Cauchy hypothesis gives us something comparable with the wave length in the geometrical dimensions of the body. Or, to take a crude matter of fact view of it, let us say the ratio of the distance from molecule to molecule (from the centre of one molecule to the centre of the next nearest molecule) to the wave length of light is the fundamental characteristic, as it were, to which we must look for the explanation of dispersion upon Cauchy's theory.

We may take this fundamental idea in connection with the two hypotheses for accounting for dispersion: that we must have, in the very essence of the ponderable medium, some relation either to wave length or to period, and it seems at first sight (although this is a proposition that may require modification) that with very long waves the velocity of propagation should be independent of the period or wave length. That, at all events, seems to be the case when the subject is only looked upon according to Cauchy's view. We are led to say then that it seems that for very long waves there should be a constant velocity of propagation. Experiment and observation now seem to be falling in very distinctly to affirm the conclusions that follow from the second hypothesis that I alluded to to account for dispersion. In this second hypothesis, instead of having a geometrical dimension in the ponderable matter which is comparable with the wave length, we have a fundamental time-relation—a certain definite interval of time somehow ingrained in the constitution of the ponderable matter, which is comparable with the period. So that, instead of a relation of length to length, we have a relation of time to time.

Now, how are we to get our time element ingrained in the constitution of matter? We need scarcely put that question now-a-days. We are all familiar with the time of vibration of the sodium atom, and the great wonders revealed by the spectroscope are all full of indications showing a relation to absolute intervals of time in the properties of matter. This is now so well understood, that it is no new idea to propose to adopt as our unit of time one of the fundamental periods—for instance, the period of vibration of light in one or other of the sodium D lines. You all have a dynamical idea of this already. You all know something about the time of vibration of a molecule. and how, if the time of vibration of light passing through any substance is nearly the same as the natural time of vibration of the molecules of the substance, this approximate coincidence gives rise to absorption. We all know of course, according to this idea, the old dynamical explanation, first proposed by Stokes, of the dark lines of the solar spectrum.

We have now this interesting point to consider, that if we would work out the idea of dispersion at all, we must look definitely to times of vibration, in connection with all ponderable matter. To get a firm hypothesis that will allow us to work on the subject, let us imagine space, otherwise full of the luminiferous ether, to be partially occupied by something different from the general luminiferous ether. That something might be a portion of denser ether, or a portion of more rigid ether; or we might suppose a portion of ether to have greater density and greater rigidity, or different density and different rigidity from the surrounding ether. We will come back to that subject in connection with the explanation of the blue sky, and, particularly, Lord Rayleigh's dynamics of the blue sky. In the meantime, I want to give something that will allow us to bring out a very crude mechanical model of dispersion.

In the first place, we must not listen to any suggestion that we are to look upon the luminiferous ether as an ideal way of putting the thing. A real matter between us and the remotest stars I believe there is, and that light consists of real motions of that matter, motions just such as are described by Fresnel and Young, motions in the way of transverse vibrations. If I knew what the electro-magnetic theory of light is, I might be able to think of it in relation to the fundamental principles of the wave theory of light. But it seems to me that it is rather a backward step from an absolutely definite mechanical notion that is put before us by Fresnel and his followers to take up the so-called Electro-magnetic theory of light in the way it has been taken up by several writers of late. In passing, I may say that the one thing about it that seems intelligible to me, I do not think is admissible. What I mean is, that there should be an electric displacement perpendicular to the line of propagation and a magnetic disturbance perpendicular to both. It seems to me that when we have an electro-magnetic theory of light, we shall see electric displacement as in the direction of propagation, and simple vibrations as described by Fresnel with lines of vibration perpendicular to the line of propagation, for the motion actually constituting light. I merely say this in passing, as perhaps some apology is necessary for my insisting upon the plain matter-of-fact dynamics and the true elastic solid as giving what seems to me the only tenable foundation for the wave theory of light in the present state of our knowledge.

The luminiferous ether we must imagine to be a substance which so far as luminiferous vibrations are concerned moves as if it were an elastic solid. I do not say it is an elastic solid. That it moves as if it were an elastic solid in respect to the luminiferous vibrations, is the fundamental assumption of the wave theory of light.

An initial difficulty that might be considered insuperable is, how can we have an elastic solid, with a certain degree of rigidity pervading all space, and the earth moving through it at the rate the earth moves around the sun, and the sun and solar system moving through it at the rate in which they move through space, at all events relatively to the other stars?

That difficulty does not seem to me so very insuperable. Suppose you take a piece of Burgundy pitch, or Trinidad pitch, or what I know best for this particular subject, Scottish shoemaker's wax. This is the substance I used in the illustration I intend

to refer to. I do not know how far the others would succeed
in the experiment. Suppose you take one of these substances, the
shoemaker's wax, for instance. It is brittle, but you could form
it into the shape of a tuning fork and make it vibrate. Take a
long rod of it, and you can make it vibrate as if it were a piece of
glass. But leave it lying upon its side and it will flatten down
gradually. The weight of a letter, sealed with sealing-wax in the
old-fashioned way, used to flatten it, will flatten it. Experiments
have not been made as to the absolute fluidity or non-fluidity of
such a substance as shoemaker's wax; but that time is all that is
necessary to allow it to yield absolutely as a fluid, is not an improbable supposition with reference to any one of the substances
I have mentioned. Scottish shoemaker's wax I have used in this
way: I took a large slab of it, perhaps a couple of inches thick,
and placed it in a glass jar ten or twelve inches in diameter. I filled
the glass jar with water and laid the slab of wax in it with a
quantity of corks underneath and two or three lead bullets on the
upper side. This was at the beginning of an Academic year. Six
months passed away and the lead bullets had all disappeared, and
I suppose the corks were half way through. Before the year had
passed, on looking at the slab I found that the corks were floating
in the water at the top, and the bullets of lead were tumbled about
on the bottom of the jar.

Now, if a piece of cork, in virtue of the greater specific gravity
of the shoemaker's wax would float upwards through that solid
material and a piece of lead, in virtue of its greater specific gravity
would move downwards through the same material, though only at
the rate of an inch per six months, we have an illustration, it seems
to me, quite sufficient to do away with the fundamental difficulty
from the wave theory of light. Let the luminiferous ether be
looked upon as a wax which is elastic and I was going to say
brittle, (we will think of that yet—of what the meaning of brittle
would be) and capable of executing vibrations like a tuning fork
when times and forces are suitable—when the times in which the
forces tending to produce distortion act, are very small indeed, and
the forces are not too great to produce rupture. When the forces
are long continued then very small forces suffice to produce unlimited change of shape. Whether infinitesimally small forces
produce unlimited change of shape or not we do not know; but
very small forces suffice to do so. All we need with respect to the

luminiferous ether is that the exceedingly small forces brought
into play in the luminiferous vibrations do not, in the times during
which they act, suffice to produce any transgression of limits of
distortional elasticity. The come-and-go effects taking place in
the period of the luminiferous vibrations do not give rise to the
consumption of any large amount of energy: not large enough an
amount to cause the light to be wholly absorbed in say its propagation from the remotest visible star to the earth.

If we have time, we shall try a little later to think of some of
the magnitudes concerned, and think of, in the first place, the magnitude of the shearing force in luminiferous vibrations of some
assumed amplitude, on the one hand, and the magnitude of the
shearing force concerned, when the earth, say, moves through the
luminiferous ether, on the other hand. The subject has not been
gone into very fully; so that we do not know at this moment
whether the earth moves dragging the luminiferous ether altogether
with it, or whether it moves more nearly as if it were through
a frictionless fluid. It is conceivable that it is not impossible that
the earth moves through the luminiferous ether almost as if it were
moving through a frictionless fluid and yet that the luminiferous
ether has the rigidity necessary for the performance of the luminiferous vibration in periods of from the four hundred million
millionth of a second to the eight hundred million millionth of
a second corresponding to the visible rays, or from the periods
which we now know in the low rays of radiant heat as recently
experimented on and measured for wave length by Abney and
Langley, to the high ultra-violet rays of light, known chiefly by
their chemical actions. If we consider the exceeding smallness of
the period from the 100 million millionth of a second to the 1600
million millionth of a second through the known range of radiant
heat and light, we need not fully despair of understanding the
property of the luminiferous ether. It is no greater mystery at
all events than the shoemaker's wax or Burgundy pitch. That is
a mystery, as all matter is; the luminiferous ether is no greater
mystery.

We know the luminiferous ether better than we know any
other kind of matter in some particulars. We know it for its
elasticity; we know it in respect to the constancy of the velocity
of propagation of light for different periods. Take the eclipses of
Jupiter's satellites or something far more telling yet, the waxings

and wanings of self-luminous stars as referred to by Prof. Newcomb in a recent discussion at Montreal on the subject of the velocity of propagation of light in the luminiferous ether. These phenomena prove to us with tremendously searching test, to an excessively minute degree of accuracy, the constancy of the velocity of propagation of all the rays of visible light through the luminiferous ether.

Luminiferous ether must be a substance of most extreme simplicity. We might imagine it to be a material whose ultimate property is to be incompressible; to have a definite rigidity for vibrations in times less than a certain limit, and yet to have the absolutely yielding character that we recognize in wax-like bodies when the force is continued for a sufficient time.

It seems to me that we must come to know a great deal more of the luminiferous ether than we now know. But instead of beginning with saying that we know nothing about it, I say that we know more about it than we know about air or water, glass or iron—it is far simpler, there is far less to know. That is to say, the natural history of the luminiferous ether is an infinitely simpler subject than the natural history of any other body. It seems probable that the molecular theory of matter may be so far advanced sometime or other that we can understand an excessively fine-grained structure and understand the luminiferous ether as differing from glass and water and metals in being very much more finely grained in its structure. We must not attempt, however, to jump too far in the inquiry, but take it as it is, and take the great facts of the wave theory of light as giving us strong foundations for our convictions as to the luminiferous ether.

To think now of ponderable matter, imagine for a moment 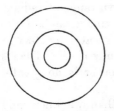 that we make a rude mechanical model. Let this be an infinitely rigid spherical shell; let there be another absolutely rigid shell inside of it, and so on, as many as you please. Naturally, we might think of something more continuous than that, but I only wish to call your attention to a crude mechanical explanation, possibly sufficient to account for dispersion. Suppose we had luminiferous ether outside, and that this hollow space is of very small diameter in comparison with the wave length. Let zig-zag springs connect the outer rigid boundary with boundary number two. I use a zig-zag, not a spiral spring possessing the helical properties for which we are not ready yet, but which must be invoked to account for such properties as sugar and quartz have in disturbing the luminiferous vibrations. 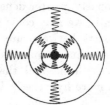 Suppose we have shells 2 and 3 also connected by a sufficient number of zig-zag springs and so on; and let there be a solid nucleus in the centre with spring connections between it and the shell outside of it. If there is only one of these interior shells, you will have one definite period of vibration. Suppose you take away everything except that one interior shell; displace that shell and let it vibrate while you hold the outer sheath fixed. The period of the vibration is perfectly definite. If you have an immense number of such sheaths, with movable molecules inside of them distributed through some portion of the luminiferous ether, you will put it into a condition in which the velocity of the propagation of the wave will be different from what it is in the homogeneous luminiferous ether. You have what is called for, viz., a definite period; and the relation between the period of vibration in the light considered, and the period of the free vibration of the molecule will be fundamental in respect to the attempt of a mechanism of that kind to represent the phenomena of dispersion.

If you take away everything except one not too massive interior nucleus, connected by springs with the outer sheath, you will have a crude model, as it were, of what Helmholtz makes the subject of his paper on anomalous dispersion. Helmholtz, besides that, supposes a certain degree or coefficient of viscous resistance against the vibration of the nucleus, relatively to the sheath.

If we had only dispersion to deal with there would be no difficulty in getting a full explanation by putting this not in a rude mechanical model form, but in a form which would commend itself to our judgment as presenting the actual mode of action of the particles, of gross matter, whatever they may be upon the luminiferous ether. It is difficult to imagine the conditions of the luminiferous ether in dense fluids or liquids, and in solids; but oxygen, hydrogen, and gases generally, must, in their detached particles, somehow or other act on the luminiferous ether, have some sort of elastic connection with it; and I cannot imagine anything that commends itself to our ideas better than this

multiple molecule which I have put before you. By taking
enough of these interior shells, and by passing to the idea of
continuous variation from the density of the ether to the enormously greater density of the molecule of grosser matter imbedded
in it, we may come as it were to the kind of mutual action that
exists between any particular atom and the luminiferous ether.
It seems to me that there must be something in this molecular
hypothesis, and that as a mechanical symbol, it is certainly not
a mere hypothesis, but a reality.

But alas for the difficulties of the undulatory theory of light;
—refraction and reflection at plane surfaces worked out by Green
differ in the most irreduceable way from the facts. They correspond in some degree to the facts, but there are differences that
we have no way of explaining. A great many hypotheses have
been presented, but none of them seems at all tenable.

First of all is the question, are the vibrations of light *perpendicular to*, or are they *in*, the plane of polarization—defining
the plane of polarization as the plane through the incident and
refracted rays, for light polarized by reflection? Think of light
polarized by reflection at a plane surface and the question is, are
the vibrations in the reflected ray perpendicular to the plane of
incidence and reflection, or are they in the plane of incidence and
reflection? I merely speak of this subject in the way of index.
We shall consider very fully, Green's theory and Lord Rayleigh's
work upon it, and come to the conclusion with absolute certainty,
it seems to me, that the vibrations must be perpendicular to the
plane of incidence and reflection of light polarized by reflection.

Now there is this difficulty outstanding—the dynamical theory
of refraction and reflection which gives this result does not give it
rigorously, but only approximately. We have by no means so
good an approach in the theory to complete extinction of the
vibrations in the reflected ray (when we have the light in the
incident ray vibrating in the plane of incidence and reflection) as
observation gives. I shall say no more about that difficulty, because it will occupy us a good deal later on, except to say that the
theoretical explanation of reflection and refraction is not satisfactory. It is not complete; and it is unsatisfactory in this, that we
do not see any way of amending it.

But suppose for a moment that it might be mended and there
is a question connected with it which is this: Is the difference

between two mediums a difference corresponding to difference of rigidity, or does it correspond to difference of density? That is an interesting question, and some of the work that had been done upon it seemed most tempting in respect to the supposition that the difference between two mediums is a difference of rigidity and not a difference of density. When fully examined, however, the seemingly plausible way of explaining the facts of refraction and reflection by difference of rigidity and no difference of density I found to be delusive, and we are forced to the view that there is difference of density and very little difference of rigidity.

In working out this subject very carefully, and endeavouring to understand Lord Rayleigh's work upon it, and to learn what had been done by others, for a time I thought it too much of an assumption that the rigidity was exactly the same and that the whole effect was due to difference of density. Might it not be, it seemed to me, that the luminiferous ether on the two sides of the interface at which the refraction and reflection takes place, might differ both in rigidity and in density. It seemed to me then by a piece of work (which I must verify, however, before I can speak quite confidently about it *) that by supposing the luminiferous ether in the commonly called denser medium to be considerably denser than it would be were the rigidities equal, and the rigidity to be greater in it than in the other medium, we might get a better explanation of the polarization by reflection than Green's result gives. Green's work ends with the supposition of equal rigidities and unequal densities. He puts the whole problem in his formulæ to begin with, but he ends with this supposition and his result depends upon it.

Not to deal in generalities, let us take the case of glass and a vacuum, say. It seemed to me that by supposing the effective rigidity of luminiferous ether in glass to be greater than in vacuum and the density to be greater, but greater in a greater proportion than the rigidity, so that the velocity of propagation is less in glass than in vacuum, we should get a better explanation of the details of polarization by reflection than Green's result gives.

It is only since I have left the other side of the Atlantic that I have worked at this thing, and going into it with keen interest, I inquired of everybody I met whether there were any observations that would help me. At last I was told that Prof. Rood had

* See Appendix A.

done what I desired to know, and on looking at his paper, I found that it settled the matter.

My question was this: Has there been any measurement of the intensity of light reflected at nearly normal incidence from glass or water showing it to be considerably greater than Fresnel's formula gives? Fresnel gives $\left(\dfrac{\mu-1}{\mu+1}\right)^2$ for the ratio of the intensity of the reflected ray to the intensity of the incident ray in the case of normal incidence, or incidence nearly normal. I wanted to find out whether that had been verified by observation. It seems that nobody had done it at all until Prof. Rood, of Columbia College, New York, took it up. His experiments showed to a rather minute degree of accuracy an agreement with Fresnel's formula, so that the explanation I was inclined to make was disproved by it. I myself had worked with the reflection of a candle from a piece of window glass, during a pleasant visit to Dr Henry Field in your beautiful Berkshire hills at the end of August; and had come to the same conclusion, even through such very crude and roughly approximate measurements. At all events, I satisfied myself that there was not so great a deviation from Fresnel's law as would allow me to explain the difficulties of refraction and reflection by assuming greater rigidity, for example, in glass than in air. We are now forced very much to the conclusion from several results, but directly from Prof. Rood's photometrical experiments, that the rigidity must be very nearly equal in the two.

There is quite another supposition that might be made that would give us the same law, for the case of normal incidence ;—the supposition that the reflection depends wholly upon difference of rigidity and that the densities are equal in the two. That gives rise to the same intensity of perpendicularly reflected light, so that the photometric measurement does not discriminate between these two extremes, but it does prevent us from pushing in on the other side of generally accepted result, a supposition of equal rigidities, in the manner that I had thought of.

We may look upon the explanation of polarization by reflection and refraction as not altogether unsatisfactory, although not quite satisfactory :—and you may see [pointing to diagrams which had been chalked on the board : see pp. 12 and 13 above] that this kind of modification of the luminiferous ether is just what would give us the virtually greater density. How this gives us precisely the same effect as a greater density I shall show when we work the thing out mathematically. We shall see that this supposition is equivalent to giving the luminiferous ether a greater density, while making the addition to virtual density according to the period of the vibration.

I am approaching an end: I had hoped to get to it sooner. We have the subject of double refraction in crystals, and here is the great hopeless difficulty.

If we look into the matter of the distortion of the elastic solid, Molar. we may consider, possibly, that that is not wonderful; but Fresnel's supposition as to the direction of the vibration of light, is that the conclusion that the plane of vibration is perpendicular to the plane of polarization proves, if it is true, that the velocity of propagation of light in uniaxial crystals depends simply on the direction of vibration and not otherwise on the plane of the distortion. In the vibrations of light, we have to consider the medium as being distorted and tending to recover its shape. Let this be a piece of uniaxial crystal, Iceland spar, for instance, a round or square column, with its length in the direction of the optic axis, which I will represent on the board by a dotted line.

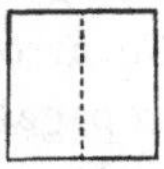

Now the relation between light polarized by passing through Iceland spar on the one hand and light polarized by reflection on the other hand, shows us that if the line of vibration is perpendicular to the plane of polarization, the velocity of propagation of light in different directions through Iceland spar must depend solely on the line of vibration and not on the plane of distortion.

Fig. 1.

There is no way in which that can be explained by the rigidity Case I. of an elastic solid. Look upon it in this way. Take a cube of Propaga-tion in Iceland spar, keeping the same direction of the axis direction of axis: as before. Let the light be passing downwards, as vibrations indicated by the arrow-head. What would be the in plane of diagram. mode of vibration, with such a direction of propaga-tion? Let us suppose, in the first place, the vibra-tions to be in the plane of the diagram. Then the dis-

Fig. 2.

tortion of that portion of matter will be of the kind, and in the direction indicated by Fig. 2. A portion which was rectangular swings into the shape represented by the dotted lines. The force tending to cause a piece of matter which has been so displaced to

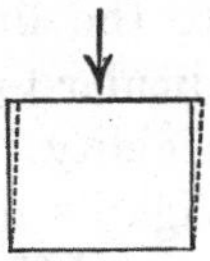

Molar.

resume its original shape depends on this kind of distortion. The mathematical expression of it would be n a constant of rigidity, multiplied into a, the amount of the distortion. How that is to be reckoned is familiar to many of you, and we will not enter into the details just now. But just consider this other case, where the

Case II. Propagation perpendicular to axis: vibrations in plane of diagram.

direction of propagation of the light is horizontal, as indicated by the arrow-head, that is to say, propagated perpendicular to the axis of the crystal (Fig. 3). What would be the nature of the distortion here, the vibration being still in the plane of the diagram?

Fig. 3.

The distortion will be in this way in which I move my two hands. A portion which was rectangular will swing into this shape, indicated by the dotted lines in Fig. 3. The return force will then depend upon a distortion of this kind. But a distortion of this kind is identical with a distortion of that kind (Fig. 2), and the result must be, if the effect depends upon the return force in an elastic solid, that we must have the same velocity of propagation in this case and in that case (Figs. 3 and 2).

Case III.

Consider similarly the distortions produced by waves consisting of vibrations perpendicular to the plane of the diagrams, the arrow-heads in Figs. 2 and 3 still representing directions of propagation. The distortion in the case of Fig. 2 will still be as by shearing perpendicular to the axis, but in the direction perpendicular to the plane of the diagram, instead of that represented by aid of the dotted lines of Fig. 2. Hence on the homogeneous elastic solid theory, and in the case of an axially isotropic crystal, the velocity of propagation in this case would be the same

Case IV. Propagation perpendicular to axis: vibrations perpendicular to plane of diagram.

as in each of the first two cases. But with vibrations perpendicular to the plane of the diagram, and propagation perpendicular to the axis, the distortion is by shearing in the plane perpendicular to the line of the arrow in Fig. 3, and in the direction perpendicular to the plane of the diagram. In the supposed crystal, the rigidity modulus for this shearing would be different from the rigidity modulus for the shearings of cases I., II. and III. Hence cases III. and IV. would correspond to the extraordinary ray, and cases I. and II. the ordinary ray.

Now observe, that light polarized in a plane through the axis, constitutes the ordinary ray, and light polarized in the plane perpendicular to the plane of the ray and the axis constitutes the extraordinary ray. There is therefore an outstanding difficulty in the assumption that the vibration is perpendicular to the plane of Molar. polarization, which is absolutely inexplicable on the bare theory of an elastic solid.

The question now occurs, may we not explain it by loading the elastic solid ? But the difficulty is, to load it unequally in different directions. Lord Rayleigh thought that he had got an explanation of it in his paper to which I have referred. He was not aware that Rankine had had exactly the same idea. Lord Rayleigh at the end of this paper puts forward the supposition that difference of effective inertias in different directions may be adduced to explain the difference of velocity of propagation in Iceland spar. But if that were the case the wave propagation would not follow Huyghens' law. It would follow the law according to which the velocity of propagation would be inversely proportional to what it is according to Huyghens' law. Huyghens' geometrical construction for the extraordinary ray in Iceland spar gives us an oblate ellipsoid of revolution according to which the velocity of propagation of light will be found by drawing from the centre of the ellipsoid a perpendicular to the tangent plane. For example (Fig. 4), CN will correspond to the velocity of propagation of the light when the front is in the direction of this (tangent) line. If the velocity is different in different directions

Fig. 4.

in virtue of an effective inertia, Lord Rayleigh's idea is that Molecular. the vibrating molecules might be like oblate spheroids vibrating in a frictionless fluid. The medium would thus have greater effective inertia when vibrating in the direction of its axis (perpendicular to its flat side), and less effective inertia when vibrating in its equatorial plane. That is a very beautiful idea, and we shall badly want it to explain the difficulty. We would be delighted and satisfied with it if the pushing forward of the conclusions from it were verified by experiment. Stokes has made the experiment. Rankine made the first suggestion in the matter, but did not push the question further than to give it as a mode of getting over the difficulty in double refraction. Stokes took away the possibility of it. He experimented on the refracting index of Iceland spar for a variety of incidences, and found with minute accuracy indeed that Huyghens' construction was verified and that therefore it was impossible to account for the unequal velocity of propagation in different directions by the beautiful suggestions of Rankine and Lord Rayleigh.

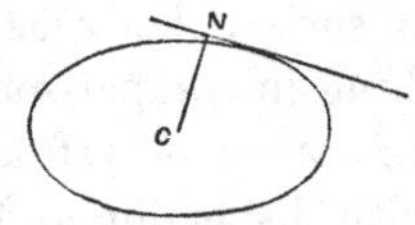

Molecular.   I have not been able to make a persistent suggestion of explanation, but I have great hopes that these spring arrangements are
going to help us out of the difficulty.   I will, just in conclusion,
give you the idea of how it might conceivably do so.

We can easily suppose these spring arrangements to have
different strengths in different directions; and their directional
law will suit exactly.   It will give the fundamental thing we want,
which is that the velocity of propagation of light shall depend on
the direction of vibration, and not merely on the plane of the
distortion.   And it will obviously do this in such a manner as to
verify Huyghens' law—giving us exactly the same shape of wave-
surface as the æolotropic elastic solid would give.

But alas, alas, we have one difficulty which seems still insuperable and prevents my putting this forward as the explanation.   It
is that I cannot get the requisite difference of propagational
velocity for different directions of the vibration *to suit different
periods.*   If we take this theory, we should have, instead of the very
nearly equal difference of refractive index for the different periods
in such a body as Iceland spar, with dispersion merely a small
thing in comparison with those differences : we should, I say, have
difference of refractive index in different directions comparable
with dispersion and modified by dispersion to a prodigious degree,
and in fact we should have anomalous dispersion coming in between
the velocity of propagation in one direction and the velocity of
propagation in another.   The impossibility of getting a difference
of wave velocity in different directions sufficiently constant for the
different periods seems to me at present a stopper.

So now, I have given you one hour and seven minutes and
brought you face to face with a difficulty which I will not say
is insuperable, but something for which nothing ever has been
done from the beginning of the world to the present* time that
can give us the slightest promise of explanation.

I shall do to-morrow, what I had hoped to do to-day, give you
a little mathematics, knowing that it is not going to explain everything.   But I think that in relation to the wave-theory of light we
really have an interest in working out the motions of an elastic
solid and obtaining a few solutions that depend on the equations of
motion of an elastic solid.   I shall first take the case of zero

_____
* [I retract this statement.   See an Appendix near the end of the present
volume.   W. T. Oct. 28, 1889.]

rigidity; that will give us sound. We shall take the most elemen- Molecular.
tary sounds possible, namely a spherical body alternately expanding and contracting. We shall pass from that to the case of a
single globe vibrating to and fro in air. We shall pass from that
to the case of a tuning-fork, and endeavour to explain the cone of
silence which you all know in the neighbourhood of a vibrating
tuning-fork. I hope we shall be able to get through that in a
short time and pass on our way to the corresponding solutions
of the motions of a wave proceeding from a centre in respect to the
wave-theory of light.

LECTURE II.

Oct. 2, 5 P.M.

PART I. In the first place, I will take up the equations of motion of an elastic solid. I assume that the fundamental principles are familiar to you. At the same time, I should be very glad if any person present would, without the slightest hesitation, ask for explanations if anything is not understood. I want to be on a conferent footing with you, so that the work shall be rather something between you and me, than something in which I shall be making a performance before you in a matter in which many of you may be quite as competent as I am, if not more so.

I want, if we can get something done in half an hour, on these problems of *molar* dynamics as we may call it, to distinguish from Molecular dynamics, to come among you, and talk with you for a few moments, and take a little rest; and then go on to a problem of molecular dynamics to prepare the way for motions depending on mutual interference among particles under varying circumstances that may perhaps have applications in physical science and particularly to the theory of light.

Molar.   The fundamental equations of equilibrium of elastic solids are, of course, included in D'Alembert's form of the equations of motion. I shall keep to the notation that is employed in Thomson and Tait's *Natural Philosophy*, which is substantially the same notation as is employed by other writers.

Let a, b, c denote components of distortion: a, is a distortion in the plane perpendicular to OX produced by slippings parallel to either or to both of the two co-ordinate planes which intersect in OX.

Let us consider this state of strain, in which, without other change, a portion of the solid in the plane yz which was a

square becomes a rhombic figure.   The measurement of that state Molar.
of strain is given fully in Thomson and Tait's geometrical preliminary for the theory of elastic solids, (*Natural Philosophy*,
§§ 169—177, and § 669, or *Elements*, §§ 148—156 and § 640).   It
is called a simple shear.   It may be measured
either by the rate of shifting of parallel planes per
unit distance perpendicular to them, or, which
comes to exactly the same thing, the change of
the angle measured in radians.   Thus I shall write
down inside this small angular space the letter a,
to denote the magnitude of the angle, measured
in radians; the particular case of strain considered
being a slipping parallel to the plane YOX.

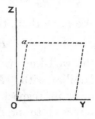

I use the word "radian"; it is not, hitherto, a very common
word; but I suppose you know what I mean.   In Cambridge
in the olden time we used to have a very illogical nomenclature,—"the unit angle"—a very absurd use of the definite
article "the".   It is illogical to talk of an angle being measured in
terms of "the" unit angle; there is no such thing as measuring
anything, except in terms of "a" unit.   The degree is a unit
angle; so is the minute; so is the second; so is the quadrant;
so is the round; all of these are units in frequent use for angular
measurement.   The unit in which it is most generally convenient
to measure angles in Analytical Mechanics is the angle whose arc
is radius.   That used to be called at Cambridge "the unit angle."
My brother, James Thomson, proposed to call it "*the radian*".

There are three principal distortions, a, b, c, relative to the
axes of OX, OY, OZ; and again, three principal dilatations—condensations of course if any one is negative, e, f, g, which are
the ratios of the augmentation of length to the length.

The general equation of energy will of course be an equation in
which we have a quadratic function of e, f, g, a, b, c, the expression for which will be

$$\tfrac{1}{2}\,(11e^2 + 12ef + 13eg + 14ea + 15eb + 16ec + 21ef + 22f^2 + 23fg + \ldots).$$

We do not here use $11, 12, \ldots$etc., as numbers but as symbols
representing the twenty-one coefficients of this quadratic subject
to the conditions $12 = 21$, etc.   If we denote this quadratic function
by E, then

$$\frac{dE}{de} = 11e + 12f + 13g + 14a + 15b + 16c.$$

Molar.   This is a component of the normal force required to produce this
compound strain e, f, g, a, b, c. According to the notation of
Thomson and Tait, let

$$P = \frac{dE}{de}, \quad Q = \frac{dE}{df}, \quad R = \frac{dE}{dg}, \quad S = \frac{dE}{da}, \quad T = \frac{dE}{db}, \quad U = \frac{dE}{dc}.$$

We have then, the relation

$$Pe + Qf + Rg + Sa + Tb + Uc = 2E,$$

the well-known dynamical interpretation of which you are of course
familiar with. A little later we shall consider these 21 coefficients,
first, in respect to the relations among them which must be
imposed to produce a certain kind of symmetry relative to the
three rectangular axes; and then see what further conditions must
be imposed to fit the elastic solid for performing the functions
of the luminiferous ether in a crystal.

Before going on to that we shall take the case of a perfectly
isotropic material. We can perhaps best put it down in tabular
form in this way:

	1	2	3	4	5	6
1	$\mathfrak{A}$	$\mathfrak{B}$	$\mathfrak{B}$	0	0	0
2	$\mathfrak{B}$	$\mathfrak{A}$	$\mathfrak{B}$	0	0	0
3	$\mathfrak{B}$	$\mathfrak{B}$	$\mathfrak{A}$	0	0	0
4	0	0	0	n	0	0
5	0	0	0	0	n	0
6	0	0	0	0	0	n

In the first place in this lower right-hand corner-square which
has to do with the distortions a, b, c, alone, if we let n represent the
rigidity-modulus the three main diagonal terms will each be n, and
those not in the diagonal will be zero. Six of the 21 coefficients
are thus determined as follows:

$$44 = 55 = 66 = n, \text{ and } 45 = 46 = 56 = 0,$$

and other nine of them, by the zeros in the upper right and lower
left corner-squares,—

$$14 = 15 = 16 = 24 = 25 = 26 = 34 = 35 = 36 = 0.$$

To verify these zeros of the upper right-hand and lower left-hand Molar. corner-squares, let us consider what possible relations there can be for an isotropic body between longitudinal strains and distortions. Clearly none. No one of the longitudinal strains can call into play a tangential force in any of the faces; and conversely, if the medium be isotropic, no distortion produced by slipping in the faces parallel to the principal planes can introduce a longitudinal stress—a stress parallel to any of the lines OX, OY, OZ. Therefore we have all zeros in these two squares. We know that $11 = 22 = 33$; and each of these will be represented by Saxon A ($\mathfrak{A}$). Now consider the effect of a longitudinal pull in the direction of OX. If the body be only allowed to yield longitudinally, that clearly will give rise to a negative pull in the directions parallel to Oy, Oz. We have then a cross connection between pulls in the directions OX, OY, OZ. Isotropy requires that the several mutual relations be all equal, so that we have just one coefficient to express these relations. That coefficient is denoted by Saxon B ($\mathfrak{B}$). Thus we fill up our 36 squares, which represent but 21 coefficients in virtue of the relations $12 = 21$, etc. We can now write down our quadratic expression for the energy,

$$E = \tfrac{1}{2}[\mathfrak{A}(e^2 + f^2 + g^2) + 2\mathfrak{B}(fg + ge + ef) + n(a^2 + b^2 + c^2)].$$

Instead of these Saxon letters $\mathfrak{A}$, $\mathfrak{B}$, which have very distinct and obvious interpretations, we may introduce the resistance of the solid to compression, the reciprocal of what is commonly called the compressibility, or, what we may call the bulk modulus, k. Then it is proved in Thomson and Tait, and in an article in the *Encyclopædia Britannica* which perhaps some of you may have*, that $\mathfrak{A} = k + \tfrac{4}{3}n$, $\mathfrak{B} = k - \tfrac{2}{3}n$. The considerations which show these relations with the bulk modulus also show us that we must have $n = \tfrac{1}{2}(\mathfrak{A} - \mathfrak{B})$. This is most important. Take a solid cube with its edges parallel to OX, OY, OZ. Apply a pull along two faces perpendicular to OX and an equal pressure on two faces perpendicular to OY; that will give a distortion in the plane xy. Find the value of that simple shear; it is done in a moment. Find the shearing force required to produce it calculated from $\mathfrak{A}$ and $\mathfrak{B}$, and equate that to the force calculated from the rigidity modulus n, and then you find this relation. The relations for complete

* Reprint of *Mathematical and Physical Papers*, Vol. III. Art. XCII. now in the Press. [W. T. Aug. 7, 1885.]

isotropy are exhibited here in this quadratic expression for the energy, if in it we take $\frac{1}{2}(\mathfrak{A} - \mathfrak{B})$ in place of n.

We shall pass on to the formation of the equations of motion. For equilibrium, the component parallel to OX of the force applied at any point x, y, z of the solid, reckoned per unit of bulk at that point must be equal to

$$\left(\frac{dP}{dx} + \frac{dU}{dy} + \frac{dT}{dz} \right),$$

if the body be held distorted in any way, by bodily forces applied all through the interior; because the resultant of the elastic force on an infinitesimal portion of matter at the point x, y, z is obviously $\left(\frac{dP}{dx} + \frac{dU}{dy} + \frac{dT}{dz} \right) dx\,dy\,dz$. To prove this remark that if the pull augments as you go forward in the direction OX there will, in virtue of that, be a resultant forward pull

$\frac{dP}{dx} dx\,.\,dy\,dz$ upon the infinitesimal element. The two tangential forces, U perpendicular to OY, and U perpendicular to OX on the one pair of forces and the pair of forces equal and in opposite directions on the other faces constitute two balancing couples, as it were. If this tangential force parallel to OX increases as we proceed in the direction y positive, there will result a positive force on the element, because it is pulled to left by the smaller and to right by the larger, and thus the force in the direction of OX receives a contribution $\frac{dU}{dy} dy\,.\,dz\,dx$. Quite similarly we find, $\frac{dT}{dz} dz\,.\,dx\,dy$ as a third and last contribution to the force parallel to OX.

Now, let there be no bodily forces acting through the material, but let the inertia of the moving part, and the reaction against acceleration in virtue of inertia, constitute the equilibrating reaction against elasticity. The result is, that we have the equation

$$\frac{dP}{dx} + \frac{dU}{dy} + \frac{dT}{dz} = \rho \frac{d^2\xi}{dt^2},$$

if by ρ we denote the density and by ξ we denote the displacement from equilibrium in the direction OX of that portion of matter having x, y, z for co-ordinates of its mean position.

I said I would use the notation of Thomson and Tait who Molar. employ α, β, γ to denote the displacements; but errors are too common when α and a are mixed up, especially in print, so we will take ξ, η, ζ instead. I have had trouble in reading Helmholtz's paper on anomalous dispersion, on this account, very frequently not being able to distinguish with a magnifying glass whether a certain letter was a or α.

The values of S, T, U, we had better write out in full, although the others may be obtained from the value of any one of the three by symmetry. The expenditure of chalk is often a saving of brains. They are:

$$S = n\left(\frac{d\eta}{dz} + \frac{d\zeta}{dy}\right), \qquad T = n\left(\frac{d\zeta}{dx} + \frac{d\xi}{dz}\right), \qquad U = n\left(\frac{d\xi}{dy} + \frac{d\eta}{dx}\right).$$

We have $P = \mathfrak{A}e + \mathfrak{B}\,(f+g)$. There are two or three other forms which are convenient in some cases, and I will put them down (writing m for $k + \frac{1}{3}n$)

$$P = (m+n)\frac{d\xi}{dx} + (m-n)\left(\frac{d\eta}{dy} + \frac{d\zeta}{dz}\right) = (m-n)\left(\frac{d\xi}{dx} + \frac{d\eta}{dy} + \frac{d\zeta}{dz}\right) + 2n\frac{d\xi}{dx}$$

$$= m\left(\frac{d\xi}{dx} + \frac{d\eta}{dy} + \frac{d\zeta}{dz}\right) + n\left(\frac{d\xi}{dx} - \frac{d\eta}{dy} - \frac{d\zeta}{dz}\right).$$

We shall denote very frequently by δ the expression

$$\frac{d\xi}{dx} + \frac{d\eta}{dy} + \frac{d\zeta}{dz},$$

so that for example, the second of these expressions is

$$P = (m-n)\,\delta + 2n\frac{d\xi}{dx}.$$

If we want to write down the equations of a heterogeneous medium, as will sometimes be the case, especially in following Lord Rayleigh's work on the blue sky, we must in taking dP/dx, dP/dy, &c., to find the accelerations, keep the symbols m, n inside of the symbols of differentiation; but for homogeneous solids, we treat m and n as constant. I forgot to say that δ is the cubic dilatation or the augmentation of volume per unit volume in the neighbourhood of the point x, y, z, which is pretty well known, and helps us to see the relations to compressibility. If we suppose zero rigidity, $P = m\delta$ is the relation between pressure and volume. In order to verify this take the preceding expression for P and make $n = 0$ and we obtain $P = m\delta$, the equation for the compression of a compressible fluid, in which m has become the bulk modulus.

Molar.   This sort of work I have called molar dynamics. It is the dynamics of continuous matter; there are in it no molecules, no heterogeneousnesses at all. We are preparing the way for dealing with heterogeneousnesses in the most analytical manner by supposing m and n to be functions of x, y, z. Lord Rayleigh studied the blue sky in that way, and very beautifully; his treatment is quite perfect of its kind. He considers an imbedded particle of water, or dust, or unknown material, whatever it is that causes the blue sky. To discover the effect of such a particle, on the waves of light, he supposes a change of rigidity and of density from place to place in the luminiferous ether; not an absolutely sudden change, but confined to a space which is small in comparison with the wave-length.

Ten minutes interval.

Molecular.   PART II. I want to take up another subject which will prepare the way to what we shall be doing afterward, which is the particular dynamical problem of the movement of a system of connected particles. I suppose most of you know the linear equations of motion of a connected system—whose integral always leads to the same formula as the cycloidal pendulum; this result being come to through a determinant equated to zero, giving an algebraical equation whose roots are essentially real for the square of the period of any one of the fundamental modes of simple harmonic motion.

As an example, take three weights, one of 7 pounds, another of 14 pounds, and another of 28 pounds, say. The lowest weight is

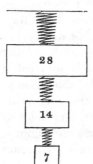

hung upon the middle weight by a spiral spring; the middle is hung upon the upper by a spiral spring, and the upper is attached to a fixed point by a spiral (or a zig-gag) spring. It is a pretty illustration; and I find it very useful to myself. I am speaking, so to say, to professors who sympathize with me, and might like to know an experiment which will be instructive to their pupils.

Just apply your finger to any one of the weights, the upper weight, for example. You soon learn to find by trial the fundamental periods. Move it up and down gently in the period which you find to be that of the

three all moving in the same direction. You will get a very Molecular.
pretty oscillation, the lowest weight moving through the greatest
amplitude, the second through a less, and the upper weight through
the smallest. That is No. 1 motion, corresponding to the greatest
root of the cubic equation which expresses the solution of the
mathematical problem. No. 2 motion will come after a little
practice. You soon learn to give an oscillation a good deal quicker
than before, the first; a second mode, in which the lowest weight
moves downward while the two upper move upwards, or the two
lower move downwards while the upper moves upwards, or it
might be that the middle weight does not move at all in this
second mode, in which case the excitation must be by putting the
finger on the upper or the lower weight. These periods depend
upon the magnitude of the weights, and the strength of the
springs that we use, and are soon learned in any particular set of
weights and springs. It might be a good problem for junior
laboratory students to find weights and springs which will in-
sure a case of the nodal point lying between the upper and
middle weights, or at the middle weight, or between the middle
and lower weights. The third mode of vibration, corresponding
to the smallest root of the cubic equation, is one in which you
always have one node in the spring between the upper and middle
weights, and another node in the spring between the middle and
lowest (the first and third weight vibrating in the same direction, and
the middle weight in an opposite direction to the first and third).

It is assumed that there is no mass in the springs. If you
want to vary your laboratory exercises, take smaller masses for the
weights, and more massive springs, and if you omit the attached
masses altogether, you pass on to a very beautiful illustration
of the velocity of sound. For that purpose a long spiral spring
of steel wire, the spiral 20 feet long, hung up, say, if you have
a lofty enough room, will answer, and you will readily get
two or three of the graver fundamental modes without any
attached weights at all. In the special problem which we have
been considering we have three separate weights and not a continuous spring; and we have three, and only three modes of vibration, when the springs are massless. We have an infinite number
of modes when the mass of the springs is taken into account. In
any convenient arrangement of heavy weights, the stiffness of the
springs is so great and their masses so small that the gravest

Molecular. period of vibration of one of the springs by itself will be very short ;
but take a long spring, a spiral of best pianoforte steel wire, if you
please, and hang it up, with a weight perhaps equal to its own,
on its lower end, and you will find it a nice illustration for getting
several of the graver of the infinite number of the fundamental
modes of the system.

I want to put down the dynamics of our problem for any
number of masses. You will see at once that that is just the case
that I spoke of yesterday, of extending Helmholtz's singly vibrating particle connected with the luminiferous ether to a multiple
vibrating heavy elastic atom imbedded in the luminiferous ether,
which I think must be the true state of the case. A solid mass
must act relatively to the luminiferous ether as an elastic body
imbedded in it, of enormous mass compared with the mass of the
luminiferous ether that it displaces. In order that the vibrations
of the ether may not be absolutely stopped by the mass, there
must be an elastic connection. It is easier to say what must
be than to say that we can understand how it comes to be. The
result is almost infinitely difficult to understand in the case of
ether in glass or water or carbon disulphide, but the luminiferous
ether in air is very easily imagined. Just think of the molecules
of oxygen and nitrogen as if each were a group of ponderable particles mutually connected by springs, and imbedded in homogeneous perfectly elastic jelly constituting the luminiferous ether.
You do not need to take into account the gaseous motions of the
particles of oxygen, nitrogen, and carbon dioxide in our atmosphere
when you are investigating the propagation of luminous waves
through the air. Think of it in this way : the period of vibration
in ultra-violet rays in luminous waves and in infra red heat
waves so far as known, is from the 1600 million-millionth of
a second to the 100 million-millionth of a second. Now think
how far a particle of oxygen or nitrogen moves, according to the
kinetic theory of gases, in the course of that exceedingly small
time. You will find that it moves through an exceedingly small
fraction of the wave-length. For example, think of a molecule
moving at the rate of 50000 centimetres per second. In the period
of orange light it crawls along 10^{-10} of a centimetre which is only
1/600000 of the wave-length of orange light. I am fully confident
that the wave motion takes place independently of the translatory
motion of the particles of oxygen and nitrogen in performing

their functions according to the kinetic theory of gases. You may Molecular.
therefore really look upon the motion of light waves through our
atmosphere as being solved by a dynamical problem such as this
before us, applied to a case in which there is so little of effective
inertia due to the imbedded molecules, that the velocity of light is
not diminished more than about one-thirty-third per cent. by it.
More difficulties surround the subject when you come to consider
the propagation of light through highly condensed gases, or transparent liquids or solids.

In our dynamical problem, let the masses of the bodies be
represented by m_1, m_2,...m_j. I am going to suppose the several
particles to be acted upon by connecting springs. I do not want
to use spiral springs here. The helicalness of the spring in these
experiments has no sensible effect; but we want to introduce
a spiral for investigating the dynamics of the helical properties, as
shown by sugar. It is usually called the rotatory property, but this
is a misnomer. The magneto-optical property which was discovered by Faraday is rotational, the property exhibited by quartz
and sugar and such things, has not the essential elements of
rotation in it, but has the characteristic of a spiral spring (a helical
spring, not a flat spiral), in the constitution of the matter that ex-
hibits it. We apply the word helical to the one and the word
rotational to the other.

I am going to suppose one other connected particle P, which is moved to and fro
with a given motion whose displacement
downwards from a fixed point O, we shall call
ξ. Let c_1 be the coefficient of elasticity of
the first spring, connecting the particle P
with the particle m_1; c_2 the coefficient of
elasticity of the next spring connecting m_1
and m_2; c_{j+1}, the coefficient of elasticity of
the spring connecting m_j to a fixed point.
We are not taking gravity into account; we
have nothing to do with it. Although in the
experiment it is convenient to use gravity,
it would be still better if we could go to the
centre of the earth and there perform the
experiment. The only difference would be,
these springs would not be pulled out by the weights hung upon

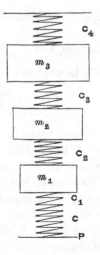

Molecular. them. In all other respects the problem would be the same, and the same symbols would apply.

We are reckoning displacements downwards as positive, the displacement of the particle m_i being x_i. The force acting upon m_1 in virtue of the spring connection between it and P is $c_1\,(\xi - x_1)$; and in virtue of the spring connection between it and m_2 is the opposing pull $- c_2\,(x_1 - x_2)$; so that the equation of motion of the first particle is

$$m_1 \frac{d^2x_1}{dt^2} = c_1\,(\xi - x_1) - c_2\,(x_1 - x_2).$$

For No. 2 particle we have

$$m_2 \frac{d^2x_2}{dt^2} = c_2\,(x_1 - x_2) - c_3\,(x_2 - x_3);\text{ and so on.}$$

Now suppose P to be arbitrarily kept in simple harmonic motion in time or period τ; so that $\xi = \text{const.} \times \cos \dfrac{2\pi t}{\tau}$. We assume that every part of the apparatus is moving with a simple harmonic motion, as will be the case if there were infinitesimal viscous resistance and the simple harmonic motion of P is kept up long enough; so that we can write $x_1 = \text{const.} \times \cos \dfrac{2\pi t}{\tau}$, etc. I am going to alter the m's so as to do away with the $4\pi^2$ which comes in from differentiation. I will let $\dfrac{m_1}{4\pi^2}$ denote the mass of the first particle, and $\dfrac{m_2}{4\pi^2}$ the mass of the second particle, etc. The result will be that the equations of motion become

$$- \frac{m_1}{\tau^2} x_1 = c_1\,(\xi - x_1) - c_2\,(x_1 - x_2),\text{ etc.}$$

Our problem is reduced now to one of algebra. There are some interesting considerations connected with the determinant which we shall obtain by elimination from these equations. To find the number of terms is easy enough; and it will lead to some remarkable expressions. But I wish particularly to treat it with a view to obtaining by very short arithmetic the result which can be obtained from the determinant in the regular way only by enormous calculation. We shall obtain an approximation, to the accuracy of which there is no limit if you push it far enough,

that will be exceedingly convenient in performing the calcu- Molecular.
lations.

In the next lecture, resuming the molar problem, we shall begin with the solution for sound, of the equations that are now before you on the board. We shall next try to go on a step further with this molecular problem, of the vibrations of our compound molecule.

LECTURE III.

Molar. WE will now go on with the problem of Molar dynamics, the propagation of sound or of light, from a source. I advise you all who are engaged in teaching, or in thinking of these things for yourselves, to make little models. If you want to imagine the strains that were spoken of yesterday, get such a box as this covered with white paper and mark upon it the directions of the forces S, T, U. I always take the directions of the axes in a certain order so that the direction of positive rotation shall be from y to z, from z to x, from x to y. What we call positive is the same direction as the revolution of a planet seen from the northern hemisphere, or opposite to the motion of the hands of a watch. I have got this box for another purpose, as a mechanical model of an elastic solid with 21 independent · moduluses, the possibility of which used to be disproved, and after having been proved, was still disbelieved for a long time.

Let us take our equations,

$$\rho \frac{d^2\xi}{dt^2} = \frac{dP}{dx} + \frac{dU}{dy} + \frac{dT}{dz}, \text{ where}$$

$$P = (m-n)\,\delta + 2n\frac{d\xi}{dx}, \quad U = n\left(\frac{d\xi}{dy} + \frac{d\eta}{dx}\right), \quad T = n\left(\frac{d\zeta}{dx} + \frac{d\xi}{dz}\right).$$

$$\left\{ \delta = \frac{d\xi}{dx} + \frac{d\eta}{dy} + \frac{d\zeta}{dz} \right\}.$$

We shall not suppose that m and n are variables, but take them constant. If we do not take them constant we shall be ready for Lord Rayleigh's paper on the blue sky, already referred to. I will do the work upon the board in full, as it is a case in which the expenditure of chalk saves brain; but it would be a waste to print such calculations, for the reason that a reader of mathematics Molar. should always have pencil and paper beside him to work the thing out. * * * * The result is that

$$\rho \frac{d^2\xi}{dt^2} = m \frac{d\delta}{dx} + n\nabla^2\xi \dots\dots\dots\dots(1).$$

We take the symbol

$$\nabla^2 = \frac{d^2}{dx^2} + \frac{d^2}{dy^2} + \frac{d^2}{dz^2}.$$

In the case of no rigidity, or $n = 0$, the last term goes out. We shall take solutions of these equations, irrespectively of the question of whether we are going to make $n = 0$ or not, and we shall find that one standard solution for an elastic solid is independent of n and is therefore a proper solution for an elastic fluid.

I have in my hand a printed report* of a Royal Institution lecture of Feb. 1883, on the Size of Atoms, containing a note on some mathematical problems which I set when I was examiner for the Smith's Prizes at Cambridge, Jan. 30, 1883. One was to show that the equations of motion of an isotropic elastic solid are what we have here obtained, and another to show that so and so was a solution. We will just take that, which is: Show that every possible solution of these three equations [(1) etc.] is included in the following:

$$\xi = \frac{d\phi}{dx} + u, \quad \eta = \frac{d\phi}{dy} + v, \quad \zeta = \frac{d\phi}{dz} + w \dots\dots\dots(2),$$

where ϕ, u, v, w, are some functions of x, y, z, t; with the condition that u, v, w are such that

$$\frac{du}{dx} + \frac{dv}{dy} + \frac{dw}{dz} = 0 \dots\dots\dots\dots\dots(3).$$

If we calculate the value of the cubic dilatation, we find

$$\delta = \nabla^2\phi + \frac{du}{dx} + \frac{dv}{dy} + \frac{dw}{dz} = \nabla^2\phi \dots\dots\dots\dots(4).$$

Again, by using $\xi = \frac{d\phi}{dx} + u$ in (1), we find (bearing in mind $\delta = \nabla^2\phi$)

$$\rho\left(\frac{d^2}{dt^2}\frac{d\phi}{dx} + \frac{d^2u}{dt^2}\right) = (m + n)\nabla^2\frac{d\phi}{dx} + n\nabla^2 u \dots\dots\dots(5).$$

---

* Reprinted in Vol. I. of Sir W. Thomson's *Popular Lectures and Addresses.* (Macmillan, 1889).

3—2

Now we may take

$$\rho \frac{d^2}{dt^2} \frac{d\phi}{dx} = (m + n) \nabla^2 \frac{d\phi}{dx} \dots\dots\dots\dots\dots (6).$$

The full justification and explanation of this procedure is reserved. [See commencement of Lecture IV. below.] Multiply (6) by dx, and the corresponding y and z equations by dy, dz, and add. We thus get a complete differential; in other words, the relation which ϕ must satisfy is

$$\rho \frac{d^2\phi}{dt^2} = (m + n) \nabla^2\phi \dots\dots\dots\dots\dots\dots (7).$$

And (2) shows that if ϕ satisfies (7) we have u, v, w, satisfying equations of the same form, but with n instead of $(m + n)$; viz.

$$\rho \frac{d^2u}{dt^2} = n\nabla^2 u, \quad \rho \frac{d^2v}{dt^2} = n\nabla^2 v, \quad \rho \frac{d^2w}{dt^2} = n\nabla^2 w \dots\dots (8).$$

By solving these four similar equations, one involving $(m + n)$, and three involving n, we can get solutions of (1), that is certain. That we get every possible solution, I shall hope to prove to-morrow. The velocity of the sound wave, or condensational wave is $\sqrt{\dfrac{m+n}{\rho}}$. The velocity of the wave of distortion in the elastic solid is $\sqrt{\dfrac{n}{\rho}}$. I shall not take this up because I am very anxious to get on with the molecular problem; but you see brought out perfectly well the two modes of waves in an isotropic homogeneous solid, the condensational wave and the distortional wave. The condensational wave follows the equations of motion of sound, which is the same as if n were null; and this gives the solution of the propagation of sound in a homogeneous medium, like air, etc. The solution is worked out ready to hand for the distortional wave because the same forms of equations give us separate components u, v, w; the same solution that gives us the velocity potential for the condensational waves, gives us the separate components of displacement for the distortional waves.

What I am going to give you to-morrow will include a solution which is alluded to by Lord Rayleigh. There is nothing new in it. I am going to pass over the parts of the solution which interpreted by Stokes explain that beautiful and curious experiment of Leslie's. Lord Rayleigh quotes from Stokes, ending his quotation of eight pages with "The importance of the subject and the masterly manner in which it has been treated by Prof. Stokes will probably be thought sufficient to justify this long quotation." I would just Molar. like to read two or three things in it. Lord Rayleigh says (*Theory of Sound*, Vol. II. p. 207), " Prof. Stokes has applied this solution to the explanation of a remarkable experiment by Leslie, according to which it appeared that the sound of a bell vibrating in a partially exhausted receiver is diminished by the introduction of hydrogen. This paradoxical phenomenon has its origin in the augmented wave length due to the addition of hydrogen in consequence of which the bell loses its hold (so to speak) on the surrounding gas." I do not like the words " paradoxical phenomenon;" " curious phenomenon," or "interesting phenomenon" would be better. There are no paradoxes in science. Lord Rayleigh goes on to say, " The general explanation cannot be better given than in the words of Prof. Stokes : ' Suppose a person to move his hand to and fro through a small space. The motion which is occasioned in the air is almost exactly the same as it would have been if the air had been an incompressible fluid. There is a mere local reciprocating motion in which the air immediately in front is pushed forward and that immediately behind impelled after the moving body, while in the anterior space generally the air recedes from the encroachment of the moving body, and in the posterior space generally flows in from all sides to supply the vacuum that tends to be created ; so that in lateral directions, the flow of the fluid is backwards, a portion of the excess of the fluid in front going to supply the deficiency behind.' " It will take some careful thought to follow it. I wish I had Green here to read a sentence of his. Green says, " I have no faith in speculations of this kind unless they can be reduced to regular analysis." Stokes speculates, but is not satisfied without reducing his speculation to regular analysis. He gives here some very elaborate calculations that are also important and interesting in themselves, partly in connection with spherical harmonics, and partly from their exceeding instructiveness in respect to many problems regarding sound. Passing by all that five or six pages of mathematics—I will not tax your brains with trying to understand the dynamics of it in the course of a few minutes; I am rather calling your attention to a thing to be read than reading it—Stokes comes more particularly to Leslie's experiments. Instead of a bell vibrating, Stokes considers the vibrations of a sphere becoming alternately prolate and

Molar.   oblate; and he shows that the principles are the same. Read all this for yourselves. I have intended merely to arouse an interest in the subject.

Ten minutes interval.

Molecular.   To return to the consideration of our molecules connected by springs, we will suppose a good fixing at the top, so firm and stiff that the changing pull of the spring does not give it any sensible motion. For any one of the springs let there be a certain change of pull, c per unit change of length. This coefficient c measures what I call the longitudinal rigidity of the spring: its effective stiffness in fact. It is not the slightest consequence whether the spring is long or short, only, if it is long, let it be so much the stiffer; but long or short, thick or thin, it must be massless. I mean that it shall have no inertia. The masses may be equal

or unequal, and are connected by springs. Let us attach here something like the handle of the bell pull of pre-electric ages;—something that you can pull by. Call it P. This, in our application to the luminiferous ether, will be the rigid shell lining between the luminiferous ether and the first moving mass.

The equation of motion for the first mass becomes, on bringing ξ to the left-hand side,

$$- c_1\xi = \left(\frac{m_1}{\tau^2} - c_1 - c_2\right) x_1 + c_2 x_2;$$

and similarly for the second mass; I shall use i to denote any integer. I find the letter i too useful for that purpose to give it up, and when I want to write the imaginary $\sqrt{-1}$, I use ι. Let us call the first coefficient on the right a_1, the similar coefficient in the next equation a_2, and so on, so that

$$a_i = \frac{m_i}{\tau^2} - c_i - c_{i+1}.$$

The ith equation will thus be

$$- c_i x_{i-1} = a_i x_i + c_{i+1} x_{i+1}.$$

Now write down all these j equations supposing the whole number of the springs to be j; form the determinant by which you find all of the others in terms of ξ, and the problem is solved.

If we had a little more time I would like to determine the number of terms in this determinant. We will come back to

that because it is exceedingly interesting; but I want at once to Molecular.
put the equations in an interesting form, taking a suggestion from
Laplace's treatment of his celebrated Tidal problem. What we
want is really the ratios of the displacements, and we shall therefore write

$$\frac{c_i x_{i-1}}{-x_i} = u_i$$

introducing the sign minus, so that when the displacements are
alternately positive and negative the successive ratios will be all
positive. We have then,

$$\frac{c_1 \xi}{-x_1} = u_1 = a_1 - \frac{c_2^2}{u_2}, \quad u_2 = a_2 - \frac{c_3^2}{u_3}, \quad \ldots u_i = a_i - \frac{c_{i+1}^2}{u_{i+1}}, \ldots$$

$$u_j = a_j \,; \; (u_{j+1} = \infty).$$

We can now form a continued fraction which, for the case that
we want, is rapidly convergent. If this be differentiated with
respect to τ^{-2}, we find a very curious law, but I am afraid we
must leave it for the present. The solution is

$$u_1 = a_1 - \cfrac{c_2^2}{a_2 - \cfrac{c_3^2}{a_3 - \cfrac{}{a_{j-1} - \cfrac{c_j^2}{a_j}}}}.$$

Thus if we are given the spring connections and the masses,
everything is known when the period is known. If you develop
this, you simply form the determinant; but the fractional form
has the advantage that in the case when the masses are larger
and larger, and the spring connections are not larger in proportion, we get an exceedingly rapid approximation to its value
by taking the successive convergents. The differential coefficient
of this continued fraction with respect to the period is essentially
negative, and thus we are led beautifully from root to root, and
see the following conditions :—First, suppose we move P to and
fro in simple harmonic motion of very short period; then when
the whole has got into periodic movement, it is necessary that P
and the first particle move in opposite directions. The vibrations
of the first particle needs to be "hurried up" (if you will allow
me an expressive American phrase) when the motion of P is
of a shorter period than the shortest of the possible independent motions of the system with P held fixed. Now if you
want to hurry up a vibrating particle, you must at each end of
its range press it inwards or towards its middle position.

Molecular. You meet this principle quite often; it is well known in the construction of clock escapements. To hurry up the vibratory motion of our system we must add to the return force of particle No. 1 by the action of the spring connected to the handle P, by moving P always in the direction opposite to the motion of m. From looking at the thing, and learning to understand it by *feeling* the experiment, if you do not understand it by brains alone, you will see that everything that I am saying is obvious. But it is not satisfactory to speak of these things in general terms unless we can submit them to a rigorous analysis.

I now set the system in motion, managing, as you see, to get it into a state of simple harmonic vibration by my hand applied to P. That, which you now see, is a specimen of the configuration in which the motion of P is of a shorter period than the shortest of the independent motions with P fixed. Suppose now, the vibration of P to be less rapid and less rapid; a state of things will come, in which, the period of P being longer and longer, the motion of the first particle will be greater and greater. That is to say, if I go on augmenting the period of P we shall find for the same range of motion of P, that the ranges of motion of m_1 and of the other particles generally will be greatly increased relatively to the range which I give to P. In analytical words, if we begin with a configuration of values corresponding to τ very small, and then, if we increase τ to a certain critical value, we shall find $\frac{x_1}{\xi}$ will become infinite. In the first place, we begin with $u_1, u_2, \ldots u_j$ all positive; and τ small enough will make them all positive as you see. Now take the differential coefficient of u_i with respect to τ and it will be found to be essentially negative. In other words, if we increase τ, we shall diminish $u_1, u_2 \ldots$ In every case u_1 will first pass through zero and become negative. When u_1 is zero we have the first infinity $\frac{x_1}{\xi} = \infty$. If we diminish τ a little further u_2 will pass through zero to negative while u_1 is still negative. Diminish τ a little further and u_3 will become zero and pass to negative, while u_2 is still negative; but in the mean time u_1 may have reached a negative maximum and passed through zero to positive, or it may not yet have done so. We shall go into this to-morrow; but I should like to have you know beforehand what is going to come from this kind of treatment of the subject.

LECTURE IV.

We found yesterday

$$\left.\begin{aligned}
\rho\,\frac{d^2\xi}{dt^2} &= (k + \tfrac{1}{3}n)\,\frac{d\delta}{dx} + n\nabla^2\xi \\[1mm]
\rho\,\frac{d^2\eta}{dt^2} &= (k + \tfrac{1}{3}n)\,\frac{d\delta}{dy} + n\nabla^2\eta \\[1mm]
\rho\,\frac{d^2\zeta}{dt^2} &= (k + \tfrac{1}{3}n)\,\frac{d\delta}{dz} + n\nabla^2\zeta
\end{aligned}\right\} \dots\dots\dots\dots(1);$$

and we saw that we get two solutions, which when fully interpreted, correspond to two different velocities of propagation, on the assumptions that were put before you as to a condensational or a distortional wave. We will approach the subject again from the beginning, and you will see at once that the sum of these solutions expresses every possible solution.

In one of our solutions of yesterday, we took, instead of ξ, η, ζ, other symbols u, v, w, which satisfied the condition,

$$\frac{du}{dx} + \frac{dv}{dy} + \frac{dw}{dz} = 0.$$

In other words, the u, v, w of yesterday express the displacements in a case in which the dilatation or condensation is zero. Now, just try for the dilatation in any case whatever, without such restriction. This we can do as follows: Differentiate (1) with respect to x (taking account of the constancy of m and n) and the corresponding equations with respect to y and z, and add. We thus find

$$\rho\,\frac{d^2\delta}{dt^2} = (m + n)\,\nabla^2\delta = (k + \tfrac{4}{3}n)\,\nabla^2\delta \dots\dots\dots\dots(2).$$

Molar. This equation, you will remember, is the same as we had yesterday
for ϕ. We shall consider solutions of this equation presently;
but now remark, that whatever be the displacements, we have
a dilatation corresponding to some solution of this equation. It
may be zero, but it must satisfy (2). Now in any actual case, δ is
a determinate function of x, y, z, t; and whether we know it or
not, we may take ϕ to denote a function such that

$$\nabla^2\phi = \delta \text{ through all space } \dots\dots\dots\dots(3).$$

This function, ϕ, is determinate. It is in fact given explicitly
(as is well known in the theory of attraction) by the equation

$$\phi = -\frac{1}{4\pi}\int_{-\infty}^{\infty}\int_{-\infty}^{\infty}\int_{-\infty}^{\infty}\frac{\delta'.dx'dy'dz'}{\sqrt{[(x-x')^2+(y-y')^2+(z-z')^2]}}\dots(4);$$

where δ' denotes the value of δ at (x', y', z'). This formula (4) is
important as giving ϕ explicitly; and exceedingly interesting on
account of the relations to the theory of attraction; but in the
wave-problem, when δ is given, (3) gives ϕ determinately and in
the easiest possible way. Putting now

$$\xi = \frac{d\phi}{dx}+u, \qquad \eta = \frac{d\phi}{dy}+v, \qquad \zeta = \frac{d\phi}{dz}+w \dots\dots(5).$$

We have
$$\delta = \nabla^2\phi + \frac{du}{dx}+\frac{dv}{dy}+\frac{dw}{dz} \dots\dots\dots\dots(6);$$

and therefore, by (3), $$\frac{du}{dx}+\frac{dv}{dy}+\frac{dw}{dz} = 0 \dots\dots\dots\dots\dots(7).$$

Now, remembering that (1) are satisfied by $d\phi/dx, d\phi/dy, d\phi/dz$
in place of ξ, η, ζ, we see, by multiplying the first by dx, or the
second by dy, or the third by dz, and integrating, that

$$\rho\,\frac{d^2\phi}{dt^2} = (k+\tfrac{4}{3}n)\,\nabla^2\phi \dots\dots\dots\dots\dots(8),$$

and we find the three equations for

$$\xi - \frac{d\phi}{dx}, \qquad \eta - \frac{d\phi}{dy}, \qquad \zeta - \frac{d\phi}{dz}$$

reduced, in virtue of (7), to the following :

$$\rho\,\frac{d^2u}{dt^2} = n\nabla^2u, \quad \rho\,\frac{d^2v}{dt^2} = n\nabla^2v, \quad \rho\,\frac{d^2w}{dt^2} = n\nabla^2w\dots\dots(9).$$

For any possible solutions of equations (1), we have a value
of δ which is a function of x, y, z; take the above volume integral

corresponding to this value of δ through all points of space $x'y'z'$, and we obtain the corresponding ϕ function which fulfils the condition $\nabla^2\phi = \delta$. Now, let us compound displacements $-\dfrac{d\phi}{dx}$, etc., with the actual displacements and denote the resultant as follows:

$$\xi - \frac{d\phi}{dx} = u, \quad \eta - \frac{d\phi}{dy} = v, \quad \zeta - \frac{d\phi}{dz} = w;$$

and remarking that therefore

$$\frac{du}{dx} + \frac{dv}{dy} + \frac{dw}{dz} = 0,$$

we see the proposition that we had before us yesterday established. Hence to solve the three equations (1) we have simply to find δ by solution of the one equation

$$\rho \frac{d^2\delta}{dt^2} = (m+n)\nabla^2\delta,$$

and to deduce ϕ from δ, by (3); or to find ϕ direct by solution of the equation

$$\rho \frac{d^2\phi}{dt^2} = (m+n)\nabla^2\phi;$$

and u, v, w from the three separate similar equations with n in the place of $(m+n)$, subject to the conditions

$$\frac{du}{dx} + \frac{dv}{dy} + \frac{dw}{dz} = 0.$$

We shall take our ϕ equation and see how we can from it obtain different forms of ϕ solutions. We can do that for the purpose of illustrating different problems in sound, and in order to familiarize you with the wave that may exist along with the wave of distortion in any true elastic solid which is not incompressible. We ignore this condensational wave in the theory of light. We are sure that its energy at all events, if it is not null, is very small in comparison with the energy of the luminiferous vibrations we are dealing with. But to say that it is absolutely null would be an assumption that we have no right to make. When we look through the little universe that we know, and think of the transmission of electrical force and of the transmission of magnetic force and of the transmission of light, we have no right to assume that there may not be something else that our philo-

Molar.   sophy does not dream of. We have no right to assume that there
may not be condensational waves in the luminiferous ether. We
only do know that any vibrations of this kind which are excited
by the reflection and refraction of light are certainly of very small
energy compared with the energy of the light from which they
proceed. The fact of the case as regards reflection and refraction
is this, that unless the luminiferous ether is absolutely incompressible, the reflection and refraction of light must generally
give rise to waves of condensation. Waves of distortion may
exist without waves of condensation, but waves of distortion cannot
be reflected at the bounding surface between two mediums without
exciting in each medium a wave of condensation. When we come
to the subject of reflection and refraction, we shall see how to
deal with these condensational waves and find how easy it is to
get quit of them by supposing the medium to be incompressible.
But it is always to be kept in mind as to be examined into, are
there or are there not very small amounts of condensational waves
generated in reflection and refraction, and may after all, the propagation of electric force be by these waves of condensation ?

Suppose that we have at any place in air, or in luminiferous
ether (I cannot distinguish now between the two ideas) a body
that through some action we need not describe, but which is conceivable, is alternately positively and negatively electrified; may
it not be that this will give rise to condensational waves ? Suppose,
for example, that we have two spherical conductors united by a
fine wire, and that an alternating electromotive force is produced in that fine wire, for instance by an "alternate current"
dynamo-electric machine; and suppose that sort of thing goes on
away from all other disturbance—at a great distance up in the
air, for example. The result of the action of the dynamo-electric
machine will be that one conductor will be alternately positively
and negatively electrified, and the other conductor negatively and
positively electrified. It is perfectly certain, if we turn the
machine slowly, that in the air in the neighbourhood of the conductors we shall have alternately positively and negatively directed
electric force with reversals of, for example, two or three hundred
per second of time with a gradual transition from negative
through zero to positive, and so on; and the same thing all
through space ; and we can tell exactly what the potential and
what the electric force is at each instant at any point. Now, does

any one believe that if that revolution was made fast enough the electro-static law of force, pure and simple, would apply to the air at different distances from each globe? Every one believes that if that process be conducted fast enough, several million times, or millions of million times per second, we should have large deviation from the electrostatic law in the distribution of electric force through the air in the neighbourhood. It seems absolutely certain that such an action as that going on would give rise to electrical waves. Now it does seem to me probable that those electrical waves are condensational waves in luminiferous ether; and probably it would be that the propagation of these waves would be enormously faster than the propagation of ordinary light waves.

I am quite conscious, when speaking of this, of what has been done in the so-called Electro-Magnetic theory of light. I know the propagation of electric impulse along an insulated wire surrounded by gutta percha, which I worked out myself about the year 1854, and in which I found a velocity comparable with the velocity of light*. We then did not know the relation between electro-static and electro-magnetic units. If we work that out for the case of air instead of gutta percha, we get simply "v," (that is, the number of electrostatic units in the electro-magnetic unit of quantity,) for the velocity of propagation of the impulse. That is a very different case from this very rapidly varying electrification I have ideally put before you : and I have waited in vain to see how we can get any justification of the way of putting the idea of electric and magnetic waves in the so-called electro-magnetic theory of light.

I may refer to a little article of mine in which I gave a sort of mechanical representation of electric, magnetic, and galvanic forces—galvanic force I called it then, a very badly chosen name. It is published in the first volume of the reprint of my papers. It is shown in that paper that the static displacement of an elastic solid follows exactly the laws of the electro-static force, and that rotatory displacement of the medium follows exactly the laws of magnetic force. It seems to me that an incorporation of the theory of the propagation of electric and magnetic disturbances with the wave theory of light is most probably to be arrived at by trying to see clearly the view that I am now indicating. In the wave theory of light, however, we shall simply suppose the

* (See an Appendix near the end of the present volume.)

Molar. resistance to compression of the luminiferous ether and the velocity of propagation of the condensational wave in it to be infinite. We shall sometimes use the words "practically infinite" to guard against supposing these quantities to be absolutely infinite.

I will now take two or three illustrations of this solution for condensational waves. Part of the problem that I referred to yesterday says:—prove that the following is a solution of (7), the equation of motion,

$$\phi = \frac{1}{r} \sin \frac{2\pi}{\lambda} \left(r - t \sqrt{\frac{k + \frac{4}{3}n}{\rho}} \right);$$

or, if we put for brevity,

$$q = \frac{2\pi}{\lambda} \left(r - t \sqrt{\frac{k + \frac{4}{3}n}{\rho}} \right) \dots\dots\dots\dots(10),$$

$$\phi = \frac{\sin q}{r} \dots\dots\dots\dots\dots\dots\dots(11).$$

The question might be put into more analytical form:—to find a solution of (7) isotropic in respect to the origin of co-ordinates; or to solve (7) on the assumption that ϕ is a function of r and t. Taking this then as our problem, remark that we now have (because ϕ is a function of r)

$$\nabla^2\phi = r^2 \frac{d}{dr} \left(r^2 \frac{d\phi}{dr} \right) = r \frac{d^2(r\phi)}{dr^2} \dots\dots\dots(12).$$

Hence (7), with both sides multiplied by r^2, becomes

$$\rho \frac{d^2(r\phi)}{dt^2} = (k + \tfrac{4}{3}n) \frac{d^2(r\phi)}{dr^2} \dots\dots\dots(13),$$

of which the general solution is

$$\left. \begin{array}{l} r\phi = F(r - vt) + f(r + vt) \\[2mm] V = \sqrt{\dfrac{k + \frac{4}{3}n}{\rho}} \end{array} \right\} \dots\dots\dots(14),$$

where

and F and f denote two arbitrary functions.

This result simply expresses wave disturbance of $r\phi$, with velocity of propagation $\sqrt{[(k + \frac{4}{3}n)/\rho]}$: and it proves (9), being merely the case of a simple harmonic wave disturbance propagated in the direction of r increasing, that is to say outwards from the origin.

Here then is the determination of a mode of motion which is possible for an elastic solid. We shall consider the nature of this motion presently. The factor $\frac{1}{r}$ in (9) prevents it from being a pure wave motion. Passing over that consideration for the present, we note that it is less and less effective, relatively to the motion considered the farther we go from the centre.

In the meantime, we remark that the velocity of propagation in an elastic solid is but little greater than in a fluid with the same resistance to compression. k is the bulk modulus and measures resistance to compression, n is the rigidity modulus. I may hereafter consider relations between k and n for real solids. k is generally several times n, so that $\frac{4}{3}n$ is small in comparison with k, and therefore in ordinary solids the velocity of propagation of the condensational wave is not *greatly* greater than if the solid were deprived of rigidity and we had an elastic fluid of the same bulk modulus.

I shall want to look at the motion in the neighbourhood of the source. That beautiful investigation of Stokes, quoted by Lord Rayleigh, has to do entirely with the region in which the change of value of the factor $\left(\frac{1}{r}\right)$ from point to point is considerable. Without looking at that now, let us find the components of the displacement and their resultant, and study carefully all the circumstances of the motion.

$\frac{d\phi}{dx}, \frac{d\phi}{dy}, \frac{d\phi}{dz}$, are the three components of the displacement.

Clearly, therefore, the displacement will be in the direction of the radius because everything is symmetrical; and its magnitude will be $\frac{d\phi}{dr}$: and from (11) and (10) we find

$$\frac{d\phi}{dr} = \frac{-1}{r^2}\sin q + \frac{2\pi}{\lambda}\frac{1}{r}\cos q \ \ldots\ldots\ldots\ldots(15).$$

Having obtained this solution of our equations, let us see what we can make of interpreting it. When r is great in comparison with $\frac{\lambda}{2\pi}$, the first term becomes very small in comparison with

Molar.   the second and we have

$$\frac{d\phi}{dr} \fallingdotseq \frac{2\pi}{\lambda} \frac{1}{r} \cos q \quad \dots\dots\dots\dots\dots(16)*.$$

Therefore, when the distance from the origin is a great many wave-lengths, the displacement is sensibly equal to $\frac{2\pi}{\lambda} \frac{1}{r} \cos q$, and is therefore approximately in the inverse proportion to the distance; and the intensity of the sound if the solution were to be applied to sound, would be inversely as the square of the distance from the source.

I want now to get a second and a third solution. Take

$$\psi = \frac{\lambda\dagger}{2\pi} \frac{d\phi}{dx} = \frac{x}{r^2}\left(\cos q - \frac{\lambda}{2\pi r} \sin q\right)\dots\dots\dots\dots(17),$$

as the velocity potential for a fresh solution. I take it that you all know that if we have one solution ϕ, for the velocity potential, we can get another solution by ψ any linear function of

$$\frac{d\phi}{dx}, \quad \frac{d\phi}{dy}, \quad \frac{d\phi}{dz}.$$

Now let us find the displacements

$$\frac{d\psi}{dx}, \quad \frac{d\psi}{dy}, \quad \frac{d\psi}{dz}.$$

Here I want to prove that though this solution is no longer symmetrical with respect to r, so that there will be motions other than radial in the neighbourhood of the source, yet still the motion is approximately radial at great distances from the source. Work it out, and you will find that

$$\frac{d\psi}{dx} = -\frac{2\pi}{\lambda} \frac{\sin q}{r}\left[\frac{x^2}{r^2} + \left(\frac{\lambda}{2\pi r}\right)^2 \frac{r^2 - 3x^2}{r^2}\right] + \frac{\cos q}{r^2} \frac{r^2 - 3x^2}{x^2} \quad \dots\dots(18).$$

The principal term here is $-\frac{2\pi}{\lambda} \frac{x^2}{r^3} \sin q$. We might go on to the third and fourth and higher differential coefficients of ϕ, with their larger and larger numbers of terms. The interpretation of this multiplicity of terms, of the terms other than those which I am now calling the "principal terms," is all-important in respect to the motion of the air in the neighbourhood of the source. It

---

* I use $\fallingdotseq$ to denote approximate equality.

† $\lambda/2\pi$ is introduced merely for convenience. The solution differs from $\psi = d\phi/dx$, only by a constant factor.

is dealt with in that splendid work of Stokes, one of the finest Molar. things ever written in physical mathematics, of which I read to you this afternoon, with reference to the effect of an atmosphere of hydrogen round a bell killing its sound. But we will drop those terms and think only of the terms which express the efficiency of the vibrator at distances great in comparison with the wave-length.

Thus, for the x-component displacement of the motion now considered, we have

$$\xi = -\frac{2\pi}{\lambda}\frac{x^2}{r^3}\sin 2\pi\left(\frac{r}{\lambda}-\frac{t}{\tau}\right) \dots\dots\dots\dots(19).$$

This approximate equality is true for distances from the centre great in comparison with the wave-length. Let me remark, it is the differentiation of $\cos q$ that gives the distantly effective terms of the displacement; and in differentiating ψ with respect to y, you have simply to differentiate $\cos q$ and to take the differential coefficient of r now with respect to y, instead of x as formerly. So that we may write down the principal terms of the y and z displacements by taking y/x and z/x of the second member of (19) as follows :—

$$\eta = -\frac{2\pi}{\lambda}\frac{xy}{r^3}\sin q, \qquad \zeta = -\frac{2\pi}{\lambda}\frac{xz}{r^3}\sin q.$$

The three component displacements being proportional to x, y, z, shows that the resultant displacement is in the direction of the radius; and its magnitude is $-\dfrac{2\pi}{\lambda}\dfrac{x}{r^2}\sin q$. If we write $x = r\cos i$, this becomes

$$-\frac{2\pi}{\lambda}\frac{\cos i}{r}\sin q \dots\dots\dots\dots\dots\dots(20),$$

or the displacement is inversely proportional to the distance. If $i = 0$ we have a maximum; if $i = \dfrac{\pi}{2}$ we have zero. The upshot of it is that the displacement is a maximum in the axis OX, zero everywhere in the plane of OY, OZ; and symmetrical all round the axis OX.

A third solution is got by taking $\dfrac{d^2\phi}{dx^2}$ as our velocity potential. At a distance from the origin, great in comparison with the wave-length, the displacement is in the direction of the radius, and its magnitude is $\dfrac{d}{dr}\dfrac{d^2\phi}{dx^2}$.

Now the interpretation of these cases is as follows:—The first solution, (velocity potential ϕ) a globe alternately becoming larger and smaller; the second solution, (velocity potential $d\phi/dx$) a globe vibrating to and fro in a straight line; the third solution, (velocity potential $d^2\phi/dx^2$) a characteristic constituent of the motion of the air produced by two globes vibrating to and fro in the line of their centres, or by the prongs of a vibrating fork.

This last requires a little nice consideration, and we shall take it up in a subsequent lecture. The third mode does not quite represent the motion in the neighbourhood of the pair of vibrating globes, or of the prongs of a vibrating fork; there must be an unknown amount of the first mode compounded with the third mode for this purpose. The expression for the vibration in the neighbourhood of a tuning fork, going so far from the ends of it that we will be undisturbed or but little disturbed, by the general shape of the whole thing, will be given by a velocity potential $A\phi + \dfrac{d^2\phi}{dx^2}$. That will be the velocity potential for the chief terms, the terms which alone have effect at great distances. The differentiation will be performed simply with reference to the r in the term $\sin q$ or $\cos q$; and will be the same as if the coefficient of $\sin q$ or $\cos q$ were constant. A differentiation of this velocity potential will show that the displacement is in the direction of the radius from the centre of the system, and the magnitude of the displacement will be $\dfrac{d}{dr}\left(A\phi + \dfrac{d^2\phi}{dx^2}\right)$.

A is an unknown quantity depending upon the tuning fork. I want to suggest this as a junior laboratory exercise, to try tuning forks with different breadths of prongs. When you take tuning forks with prongs a considerable distance asunder you have much less of the ϕ in the solution: try a tuning fork with flat prongs, pretty close together, and you will find much more of the ϕ. The ϕ part of the velocity potential corresponds to the alternate swelling and shrinking of the air between the two prongs of the tuning fork. The larger and flatter the prongs are the greater is the proportion of the ϕ solution, that is to say, the larger is the

value of A in that formula; and the smaller the angle of the cone   Molar.
of silence.

The experiment that I suggest is this: take a vibrating tuning fork and turn it round until you find the cone of silence, or find the angle between the line joining the prongs and the line going to the place where your ear must be to hear no sound. The suddenness of transition from sound to no sound is startling. Having the tuning fork in the hand, turn it slowly round near one ear until you find its position of silence. Close the other ear with your hand. A very small angle of turning round the vertical axis from that position gives you a startlingly loud sound. I think it is very likely that the place of no sound will, with one and the same fork, depend on the range of vibration. If you excite it very powerfully, you may find less inclination between the line of vibration of the prongs and the line to the place of silence; less powerfully, greater inclination. It will certainly be different with different tuning forks.

LECTURE V.

SATURDAY, *Oct.* 4, 5 P.M.

Molar:
Recapitu-
lation.

I STATED in the last lecture that the second solution, corresponding to the velocity potential $\frac{d\phi}{dx}$, would represent the effect,
at a great distance from the mean position, of a single body vibrating to and fro in a straight line. I said a sphere, but we may take
a body of any shape vibrating to and fro in a straight line; and at
a very great distance from the vibrator, the motion produced will
be represented by the velocity potential $\frac{d\phi}{dx}$, provided the period
of the vibration is great in comparison with the time taken by
sound to travel a distance equal to the greatest diameter of the
body. Then the velocity potential $\frac{d^2\phi}{dx^2}$, in the third solution,
would, I believe, represent (without an additional term $A\phi$) the
motion at great distances, when the origin of the sound consists
in two globes, let us say, for fixing the ideas, placed at a distance
from one another very great in comparison with their diameters
and set to vibrate to and fro through a range small in comparison
with the distance between them, but not necessarily small in
comparison with their diameters: provided always that the period
of the vibration is great in comparison with the time taken by
sound to travel the distance between the vibrators. Suppose
this is a globe in one hand, and this is one in the other. I now
move my hands towards and from each other—the motion of the
air produced by that sort of motion of the exciting bodies would, at
a very great distance, be expressed exactly by the velocity potential
$\frac{d^2\phi}{dx^2}$.

But when you have two globes, or two flat bodies, very near
one another, you need an unknown amount of the ϕ vibration to

represent the actual state of the case. That unknown amount might be determined theoretically for the case of two spheres. The problem is analogous to Poisson's problem of the distribution of electricity upon two spheres, and it has been solved by Stokes for the case of. fluid motion (see *Mem. de l'Inst.*, Paris, 1811, pp. 1, 163; and Stokes' Papers, Vol. I., p. 230—"On the resistance of a fluid to two oscillating spheres"). You can thus tell the motion exactly in the neighbourhood of two spheres vibrating to and fro provided the amplitudes of their vibrations are small in comparison with the distance between them; and you can find the value of A for two spheres of any given radii and any given distance between them. For such a thing as a tuning fork, you could not, of course, work it out theoretically; but I think it would be an interesting subject for junior laboratory work, to find it by experiment.

I suppose you are all now familiar with the zero of sound in the neighbourhood of a tuning fork; but I have never seen it described correctly anywhere. We have no easy enough theoretical means of determining the inclination of the line going to the position of the ear for silence to the line joining the prongs; but we readily see that it is dependent upon the proportions of the body. In turning the tuning fork round its axis, you can get with great nicety the position for silence; and a surprisingly small turning of the tuning fork from the position of silence causes the motion to be heard. It would be very curious to find whether the position of zero sound varies perceptibly with the amplitude of the vibrations. I doubt whether any perceptible difference will be found in any ordinary case however we vary the amplitude of the vibrations. But I am quite sure you will find considerable difference, according as you take tuning forks with cylindrical prongs, or with rectangular prongs of such proportions as old Marlowe used to make, or tuning forks like the more modern ones that Koenig makes, with very broad flat prongs.

Now for our molecular problem.

I want to see how the variable quantities vary, when we vary the period. Remember that

$$a_i = \frac{m_i}{\tau^2} - c_i - c_{i+1} \dots\dots\dots\dots(1);$$

so that
$$\frac{da_i}{d\left(\tau^{-2}\right)} = m_i \quad\dots\dots\dots\dots\dots\dots\dots(2).$$

Write for the moment ∂ for $\dfrac{d}{d\left(\tau^{-2}\right)}$, and differentiate the equation for u_i; we find

$$\partial u_i = m_i + \left(\frac{c_{i+1}}{u_{i+1}}\right)^2 \partial u_{i+1}, \quad \partial u_{i+1} = m_{i+1} + \left(\frac{c_{i+2}}{u_{i+2}}\right)^2 \partial u_{i+2}, \dots \partial u_j = m_j.$$

Substitute successively, and we find,

$$\partial u_i = m_i + \left(\frac{c_{i+1}}{u_{i+1}}\right)^2 m_{i+1}$$
$$+ \left(\frac{c_{i+1}c_{i+2}}{u_{i+1}u_{i+2}}\right)^2 m_{i+2} + \dots \left(\frac{c_{i+1}\dots c_j}{u_{i+1}\dots u_j}\right)^2 m_j \quad\dots\dots\dots\dots(3).$$

This is our expression, and remark the exceedingly important property of it that it is essentially positive, i.e. the variation of u_i with increase of τ^{-2} is essentially positive. Now

$$\frac{du_i}{d\tau} = -2\tau^{-3}\partial u_i:$$

also $\quad \dfrac{c_{i+1}}{u_{i+1}} = -\dfrac{x_{i+1}}{x_i}, \quad \dfrac{c_{i+2}}{u_{i+2}} = -\dfrac{x_{i+2}}{x_{i+1}}$, and so on.

The result (3) therefore is equivalent to the following expression (4) for the differential coefficient of u_i with respect to the period.

$$\frac{du_i}{d\tau} = -\frac{2}{\tau^3}\cdot\frac{1}{x_i^2}\left(m_i x_i^2 + m_{i+1}x_{i+1}^2 + \dots m_j x_j^2\right)\dots\dots\dots(4).$$

This is certainly a very remarkable theorem, and one of great importance with reference to the interpretation of the solution of our problem. Remember that x_i is the displacement of m_i at any time of the motion. You may habitually think of the maximum values of the displacements, but it is not necessary to confine yourselves to the maximum values. Instead of $x_1, x_2, \dots x_j$ we may take constants equal to the maximum values of the x's, each multiplied into $\sin\dfrac{2\pi t}{\tau}$, because the particles vibrate according to the simple harmonic law, all in the same period and the same phase, that is all passing through zero simultaneously, and reaching maximums simultaneously, every vibration. The masses are positive, and we have squares of the displacements in the several terms of (4); so that the second member of (4) is essentially negative. Hence, as we augment the period, each one of the ratios Molecular. u_i decreases.

Let us now consider the configurations of motion in our spring arrangement, for different given periods of the exciting vibrator, P. I am going to suppose, in the first place, that the period of vibration is very small, and is then gradually increased. As you increase the period, we have seen that the value of each one of the quantities $u_1, u_2, \ldots$ decreases. It is interesting to remark that this is so continuously throughout each variety of the configurations found successively by increasing τ from 0 to ∞. But we shall find that there are critical values of τ, at which one or other of the u's, having become negative decreases to $-\infty$; then suddenly jumps to $+\infty$ as τ is augmented through a critical value; and again decreases, possibly again coming to $\mp \infty$, possibly not, while τ is augmented farther and farther, to infinity. In the first place, τ may be taken so small that the u's are all very large positive quantities; for u_i being equal to $\dfrac{m_i}{\tau^2} - c_i - c_{i+1}, \ -\dfrac{c_{i+1}^2}{u_{i+1}}$ may be certainly made as very large positive as we please by taking τ small enough, if, at the same time the succeeding quantity, u_{i+1}, is large, a condition which we see is essentially fulfilled where τ is very small, because we have $u_j = m_j/\tau^2 - c_j - c_{j+1}$, which makes u_j very great.

Observe that the u's all positive implies that $\xi, x_1, x_2, x_3, \ldots x_j$ are alternately positive and negative. In other words the handle P and the successive particles $m_1, m_2, \ldots m_j$, are each moving in a direction opposite to its neighbour on either side. Since the magnitudes of the ratios $u_1, u_2, \ldots u_j$ of the successive amplitudes decrease with the increase of the period, the amplitude of particle m_i is becoming smaller in proportion to the amplitude of the succeeding particle m_{i+1}, as long as the vibrations of the successive particles are mutually contrary-wards. I am going to show you that as every one of these quantities u_i decreases, the first that passes through zero is necessarily u_1; corresponding to a motion of each particle of the system infinitely great in comparison with the motion of the handle P; that is to say, finite simple harmonic motion of the system with P held fixed. This is our first critical case. It is the only one of the j fundamental modes of vibration of the system with P fixed, in which the directions of vibration of the successive particles are all mutually contrariwise, and it is the one

Molecular. of them of which the period is shortest.   After that, as we increase τ, u_1 becomes negative, and the motion of P comes to be in
the direction of the motion of the first particle.   As we go on
increasing the period we shall find that the next critical case that
comes is one in which particle m_1 has zero motion, or

$$u_1 = \frac{c_1 \xi}{-x} = -\infty .$$

To prove this, and to investigate the further progress, let us look
at the state of things when a positive decreasing u_i has approached
very near to zero.   We shall have u_{i-1}, being equal to $a_{i-1} - \dfrac{c_1^{\,2}}{u_i}$, a
very large negative quantity.   This alone shows that u_{i-1} must
have preceded u_i in becoming zero, since it must have passed
through zero before becoming negative.   Therefore, as we augment
τ, the first of the u's to become zero is $u_1 = \dfrac{c_1 \xi}{-x}$; or, as I said
before, the motion of particle m_1 and also of each of the other
particles is infinite in comparison with the motion of P.   Just
before this state of things all the particles P, m_1, ... m_j are, as we
saw, moving each contrary-wards to its neighbour ; just after it, P
has reversed its motion with reference to the first particle, and is
moving in the same direction with it.

This continues to be the configuration, till just before the
second critical case, in which we have u_1 large negative, u_2 small
positive, u_3, ... u_j, all positive.   At this critical case, we have

$$u_1 = \frac{c_1 \xi}{-x_1} = \mp \infty \; ; \; \text{or } x_1 = \pm\, 0 \times \xi.$$

The period of motion of P that will produce this state of things is
equal to the period of the free vibration of the system of particles,
with mass m_1 held at rest, and each of the other masses moving
contrary-wards to its neighbour on each side.   When the period of
the simple harmonic motion of P is equal to a period of motion
of the system with the first particle held at rest, then the only
simple harmonic motion which the system with all the particles
unconstrained can have is in that period, and with the amplitude
of vibration of the second particle in one direction just so great as
to produce by spring No. 2, a pull in that direction, on m_1, equal
to the pull exercised on it through spring No. 1, by P in the
opposite direction ; so as to let the first particle be at rest.   Sup-

pose now τ to be continuously increased through this critical value. Molecular. Immediately after the critical case, u_1 has changed from large negative, through $\mp \infty$, to large positive, and u_2 from small positive, through ± 0, to small negative; or the first particle has reversed the direction of its motion and come to move same-wards with P and contrary-wards to the second particle.

The third critical case might be that of the second particle coming to rest, $(u_2 = \mp \infty, u_3 = \pm 0)$; or it might be $u_1 = 0$ a second time: it must be either one or other of these two cases. But we must not stop longer on the line of critical cases at present*. I will just jump over the remaining critical cases to the final condition.

It would be curious to find the solution when the period is infinitely great out of our equations. When τ is infinite, $\dfrac{m_i}{\tau^2}$

[Note added; Jan. 11, 1886, Netherhall, Largs.]

* As we go on increasing τ from the first critical value (that which made $u_1 = 0$), the essential decreasings of u_1 (negative) and u_2 (still positive) bring u_1 to $-\infty$ and u_2 to 0 simultaneously. With farther increase of τ, the decreasings of u_2 (now negative) and of u_3 (still positive) bring u_2 to $-\infty$ and u_3 to 0, simultaneously; and so on, in succession from u_1 to u_j which passes through zero to negative, but cannot become $-\infty$ and remains negative for all greater values of τ, diminishing to the value $-(c_j + c_{j+1})$ as τ is augmented to infinity.

But u_1, after decreasing to $-\infty$, must pass to $+\infty$ and again become decreasing positive. It must again pass through zero; and thus there is started another procession of zeros along the line from P, through $m_1, m_2, \ldots$ successively, but ending in m_{j-2}: not in m_{j-1} whose amplitude $(u_j a_j / c_j)$ is made zero and negative by the conclusion of the first procession, before the second procession can possibly reach it. Thus u_{j-1} passes a second and last time through zero and diminishes to the limiting values $-c_{j-1}\left(\dfrac{1}{c_{j+1}} + \dfrac{1}{c_j} + \dfrac{1}{c_{j-1}}\right) \Big/ \left(\dfrac{1}{c_{j-2}} + \dfrac{1}{c_{j-1}}\right)$, as τ is augmented to ∞.

A third procession of zeros, similarly commencing with P, passing along the line, $m_1, m_2, \ldots$ and ending with m_{j-3}, makes u_{j-2} zero for a third and last time, and leaves it to diminish to its limit (shown by the formula reported in the text below, from the lecture,) as τ augments to ∞. Similarly procession after procession, in all j processions, commence with m_1. The last begins and ends in m_1, and leaves u_1 to go from zero to its negative limiting value

$$-c_1\left(\frac{1}{c_{j+1}} + \frac{1}{c_j} + \ldots + \frac{1}{c_2} + \frac{1}{c_1}\right) \Big/ \left(\frac{1}{c_{j+1}} + \frac{1}{c_j} + \ldots + \frac{1}{c_2}\right),$$

as τ augments to ∞. There is no general rule of precedence in respect to magnitude of τ, of the different transitional zeros of the different processions. For example, there is no general rule as to order of commencement of one procession, and termination of its predecessor. The one essential limitation is that no collision can take place between the front of one procession and the rear of its predecessor.

Molecular. vanishes, and $a_j = -c_i - c_{i+1}$. That applied to the equations for the u's ought to find the solution quite readily.

You know, when you think of the dynamics of this case, that when τ is infinitely great, P is moving infinitely slowly, so that the inertia of each particle has no sensible effect; and all the particles are in equilibrium. Let F be the force, then, on the spring; that is to say, pull P slowly down with a force F and hold it at rest. What will be the displacements of the different particles? Answer, $x_j = \dfrac{F}{c_{j+1}}$, $x_{j-1} = \dfrac{F}{c_{j+1}} + \dfrac{F}{c_j}$, and so on. Particle number j is displaced to a distance equal to the force, divided by the coefficient of elongation of the spring. To obtain the displacement of particle $j-1$, we have to add the displacement resulting from the elongation of the next spring c_j, and so on. The general equation then is

$$x_i = \left(\frac{1}{c_{j+1}} + \frac{1}{c_j} + \dots \frac{1}{c_{i+1}} \right) F.$$

$$\therefore \quad u_i = -c_i \left(\frac{1}{c_{j+1}} + \dots \frac{1}{c_i} \right) \bigg/ \left(\frac{1}{c_{j+1}} + \dots \frac{1}{c_{i+1}} \right).$$

It is a curious but of course a very simple and easy problem to substitute the value of $a_i = -c_i - c_{i+1}$ in the continued fraction which gives u_i, and verify this solution. No more of this now however.

It is fiddling while Rome is burning; to be playing with trivialities of a little dynamical problem, when phosphorescence is in view, and when explanation of the refraction of light in crystals is waiting. The difficulty is, not to explain phosphorescence and fluorescence, but to explain why there is so little of sensible fluorescence and phosphorescence. This molecular theory brings everything of light, to fluorescence and phosphorescence. The state of things as regards our complex model-molecule would be this: Suppose we have this handle P moved backwards and forwards until everything is in a perfectly periodic state. Then suddenly stop moving P. The system will continue vibrating for ever with a complex vibration which will really partake something of all the modes. That I believe is fluorescence.

But now comes Mr Michelson's question, and Mr Newcomb's question, and Lord Rayleigh's question;—the velocity of groups of waves of light in gross matter. Suppose a succession of luminous vibrations commences. In the commencement of the luminous vi- Molecular. brations the attached molecules imbedded in the luminiferous ether, do not immediately get into the state of simple harmonic vibration which constitutes a regular light. It seems quite certain that there must be an initial fluorescence. Let light begin shining on uranium glass. For the thousandth of a second, perhaps, after the light has begun shining on it, you should find an initial state of things, which differs from the permanent state of things somewhat as fluorescence differs from no light at all.

There is still another question. which is of profound interest, and seems to present many difficulties, and that is, the actual condition of the light which is a succession of groups. Lord Rayleigh has told us in his printed paper in respect to the agitated question of the velocity of light, and then again, at the meeting of the British Association at Montreal, he repeated very peremptorily and clearly, that the velocity of a group of waves must not be confounded with the wave velocity of an infinite succession of waves, and is of necessity largely different from the velocity of an infinite succession of waves, in every dispersively refracting medium, that is to say, medium in which the velocity is different for lights of different period. It seems to be quite certain that what he said is true. But here is a difficulty which has only occurred to me since I began speaking to you on the subject; and I hope, before we separate, we shall see our way through it. All light consists in a succession of groups. We are all, already, familiar with the question;—Why is all light not polarized? and we are all familiar with its answer. We are now going to work our way slowly on until we get expressions for sequences of vibrations of existing light. Take any conceivable supposition as to the origin of light, in a flame, or a wire made incandescent by an electric current, or any other source of light; we shall work our way up from these equations which we have used for sound, to the corresponding expression for light from any conceivable source. Now, if we conceive a source consisting of a motion kept going on with perfectly uniform periodicity; the light from that source would be plane polarized, or circularly polarized, or elliptically polarized, and would be absolutely constant. In reality, there is a multiplicity of successions of groups of waves, and no constant periodicity. One molecule, of enormous mass in comparison with the luminiferous ether that

Molecular. it displaces, gets a shock and it performs a set of vibrations until it comes to rest or gets a shock in some other direction; and it is sending forth vibrations with the same want of regularity that is exhibited in a group of sounding bodies consisting of bells, tuning forks, organ pipes, or all the instruments of an orchestra played independently in wildest confusion, every one of which is sending forth its sound which, at large enough distances from the source, is propagated as if there were no others. We thus see that light is essentially composed of groups of waves; and if the velocity of the front or rear of a group of waves, or of the centre of gravity of a group, differs from the wave-velocity of absolutely continuous sequences of waves, in water, or glass, or other dispersively refracting mediums, we have some of the ground cut from under us in respect to the velocity of waves of light in all such mediums.

I mean to say, that all light consists of groups following one another, irregularly, and that there is a difficulty to see what to make of the beginning and end of the vibrations of a group: and that then there is the question which was talked over a little in Section A at Montreal,—will the mean of the effects of the groups be the same as that of an infinite sequence of uniform waves, and will the deviation from regular periodicity at the beginning and end of each group have but a small influence on the whole? It seems almost certain that it must have but a small influence from the known facts regarding the velocity of light proved by the known, well-observed, and accurately measured, phenomena of refraction and interference. But I am leading you into a muddle, not however for you, I hope, a slough of despond; though I lead you into it and do not show you the way out. You will all think a good deal along with me about the connections of this subject.

LECTURE VI.

I WANT to ask you to note that when I spoke of $k + \frac{1}{3}n$ not   Molar. differing very much from k for most solids, I was rather under the impression for the moment that the ratio of n to k was smaller than it is; and also you will remember that we had $k + \frac{1}{3}n$ on the board by mistake for $k + \frac{4}{3}n$. The square of the velocity of a condensational wave in an elastic solid is $(k + \frac{4}{3}n)/\rho$. For solids fulfilling the supposed relation of Navier and Poisson between compressibility and rigidity we have $n = \frac{3}{5}k$; and for such cases the numerator becomes $\frac{9}{5}k$. It would be k if there were no rigidity; it is $\frac{9}{5}k$ if the rigidity is that of a solid for which Poisson's ratio has its supposed value.

Metals are not enormously far from fulfilling this condition, but it seems that for elastic solids generally n bears a less proportion to k than this. It is by no means certain that it fulfils it even approximately for metals; and for india rubber, on the other hand, and for jellies, n is an exceedingly small fraction of k, so that in these cases the velocity of the condensational wave is but very little in excess of $\sqrt{\dfrac{k}{\rho}}$. The velocity of propagation of a distortional wave is $\sqrt{\dfrac{n}{\rho}}$; so that for jellies, the velocity of propagation of condensational wave is enormously greater than that of distortional waves.

I am asked by one of you to define velocity potential. Those who have read German writers on Hydrodynamics already know the meaning of it perfectly well. It is a purely technical expression which has nothing to do with potential or force. "Velocity potential" is a function of the co-ordinates such that its rate of variation per unit distance in any direction is equal to the

Molar.　component of velocity in that direction. A velocity potential exists when the distribution of velocity is expressible in this way; in other words when the motion is irrotational. The most convenient analytical definition of irrotational motion is, motion such that the velocity components are expressed by the differential coefficients of a function. That function is the velocity potential. When the motion is rotational there is no velocity potential.

This is the strict application of the words "velocity potential" which I have used. A corresponding language may be used for displacement potential. It is not good language, but it is convenient, it is rough and ready. And when we are speaking of component displacements in any case, whether of static displacement in an elastic solid or of vibrations, in which the components of displacement are expressible as the differential coefficients of a function, we may say that it is an irrotational displacement. If from the differentiation of a function we obtain components of velocity, we have velocity potential; whereas, if we so get components of displacement, we have displacement potential. The functions ϕ, that we used, are not then, strictly speaking, velocity potentials but displacement potentials.

I want you in the first place to remark what is perfectly well known to all who are familiar with Differential equations, that taking the solution $\phi = \dfrac{1}{r} \cdot \sin q$ as a primary, where

$$q = \frac{2\pi}{\lambda} \left(r - t \sqrt{\frac{k + \tfrac{4}{3}n}{\rho}} \right),$$

we may derive other solutions by differentiations with respect to the rectangular co-ordinates. The first thing I am going to call attention to is that at a distance from the origin, whatever be the solution derived from this primary by differentiation, the corresponding displacement is nearly in the direction through the origin of co-ordinates.

Take any differential coefficient whatever, $\dfrac{d^{i+j+k}}{dx^i dy^j dz^k}$; the term of this which alone is sensible at an infinitely great distance is that which is obtained by successive differentiation of $\sin q$. That distance term in every case is as follows:

$$\left(\frac{2\pi}{\lambda}\right)^{i+j+k} \cdot \left(\frac{dr}{dx}\right)^i \cdot \left(\frac{dr}{dy}\right)^j \cdot \left(\frac{dr}{dz}\right)^k \cdot \frac{1}{r} \frac{\sin}{\cos} q$$

It will be $\sin q$ or $\cos q$, according as $i + j + k$ is even or odd. We do not need to trouble ourselves about the algebraic sign, because we shall make it positive, whether the differential coefficient is positive or negative. Now $\dfrac{dr}{dx} = \dfrac{x}{r}$, $\dfrac{dr}{dy} = \dfrac{y}{r}$, $\dfrac{dr}{dz} = \dfrac{z}{r}$. Thus our type solution becomes, $\dfrac{x^i y^j z^k}{r^{i+j+k+1}} \dfrac{\sin}{\cos} q$. This expresses the most general type of displacement potential for a condensational wave proceeding from a centre, and having reached to a distance in any direction from the centre, great in comparison with the wave-length. I have not formally proved that this is the most general type, but it is very easy to do so. I am rather going into the thing synthetically. It is so thoroughly treated analytically by many writers that it would be a waste of your time to go into anything more, at present, than a sketch of the manner of treatment, and to give some illustrations.

But now to prove that the displacement at a distance from the origin of the disturbance is always in the direction of the radius vector. Once more, the differential coefficient of this displacement potential, which has several terms depending upon the differentiation of the r's, x's, etc. has one term of paramount importance, and that is the one in which you get $\dfrac{2\pi}{\lambda}$ as a factor. The smallness of λ in proportion to the other quantities makes the factor $\dfrac{2\pi}{\lambda}$ give importance to the term in which it is found. The distance terms then for the components of the displacement are

$$\xi = \frac{x^i y^j z^k}{r^{i+j+k+1}} \cdot \frac{2\pi}{\lambda} \cdot \frac{x}{r} \frac{\cos}{\sin} q = R \frac{x}{r} \frac{\cos}{\sin} q,$$

$$\eta = R \frac{y}{r} \frac{\cos}{\sin} q, \quad \zeta = R \frac{z}{r} \frac{\cos}{\sin} q.$$

These are then the components of a displacement which is radial; and the expression for the amplitude of the radial displacement is

$$R = \frac{2\pi}{\lambda} \cdot \frac{x^i y^j z^k}{r^{i+j+k+1}}.$$

The sum of any number of such expressions will express the distance effect of sound proceeding from a source. It is interest-

Molar.

ing to see how, simply by making up an algebraic function in the numerator out of the x's, y's and z's, we can get a formula that will express any amount of nodal subdivision where silence is felt. The most general result for the amplitude of the radial displacement is $R = \Sigma \frac{cx^i y^j z^k}{r^{i+j+k+1}}$. Remark that $\frac{x}{r}, \frac{y}{r}, \frac{z}{r}$ are merely angular functions and may be expressed at once as $\sin\theta\cos\psi$, $\sin\theta\sin\psi$, $\cos\theta$; and therefore R is an integral algebraic function of $\sin\theta\cos\psi$, $\sin\theta\sin\psi$, $\cos\theta$. It is thus easy to see that you can vary indefinitely the expressions for sound proceeding from a source with cones of silence and corresponding nodes or lines in which those cones cut the spherical wave surface. It is interesting to see that even in the neighbourhood of the nodes the vibration is still perpendicular to the wave surface; so that we have realized in any case a gradual falling off of the intensity of the wave to zero and a passing through zero, which would be equivalent to a change of phase, without any motion perpendicular to the radius vector.

The more complicated terms that I have passed over are those that are only sensible in the neighbourhood of the source. Suppose, for instance, that you have a bell vibrating. The air slipping out and in over the sides of the bell and round the opening gives rise to a very complicated state of motion close to the bell; and similarly with respect to a tuning fork. If you take a spherical body, you can very easily express the motion in terms of spherical harmonics. You see that in the neighbourhood of the sounding body there will be a great deal of vibration in directions perpendicular to the radius vector, compounded with motions out and in; but it is interesting to notice that all except the radial component motions become insensible at distances from the centre large in comparison with the wave-length. It is the consideration of the motion at distances that are moderate in comparison with the wave-length that Stokes has made the basis of that very interesting investigation with reference to Leslie's experiment of a bell vibrating in a vacuum, to which I have already referred. (Lecture III., pp. 36, 37 above.)

We may just notice, before I pass away from the subject, two or three points of the case, with reference to a tuning fork, a bell, and so on. Suppose the sounding body to be a circular bell. In that case clearly, if the bell be held with its lip horizontal, and if it be kept vibrating steadily in its gravest ordinary mode, the kind of vibration will be this: a vibration from a circular figure

 into an elliptic figure along one diameter, and

a swinging back through the circular figure into an

elliptic figure along the diameter at right angles to the

first. Clearly there would be practically a plane of silence here and another at right angles to it here (represented on the diagrams by dotted lines). Hence the solution for the radial component corresponding to this case, at a considerable distance from the bell, is $R = (\frac{1}{2} - \cos^2 A)\dfrac{\cos q}{r}$, in order that the component may vanish when $\cos^2 A = \frac{1}{2}$, or $A = \pm 45°$; A being an azimuthal angle if the axis of the bell is vertical.

On the other hand, consider a tuning-fork vibrating to-and-fro or an elongated (elliptic) bell, which I got from that fine old Frenchman, Koenig's predecessor, Marloye. It makes an exceedingly loud sound and has an advantage in acoustic experiments over a circular bell. If you set a circular bell vibrating and leave it to itself you always hear a beating sound, because the bell is approximately but not accurately symmetrical. Excite it with a bow, and take your finger off, and leave it to itself: and if you do not choose a proper place to touch it, for a fundamental mode, when you take your finger off it will execute the resultant of two fundamental modes.

I do not know whether the corresponding experiment with circular plates is familiar to any of you. I would be glad to know whether it is. I make it always before my own classes, in illustrating the subject. Take a circular plate—just one of the ordinary circular plates that are prepared for showing vibration in acoustic illustrations. Excite it in the usual way with a violoncello bow, and putting a finger, or two fingers, to the edge to make the quadrantal vibration. If sand is sprinkled on the plate, the vibrations toss it into sand-hills with ridges lying along two diameters of the disc, perpendicular to one another, one of them through the point or

Molar. two of them through the points of the edge touched by finger. Now cease bowing and take your finger off the edge of the disc.   The sand-hills are tossed up in the air and the sand is scattered to considerable distances on each side, and is continually tossed up and not allowed to rest anywhere.   At the same time you hear a beating sound.   But by a little trial, I find one place where if I touch with the finger, and ply the bow so as to make a quadrantal vibration, and if I then cease bowing and take off my finger, the sand remains undisturbed on two diameters at right angles, and no beat is heard.   Then having found one pair of nodal diameters, I know there will be another pair got by touching the plate here, 45° from the first place.   Now touch therefore with two fingers, at two points 90° from one another midway between the first and second pairs of nodal diameters: cause the plate to vibrate and then take off your fingers, and stop bowing: you will hear very marked beats; a sound gradually waxing from absolute silence to loudest sound, then gradually waning to silence again, and so on, alternating between loudest sound and absolute silence with perfect gradualness and regularity of waxing and waning.

Take a division of the circumference into six equal parts by three diameters, and you find the same thing over again.   Go on by trial touching the plate at two points 60° or 120° asunder, and bowing it 30° from either; and you will see the sand resting on the three diameters determined by your fingers.   Take off your fingers and you will in general see the sand scattered and hear a beat.   Follow your way around, little by little—it is very pretty when you come near a place of no beat.   The moment you take off your fingers you see the lines of nodes swaying to-and-fro on each side of a mean position, with a slow oscillation; and you hear a very distinct beat, though of a soft but perfectly regular character from loudest to least loud sound.   Get exactly the mean position, steadying the nodal sand-hills while still plying the bow: then cease bowing and suddenly remove your finger or fingers from the plate; you will see the nodal lines remaining absolutely still and you will hear a pure note without beats.   If you touch at exactly 30° from the nodal lines first found you will have the strongest beat possible, which is a beat from loud sound to silence.   Advance your fingers another 30° and you will again find the sand-hills remain absolutely still when you remove your fingers. You may go on in this way with eight and ten subdivisions, and so on; but you must not expect that the places for the sextantal, _{Molar.} octantal, and higher, subdivisions correspond to the places for the quadrantal subdivisions. The places for quadrantal subdivision will not in general be places for octantal subdivision. You must experiment separately for the octantal places, and you will find generally that their diameters are oblique to the quadrantal.

The reason for all this is quite obvious. In each case, the plate being only approximately circular and symmetrical, the general equation for the motion has two approximately equal roots corresponding to the nodes or divisions by one, two, three, or four diameters, and so on. These two roots always correspond to sounds differing a little from one another. The effect of putting the finger down at random is to cause the plate, as long as your finger is on it, to vibrate forcedly in a simple harmonic vibration of period greater than the one root and less than the other. But as soon as you take your finger off, the motion of the plate follows the law of superposition of fundamental modes; each fundamental mode being a simple harmonic vibration. I have often, in showing this experiment, tried musicians with two notes which were very nearly equal, and said to them, "Now, which of the two notes is the graver?" Rarely can they tell. The difference is generally too small for a merely musical ear, and the verdict is that the notes are "the same;" musicians are not accustomed to listen to sounds with scientific ears and do not always say rightly which is the graver note, even when the difference is perceptible. Any person can tell, after having made a few experiments of the kind, that this is the graver and that the less grave note, even though he may have what, for musicians, is an uncultivated ear, or truly a very bad ear for music, not good enough in fact to guide him in sounding a note with his voice, or to make him sing in tune if he tries to sing. It is very curious, when you have two notes which you thoroughly know are different, that if you sound first one and then the other, most people will say they are about the same. But sound them both together, and then you hear the discord of the two notes in approximate unison.

In every case of a circular plate vibrating between diametral lines of nodes, there is an even number of planes of silence in the surrounding air; being the planes perpendicular to the plate, through the nodal diameters. If you take a square plate or bell vibrating in a quadrantal mode, for instance, then you have two

Molar.   vertical planes of silence at right angles to one another.   If you
             make it vibrate with six or more subdivisions, you will have a
             corresponding number of planes of silence.

                 With reference to the motions in the neighbourhood of the
             tuning-fork, you get this beautiful idea, that we have essentially
             harmonic functions to express them.   Essentially algebraic func-
             tions of the co-ordinates appear in these distant terms, but in the
             other terms which Prof. Stokes has worked out, and which have
             been worked out in Prof. Rowland's paper on Electro-Magnetic
             Disturbances*, quite that kind of analysis appears, and it is most
             important.   I have not given you a detailed examination of that
             part of our general solution, but only called your attention specially
             to the " distance terms," partly because of their interest for sound
             and partly because the consideration of them prepares us for our
             special subject, waves of light.

                 To-morrow we shall begin and try to think of sources of waves
             of light.   I want to lead you up to the idea of what the simplest
             element of light is.   It must be polarized, and it must consist of
             a single sequence of vibrations.   A body gets a shock so as to
             vibrate; that body of itself then constitutes the very simplest
             source of light that we can have; it produces an element of light.
             An element of light consists essentially in a sequence of vibrations.
             It is very easy to show that the velocity of propagation of
             sequences in the pure luminiferous ether is constant.   The sequence
             goes on, only varying with the variation of the source.   As the
             source gradually subsides in giving out its energy the amplitude
             evidently decreases; but there will be no throwing off of wavelets
             forward, no lagging in the rear, no ambiguity as to the velocity of
             propagation.   But when light, consisting as it does of sequences
             of vibrations, is propagated through air or water, or glass, or
             crystal, what is the result ?   According to the discussion to which
             I have referred, the velocity should be quite uncertain, depending
             upon the number of waves in the sequence, and all this seems
             to present a complicated problem.

                 But I am anticipating a little.   We shall speak of this here-
             after.   One of you has asked me if I was going to get rid of the
             subject of groups of waves.   I do not see how we can ever get rid
             of it in the wave theory of light.   We must try to make the best
             of it, however.

                          * *Phil. Mag.* xvii., 1884, p. 413.   *Am. Jour. Math.* vi., 1884, p. 359.

Ten minutes' interval.

This question of the vibration of connected particles is a Molecular. peculiarly interesting and important problem. I hope you are not tired of it yet. You see that it is going to have many applications. In the first place remark that it might be made the base of the theory of the propagation of waves. When we take our particles uniformly distributed and connected by constant springs we may pass from the solution of the problems for the mutual influence of a group of particles to the theory, say, of the longitudinal vibrations of an elastic rod, or, by the same analysis, to the theory of the transverse vibrations of a cord.

I am going to refer you to Lagrange's *Mécanique Analytique*, [Part II. p. 339]. The problem that I put before you here is given in that work under the title of vibrations of a linear system of bodies. Lagrange applies what he calls the algorithm of finite differences to the solution. The problem which I put before you is of a much more comprehensive kind; but it is of some little interest to know that cases of it may be found, ramifying into each other.

I wish to put before you some properties of the solution which are of very great importance. I want you to note first the number of terms.

We have:

$$c_j x_{j-1} = -a_j x_j,$$

$$c_{j-1} x_{j-2} = -a_{j-1} x_{j-1} - c_j x_j = \frac{a_j a_{j-1}}{c_j} x_j - c_j x_j,$$

$$c_{j-2} x_{j-3} = \text{etc.}$$

All the x's being expressed in this way successively in terms of x_j. Let N_i be the number of terms in x_{j-i}. These terms are obtained by substituting the values of x_{j-i+1}, x_{j-i+2} in the formula

$$-c_{j-i+1} x_{j-i} = a_{j-i+1} x_{j-i+1} + c_{j-i+2} x_{j-i+2}.$$

None of the terms can destroy one another except for special values, and the conclusion is that we have the following formula for obtaining the number of terms:

$$N_i = N_{i-1} + N_{i-2}.$$

This is an equation of finite differences. Apply the algorithm

Molecular. of finite differences, as Lagrange says; or, which is essentially the same, we may try for solutions of this equation by the following formula: $N_i = zN_{i-1}$. We thus find

$$z^2 = z + 1, \text{ or } z = \frac{1 \pm \sqrt{5}}{2}.$$

We can satisfy our equation by taking either the upper or the lower sign.   The general solution is, of course,

$$N_i = C \left(\frac{1 + \sqrt{5}}{2}\right)^i + C' \left(\frac{1 - \sqrt{5}}{2}\right)^i$$

where C, C' are to be determined by the equation $N_0 = 1$, $N_1 = 1$. It is rather curious to see an expression of this kind for the number of terms in a determinant. You will find that, of the more general equation

$$N_i = aN_{i-1} + bN_{i-2},$$

the following is a solution:

$$N_i = N_1 \frac{(r^i - s^i)}{r - s} + bN_0 \frac{(r^{i-1} - s^{i-1})}{r - s},$$

where r, s are the two roots of the equation $x^2 = ax + b$.   Remark that the coefficients of N_0, N_1, being symmetrical functions of the two roots, are, as they must of course be, integral functions of a and b.

If one of the roots, s, for example, be less than unity, we may omit the large powers of s, and therefore for large values of i we may be sure of obtaining N_i to within a unit, and therefore the absolutely correct value, by calculating the integral part of

$$\frac{N_1 r^i + N_0 b r^{i-1}}{r + \frac{b}{r}}.$$

It is interesting to remark that the numerical value of this formula differs less and less from an integer the greater is i, and differs infinitely little from an infinitely large integer when i is infinitely great.

The values of N_i up to $i = 12$ for the case of our problem $(a = b = N_0 = N_1 = 1)$ are,

$$i = 2, \quad 3, \quad 4, \quad 5, \quad 6, \quad 7, \quad 8, \quad 9, \quad 10, \quad 11, \quad 12.$$

$$N_i = 2, \quad 3, \quad 5, \quad 8, \quad 13, \quad 21, \quad 34, \quad 55, \quad 89, \quad 144, \quad 233.$$

LECTURE VII.

TUESDAY, *Oct.* 7, 3.30 P.M.

LAGRANGE, in the second section of the second part of his *Mé-* Molecular.
canique Analytique on the Oscillation of a Linear System of Bodies,
has worked out very fully the motion in the first place for bodies
connected in series, and secondly for a continuous cord. The case
that we are working upon is not restricted to equal masses and
equal connecting springs, but includes the particular linear system
of Lagrange, in which the masses and springs are equal. I hope to
take up that particular case, as it is of great interest. We shall
take up this subject first to-day, and the propagation of disturbances
in an elastic solid second.

It was pointed out by Dr Franklin that, for the particular case
$N_1 = aN_0$, which is the case of our particular question as to the
number of terms in our determinant, the formula becomes

$$\left(a \frac{r^i - s^i}{r - s} + b \frac{r^{i-1} - s^{i-1}}{r - s} \right) N_0$$

and may be thus simplified.

We have $r^2 = ar + b$, or multiplying by r^{i-1}, $r^{i+1} = ar^i + br^{i-1}$. So
that the expression simplifies down to

$$N_i = N_0 \frac{r^{i+1} - s^{i+1}}{r - s}.$$

This may be obtained directly, by determining C, C', in terms of
N_0, with $N_{-1} = 0$, to make $N_i = Cr^i + C's^i$.

We have, in our case, $a = b = 1$: whence

$$r - s = \sqrt{5}, \; r = \frac{1 + \sqrt{5}}{2} \doteqdot 1\cdot 618, \; s \doteqdot - \cdot 618.$$

Molecular. If we work this out by very moderate logarithms for the case

$$N_{12} = \frac{r^{13}}{r-s},$$

dropping s^{13}, we find

$$13 \log 1{\cdot}618 - \log \sqrt{5} = 13 \times {\cdot}209 - {\cdot}3495 = 2{\cdot}3675 = \log 233,$$

which comes out exact; and this working with only 4-place-logarithms.

I want to call your attention to something far more important than this. The dynamical problem, quite of itself, is very interesting and important, connected as it is with the whole theory of modes and sequences of vibration; but the application to the theory of light, for which we have taken this subject up, gives to it more interest than it could have as a mere dynamical problem. I want to justify a fundamental form into which we can put our solution, which is of importance in connection with the application we wish to make.

Algebra shows that we must be able to throw $\dfrac{-x_1}{\xi}$ into the form

$$\frac{q_1}{\dfrac{\kappa_1^{2}}{\tau^{2}} - 1} + \frac{q_2}{\dfrac{\kappa_2^{2}}{\tau^{2}} - 1} + \cdots \frac{q_j}{\dfrac{\kappa_j^{2}}{\tau^{2}} - 1},$$

where $q_1, q_2, \ldots q_j$ are determinate constants, and $\kappa_1, \kappa_2, \ldots \kappa_j$ are the values of the period τ for which $\dfrac{-x_1}{\xi}$ becomes infinite. We can put it into this form certainly, for if x_1 and ξ be expressed in terms of x_j, they will be functions of the $(j-1)^{\text{st}}$ and j^{th} degrees, respectively, in $\dfrac{1}{\tau^2}$. This is easily seen if we notice that $x_{j-1} = -\dfrac{a_j}{c_j} x_j$ is of the first degree in $\dfrac{1}{\tau^2}$, and that the degree of each x is raised a unit above that of the succeeding x by the factor $a_j = \dfrac{m_i}{\tau^2} - c_i - c_{i+1}$ in the equation $-c_i x_{i-1} = a_i x_i + c_{i+1} x_{i+1}$. Therefore, writing z for $\dfrac{1}{\tau^2}$, we have

$$-\frac{x_1}{c_1 \xi} = \frac{A z^{j-1} + A' z^{j-2} + \cdots}{B z^j + B' z^{j-1} + \cdots}.$$

This, on being expanded into partial fractions, becomes Molecular.

$$\frac{K_1}{z-\dfrac{1}{\kappa_1^{\,2}}}+\frac{K_2}{z-\dfrac{1}{\kappa_2^{\,2}}}+\ldots\frac{K_j}{z-\dfrac{1}{\kappa_j^{\,2}}},$$

which takes the previously written form if we put $K_i\kappa_i^{\,2}=q_i.$

We know that the roots of the equation of the ith degree in z which makes $\dfrac{-x_1}{c_1\xi}$ become infinite are all real; they are the periods of vibration of a stable system of connected bodies. We have formal proof of it in the work which we have gone through in connection with such a system. I am putting our solution in this form, because it is convenient to look upon the characteristic feature of the ratio of τ to one or other of the fundamental periods. In the first place it is obvious that if we know the roots $\kappa_1,\ \kappa_2,\ldots$, the determination of $q_1,\ q_2,\ldots$ is algebraic. Another form which I shall give you is an answer to that algebraic question, what are the values of $q_1,\ q_2\ldots$? It is an answer in a form that is particularly appropriate for our consideration because it introduces the energy of the vibrations of the several fundamental modes in a remarkable manner. We will just get that form down distinctly.

Take the differential coefficients of $\dfrac{c_1\xi}{-x_1}$ with respect to $\dfrac{1}{\tau^2}$, and denoting $\dfrac{\kappa_1}{\tau^2}-1,\ \dfrac{\kappa_2}{\tau^2}-1,\ldots\ldots$, for brevity, by $D_1,\ D_2,\ldots\ldots$we find

$$\frac{d}{d\tau^{-2}}\frac{c_1\xi}{-x_1}=\frac{\kappa_1^{\,2}q_1/D_1^{\,2}+\kappa_2^{\,2}q_2/D_2^{\,2}+\ldots}{(q_1/D_1+q_2/D_2+\ldots)^2}.$$

For the case $\tau=\kappa_1$, our differential coefficient becomes $\dfrac{\kappa_1^{\,2}}{q_1}$, which determines

$$q_1=\kappa_1^{\,2}\Big/\frac{d}{d\tau^{-2}}\frac{c_1\xi}{-x_1}.$$

Now you will remember that we had

$$\frac{d}{d\tau^{-2}}\frac{c_1\xi}{-x_1}=m_1+m_2\left(\frac{x_2}{x_1}\right)^2+\ldots m_j\left(\frac{x_j}{x_1}\right)^2.$$

For the moment, take the expression for the simple harmonic motion, and you see at once that that comes out in terms of the energy. Adopt the temporary notation of representing the

Molecular. maximum value by an accented letter. Then we have at any time

of the motion $x_1 = x'_1 \sin \dfrac{2\pi\epsilon}{\tau}$, if we reckon our time from an era of

each particle passing through its middle position, remembering
that all the particles pass the middle position at the same instant.
We have therefore for the velocity of particle No. 1,

$$\dot{x}_1 = \frac{2\pi}{\tau} x'_1 \cos \frac{2\pi\epsilon}{\tau}.$$

The energy which at any time is partly kinetic and partly
potential, will be all kinetic at the instant of passing through the
middle position. Take then the energy at that instant. For $t = 0$

we have $x_1 = 0$, $\dot{x}_1 = \dfrac{2\pi}{\tau} x'_1$. Denoting the whole energy by E

$\left(\text{and remembering that the mass} = \dfrac{m_1}{4\pi^2}\right)$ we have

$$E = \tfrac{1}{2} \left(m_1 x_1'^2 + m_2 x_2'^2 + \dots m_j x_j'^2 \right) \frac{1}{\tau^2}.$$

Thus, the ratio of the whole energy to the energy of the first

particle $\left(\dfrac{E}{\frac{1}{2}\dfrac{m_1 x_1'^2}{\tau^2}} \right)$ being denoted by R^{-1}, we have

$$m_1 R^{-1} = \frac{d}{d\tau^{-2}} \frac{c_1 \xi}{-x_1}.$$

This is true for any value of τ whatever. From this equation find
then the ratios of the whole energy to the energy of the first
particle when $\tau = \kappa_1,\ \kappa_2, \dots$. Denoting these several ratios by R_1^{-1},

$R_2^{-1}\dots$, we find $q_1 = \dfrac{\kappa_1^2 R_1}{m_1}$, $q_2 = \dfrac{\kappa_2^2 R_2}{m_1}$, $\dots$ Our solution becomes

then

$$\frac{-x_1}{c_1 \xi} = \frac{\tau^2}{m_1} \left(\frac{\kappa_1^2 R_1}{\kappa_1^2 - \tau^2} + \frac{\kappa_2^2 R_2}{\kappa_2^2 - \tau^2} + \dots \right).$$

This is a very convenient form, as it shows us everything in terms
of quantities whose determinations are suitable, and intrinsically
important and interesting, viz. the periods and the energy-ratios.

It remains, lastly, to show how, from our process without calculating the determinants, we can get everything that is here
concerned. Our process of calculating gives us the u's in order,
beginning with u_{j-1}. That gives us the x's in order, and thus
we have all that is embraced in the differential coefficient with

respect to $\dfrac{1}{T^2}$. Everything is done, if we can find the roots. I Molecular. will show how you can find the roots from the continued fraction, without working out the determinant at all. The calculation in the neighbourhood of a root gives us the train of x's corresponding to that root, and then by multiplying the squares of the ratios of the x's to x_1, by the masses, and adding, we have the corresponding energy.

The case that will interest us most will be the successive masses greater and greater; and the successive springs stronger and stronger, but not in proportion to the masses so that the periods of vibration of limited portions of the higher numbered particles of the linear system shall be very large: so that if we hold at rest particles 4 and 6, the natural time of vibration of particle 5 will be longer than No. 2's would be if we held Nos. 1 and 3 at rest and set No. 2 to vibrate, and so on.

We will just put down once more two or three of our equations :

$$\frac{c_1 \xi}{-x_1} = a_1 - \frac{c_2^{\;2}}{u_2}; \quad \dots; \quad u_i = a_i - \frac{c_{i+1}^{\;\;2}}{u_{i+1}}: \quad a_i = \frac{m_i}{T^2} - c_i - c_{i+1}.$$

Without considering whether u_{i+1} is absolutely large or small, let us suppose that it is large in comparison with c_{i+1}; u_i will then be of the order a_i; u_{i-1} of the order a_{i-1}; and so on. We are to suppose that $a_1, a_2, \dots a_i$ are in ascending order of magnitude. Now, $u_1 u_2 \dots u_i = (-)^i c_1 \dots c_i \dfrac{\xi}{x_i}$. We thus have this important proposition, that the magnitudes of the vibrations of the successive particles decrease from particle No. 1 towards No. j; and x_i is exceedingly small in comparison with ξ, even though there is only a moderate proportion of smallness with respect to the ratios $\dfrac{c_1}{u_1}, \dfrac{c_2}{u_2}, \dots \dfrac{c_i}{u_i}$. Thus see how small is the motion at a considerable distance from the point at which the excitation is applied, under the suppositions that we have been making.

Now, as to the calculations. I do not suppose anybody is going to make these calculations*, but I always feel in respect to arithmetic somewhat as Green has expressed in reference to

---

* Happily this negative prognostication was not fulfilled. See (Appendix C) Numerical Solution with Illustrative Curves, by Prof. E. W. Morley, of the case of seven connected masses, proposed in Lecture IX.

Molecular. analysis.   I have no satisfaction in formulas unless I feel their arithmetical magnitude—at all events when formulas are intended for definite dynamical or physical problems.   So that if I do not exactly calculate the formulas, I would like to know how I could calculate them and express the order of the magnitudes concerned in them.   We are not going to make the calculations, but you will remark that we have every facility for doing so.   In the first place, is the exceeding rapidity of convergence of the formulas. The question is to find $\dfrac{c_i \xi}{-x_1}$; everything, you will find, depends upon that.   The exceeding rapidity of the convergence is manifest.   Since u_2 is large, u_1 is equal to a_1 with a small correction; similarly $u_2 = a_2$ with a small correction, and so on; so that two or three terms of the continued fraction will be sufficient for calculating the ratio denoted by u_1.   The continued fractions converge with enormous rapidity upon the suppositions we have been making.   We thus know the value of the differential coefficient $\dfrac{du_1}{d(\tau^{-2})}$.   We can in this way obtain several values of u_1 and begin to find it coming near to zero.   Then take the usual process. Knowing the value of the differential coefficient allows you to diminish very much the number of trials that you must make for calculating a root.   The process of finding the roots of this continued fraction will be quite analogous to Newton's process for finding the roots of an algebraic equation; and I tell any of you who may intend to work at it, that if you choose any particular case you will find that you will get at the roots very quickly.

I should think something like an arithmetical laboratory would be good in connection with class work, in which students might be set at work upon problems of this kind, both for results, and in order to obtain facility in calculation.

I hinted to you in the beginning about the kind of view that I wanted to take of molecules connected with the luminiferous ether, and affecting by their inertia its motions.   I find since then that Lord Rayleigh really gave in a very distinct way the first indication of the explanation of anomalous dispersion.   I will just read a little of his paper on the Reflection and Refraction of Light by intensely Opaque Matter (*Phil. Mag.*, May, 1872).   He commences, " It is, I believe, the common opinion, that a satisfactory mechanical theory of the reflection of light from metallic surfaces

has been given by Cauchy, and that his formulæ agree very well Molecular.
with observation. The result, however, of a recent examination of
the subject has been to convince me that, at least in the case of
vibrations performed in the plane of incidence, his theory is
erroneous, and that the correspondence with fact claimed for it is
illusory, and rests on the assumption of inadmissible values for the
arbitrary constants. Cauchy, after his manner, never published
any investigation of his formulæ, but contented himself with a
statement of the results and of the principles from which he
started. The intermediate steps, however, have been given very
concisely and with a command of analysis by Eisenlohr (Pogg.
Ann. Vol. CIV. p. 368), who has also endeavoured to determine
the constants by a comparison with measurements made by Jamin.
I propose in the present communication to examine the theory of
reflection from thick metallic plates, and then to make some
remarks on the action on light of a *thin* metallic layer, a subject
which has been treated experimentally by Quincke.

" The peculiarity in the behaviour of metals towards light is
supposed by Cauchy to lie in their *opacity*, which has the effect of
stopping a train of waves before they can proceed for more than a
few lengths within the medium. There can be little doubt that
in this Cauchy was perfectly right; for it has been found that
bodies which, like many of the dyes, exercise a very intense
selective absorption on light, reflect from their surfaces in excessive proportion just those rays to which they are most opaque.
Permanganate of potash is a beautiful example of this given by
Prof. Stokes. He found (*Phil. Mag.*, Vol. VI. p. 293) that when
the light reflected from a crystal at the polarizing angle is examined through a Nicol held so as to extinguish the rays polarized
in the plane of incidence, the residual light is green; and that,
when analyzed by the prism, it shows bright bands just where
the absorption-spectrum shows dark ones. This very instructive
experiment can be repeated with ease by using sunlight, and
instead of a crystal a piece of ground glass sprinkled with a little
of the powdered salt, which is then well rubbed in and burnished
with a glass stopper or otherwise. It can without difficulty be so
arranged that the two spectra are seen from the same slit one
over the other, and compared with accuracy.

" With regard to the chromatic variations it would have seemed
most natural to suppose that the opacity may vary in an arbitrary

Molecular. manner with the wave-length, while the optical density (on which
alone in ordinary cases the refraction depends) remains constant,
or is subject only to the same sort of variations as occur in transparent media.   But the aspect of the question has been materially
changed by the observations of Christiansen and Kundt (Pogg.
Ann. vols. CXLI., CXLIII., CXLIV.) on anomalous dispersion in
Fuchsin and other colouring-matters, which show that on either
side of an absorption-band there is an abnormal change in the
refrangibility (as determined by prismatic deviation) of such a
kind that the refraction is *increased* below (that is, on the red
side of) the band and *diminished* above it.   An analogy may be
traced here with the repulsion between two periods which frequently
occurs in vibrating systems.   The effect of a pendulum suspended
from a body subject to horizontal vibration is to increase or diminish
the virtual inertia of the mass according as the natural period of
the pendulum is shorter or longer than that of its point of suspension.   This may be expressed by saying that if the point of support
tends to vibrate more rapidly than the pendulum, it is made to go
faster still, and *vice versâ*"—I cannot understand the meaning of the
next sentence at all.   There is a terrible difficulty with writers in
abstruse subjects to make sentences that are intelligible.   It is
impossible to find out from the words what they mean; it is only
from knowing the thing* that you can do so—"Below the absorption-band the material vibration is naturally the higher, and hence
the effect of the associated matter is to increase (abnormally) the
virtual inertia of the aether, and therefore the refrangibility.   On
the other side the effect is the reverse."   Then follows a note, "See
Sellmeier, Pogg. Ann. vol. CXLIII. p. 272."   Thus Lord Rayleigh
goes back to Sellmeier, and I suppose he is the originator of all
this.   "It would be difficult to exaggerate the importance of these
facts from the point of view of theoretical optics, but it lies
beside the object of the present paper to go further into the question here."

There is the first clear statement that I have seen.   Prof.
Rowland has been kind enough to get these papers of Lord
Rayleigh for me.   I am most grateful to him and others among
you, by whom, with great trouble kindly taken for me, an

* In the next sentence, for "the refrangibility," substitute *its refractivity*.
W. T. Feb. 9, 1892.

immense number of books have been brought to me, in every one Molecular. of which I have found something very important.

Sellmeier, Lord Rayleigh, Helmholtz, and Lommel seem to be about the order. Lommel does not quote Helmholtz. I am rather surprised at this, because Lommel comes three or four years after Helmholtz: 1874 and 1878 are the respective dates. Lommel's paper is published in Helmholtz's Journal (*Ann. der Physik und Chemie*, 1878, vol. III. p. 339), Helmholtz's paper is excellent. Lommel goes into the subject still further, and has worked out the vibrations of associated matter to explain ordinary dispersion.

I only found this forenoon that Lommel (*Ann. der Ph. und Chem.* 1878, vol. IV. p. 55) also goes on to double refraction of light in crystals—the very problem I am breaking my head against. He is satisfied with his solution, but I do not think it at all satisfactory. It is the kind of thing that I have seen for a long time, but could not see that it was satisfactory; and I do see reason for its not being satisfactory. He goes on from that and obtains an equation which would approximately give Huyghens' surface. I have not had time to determine how far it may be correct, but I believe it must essentially differ from Huyghens' wave-surface to an extent comparable with that experimentally disproved by Stokes in his experimental disproof of Rankine's theory explaining optical aeolotropy by difference of inertia in different directions. The exceedingly close agreement of Huyghens' surface with the facts of the case which Stokes has found, absolutely cuts the ground from under a large number of very tempting modes of explaining double refraction.

Molar.

WE shall take some fundamental solutions for wave motion such as we have already had before us, only we shall consider them as now applicable to non-condensational distortional waves, instead of condensational waves. We can take our primary solution in the form $\phi = \dfrac{1}{r} \sin \dfrac{2\pi}{\lambda} (r - ct)$, where $c = \sqrt{\dfrac{k + \frac{4}{3}n}{\rho}}$ if the wave is condensational, and $\sqrt{\dfrac{n}{\rho}}$ if the wave is distortional. But for a distortional wave we must also have $\delta = 0$.

In the first place, if our value of c is $\sqrt{\dfrac{n}{\rho}}$, we know that ϕ satisfies $\rho \dfrac{d^2\phi}{dt^2} = n\nabla^2\phi$. [I want very much a name for that symbol ∇^2 (delta turned upside down). I do not know whether, Prof. Ball, you have any name for it or not; your predecessor, Sir William Hamilton, used it a great deal, and I think perhaps you may know of a name for it.] The conditions to be fulfilled by the three components of displacement, ξ, η, ζ, of a distortional wave are, in the first place,

$$\rho \frac{d^2\xi}{dt^2} = n\nabla^2\xi, \quad \rho \frac{d^2\eta}{dt^2} = n\nabla^2\eta, \quad \rho \frac{d^2\zeta}{dt} = n\nabla^2\zeta;$$

and we must have besides

$$\frac{d\xi}{dx} + \frac{d\eta}{dy} + \frac{d\zeta}{dz} = 0.$$

Thus ξ, η, ζ must be three functions, each fulfilling the same equation. There is a fulfilment of this equation by functions ϕ; and as we have one solution, we can derive other solutions from that by differentiation. Let us see then, if we can derive three Molar. solutions from this value of ϕ which shall fulfil the remaining condition. It is not my purpose here to go into an analytical investigation of solutions; it is rather to show you solutions which are of fundamental interest. Without further preface then, I will put before you one, and another, and then I will interpret them both.

Take for example the following, which obviously fulfils the

equation $\qquad \dfrac{d\xi}{dx} + \dfrac{d\eta}{dy} + \dfrac{d\zeta}{dz} = 0 :$

$$\xi = 0, \quad \eta = -\frac{d\phi}{dz}, \quad \zeta = \frac{d\phi}{dy}.$$

In each case the distance-terms only of our solution are what we wish. Thus

$$\eta = -\frac{d\phi}{dz} \doteqdot -\frac{2\pi}{\lambda} \cdot \frac{z}{r^2} \cdot \cos q, \quad \zeta = \frac{d\phi}{dy} \doteqdot \frac{2\pi}{\lambda} \cdot \frac{y}{r^2} \cdot \cos q.$$

Remark that in this solution the displacement at a distance from the source is perpendicular to the radius vector; i.e. we have

$$x\xi + y\eta + z\zeta = -y\frac{d\phi}{dz} + z\frac{d\phi}{dy} \doteqdot 0.$$

Before going further, it will be convenient to get the rotation. It is an exceedingly convenient way of finding the direction of vibration in distortional displacements. The rotations about the axes of x, y, z will be:

$$\tfrac{1}{2}\left(\frac{d\zeta}{dy} - \frac{d\eta}{dz}\right) \doteqdot -\frac{2\pi^2}{\lambda^2}\frac{y^2 + z^2}{r^3}\sin q;$$

$$\tfrac{1}{2}\left(\frac{d\xi}{dz} - \frac{d\zeta}{dx}\right) \doteqdot \frac{2\pi^2}{\lambda^2}\frac{xy}{r^3}\sin q;$$

$$\tfrac{1}{2}\left(\frac{d\eta}{dx} - \frac{d\xi}{dy}\right) \doteqdot \frac{2\pi^2}{\lambda^2}\frac{xz}{r^3}\sin q.$$

These rotations are proportional to $\dfrac{x^2}{r^3} - \dfrac{1}{r}$, $\dfrac{xy}{r^3}$, $\dfrac{xz}{r^3}$; that is

to say, besides an x component equal to $-\dfrac{1}{r}$, we have an r com-

ponent equal to $\dfrac{x}{r^2}$. We have a rotation around the radius vector

Molar. r, and a rotation around the axis of x, whose magnitudes are proportional to $\dfrac{x}{r^2}$ and $\dfrac{1}{r}$.

If you think out the nature of the thing, you will see that it is this : a globe, or a small body at the origin, set to oscillating rotationally about Ox as axis. You will have turning vibrations everywhere; and the light will be everywhere polarized in planes through Ox. The vibrations will be everywhere perpendicular to the radial plane through Ox.

In the first place we have $\left(\text{omitting the constant factor } \dfrac{2\pi}{\lambda}\right)$

$$\xi = 0, \quad \eta = -\frac{z}{r^2}\cos q, \quad \zeta = \frac{y}{r^2}\cos q.$$

Hence for $(y = 0,\ z = 0,)$ the displacements are zero, or we have zero vibration in the axis of x. Everywhere else the displacements are not null and are perpendicular to Ox (since we always have $\xi = 0$); and being perpendicular also to the radius vector, they are perpendicular to the radial plane through the axis of x.

Let us consider the state of things in the plane yz. Suppose we have a small body here at the origin or centre of disturbance, and that it is made to turn forwards and backwards in this way (indicating a turning motion about an axis perpendicular to the plane of the paper) in any given period. What is the result? Waves will proceed out in all directions from the source, and the intersections of the wave fronts with the plane (yz) of the paper will be circles. We shall have vibrations perpendicular to the radius vector; of magnitude $\dfrac{2\pi}{\lambda}\dfrac{\cos q}{r}$, which is the same in all directions. The rotation (molecular rotation about the axis of x, or in the plane yz,) is $\dfrac{2\pi^2}{\lambda^2}\dfrac{\sin q}{r}$.

There is therefore zero displacement where the rotation or the distortion is a maximum (positive or negative): and *vice versa*, at a place of maximum displacement (positive or negative) there is zero rotation and zero distortion. The rotation is clearly equal to half the distortion in the plane yz by shearing (or differential motion) perpendicular to rx, of infinitesimal planes perpendicular to r. The result is polarized light consisting of vibrations in the Molar. plane yz and perpendicular to the radius vector, and therefore the plane of polarization is the radial plane through OX.

Here we have a simple source of polarized light; it is the simplest form of polarization and the simplest source that we can have. Every possible light consists of sequences of light from simple sources. Is it probable that the shocks to which the particles are subjected in the electric light, or in fire, or in any ordinary source of light, would give rise to a sequence of this kind? No, at all events not much; because there cannot be much tendency in these shocks or collisions, to produce rotatory oscillations of the gross molecules. We can arbitrarily do it, for we can do what we will with the particle. That privilege occurred to me in Philadelphia last week, and I showed the vibrations by having a large bowl of jelly made with a ball placed in the middle of it. I really think you will find it interesting enough to try it for yourselves. It allows you to see the vibrations we are speaking of. I wish I had it to show you just now, so that you might see the idea realized. It saves brain very much.

I had a large glass bowl quite filled with yellowish transparent jelly, and a red-painted wooden ball floating in the middle of it.

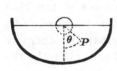

Try it, and you will find it a very pretty illustration. Apply your hand to the ball, and give it a turning motion round its vertical diameter, and you have exactly the kind of motion expressed by our equations. The motion in any oblique direction, such as at this point P (x, y, z) represents that of polarized light vibrating perpendicularly to the radial plane (or plane through the vertical central axis). The amplitude of the vibration here (in the vertical axis) is zero; here at the surface (in the plane yz) it is $\frac{1}{r}\cos q$; and if you use polar coordinates, calling this angle θ (indicating on the diagram) then the amplitude here (at P) is $\frac{1}{r}\cos q \sin\theta$, giving when θ is a right angle the previous expression.

I say that this is the simplest source and the simplest system of polarized light that we can imagine. But it can scarcely be induced naturally. The next simplest is a globe or small body vibrating to-

Molar.

and-fro in one line. We will take the solution for that presently. Still we have not got up to the essential complexity of the vibration produced naturally by the simplest natural vibrator with unmoved centre of inertia. I may take my hand and, instead of the torsional oscillations which we have been considering, I may give vertical oscillations to the globe in the jelly (and that makes a very pretty modification of the experiment), and people all call out, "O, there is the natural time of the vibration, if you only leave the globe to itself," oscillating up and down in the jelly. But the case is not proper for an *illustration* of undulatory vibrations spreading out from a centre. We are troubled here also by reflection back, as it were, from the containing bowl, just as in experiments on a stretched rope to show waves running along it, we are troubled by the rope not being infinitely long. You can always see sets of vibrations running along the rope, and reflected back from the ends. But in this experiment with the jelly in the bowl, you do not see the waves travelling out at all because the distance to the boundary is not large enough in comparison with the wave-length; and what you really see is a certain set of standing vibrations, depending on the finiteness of the bowl. But just imagine the bowl to be infinitely large, and that you commence making torsional oscillations; what will take place? A spreading outwards of this kind of vibrations, the beginning being, as we shall see, abrupt. We shall scarcely reach that to-day, but we shall perhaps another day consider if the motion in the source begins and ends abruptly, the consequent abruptness* of the beginnings and endings of the vibrations throughout an elastic solid ; in every case in which the velocity of propagation is independent of the wave-length.

When you apply your hands and force the ball to perform those torsional vibrations, you have waves proceeding from it ; but if you then leave it to itself, there is no vibrating energy in it at all, except the slight angular velocity that you leave it with. A vibrator which can send out a succession of impulses independently of being forced to vibrate from without, must be a vibrator with the means of conversion of potential into kinetic energy in itself. A tuning-fork, and a bell, are sample vibrators for sound. The simplest sample vibrator that we can imagine to represent the

---

* This is in fact proved by the solution of Lecture IV. expressed by (14) in terms of an arbitrary function.

origin of an independent sequence of light may be like a tuning-fork. Molar.
Two bodies, joined by a spring would be more symmetrical than a
tuning-fork.   Two rigid globes joined by a spring—that will give
you the idea ; or (which will be a vibration of the same type still)
one elastic spherical body vibrating from having been drawn into
an oval shape, and let go.

I will look, immediately, at a set of vibrations produced in an
elastic solid by a sample vibrator.   But suppose you produce
vibrations in your jelly elastic solid by taking hold of this ball
and moving it to-and-fro horizontally, or again moving it up and
down vertically and think of the kinds of vibrations it will make
all around.   Think of that, in connection with the formulas, and
it will help us to interpret them.   This is in fact the kind of
vibration produced in the ether by the rigid spherical containing
shell of our complex molecule (Lecture II.).   Or think of the
vibrations due to the higher though simpler order of vibrator, of
which we have taken as an example a very dense elastic globe
vibrating from prolate to oblate and back periodically.   We might
also have those torsional vibrations; but among all the possible
vibrations of atoms in the clang and clash of atoms that there is
in a flame, or other source of light, a not very rare case I think
would be that which I am going to speak of now.   It consists of
opposite torsional vibrations at the two ends of an elongated mass.
To simplify our conception for a moment, imagine two globes connected by a columnar spring; twist them in opposite directions, and
let them go.   There you have an imaginable source of vibrations.
If in any one of our cases the potential energy of the spring is very
large in comparison with the energy that is carried off in a
thousand, or a hundred thousand vibrations, you will have a
nearly uniform sequence of vibrations such as those we have been
considering, but gradually dying down.

Before passing on to the to-and-fro vibrator we will think of
this motion for a moment, but we will not work it out, because it
is not so interesting.   To suit our drawing we shall suppose one
globe here, and another upon the opposite side on a level with the
first, so that the line of the two is perpendicular to the board.
Give these globes opposite torsional vibrations about their common
axis, and what will the result be ?   A single one produces zero
light in the axis and maximum light in the equatorial plane.   The
two going in opposite directions will produce zero light in the

Molar.

equatorial plane and zero light in the axis; so that you will proceed from zero in the equatorial plane to a maximum between the equatorial plane and the poles, and zero at the poles; and you will have opposite vibrations in each hemisphere. That constitutes a possible case of vibrations of polarized light, proceeding from a possible independent vibrator.

One of the most simple and natural suppositions in respect to an independent vibrator is afforded by the illustration of a bell, or a tuning-fork, or an elastic body deformed from its natural shape and left to vibrate. In all these cases, as also in the case of our supposed complex molecule (Lecture I.), remark that the centre of gravity of the vibrator is at rest; except for the comparatively very small reaction of the ether upon it; and this is essential to an independently acting vibrator. The vibrator must have potential energy in itself, for many thousand vibrations generating waves travelling outwards through ether; and its centre of gravity must be at rest, except in so far as the reaction of the medium upon it causes a slight motion of the centre of gravity.

I will put down the solution which corresponds to a to-and-fro vibration in the axis of x, viz.:

$$\xi = \frac{4\pi^2}{\lambda^2}\,\phi + \frac{d^2\phi}{dx^2}, \qquad \eta = \frac{d^2\phi}{dydx}, \qquad \zeta = \frac{d^2\phi}{dzdx}.$$

ϕ is our old friend,

$$\frac{1}{r}\sin\frac{2\pi}{\lambda}\left(r - t\sqrt{\frac{n}{\rho}}\right),$$

but with n now in place of the $n + \frac{4}{3}k$ which we had formerly when we were dealing with condensational waves. First remark that we know that

$$\rho\frac{d^2\xi}{dx^2} = n\nabla^2\xi, \text{ etc.,}$$

are satisfied, because ϕ and all its differential coefficients satisfy this relation. We have therefore only to verify that the dilatation is zero. Instead of merely going through the verification, I wish to help you to make the solution your own by showing you how I obtained it. I will not say that there is anything novel in it, but it is simply the way it occurred to me. I obtained it to illustrate Stokes' explanation of the blue sky. I afterwards found that Lord Rayleigh had gone into the subject even more searchingly than Stokes, and I read his work upon it.

The way I found the solution was this : $\dfrac{d\phi}{dx}$ is clearly the dis- Molar

placement-potential for an elastic fluid, corresponding to a source of the kind, constituted by an immersed solid moving to-and-fro along the axis of x. The displacement function of which the displacements are the differential coefficients would take simply that form if the question were of sound in air or other compressible fluid, and not of light, or of waves in an incompressible solid, or of waves of distortion in a compressible solid. It was a question of condensational vibrations with us several days ago. I did not go into the matter in detail then, but we saw that for condensational vibrations proceeding from a vibrator vibrating to-and-fro along the axis of x that $\dfrac{d\phi}{dx}$ was the displacement potential; and it is obvious, if we start from the very root of the matter that it must be so.

Hence we may judge that the differential coefficients $\dfrac{d}{dx}, \dfrac{d}{dy}, \dfrac{d}{dz}$

of $\dfrac{d\phi}{dx}$, with n in place of $n + \tfrac{4}{3}k$, must therefore be at all events constituents of the components of displacement in the case of light, or of distortional waves, from such a source : but neither they nor the differential coefficients of any function can be simply equal to the displacement-components of our present problem, in which the motion is essentially rotational. The irrotational displacements in the condensational wave problem are displacements which fulfil certain of the conditions of our present problem : but they do not fulfil the condition of giving us a purely distortional wave, unless we add a term or terms in order to make the dilatation zero. This is done in fact, as I found, by the addition of the spherically symmetric term $\dfrac{4\pi}{\lambda^2}\phi$ to $\dfrac{d}{dx}\dfrac{d\phi}{dx}$, for the x-component of displacement. Just try for the dilatation. We have

$$\nabla^2\phi - \frac{4\pi^2}{\lambda^2}\phi,$$

in which we may substitute $\dfrac{d\phi}{dx}$ for ϕ. Thus

$$\nabla^2 \frac{d\phi}{dx} = -\frac{4\pi^2}{\lambda^2} \cdot \frac{d\phi}{dx}.$$

We verify therefore in a moment that the displacements given by the formulas satisfy

$$\frac{d\xi}{dx} + \frac{d\eta}{dy} + \frac{d\zeta}{dz} = 0;$$

and thus we have made up a solution which satisfies the conditions of being rigorously non-condensational, (no condensation or rarefaction anywhere,) and of being symmetrical round the axis of x.

In the first place, taking the distant terms only, we have

$$\xi \doteqdot \frac{4\pi^2}{\lambda^2} \frac{r^2 - x^2}{r^3} \sin q,$$

$$\eta \doteqdot - \frac{4\pi^2}{\lambda^2} \frac{xy}{r^3} \sin q,$$

$$\zeta \doteqdot - \frac{4\pi^2}{\lambda^2} \frac{xz}{r^3} \sin q.$$

It is easy to verify that these displacements are perpendicular to the radius vector, i.e. that we have $x\xi + y\eta + z\zeta \doteqdot 0$. Just look at the case along the axis of x, and again in the plane yz. It is written down here in mathematical words painting as clearly and completely as any non-mathematical words can give it. Take $y = 0$, $z = 0$, and that makes $\xi = 0$, $\eta = 0$, $\zeta = 0$. Therefore, in the direction of the axis of x there is no motion. That is a little startling at first, but is quite obviously a necessity of the fundamental supposition. Cause a globe in an elastic solid to vibrate to-and-fro. At the very surface of the globe the points in which it is cut by Ox have the maximum motion; and throughout the whole circumference of the globe, the medium is pulled, by hypothesis, along with the globe. But this is not a solution for that comparatively complex, though not difficult, problem. I am only asking you to think of this as the solution for the motion at a great distance. It may not be a globe, but a body of any shape moved to-and-fro. To think of a globe will be more symmetrical. In the immediate neighbourhood of the vibrator there is a motion produced in the line of vibration; the motion of the elastic solid in that neighbourhood consists in a somewhat complex, but very easily imagined state of things, in which we have particles in the axis of x, moving out and in directly along the radius vector; in all other places except the plane of yz, slipping around with motions oblique to the radius vector; and in the plane of yz moving exactly perpendicular to the radius vector. All, however, except motions perpendicular Molar. to the radius vector, become insensible at distances very great in comparison with the wave-length. We have taken, simply, the leading terms of the solution. These represent the motion at great distances, quite irrespectively of the shape of the body, and of the comparatively complicated motion in the neighbourhood of the vibrating body.

Take now $x = 0$, and think of the motions in the plane yz. The vibrator is supposed to be vibrating perpendicular to this plane. We have

$$\xi \fallingdotseq \frac{4\pi^2}{\lambda^2} \frac{1}{r} \sin q, \quad \eta = 0, \quad \zeta = 0.$$

What does that mean? Clearly, that the vibrations are perpendicular to the plane yz. We have the wave spreading out uniformly in all directions in that plane, and " polarized in " that plane, the vibrations being perpendicular to it. That is exactly what Stokes supposed was of necessity the dynamical theory of the blue light of the sky. Lord Rayleigh showed that it was not so obvious as Stokes had supposed. He elaborately investigated the question, " Whether is the blue light of the sky, (which we assume to be owing to particles in the air,) due to the particles being of density different from the surrounding luminiferous ether, or being of rigidity different from the surrounding luminiferous ether ? " The question would really be, If the particles are water, what is the theory of waves of light in water; does it differ from air in being, as it were, a denser medium with the same effective rigidity, or is it a medium of the same density and less effective rigidity, or does it differ from ether both as to density and as to rigidity ?

Lord Rayleigh examined that question very thoroughly, and finds, if the fact that the cause were, for instance, little spherules of water, and if in the passage of light through water the propagation is slower than in air were truly explained by less rigidity and the same density we should have something quite different in the polarization of the sky from what we would have on the other supposition. On the other hand, the observed polarization of the sky supports the other supposition (as far as the incertitude of the experimental data allows us to judge) that the particles, whether they be particles of water, or motes of dust, or whatever

Molar.   they may be, act as if they were little portions of the luminiferous
ether of greater density than, and not of different rigidity from,
the surrounding ether. Hence our present solution, which has for
us such special interest as being the expression of the disturbance
produced in the ether by our imbedded spring-molecule, acquires
farther and deeper interest as being the solution for dynamical
action which according to Stokes and Rayleigh is the origin cause
of the blue light coming from the sky. I will call attention a
little more to Lord Rayleigh's dynamics of the blue sky in a
subsequent lecture.   Meantime, returning to our solution, we
may differentiate once more with respect to x, in order to get
a proper form of function to express the motion from a double
vibrator vibrating to-and-fro like this—vibrating my hands towards and from each other.   Then we shall have a solution which
will express another important species of single sequence of vibrations, of which multitudes may constitute the whole, or a large
part, of the light of any ordinary source.

A question is now forced upon us,—what is the velocity of a
group of waves in the luminiferous ether disturbed by ordinary
matter? With a constant velocity of propagation, as in pure ether,
each group remains unchanged. But how about the propagation
of light-sequences in a transparent medium like glass? It is a
question that is more easily put than answered. We are bound
to consider it most carefully. I do not despair of seeing the
answer. I think, if we have a little more patience with our
dynamical problem we shall see something towards the answer.

Molecular.   Here is a perfectly parallel problem.   Commence suddenly to
give a simple harmonic motion through the handle P to our
system of particles $m_1, m_2,...m_j$, which play the part of a molecule.
If you commence suddenly imparting to the handle a motion of
any period whatever, only avoiding every one of the fundamental
periods, if there be a little viscosity it will settle into a state of
things in which you have perfectly regular simple harmonic vibration. But if there be no viscosity whatever, what will the result
be? It will be composed of simple harmonic motions in the
period of our applied motion at the bell-handle P; with the amplitude of each calculated from our continued fraction; and superimposed upon it, a jangle as it were, consisting of coexistent simple
harmonic vibrations of all the fundamental periods. If there is no
viscosity, that state of things will go on for ever. I cannot satisfy

myself with viscous terms in these theories not only because the Molecular. assumption of viscosity, *in molecular dynamics*, is a theoretic violation of the conservation of energy ; but because the smallest degree of viscosity of ether sufficing. to practically rid us of any of this jangling, or to have any sensible influence in any of the motions we have to do with in sources or waves of light, would not allow a light-sequence through ether to last, as we know it lasts, through millions of millions of millions of millions of vibrations. But if we have no viscosity at all, whatever energy of any vibrations, regular or irregular, we have at any time in our complex molecule must show in the vibrations of something else, and that is what ?   In studying that sort of vibration with which we have been occupied in the molecular part of our course, we must account for these irregular vibrations somehow or other.   The viscous terms which Helmholtz and others have introduced represent merely an integral effect, as it were, of actions not followed in detail, not even explained, in the theory.   By viscous terms, I mean terms that assume a resistance in simple proportion to velocity.

But the state of things with us is that that jangling will go on for ever, if there is no loss of energy ; and we want to coax our system of vibrators into a state of vibration with an arbitrarily chosen period without viscous consumption of energy.   Begin thus: commence suddenly acting on P just as we have already supposed, but with only a very small range of motion of P.   The result will be just as I have said, only with very small ranges of all the constituent motions.   After waiting a little time increase the range of the motion of P; after waiting a little longer, increase the range farther, and so go on, increasing the range by successive steps.   Each of those will superimpose another state of vibration.   There would be, I believe, virtually an addition of the energies, not of the amplitudes, of the several janglings if you make these steps quite independent of one another.

For example, suppose you proceed thus: In the first place, start right off into vibrations of your handle P through a space, say of 30 inches.   You will have a certain amount of energy, J, in the irregular vibrations (the " jangling ").   In the second place, commence with a range of three inches.   After you have kept P vibrating three inches through many periods, suddenly increase its range by three inches more, making it six inches.   Then, some-

Molecular. time after, suddenly increase the range to nine inches ; and so on
in that way by ten steps.   The energy of the jangle produced by
suddenly commencing through the range of three inches, which
is one-tenth of 30 inches, will be exactly one-hundredth of J, the
energy of jangle which you would have if you commenced right
away with the vibration through 30 inches.   Each successive
addition of three inches to the range of P will add an amount
of energy of jangling of which the most probable value is the one-
hundredth of J; and the result is that if you advance by these
steps to the range of 30 inches, you will have in the final jangle
ten-hundredths, that is to say one-tenth, of the energy of jangle
which you would get if you began at that range right away.
Thus, by very gradually increasing the range, the result will be
that, without any viscosity at all there will be infinitely little of
the irregular vibrations.

But there are cases in which we have that tremendous jangling
of the molecules concerned in luminous vibrations; for instance,
the fluorescence of such a thing as uranium glass or sulphate
of quinine which lasts for several thousandths of a second after
the exciting light is taken away, and then again in phosphorescence
that lasts for hours and days.   There have been exceedingly
interesting beginnings, in the way of experiments already made,
in these subjects, but nobody has found whether initial refraction
is exactly the same as permanent refraction.   For this purpose
we might use Becquerel's phosphoroscope or we might use methods
such as those of Fizeau or Foucault, or take such an appliance as
Prof. Michelson has been recently using, for finding the velocity of
light, and so get something enormously more searching than even
Becquerel's phosphoroscope, and try whether in the first hundredth
of a second, or the first millionth of a second, there is any
indication of a different wave velocity from that which we find
from the law of refraction, when light passes continuously through
a transparent liquid or solid.   If, with the methods employed for
ascertaining the velocity of light in a transparent body (to take
account of the criticisms that they have received at the British
Association meeting, to which I have referred several times), we
combine a test for instantaneous refraction, it seems likely that
we should not get negative results, but rather find phenomena
and properties of ultimate importance.   We might take not only
ordinary transparent solids and liquids, but also bodies in which,

like uranium glass, the phosphorescence lasts only a few Molecular.
thousandths of a second; and then again bodies in which phosphorescence lasts for minutes and hours. With some of those
we should have anomalous dispersion, gradually fading away after
a time. I cannot but think that by experimenting, in some such
way, we should find some very interesting and instructive results
in the way of initial fluorescence.

LECTURE IX.

WEDNESDAY, *Oct.* 8, 5 P.M.

Molar.   WE shall go on for the present with the subject of the
propagation of waves from a centre.   Let us pass to the case of
two bodies vibrating in opposite directions, by superposition of
solutions such as that which we have already found for a single
to-and-fro vibrator, which was expressed by

$$\xi = \frac{4\pi^2}{\lambda^2}\,\phi + \frac{d}{dx}\frac{d\phi}{dx}, \quad \eta = \frac{d}{dy}\frac{d\phi}{dx}, \quad \zeta = \frac{d}{dz}\frac{d\phi}{dx}.$$

We verified that

$$\frac{d\xi}{dx} + \frac{d\eta}{dy} + \frac{d\zeta}{dz} = 0,$$

so that this expresses rigorously a distortional wave.   It is obvious
that this expresses the result of a to-and-fro motion through the
origin in the line *OX*.   Remark, for one thing, that in the
neighbourhood of the origin, at such moderate distance from it
that the component motion in the direction *OX* is not insensible,
we have on the two sides of the origin simultaneously positive
values. ξ is the same for a positive value of *x* as for the negative
of that value.   At distances from the origin in the line *OX* which
are considerable in comparison with the wave-length the motion
vanishes as we have seen.

Pass on, now, to this case: a positive to-and-fro motion on the
one side of the origin, and a simultaneous negative to-and-fro
motion on the other side of the origin; that is to say, two simultaneous co-periodic vibrations of portions of matter on the two
sides of the origin moving simultaneously in opposite directions.
I will indicate these motions by arrow-heads, continuous arrow-
heads to indicate directions of motion at one instant of the period,

and dotted at the instant half a period earlier or later. The first Molar. case already considered

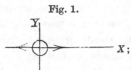

Fig. 1.

the second case

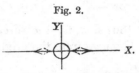

Fig. 2.

The effect in the first case being expressed by the displacements ξ, η, ζ, already given, the effect in the second case will be expressed by displacement-coefficients respectively equal to

$$\frac{d\xi}{dx}, \quad \frac{d\eta}{dx}, \quad \frac{d\zeta}{dx}.$$

This configuration of displacement clearly implies a motion of which the component parallel to OX has opposite signs, and the components perpendicular to OX are equal with the same sign, for equal positive and negative values of x; it is a simultaneous out and in vibration on the two sides of the origin in the line OX, and a simultaneous in and out vibration perpendicular to OX, every-where in the plane yz. A motion of the matter at distances from the origin moderate in comparison with the wave-length will be accurately expressed by these functions. Passing now from Fig. 2 which shows the germ from which we have developed the idea of this configuration of motion, and the functions expressing it; look

Fig. 3.

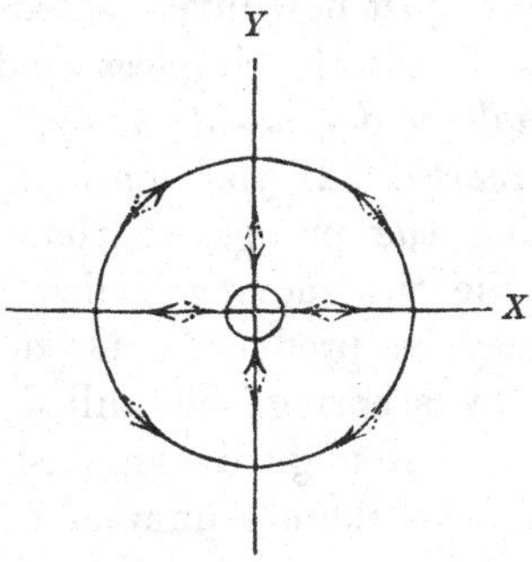

Molar.   to Fig. 3 illustrating the in and out vibration perpendicular to OX which accompanies the out and in vibration along OX from which we started. The configuration of arrow-heads on the circle in Fig. 3 shows the component motions perpendicular to the radius vector at any distance, small or great, from the origin; which constitute sensibly the whole motion at the great distances. To express this motion, take only the "distance-terms" (as in previous cases,) and drop the factor $\dfrac{-8\pi^3}{\lambda^3}$, from the differential coefficients indicated above. We thus find, for the three components of the displacement at great distances from the origin,

$$\xi \doteqdot x\,\frac{(x^2 - r^2)}{r^4}\cos q, \quad \eta \doteqdot \frac{x^2 y}{r^4}\cos q, \quad \zeta \doteqdot \frac{x^2 z}{r^4}\cos q.$$

To satisfy ourselves that the radial component of the displacement is zero verify that we have $x\xi + y\eta + z\zeta = 0$.

To think of the kind of "polarization" that will be found, when the case is realized in a sequence of waves of light, remark that the motion is everywhere symmetrical around the axis of x, and is in the radial plane through OX. Therefore, we have light polarized in the plane through the radius to the point considered and perpendicular to the plane through OX.

This is, next to the effect of a single to-and-fro rigid vibrator, the simplest set of vibrations that we can consider as proceeding from any natural source of light. As I said, we might conceive of a pair of equal and opposite torsional motions at the two ends of a vibrating molecule. That is one of the possibilities, and it would be rash to say that any one possible kind of motion does not exist in so remarkably complex a thing as the motion of the particles from which light originates.

The motion we have just now investigated is perhaps the most interesting, as it is obviously the simplest kind of motion that can proceed from a *single independent non-rotating vibrator with unmoved centre of inertia*. If you consider the two ends of a tuning-fork, neglecting the prongs, so that everything may be symmetrical around the two moving bodies, you have a way by which the motion may be produced. Or our source might be two balls connected by a spring and pulled asunder and set to vibrating in and out; or it might be an elastic sphere which has experienced a shock. An infinite number of modes of vibration are generated when an elastic ball is struck a blow, but the gravest mode, which is also no doubt the one in which the energy is Molar. greatest, if the impinging body be not too hard, consists of the globe vibrating from an oblate to a prolate figure of revolution; and this originates in the ether the motion with which we have just now been occupied.

The kind of thing that an elemental source of light in nature consists of, seems to me to be a sudden initiation of a set of vibrations and a sequence of vibrations from that initiation which will naturally become of smaller and smaller amplitude. So that the graphic representation of what we should see if we could see what proceeds from one element of the source, the very simplest conceivable element of the source, would consist of polarized waves of light spreading out in all directions according to some such law as we have here. In any one direction, what will it be? Suppose that the wave advances from left to right; you will then see what is here represented on a magnified scale.

I have tried to represent a sudden start, and a gradual falling off of intensity. Why a sudden start? Because I believe that the light of the natural flame or of the arc-light, or of any other known source of light, must be the result of sudden shocks upon a number of vibrators. Take the light obtained by striking two quartz pebbles together. You have all seen that. *There* is one of the very simplest sources of light. Some sort of a chemical or ozoniferous effect connected with it which makes a smell, there must be. As to what the cause of this peculiar smell may be, I suppose we are almost assured, now, that it proceeds from the generation of ozone. What sort of a thing can the light be that proceeds from striking two quartz pebbles together? Under what circumstances can we conceive a group of waves of light to begin gradually and to end gradually? You know what takes place in the excitation of a violin string or a tuning fork by a bow. The vibrations gradually get up from zero to a maximum and then, when you take the bow off, gradually subside. I cannot see anything like that in the source of light. On the contrary, it seems to me to be all shocks, sudden beginnings and gradual subsidences; rather like the excitation of a harp string *plucked* in the usual manner, or of a pianoforte string struck by the

Molar.

hammer; and left to itself to give away all its energy gradually in waves of sound.

I say this, because I have just been reading very interesting papers by Lommel and Sellmeier*, both touching upon this subject. Helmholtz remarks that Sellmeier gets into a difficulty in his dynamics and does not show clearly what becomes of the energy in a certain case; but it seems to me that Sellmeier really takes hold of the thing with great power. He goes into this case very fully, and in a way with which we are all now more or less familiar. He remarks that Fizeau obtained a suite of 50,000 vibrations interfering with one another, and judges from that that, though ordinary light consists of polarized light, circularly-, or elliptically-, or plane-polarized as I said to you myself, one or two days ago, with (what I did not say) the plane of polarization, or one or both axes of the ellipse if it be elliptically polarized, gradually varying, and the amplitude gradually changing, the changing must be so gradual that the whole amount of the change, whether of amplitude or of mode of polarization or of phase, in the course of 50,000 or 100,000, or perhaps several million vibrations cannot be so great as to prevent interference. In fact, I suppose there is no perceptible difference between the perfectness of the annulments in Fizeau's experiment, with 50,000 vibrations and with 1,000; although I speak here not with confidence and I may be corrected. You have seen that with your grating, have you not, Prof. Rowland?

PROF. ROWLAND. Yes; but it is very difficult to get the interferences.

SIR WM. THOMSON. But when you do get them, the black lines are very black, are they not?

PROF. ROWLAND. I do not know. They are so very faint that you can hardly see them.

SIR WM. THOMSON. What do you infer from that?

PROF. ROWLAND. That there is a large number. The narrowness of the lines of the spectrum indicates how perfectly the light interferes; and with a grating of very fine lines I *find exceedingly perfect interference for at least* 100,000 *periods* I should think.

SIR WM. THOMSON. That goes further than Fizeau. Sellmeier says that probably a great many times 50,000 waves must pass before there can be any great change. He goes at the thing very

* Sellmeier; *Ann. der Phy. u. Chem.* 1872, Vols. CXLV., CXLVII.

admirably for the foundation of his dynamical explanation of Molar. absorption and anomalous refraction.   The only thing that I do not fully agree with him in his fundamentals is the gradualness of the initiation of light at the source.   I believe, in the majority of cases at all events, in sudden beginnings and gradual endings. Prof. Rowland has just told us how gradual the endings are. Fizeau could infer that the amplitude does not fall off greatly in 50,000 vibrations.   It is quite possible from all we know, that the amplitude may fall off considerably in 100,000 vibrations, is it not?

PROF. ROWLAND.   The lines are then very sharp.

SIR WM. THOMSON.   It would not depend on the sharpness of the lines, would it?

PROF. ROWLAND.   O, yes.   It would draw them out of fineness.

SIR WM. THOMSON.   Would it broaden them out, or would it leave them fine, but throw a little light over a place that should be dark?

PROF. ROWLAND.   It would broaden them out.

SIR WM. THOMSON.   It is a very interesting subject; and from the things that have been done by Prof. Rowland and others, we may hope to see, if we live, a conquering of the difficulties quite incomparably superior to what we have now.   I have no doubt, however, but that some now present will live to see knowledge that we can have hardly any conception of now, of the way of the extinction of vibrations in connection with the origin and the propagation of light.   We are perfectly certain that the diminution of amplitude in the majority of sequences in any ordinary source, must be exceedingly small—practically nil—in 1,000 vibrations; we can say that probably it is practically nil in 50,000 vibrations; we know that it is nearly nil in 100,000 vibrations.   Is it practically nil in two or three hundred thousand vibrations, or in several million vibrations?   Possibly not.   Dynamical considerations come into play here.   We shall be able to get a little insight into these things by forming some sort of an idea of the total amount of energy there can possibly be in one elemental vibrator, in a source of light, and what sequences of waves it can supply.   That the whole energy of vibration of a single freshly excited vibrator in a source of light is many times greater than what it parts with in the course of 100,000

Molar.

molar-vibrations, is a most interesting experimental conclusion, drawn from Fizeau's and Rowland's grand observations of interference.

In speaking of Sellmeier's work, and Helmholtz's beautiful paper which is really quite a mathematical gem, I must still say that I think Helmholtz's modification is rather a retrograde step. It is not so perhaps in the mathematical treatment; and at the same time Helmholtz is perfectly aware of the kind of thing that is meant by viscous consumption of energy. He knows perfectly well that that means, conversion of energy into heat ; and in introducing viscosity he is throwing up the sponge, as it were, so far as the fight with the dynamical problem is concerned.

Mr Mansfield brought me another quarter hundred weight of books on the subject last night. I have not read them all through. I opened one of them this forenoon, and exercised myself over a long mathematical paper. I do not think it will help us very much in the mathematics of the subject. What we want is to try and see if we cannot understand more fully what Sellmeier has done, and what Lommel has done. I see that both stick firmly to the idea that we must in the particles themselves account for the loss of energy from the transmitted wave. That is what I am doing; and we shall never have done with it until we have explained every line in Prof. Rowland's splendid spectrum. If we are tired of it, we can rest, and go at it again.

Lommell and Sellmeier do not go very fully into these multiple modes of vibrations, although they take notice of them. But they do indicate that we must find some way of distributing the energy without supposing annulment of it. That is the reason why I do not like the introducing of viscous terms in our equations. It is very dangerous, in an ideal sense, to introduce them at all. This little bit of viscosity in one part of the system might run away with all our energies long before 50,000 vibrations could be completed. If there were any sensibly effective viscosity in any of the material connected with the moving particle it might be impossible to get a sequence of one-hundred thousand or a million vibrations proceeding from one initial vibration of one vibrator.

What the dynamical problem has to do for us is to show how we can have a system capable of vibrations in itself and acted upon by the luminiferous ether, that under ordinary circumstances does not absorb the light in millions of vibrations, as for transparent

liquids or solids, or in hundreds of thousands of millions of vibra- Molar.
tions as in our terrestrial atmosphere. That is the case with
transparent bodies; bodies that allow waves to pass through them
one-hundred feet or fifty miles, or greater distances; transparent
bodies with exceedingly little absorption. If we take vibrators,
then, that will perform their functions in such a way as to give a
proper velocity of propagation for light in a highly transparent
body, and yet which, with a proper modification of the magnitudes
of the masses or of the connecting springs, will, in certain complex
molecules, such as the molecules of some of those compounds that
give rise to fluorescence and phosphorescence, take up a large
quantity of the energy, so that perhaps the whole suite of
vibrations from a single initiation may be absolutely absorbed
and converted into vibrations of a much lower period, which will
have, lastly, the effect of heating the body, I think we shall see
a perfectly clear explanation of absorption without introducing
viscous terms at all; and that idea we owe to Sellmeier.

I would like, in connection with the idea of explaining
absorption and refraction, and lastly, anomalous refraction and
dispersion, to just point out as a matter of history, the two
names to which this is owing,—Stokes and Sellmeier. I would
be glad to be corrected with reference to either, if there is any
evidence to the contrary; but so far as I am aware, the very first
idea of accounting for absorption by vibrating particles taking up,
in their own modes of natural vibration, all the energy of those
constituents of mixed light trying to pass through, which have
the same periods as those modes, was from Stokes. He taught
it to me at a time that I can fix in one way indisputably.
I never was at Cambridge once from about June 1852 to May
1865; and it was at Cambridge walking about in the grounds
of the colleges that I learned it from Stokes. Something was
published of it from a letter of mine to Helmholtz, which he
communicated to Kirchhoff and which was appended by Kirchhoff
in his postscript to the English translation (published in *Phil.
Mag.*, July 1860) of his paper on the subject which appeared in
Poggendorff's *Annalen*, Vol. CIX. p. 275.

In the postscript you will find the following statement taken
from my letter :—

" Prof. Stokes mentioned to me at Cambridge some time ago,
probably about ten years, that Prof. Miller had made an experiment

Molar.

testing to a very high degree of accuracy the agreement of the
double dark line D of the solar spectrum, with the double bright
line constituting the spectrum of the spirit lamp with salt.   I
remarked that there must be some physical connection between
two agencies presenting so marked a characteristic in common.
He assented, and said he believed a mechanical explanation of
the cause was to be had on some such principle as the following:—
Vapour of sodium must possess by its molecular structure a tendency to vibrate in the periods corresponding to the degrees of
refrangibility of the double line D.   Hence the presence of sodium
in a source of light must tend to originate light of that quality.
On the other hand, vapour of sodium in an atmosphere round a
source must have a great tendency to retain in itself, i.e. to absorb,
and to have its temperature raised by, light from the source, of
the precise quality in question.   In the atmosphere around the
sun, therefore, there must be present vapour of sodium, which,
according to the mechanical explanation thus suggested, being
particularly opaque for light of that quality prevents such of it as
is emitted from the sun from penetrating to any considerable
distance through the surrounding atmosphere.   The test of this
theory must be had in ascertaining whether or not vapour of
sodium has the special absorbing power anticipated.   I have the
impression that some Frenchman did make this out by experiment,
but I can find no reference on the point.

"I am not sure whether Prof. Stokes' suggestion of a mechanical theory has ever appeared in print.   I have given it in
my lectures regularly for many years, always pointing out along
with it that solar and stellar chemistry were to be studied by
investigating terrestrial substances giving bright lines in the
spectra of artificial flames corresponding to the dark lines of the
solar and stellar spectra*."

* [The following is a note appended by Prof. Stokes to his translation of a paper
by Kirchhoff in *Phil. Mag.*, March 1860, p. 196:—"The remarkable phenomenon
discovered by Foucault, and rediscovered and extended by Kirchhoff, that a body
may be at the same time a source of light giving out rays of a definite refrangibility,
and an absorbing medium extinguishing rays of the same refrangibility which
traverse it, seems readily to admit of a dynamical illustration borrowed from sound.
We know that a stretched string which on being struck gives out a certain note
(suppose its fundamental note) is capable of being thrown into the same state of
vibration by aërial vibrations corresponding to the same note.   Suppose now a portion of space to contain a great number of such stretched strings forming thus the

What I have read thus far is with reference not to the Molar. origin of spectrum analysis, but to the definite point, of Stokes' suggested dynamics of absorption.   There is no hint there of the effect of the reaction of the vibrating particles in the luminiferous ether in the way of affecting the velocity of the propagation of the light through it.   Sellmeier's first title has reference to that effect; he explains ordinary refraction through the inertia of these particles and he shows how, when the light is nearly of the period corresponding to any of the fundamental periods of the embedded vibrators, there will be anomalous dispersion.   He gives a mathematical investigation of the subject, not altogether satisfactory, perhaps, but still it seems to me to formulate a most valuable step towards a wholly satisfactory treatment of the thing. Lord Rayleigh, Helmholtz and others have quoted Sellmeier. Lommel begins afresh, I think, but he notices Sellmeier also, so the idea must have originated with Sellmeier, and it seems to me a very important new departure with respect to the dynamical explanation of light.

Ten minutes interval.

Now, let us look at this problem of vibrating particles once Molecular. more.   I have a little exercise to propose for the ideal arithmetical laboratory.   Just try the arithmetical work for this problem for 7 particles.   I do not know whether it will work out well or not. I have not the time to do it myself, but perhaps some of you may find the time, and be interested enough in the thing, to do it. Take the m's in order, proceeding by ratios of 4; and the c's in order, proceeding by differences of 1:

$$m_1, m_2, m_3, m_4, m_5, m_6, m_7 = 1, 4, 16, 64, 256, 1024, 4096,$$
$$c_1, c_2, c_3, c_4, c_5, c_6, c_7, c_8 = 1, 2, 3, 4, 5, 6, 7, 8.$$

There will be 7 roots to find by trial.   I would like to have some of you try to find some of these, if not all; also the energy ratios.   You will probably find it an advantage in the calculation if you proceed thus: put $\dfrac{1}{r^2} = z$, and by "roots" let us understand

analogue of a "medium."   It is evident that such a medium, on being agitated, would give out the note above mentioned, while on the other hand, if that note were sounded in air at a distance, the incident vibrations would throw the strings into vibration and consequently would themselves be gradually extinguished since otherwise there would be a creation of vis viva.   The optical application of this illustration is too obvious to need comment.—G. G. S."   H.]

Molecular. values of z, making $\xi = 0$, which implies, and is secured by, $u_1 = 0$. We have

$$a_1 = z - 3, \quad a_2 = 4z - 5, \quad a_3 = 16z - 7, \quad a_4 = 64z - 9,$$

$$a_5 = 256z - 11, \quad a_6 = 1024z - 13, \quad a_7 = 4096z - 15.$$

You will have to take values of z by trial until you get near a root. The convergence of the continued fraction (p. 39) will be so rapid that you will have very little trouble in getting the largest roots. Begin then with the largest z-root, corresponding to the shortest of the critical periods τ, and proceed downwards, according to the indications of pp. 55—58. In the course of the process, you will have the whole series of the u's for each root; by multiplying these in order, you have the x's for each particular root, and then you can calculate the energy ratios for each root. We shall then be able to put our formula into numbers; and I feel that I understand it much better when I have an example of it in numbers than when it is merely in a symbolic form.

I want to show you now the explanation of ordinary refraction. Let us go back to our supposition of spherical shells, or, if you like, our rude mechanical model. Suppose an enormous number of spherical cavities distributed equally through the space we are concerned with. Let the quantity of ether thus displaced be so exceedingly small in proportion to the whole volume that the elastic action of the residue will not be essentially altered by that. These suppositions are perfectly natural. Now, what is unnatural mechanically, is, that we suppose a massless rigid spherical lining to this spherical cavity in the luminiferous ether connected with an interior rigid massive shell, m_1, by springs—in the first place symmetrical. We shall try afterwards to see if we cannot do something in the way of aeolotropy; but as I have said before I do not see the way out of the difficulties yet. In the meantime, let us

Massless rigid shell lining to spherical cavity in the luminiferous ether. ....

Shell No. 1, m_1 ....
Shell No. 2, m_2 ....

suppose this first shell m_1 to be isotropically connected by springs with the rigid shell lining of the spherical cavity in the ether. When I say isotropically connected I mean distinctly this: that if you draw this first shell m_1 aside through a certain distance in any direction, and hold it so, the

required force will be independent of the direction of the displace- Molecular.
ment. Certain springs in the drawing—the smallest number
would be three—placed around in proper positions will rudely
represent the proper connections for us. Similarly, let there
be another shell here, m_2, in the interior of m_1, isotropically
connected with it by springs; and so on.

This is the simplest mechanical representation we can give of a
molecule or an atom, imbedded in the luminiferous ether, unless
we suppose the atom to be absolutely hard, which is out of the
question. If we pass from this problem to a problem in which we
shall have continuous elastic denser matter instead of a series of
connections of associated particles, we shall be, of course, much
nearer the reality. But the consideration of a group of particles
has great advantage, for we are more familiar with common algebra
than with the treatment of partial differential equations of the
second order with coefficients not constant, but functions of the
independent variable,—which are the equations we have to deal
with if we take a continuous elastic molecule, instead of one made
up of masses connected by springs as we have been supposing.

Let us suppose the diameters of these spherical cavities to be
exceedingly small in comparison with the wave length. Practically
speaking, we suppose our structure to be infinitely fine-grained.
That will not in the least degree prevent its doing what we want.
The distance also from one such cavity, containing within it a
series of shells, to another such cavity, in the luminiferous ether, is
to be exceedingly small in comparison with the wave length, so
that the distribution of these molecules through the ether leaves
us with a body which is homogeneous when viewed on so coarse a
scale as the wave length; but it is, if you like, heterogeneous
when viewed with a microscope that will show us the millionth or
million-millionth of a wave length. This idea has a great advantage over Cauchy's old method, in allowing an infinitely fine-
grainedness of the structure, instead of being forced to suppose
that there are only several molecules, ten or twelve, to the wave
length, as we are obliged to do in getting the explanation of
refraction by Cauchy's method.

I wish to show you the effect of molecules of the kind now
assumed upon the velocity of light passing through the medium.

Let $\dfrac{m_1}{4\pi^2}$ denote the sum of all the masses of shells No. 1 in any

Molecular. volume divided by the volume; let $\dfrac{m_2}{4\pi^2}$ denote the sum of the masses of No. 2 interior shell in any volume divided by the volume; and so on. We will not put down the equations of motion for all directions, but simply take the equations corresponding to a set of plane waves in which the direction of the vibration is parallel to OX, and the direction of the propagation is parallel to OY.

If we denote by $\dfrac{\rho}{4\pi^2}$ the density of the vibrating medium, (I am taking $\dfrac{\rho}{4\pi^2}$ instead of the usual ρ for the reason you know, viz.: to get rid of the factor $4\pi^2$ resulting from differentiation). Let $\dfrac{l}{4\pi^2}$, (instead of n as formerly,) denote the rigidity of the luminiferous ether. The dynamical equation of motion of the ether and embedded cavity-linings will clearly be

$$\frac{\rho}{4\pi^2}\frac{d^2\xi}{dt^2} = \frac{l}{4\pi^2}\frac{d^2\xi}{dy^2} + c_1\,(x_1 - \xi).$$

For waves of period T, we have $\xi = \text{const.}\,X \sin 2\pi\left(\dfrac{x}{\lambda} - \dfrac{t}{T}\right)$. The second differential coefficients of this with respect to t and x will be $-\dfrac{4\pi^2}{T^2}\xi,\ -\dfrac{4\pi^2}{\lambda^2}\xi$ respectively. Therefore our equation becomes $\dfrac{\rho}{T^2} = \dfrac{l}{\lambda^2} + c_1\left(1 - \dfrac{x_1}{\xi}\right)$. Let us find $\dfrac{T^2}{\lambda^2}$, which is the reciprocal of the square of the velocity of propagation. You may write it $\dfrac{1}{v^2}$ if you like, or μ^2, the square of the refractive index. We have,

$$\frac{T^2}{\lambda^2} = \frac{1}{l}\left\{\rho - c_1 T^2\left(1 - \frac{x_1}{\xi}\right)\right\}.$$

Substitute our value (Lecture VII.) for $-x_1/\xi$;

$$-\frac{x_1}{\xi} = \frac{c_1 T^2}{m_1}\left(\frac{\kappa_1^{\,2} R_1}{\kappa_1^{\,2} - T^2} + \frac{\kappa_2^{\,2} R_2}{\kappa_2^{\,2} - T^2} + \dots\right)$$

and this becomes

$$\frac{T^2}{\lambda^2} = \frac{1}{l}\left[\rho - c_1 T^2\left\{1 + \frac{c_1 T^2}{m_1}\left(\frac{\kappa_1^{\,2} R_1}{\kappa_1^{\,2} - T^2} + \frac{\kappa_2^{\,2} R_2}{\kappa_2^{\,2} - T^2} + \dots\right)\right\}\right].$$

This is the expression for the square of the refractive index as it is affected by the presence of molecules arranged in the way we hàve supposed. It is too late to go into this for interpretation Molecular. just now, but, I will tell you that if you take T considerably less than κ_1, and very much greater than κ_2, you will get a formula with enough of disposable constants to represent the index of refraction by an empirical formula, as it were; which, from what we know, and what Sellmeier and Ketteler have shown, we can accept as ample for representing the refractive index of ordinary transparent substances.

We shall look into this farther, a little later, and I will point out the applications to anomalous dispersion. We must think a good deal of what can become of vibrations in a system of that kind when the period of the vibration of the luminiferous ether is approximately equal to any one of the fundamental periods that the internal complex molecule could have were the shell lining in the ether held absolutely at rest.

LECTURE X.

Molar.
WE shall now think a little about the propagation of waves with a view to the question, what is the result as regards waves at a distance from the source, the source itself being discontinuous in its action. In the first place, we will take our expression for a plane wave. The factor in our formulas showing diminution of amplitude at a distance from a source does not have effect when we come to consider plane waves. So we just take the simple expression for plane harmonic waves propagated along the axis of y with velocity v;

$$\xi = a \cos \frac{2\pi}{\lambda} (y - vt).$$

Let us consider this question:—what is the work done per period by the elastic force in any plane perpendicular to the line of propagation of the wave. We shall think of the answer to that question with the view to the consideration of the possibility of a series of waves advancing through space previously quiescent. Suppose I draw a straight line here for the line of propagation and let this curve represent a succession of waves travelling from left to right and penetrating into an elastic solid previously

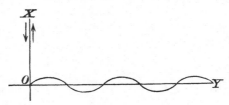

quiescent. Take a plane perpendicular to the line of propagation of the waves, and think of the work done by the elastic solid upon one side of this plane upon the elastic solid on the

other side, in the course of a period in the vibration. We shall Molar.
take an expression for the tangential force in the plane XOZ,
and in the direction OX, which we denote by T (according
to our old notation of S, T, U, P, Q, R). We shall virtually investigate here the formula for the propagation of the wave independently of our general formula in three dimensions. Taking
T to denote the tangential force of the elastic medium on the one
side of the plane XOZ, the downward direction of the arrow-head
which I draw being that direction in which the medium on the
left pulls the medium on the right, I put infinitely near that in
the medium on the right another arrow-head. Imagine for the
moment a split in the medium to indicate the reaction which the
medium on the right exerts on the medium on the left by this
plane ; and imagine the medium on the left taken away, and that
you act upon the plane boundary of the medium on the right,
with the same force as in the continuous propagation of waves.
The medium upon the left acts in this way upon the plane interface :—that is an easy enough conception. I correctly represent
it in my diagram by an arrow-head pointing down infinitely near
to the plane on the left-hand side. The displacement of the
medium is determined by a distortion from a square figure to an
oblique figure, and there is no inconsistency in putting into this
little diagram an exaggeration of the obliquity, so as to
show the direction of it. The force required to do that
is clearly as our diagram lies, upward on the right and
downward on the left.

Let us consider now the work done by that force. Calling
ξ the displacement of a particle from its mean position, $T \cdot \dot{\xi}$ is
the work done by that tangential force per unit of time. $\dfrac{d\xi}{dy}$ is
the shearing strain experienced in the medium so that

$$n \frac{d\xi}{dy} = -T.$$

In this particular position which we have taken, ξ increases with y
so that the sign minus is correct according to the arrow heads.

Let there be simple harmonic waves propagated from left to
right with velocity v. This is the expression for it

$$\left[\text{indicating } \xi = A \cos \frac{2\pi}{\lambda}\,(y - vt) \right].$$

Hence,

$$\dot{\xi}=\frac{2\pi}{\lambda}\, va \sin q \; ; \quad \frac{d\xi}{dy}=-\frac{2\pi}{\lambda}\, a \sin q \; ;$$

and the rate of doing work is

$$\frac{4\pi^2}{\lambda^2}\, a^2 vn \sin^2 q.$$

That is the rate at which this plane, working on the elastic solid on the right-hand side of it, does work (" per unit area of the plane" understood). Multiply this by dt and integrate through a period $\tau = \lambda/v$. Now

$$\int_0^\tau \sin^2 q\, dt = \int_0^\tau \frac{1}{2}\,(1-\cos 2q)\, dt = \int_0^\tau \frac{1}{2}\, dt = \frac{1}{2}\,\tau.$$

The rate of doing work then, per period, is

$$\frac{2\pi^2}{\lambda^2}\, a^2 vn\tau = \frac{2\pi^2 a^2 n}{\lambda}\,.$$

If it is possible for a set of waves to advance uniformly into space previously undisturbed, then it is certain that the work done per period must be equal to the energy in the medium per wave length. Let us then work out the energy per wave length.

It is easily proved that, in waves in a homogeneous elastic solid, the energy is half potential of elastic stress, and half kinetic energy; and it will shorten the matter, simply to calculate the kinetic energy and double it, taking that as the energy in the medium per wave length. In our notation of yesterday, we took $\frac{\rho}{4\pi^2}$ as the density. Multiply this by dy, to get the mass of an infinitesimal portion (per unit of area in the plane of the wave). The kinetic energy of this mass is

$$\frac{1}{2}\frac{\rho}{4\pi^2}\, dy\, \dot{\xi}^2 = \frac{1}{2}\frac{\rho v^2 a^2}{\lambda^2}\sin^2 q \cdot dy.$$

Integrating this through a wave length, and doubling it so as to get the whole energy, we have $\frac{1}{2}\frac{\rho v^2 a^2}{\lambda}$. Compare that with the work done per period, viz. $\frac{1}{2}\frac{a^2}{\lambda}\, l$, if $\frac{l}{4\pi^2}$ denote as yesterday the rigidity instead of n. We see that they are equal, because (velocity

of propagation) $v = \sqrt{\dfrac{l}{\rho}}$ as we find from the elementary equation Molar.
of the wave motion,

$$\frac{d^2\xi}{dt^2} = \frac{l}{\rho}\frac{d^2\xi}{dy^2}.$$

Thus the work done per period is equal to the energy per wave length.

This agrees with what we know from the ordinary general solution of the equation of motion by arbitrary functions that it is possible for a discontinuous series of waves to be propagated into the elastic medium, previously quiescent:—and is coextensive with the case of velocity of propagation independent of wave length, for a regular simple harmonic endless succession of waves. But if our present energy equation did not verify, it would be impossible to have a discontinuous series of waves propagated forward without change of form into a medium previously quiescent. I wanted to verify the energy equation for the case of the homogeneous elastic solid, because we are concerned with a case in which this is not verified; that is to say, when we put in our molecules. In this case, the work done per period is less than the energy in the medium per wave length, and therefore it is impossible for the waves to advance without change of form.

Before we go on to that, let us stay a little longer in a homogeneous elastic solid, and look at the well-known solution by discontinuous functions. The equation of motion is

$$\rho\,\frac{d^2\xi}{dt^2} = l\,\frac{d^2\xi}{dy^2}.$$

Although I said I would not formally prove this now, it is in reality proved by our old equation

$$\rho\,\frac{d^2\xi}{dt^2} = l\nabla^2\xi.$$

I took the liberty of asking Professor Ball two days ago whether he had a name for this symbol ∇^2; and he has mentioned to me *nabla*, a humorous suggestion of Maxwell's. It is the name of an Egyptian harp, which was of that shape. I do not know that it is a bad name for it. Laplacian I do not like for several reasons both historical and phonetic. [Jan. 22, 1892. Since 1884 I have found nothing better, and I now call it Laplacian.]

I should have told you that this is the case of a plane wave propagated in the direction of OY, with the plane of the wave parallel to XZ; for which case, *nabla* of ξ (that is to say $\nabla^2 \xi$) becomes simply $\dfrac{d^2\xi}{dy^2}$. The time-honored solution of this equation is

$$\xi = f(y - vt) + F(y + vt),$$

where f and F are arbitrary functions. You can verify that by differentiation. This solution in arbitrary functions proves that a discontinuous series of waves is possible; and knowing that a discontinuous series is possible, you could tell without working it out, that the work done per period by the medium on the one side of the plane which you take perpendicular to the line of propagation must be equal to the energy of the medium per wave length.

Before passing on to the energy solution for the case in which we have attached molecules, in which this equality of energy and work does not hold, with the result that you cannot get the discontinuous single pulse or sequence of pulses, I want to suggest another elementary exercise for the anticipated arithmetical laboratory. It is to illustrate the propagation of waves in a medium in which the velocity is not independent of the wave length, and to contrast that with the propagation of waves when the velocity is independent of the wave length in order that you may feel for yourselves what these two or three symbols show us, but which we need to look at from a good many points of view before we can make it our own, and understand it thoroughly. To realize that this equation $\dfrac{d^2\xi}{dt^2} = \text{const.} \times \dfrac{d^2\xi}{dy^2}$ gives us constant velocities for all wave lengths, and that constant velocities for all wave lengths implies this equation, and to see that that goes along with the propagation of a discontinuous pulsation without change of figure, or a discontinuous succession of pulsations without change of character, I want an illustration of it, by the consideration of a case in which the condition of constancy of velocity for different wave lengths is not fulfilled.

I ask you first to notice the formula

$$S = \frac{\frac{1}{2}(1 - e^2)}{1 - 2e \cos q + e^2} = \frac{1}{2} + e \cos q + e^2 \cos 2q + \ldots$$

which is familiar to all mathematical readers as leading up to Molar. Fourier's harmonic series of sines and cosines. It is proved by taking

$$2 \cos q = \epsilon^{\iota q} + \epsilon^{-\iota q},$$

and resolving S into two partial fractions. Poisson and others* make this series the foundation of a demonstration of Fourier's theorem. If $e < 1$ the series is convergent; when $e = 1$ it ceases to converge. If we take $q = \dfrac{2\pi y}{a}$ and draw the curve whose dependent coordinate is $x = S$, what have we?

Take $t = 0$ and measure off lengths from the origin

$$y = a, \quad 2a, \ldots$$

The curve represented will be this (heavy curve).

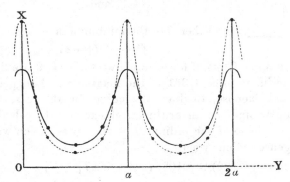

The heavy curve is

$$x = \frac{2\delta}{5 - 3 \cos 2\pi y} \quad (a = 1, \ e = \tfrac{1}{3}).$$

It is here drawn by the points

$$(y, x) = (0, 1), \ (\tfrac{1}{8}, \tfrac{7}{10}), \ (\tfrac{1}{4}, \tfrac{4}{10}), \ (\tfrac{1}{2}, \tfrac{1}{4}),$$

and symmetrical continuation.

The dotted curve is

$$x = \frac{3/2}{5 - 4 \cos 2\pi y} \quad (a = 1, \ e = \tfrac{1}{2}).$$

It is here drawn by the points

$$(y, x) = (0, \tfrac{3}{2}), \ (\tfrac{1}{8}, \tfrac{7}{10}), \ (\tfrac{1}{4}, \tfrac{3}{10}), \ (\tfrac{1}{2}, \tfrac{1}{6}),$$

and symmetrical continuation.

I want the arithmetical laboratory to work this out and give

---

* See Thomson and Tait's *Natural Philosophy*, § 77.

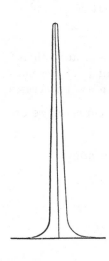

Molar.   graphic representations of the periodic curves for several different values of *e*. The particular numerical case that I am going to suggest is one in which the curve will be more like this second curve which I draw ; it is much steeper and comes down more nearly to zero. Take the extreme case of $e = 1$, and what happens ? S is infinitely great for q infinitely small, and is infinitely small for all other values of q less than a. For any value of e, the maximum and minimum ordinates of the curve (corresponding respectively to $q = 0°$, and $q = 180°$) are

$$\tfrac{1}{2} \cdot \frac{1+e}{1-e}, \text{ and } \tfrac{1}{2} \cdot \frac{1-e}{1+e};$$

and therefore the minimum is

$$(1-e)^2/(1+e)^2$$

of the maximum. Thus, if for example we take $e = \cdot9$, we find the minimum ordinate to be $1/361$ of the maximum. I suggest, as a mathematical exercise, to draw the curve for this case by the finite formula: and, as an arithmetical exercise, to calculate as many as you please of the ordinates by the series. You will find its convergence tediously slow.

$$[(\cdot9)^{43} \fallingdotseq \cdot0108, \ (\cdot9)^{44} \fallingdotseq \cdot0097]:$$

you must take more than 43 or 44 terms to reach an accuracy of one per cent. in the result. So I do not think you will be inclined to calculate very many of the ordinates by the series.

Molar: Digression, Deep-sea Waves.   I would also advise those who have time to read Poisson's and Cauchy's great papers on deep-sea waves. (Poisson's Mémoire sur la théorie des ondes. Paris, *Mém. Acad. Sci.* I., 1816, pp. 71—186 ; *Annal. de Chémie*, V., 1817, pp. 122—142. *Cauchy*, Mémoire sur la théorie de la propagation des ondes à la surface d'un fluide pesant d'une profondeur indéfinie [1815], Paris, *Mém. Sav. Étrang.* I., 1827, pp. 3—312.) Those papers are exceedingly fine pieces of true mathematics; *and they are very strong.* But you might have the hydrodynamical beginnings presented much more fascinatingly. If you know the elementary theory of deep-sea waves, well and good: then take Poisson and Cauchy for the higher analytical treatment. Those who do not know the theory of deep-sea waves may read it up in elementary books. The best text-books I know for Hydrokinetics are Besant's and Lamb's.

The great struggle of 1815, *not* that fought out on the plains of Belgium, was, who was to rule the waves, Cauchy or Poisson. Their memoirs seem to me of very nearly equal merit. I have no doubt the judges had good reason for giving the award to Cauchy, but Poisson's paper also is splendid. I can see that the two writers respected each other very much, and I suppose each thought the other's work as good as his own (? and sometimes better !).

The problem which they solve is this, in their high analytical style : Every portion of an infinite area of water is started initially with an arbitrarily stated infinitesimal displacement from the level and an arbitrarily stated velocity up or down from the level, and the inquiry is, what will be the result ? It is obvious that you have the solution of that problem from the more elementary problem, what is the result of an infinitesimal displacement at a single point, such as may be produced by throwing a stone into water ? Let a solid, say, cause a depression in any place, the velocity of the solid performing the part of giving velocity and displacement to the surface of the water : then consider the solid suddenly annulled. The same thing in two dimensions is exceedingly simple. Take, for example, waves in an infinitely deep canal with vertical sides. Take a sudden disturbance in the canal, equal over all its breadth, and inquire what will the result be ?

I wish now to help you toward an understanding of Cauchy's and Poisson's solutions. They only give symbols and occasionally numerical results : they do not give any diagrams or graphical representations ; and I think it would repay any one who is inclined to go into the subject to work it out with graphic representations thus :— first I must tell you that the elementary Hydrokinetic solution for deep-sea waves is simply a set of waves, or a set of standing vibrations (take which you please): the propagational velocity of the waves being $\sqrt{\dfrac{g\lambda}{2\pi}}$, and therefore for different waves directly proportional to the square root of the wave-length ; and the vibrational period of waves or of standing vibrations being $\sqrt{\dfrac{2\pi\lambda}{g}}$, also directly proportional to the square root of the wave-length. Thus, so far as concerns only the varying shape of the disturbed water surface, the whole result of the elementary Hydrokinetics of deep-sea wave-motion is expressed by one or other of these equations which I write down ;—

$$x = h \cos \frac{2\pi}{\lambda}\left(y \mp t \sqrt{\frac{g\lambda}{2\pi}}\right); \quad \text{or} \quad x = h \frac{\sin}{\cos} \frac{2\pi y}{\lambda} \frac{\sin}{\cos} t \sqrt{\frac{2\pi g}{\lambda}}.$$

Superposition of two motions represented by either formula gives a specimen of single motion represented by the other, as you all know well by your elementary trigonometry.

And now, in respect to Poisson's and Cauchy's great mathematical work on deep-sea waves, it will be satisfactory to you to know that it consists merely in the additions of samples represented by whichever of these formulas you please to take. The simple formula, or summation of it with different values of h and λ, represents waves with straight ridges, or generally straight lines of equal displacement: or, as we may call it, two-dimensional wave-motion. Every possible case of three-dimensional wave-motion (including the circular waves produced by throwing a stone into water) is represented by summation of the samples of the formula, as it stands, and with z substituted for y; y and z representing Cartesian coordinates in the horizontal plane of the undisturbed water-surface, and x representing elevation of the disturbed water-surface above this plane.

And now, confining ourselves to the two-dimensional wave-motion, I suggest to our arithmetical laboratory, to calculate and draw the curve represented by

$$s = \tfrac{1}{2} + e \cos q_1 + e^2 \cos 2q_2 + e^3 \cos 3q_3 + \ldots e^i \cos iq_i + \ldots$$

where
$$q_i = \frac{2\pi}{a}(y - v_i t); \quad \text{with } v_i = \frac{1}{\sqrt{i}}.$$

Calculate the curve corresponding to any values you please of e and of t. Also calculate and draw curves representing the sum of values of s with equal positive and negative values of v_i.

I suggest particularly that you should perform this last described calculation for the cases, $e = \tfrac{1}{3}$ and $e = \tfrac{1}{2}$. You have already the curves for $t = 0$ in these cases shown on the diagram of page 113, and you will find it interesting to work out, in considerable detail, curves for other values of t which you can do without inordinately great labour, as the series are very rapidly convergent. You will thus have graphic representations of the two-dimensional case of Cauchy and Poisson's problem for infinitely deep water between two fixed parallel vertical planes.

Lastly, I may say that if there are some among you who Molar: Digression, Deep-sea Waves. will not shrink from the labour of calculating and adding forty or fifty terms of the series, I advise you to do the same for $e = ·9$; and you will have splendid graphical illustrations of the two-dimensional problem of deep-sea waves initiated by a single disturbance along an endless straight line of water. If you do so, or if you spend a quarter-of-an-hour in planning to begin doing so, you will learn to thank Cauchy and Poisson for their magnificent mathematical treatment of their problem by definite integrals, and for their results from which, with very moderate labour, you may calculate the answer to any particular question that may be reasonably put with reference to the subject; and may work out very thorough graphical illustrations of all varieties of the problem of deep-sea waves *.

We are going to take our molecules again, and put them in the Molecular. ether; and look at the question, what is the velocity of propagation of waves through it under some suppositions which we shall make as to the masses of these embedded molecules, and how much they will modify the velocity of propagation from what it is in pure ether. Then we shall look at the matter, with respect to the question of the work done upon a plane perpendicular to the line of propagation, and we shall see that the energy per wave-length is greater than the work done per period, and that therefore it is impossible under these conditions for waves to advance uniformly into space previously occupied by quiescent matter.

You will find, in Lord Rayleigh's book on sound, the question of the work done per period, and the energy per wave length, gone into: and the application of this principle, with respect to the possibility of independent suites of waves travelling without change of form, thoroughly explained.

To-morrow we shall consider investigations respecting the difference of velocity of propagation in different directions in an aeolotropic elastic solid, for the foundation of the explanation of double refraction on mere elastic solid idea. The thing is quite

---

* [Note of May, 1898. For some of these see my papers:—"On Stationary Waves in Flowing Water," *Phil. Mag.* 1886, Vol. 22, pp. 353, 445, 517; 1887, Vol. 23, p. 52. "On the Front and Rear of a Free Procession of Waves in Deep Water," *Phil. Mag.* 1887, Vol. 23, p. 113; and, "On the Waves produced by a Single Impulse in Water of any depth or in a Dispersive Medium," *Phil. Mag.* 1887, Vol. 23, p. 252.]

Molecular. familiar to many of you, no doubt, and you also know that it is a
failure in regard to the explanation of the propagation of light in
biaxal crystals. It is, however, an important piece of physical
dynamics, and I shall touch upon it a little, and try to show it in
as clear a light as I can.

Ten minutes interval.

Now for our proper molecular question. The distance from
cavity to cavity in the ether is to be exceedingly small, in
comparison with the wave-length, and the diameter of each
cavity is to be exceedingly small, in comparison with the distance
from cavity to cavity. Let the lining of the cavity be an
ideal rigid massless shell. Let the next shell within be a rigid

shell of mass $\frac{m_1}{4\pi^2}$.   I represent the thing in this diagram as

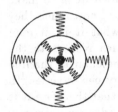

if we had just two of these massive shells
and a solid nucleus. The enormous mass of
the matter of the grosser kind which exists
in the luminiferous ether when permeated
by even such a comparatively non-dense
body as air, would bring us at once to very
great numbers in respect to the masses which
we will suppose inside this cavity, in comparison with the masses
of comparable bulks of the luminiferous ether. If there is time
to-morrow, we shall look a little to the possible suppositions as to
the density of the luminiferous ether, and what limits of greatness
or smallness are conceivable in respect to it. At present we have
enough to go upon to let us see that, even in air of ordinary
density, the mass of air per cubic centimetre must be enormously
great, in comparison with the mass of the luminiferous ether per
cubic centimetre. We must therefore have something enormously
massive in the interior of these cavities. We shall think a good
deal of this yet, and try to find how it is we can have the large
quantity of energy that is necessary to account for the heating of
a body such as water by the passage of light through it, or for
the phosphorescence of a body which is luminous for several days
after it has been excited by light. I do not think we shall have
the slightest difficulty in explaining these things. These are not

the difficulties.   The difficulties of the wave theory of light are Molecular. difficulties which do not strike the popular imagination at all. They are the difficulties of accounting for polarization by reflection with the right amount of light reflected; and of accounting for double refraction with the form of wave-surface guessed by Huyghens and proved experimentally by Stokes.   With the general character of the phenomena we have no difficulty whatever; the great difficulty, in respect to the wave theory of light, is to bring out the proper quantities in the dynamical calculation of these effects.

There is no difficulty in explaining the energy required for heating a body by radiant heat passing through it, nor how it is that it sometimes comes out as visible light and, it may be, so slowly that it may continue appearing as light for two or three days.   All these properties, wonderful as they are, seem to come as a matter of course from the dynamical consideration.   So much so that any one not knowing these phenomena would have discovered them on working out the subject dynamically.   He would discover anomalous dispersion, fluorescence, phosphorescence, and the well-known visible and invisible radiant heat of longer periods emitted by a body which has been heated and left to cool.   All these phenomena might have been discovered by dynamics; and a dynamical theory that discovers what is afterwards verified by experiment is a very estimable piece of physical dynamics.

I speak with confidence in this subject because I am ashamed to say that I never heard of anomalous dispersion until after I found it lurking in the formulas.   And, when I looked into the matter, I found to my shame that a thing which had been known by others for fifteen or twenty years* I had not known until I found it in the dynamics.

Take our concluding formula of yesterday (p. 106 above), with some changes of notation

$$\frac{1}{\zeta^2} = \frac{\rho}{n} + \frac{c_1\tau^2}{n} \left\{ -1 + \frac{c_1\tau^2}{m_1} \left(\frac{\kappa^2 R}{\tau^2 - \kappa^2} + \frac{\kappa_{\prime}^2 R_{\prime}}{\tau^2 - \kappa_{\prime}^2} + \&c. \right) \right\},$$

where ζ denotes the propagational velocity of waves of period τ ;

* Leroux, "Dispersion anomale de la vapeur d'iode," *Comptes Rendus*, LV., 1862, pp. 126—128: *Pogg. Ann.* CXVII., 1862, pp. 659, 660.   Christiansen, "Ueber die Brechungsverhältnisse einer weingeistigen Lösung des Fuchsins," *Ann. Phys. Chem.* CXLI., 1870, pp. 479, 480: *Phil. Mag.*, XLI., 1871, p. 244; *Annales de Chimie*, XXV., 1872, pp. 213, 214.

Molecular. n, ρ, and m_1 denote now respectively the rigidity of the ether, the mass of the ether in unit volume of space, and the sum of the masses of the first interior shells of the embedded molecules in unit volume of space;

c_1 the force of the first spring, per unit elongation, multiplied by the number of molecules embedded in the ether per unit volume of space;

κ, $\kappa_{\prime}$, $\kappa_{\prime\prime}$, &c., in order of ascending magnitude, the fundamental periods of the molecule when the outer shell is held fixed;

R, $R_{\prime}$, $R_{\prime\prime}$, &c., denote for the separate fundamental vibrations the ratio of the energy of the first interior shell to the whole energy of the complex vibrator.

Let us consider what the wave-period τ may be relatively to the fundamental periods κ, $\kappa_{\prime}$, $\kappa_{\prime\prime}$, ... of the vibrator on the supposition of the bounding shell held fixed, to give us a good reasonable explanation of dispersion, in accordance with the facts of observation with respect to the difference of velocity for different periods. To help us with this consideration, take our previous auxiliary formulas

$$\frac{1}{\zeta^2} = \frac{\rho}{n} + \frac{c_1 \tau^2}{4\pi^2 n}\left(\frac{x_1}{\xi} - 1\right),$$

$$\frac{x_1}{\xi} = \frac{c_1}{4\pi^2 m_1}\left(\frac{\kappa^2 R \tau^2}{\tau^2 - \kappa^2} + \frac{\kappa_{\prime}^2 R_{\prime} \tau^2}{\tau^2 - \kappa_{\prime}^2} + \&c.\right),$$

where ξ and x_1 denote respectively the simultaneous maximum displacements of the outermost massless shell and the first of the massive shells within it.

If τ were less than the smallest of the fundamental periods, $\frac{x_1}{\xi}$ would be negative, the wave-velocity would be greater than in free ether, and the refractive index would be less than unity. But in all known cases the refractive index is greater than unity; and when this is so, $\frac{x_1}{\xi} - 1$ must be positive. I want to see if we can get our formula to cover a range, including all light from the highest ultra-violet photographic light of about half the wave-length of sodium light down to the lowest we know of, which is the radiant heat from a Leslie cube with a wave-length that I hear from Prof. Langley since I spoke to you on the Molecular. subject a week ago is about $\frac{1}{1000}$ of a centimetre or 17 times the wave-length of sodium light. That will be a range of about forty to one. The highest invisible ultra-violet light hitherto determined, by its photographic action, has a period about 1/40 of the period of the lowest invisible radiation of radiant heat that has yet been experimented upon.

It is probable that all or many colourless transparent liquids and solids are mediums for which throughout every part of that range there are no anomalous dispersions. I think it is almost certain that for rock-salt, in the lower part of the range, there are no anomalous dispersions at all. In fact Langley's experiments on radiant heat are made with rock-salt; and in all experiments made with rock-salt, it seems as if little or no radiant heat is absorbed by it. At all events, we could not be satisfied unless we can show that this kind of supposition will account for dispersion through a range of period from one to forty. It is obvious that if we are to have continuous refraction without anomalous dispersion through that wide range of periods, there cannot be any of the periods κ, $\kappa_{,}$, $\kappa_{,,}$, ... within it.

To-morrow we shall consider the case in which the wave-period is longer than the longest of the molecular periods; and we shall find that on this supposition our formula serves well to represent all we have hitherto known by experiment regarding ordinary dispersion. [*Added July* 7, 1898. This was true in October 1884; but measurements by Langley of the refractivity of rock-salt for radiant heat of wave-lengths (in air or ether) from ·43 of a mikron to 5·3 mikrons, (the " mikron " being 10^{-6} of a metre or 10^{-4} of a centimetre), published in 1886 (*Phil. Mag.* 1886, 2nd half-year), showed that there must be a molecular period longer than that corresponding to wave-length 5·3. In an addition to Lecture XII, Part II, we shall see that subsequent measurements of refractivity by Rubens, Paschen, and others, extend the range of ordinary dispersion by rock-salt to wave-lengths of 23 mikrons; and give results in splendid agreement with our formula, which is identical with Sellmeier's expression of his own original theory, through a range of from ·4 of a mikron to 23 mikrons, and indicate 56 mikrons as being the probable wave-length of radiant heat in ether of which the period is a critical period for rock-salt.]

LECTURE XI.

Molar.

WE shall now take up the subject of an elastic solid which is not isotropic. As I said yesterday, we do not find the mere consideration of elastic solid satisfactory or successful for explaining the properties of crystals with reference to light. It is, however, to my mind quite essential that we should understand all that is to be known about homogeneous elastic solids and waves in them, in order that we may contrast waves of light in a crystal with waves in a homogeneous elastic solid.

Aeolotropy is in analogy with Cauchy's word isotropy which means equal properties in all directions. The formation of a word to represent that which is not isotropic was a question of some interest to those who had to speak of these subjects. I see the Germans have adopted the term anisotropy. If we used this in English we should have to say : " An anisotropic solid is not an isotropic solid "; and this jangle between the prefix *an* (privative) and the article *an*, if nothing else, would prevent us from adopting that method of distinguishing a non-isotropic solid from an isotropic solid. I consulted my Glasgow University colleague Prof. Lushington and we had a good deal of talk over the subject. He gave me several charming Greek illustrations and wound up with the word aeolotropy. He pointed out that αἰόλος means variegated; and that the Greeks used the same word for variegated in respect to shape, colour and motion; example of this last, our old friend "κορυθαίολος Ἕκτωρ." There is no doubt of the classical propriety of the word and it has turned out very convenient in science. That which is different in different directions, or is variegated according to direction, is called aeolotropic.

The consequences of aeolotropy upon the motion of waves, or the equilibrium of particles, in an elastic solid is an exceedingly interesting fundamental subject in physical science; so that there Molar. is no need for apology in bringing our thoughts to it here except, perhaps, that it is too well known. On that account I shall be very brief and merely call attention to two or three fundamental points. I am going to take up presently, as a branch of molar dynamics, the actual propagation of a wave; and in the mathematical investigation, I intend to give you nothing but what is true of the propagation of a plane wave in an elastic solid, not limited to any particular condition of aeolotropy; in an elastic solid, that is to say, which has aeolotropy of the most general kind.

Before doing that, which is strictly a problem of continuous or molar dynamics, I want to touch upon the somewhat cloud-land molecular beginning of the subject, and refer you back to the old papers of Navier and Poisson, in which the laws of equilibrium or motion of an elastic solid were worked out from the consideration of points mutually influencing one another with forces which are functions of the distance. There can be no doubt of the mathematical validity of investigations of that kind and of their interest in connection with molecular views of matter; but we have long passed away from the stage in which Father Boscovich is accepted as being the originator of a correct representation of the ultimate nature of matter and force. Still, there is a never-ending interest in the definite mathematical problem of the equilibrium and motion of a set of points endowed with inertia and mutually acting upon one another with any given forces. We cannot but be conscious of the one splendid application of that problem to what used to be called physical astronomy but which is now more properly called dynamical astronomy, or the motions of the heavenly bodies. But it is not of these grand motions of mutually attracting particles that we must now think. It is equilibriums and infinitesimal motions which form the subject of the special molecular dynamics now before us.

Many writers [Navier (1827), Poisson (1828), Cauchy, F. Neumann, Saint-Venant, and others] who have worked upon this subject have come upon a certain definite relation or set of relations between moduluses of elasticity which seemed to them essential to the hypothesis that matter consists of particles acting upon one another with mutual forces, and that the elasticity of a solid is the manifestation of the forcive required to hold the particles displaced infinitesimally from the position in which the

Molar.   mutual forces will balance. This, which is sometimes called
Navier's relation, sometimes Poisson's relation, and in connection
with which we have the well-known "Poisson's ratio," I want to
show you is not an essential of the hypothesis in question. Their
supposed result for the case of an isotropic body is interesting,
though now thoroughly disproved theoretically and experimentally.
Doubtless most of you know it; it is in Thomson and Tait, and
I suppose in every elementary book upon the subject. I will just
repeat it.

An isotropic solid, according to Navier's or Poisson's theory,
would fulfil the following condition : if a column of it were pulled
lengthwise, the lateral dimensions would be shortened by a
quarter of the proportion that is added to the length; and the
proportionate reduction of the cross-sectional area would therefore
be half the proportion of the elongation. Stokes called attention
to the viciousness of this conclusion as a practical matter in
respect to the realities of elastic solids. He pointed out that jelly
and india-rubber and the like, instead of exhibiting lateral
shrinkage only to the extent of one quarter of the elongation, give
really enough of shrinkage to cause no reduction in volume at all.
That is to say, india-rubber and such bodies vary the area of the
cross-section in inverse proportion to the elongation so that the
product of the length into the area of the cross-section remains
constant. Thus the proportionate linear contraction across the
line of pull is half the elongation instead of only quarter as
according to Navier and Poisson.

Stokes* also referred to a promise that I made, I think it
was in the year 1856, to the effect that out of matter fulfilling
Poisson's condition a model may be made of an elastic solid,
which, when the scale of parts is sufficiently reduced, will be a
homogeneous elastic solid not fulfilling Poisson's condition. That
promise of mine which was made 30 years ago, I propose this
moment to fulfil, never having done so before.

Let this box help us to think of 8 atoms placed at its 8
corners, with the box annulled. The kind of elastic model I am
going to suppose is this : particles or atoms arranged equidistantly
in equidistant parallel rows and connected by springs in a certain

* [Report to *Brit. Association*, Cambridge, 1862. " On Double Refraction,"
p. 262, at bottom.]

definite way. I am going to show you that we can connect Molar.
neighbouring particles of a Boscovich elastic solid with a special
appliance of cord, and a sufficient number of springs, to fulfil the
condition of giving 18 independent moduluses; then by trans-
forming the coordinates to an orientation in the solid taken at
random, we get the celebrated 21 coefficients, or moduluses, of
Green's theory. I suppose you all know that Green took a short-
cut to the truth; he did not go into the physics of the thing at
all, but simply took the general quadratic expression for energy in
terms of the 6 strain components, with its 21 independent coefficients, as the most general supposition that can be made with
regard to an elastic solid.

To make a model of a solid having the 21 independent coefficients of Green's theory, think of how many disposable springs we
have with which to connect 8 particles at the corners of a parallelepiped. Let them be connected by springs first along the 12
edges of the parallelepiped. That clearly will not be sufficient to
give any rigidity of figure whatever, so far as distortions in the
principal planes are concerned. These 12 springs connecting in
this way the 8 particles would give resistances to elongations
in the directions of the edges; but no resistance whatever to
obliquity; you could change the configuration from rectangular,
if given so as in the box before you, into an oblique parallelepiped,
and alter the obliquity indefinitely, bringing if you please all
the 8 atoms into one plane or into one line, without calling any
resisting forces into play. What then must we have, in order
to give resistance against obliquity? We can connect particles
diagonally. We have in the first place, the two diagonals in each
face although we shall see that the two will virtually count as but
one; and then we have the four body diagonals.

Now let me see how many disposables we have got. Remark
that each edge is common to four parallelepipeds of the Boscovich
assemblage. Hence we have only a quarter of the number of
the twelve edge springs independently available. Thus we have
virtually three disposables from the edge springs. Each face is
common to two parallelepipeds; therefore from the two diagonals
in each face we have only one disposable, making in the six faces,
six disposables. We have the four body diagonals not common to
any other parallelepipeds and therefore four disposables from them.
We have thus now 13 disposables in the stiffnesses of 13 springs.

Molar. And we have two more disposables in the ratios of edges of the parallelepiped. Lastly we have three angles of three of the oblique parallelograms constituting the faces of our oblique parallel-epiped. Thus we have in all 18 disposables. But these 18 disposables cannot give us 18 independent moduluses because it is obvious that they cannot give us infinite resistance to compression with finite isotropic rigidity, a case which is essentially included in 18 independent moduluses. Hence I must now find some other disposable or disposables that will enable me to give any compressibility I please in the case of an isotropic solid, and to give Green's 21 independent coefficients, for an aeolotropic solid. For this purpose we must add something to our mechanism that can make the assemblage incompressible or give it any compres-sibility we please; so that, for example, we can make it represent either cork or india-rubber, the extremes in respect to elasticity of known natural solids.

I must confess that since 1856 when I promised this result I have never seen any simple definite way of realising it until a few months ago when in making preparations for these lectures I found I could do it by running a cord twice round the edges of our parallelepiped of atoms as you see me now doing on the model before you. It is easier to do this thus with an actual cord and with rings fixed at the eight corners of a cubic box, than to imagine it done. There is a vast number of ways of doing it; I cannot tell you how many, I wish I could. It is a not unin-teresting labyrinthine puzzle to find them all and to systematise the finding. You see now we are finding *one* way.

Here it is expressed in terms of the coordinates of the corners as we have taken them in succession.

(000) (001) (011) (010) (000) (001) (011)(010)(000)(100)(110)(010)
(110) (111) (011) (111) (101) (001) (101)(111)(110)(100)(101)(100)

[*April* 14, 1898. For the accompanying very clear diagram (fig. 1) representing another of the vast number of ways of laying a cord round the edges of a parallelepiped, I am indebted to M. Brillouin who has added abstracts of some of the present lectures to a translation by M. Lugol of Vol. I. of my *Popular Lectures and Addresses*. M. Brillouin describes his diagram as follows:—" J'ai modifié l'ordre indiqué par Thomson en permutant le 6e et le 8e sommet, pour que la corde suive chaque côté en sens

opposé à ses deux trajets. Dans l'ordre primitif, la face (000) Molar.
(001) (011) (010) était parcourue deux fois dans le même sens."]

You see I have drawn the cord just three times through each
corner ring and I now tighten it over its whole length and tie the
ends together. Remark now that if the cord is inextensible it
secures that the sum of the lengths of the 12 edges of the parallelepiped remains constant, whatever change be given to the relative
positions of the 8 corners. This condition would be fulfilled for
any change whatever of the configuration; but it is understood
that it always remains a parallelepiped, because our application of
the arrangement is to a homogeneous strain of an elastic solid
according to Boscovich. A cord must similarly be carried twice
round the 4 edges of every one of the contiguous parallelepipeds ;
and eight of these have a common corner, at which we suppose
placed a single ring. Hence every ring is traversed three times

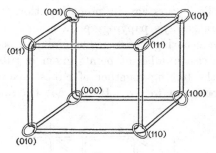

Fig. 1.

by each one of eight endless cords. Each edge is common to four
contiguous parallelepipeds and therefore it has two portions of
each of four endless cords passing along it.

Suppose now for example the parallelepiped to be a cube.
Inextensible cords applied in the manner described between
neighbouring atoms, keep constant the sum of the lengths of the
12 edges of each cube, and therefore secure that its volume is
constant for every infinitesimal displacement. Consider for a
moment the assemblage of ring atoms thus connected by endless
cords. In itself and without further application to molecular
theory we see a very interesting structure which, provided the
cords are kept stretched, occupies a constant volume of space and
yet is perfectly without rigidity for any kind of distortion. You

Molar.

see if I elongate it in one of the three directions of the parallel edges of the cubes, it necessarily shrinks in the perpendicular directions so as to keep constant volume. This kind of deformation gives us two of the five components of distortion without change of bulk ; and the other three are given by the shearings which we have called a, b, c ; of which, for instance, a is a distortion such that the two square parallel faces of the cube perpendicular to OX become rhombuses while the other four remain square.

[*April* 14, 1898. The cube (fig. 1) with an endless cord twice along each edge is, at least mechanically, somewhat interesting in realising an isolated mechanism for securing constancy of volume of a hexahedron, without other restriction of complete liberty of its eight corners than constancy of volume requires: provided only that the hexahedron is infinitely nearly cubic, and that each face is the (plane or curved) surface of minimum area bounded by four straight lines. Two years ago in preparing this lecture for the press from Mr Hathaway's papyrograph, a much simpler mechanism of cords for securing constancy of volume, occupied by an assemblage of a vast number of points given in cubic order, than is provided by the linking together of cubes separately fulfilling this condition, occurred to me. It was not till some time later

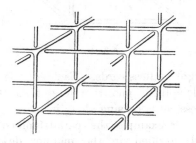

Fig. 2.

that I found myself anticipated by M. Brillouin. I am much pleased to find that he has interested himself in the subject and has introduced a new idea, by which the condition of constant volume is realised in the exceedingly simple and beautiful mechanism of endless cords represented in the annexed diagram (fig. 2), copied from one of his articles in the volume already referred to. This diagram represents endless cords of which just one is shown

complete. These cords pass through rings not shown in fig. 2. Molar.
Fig. 3 shows one of the rings viewed in the direction of a diagonal
of the cube and seen to be traversed twelve times by four endless
cords of which one is shown complete. In an assemblage of a vast
number of rings thus connected by endless cords and arranged in
cubic, or approximately cubic, order, one set of three conterminous
edges of each cube is kept constant, and therefore for the exactly

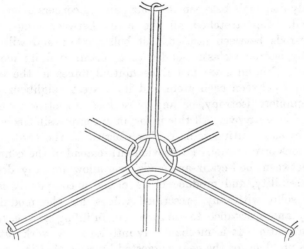

Fig. 3.

cubic order the volume is a maximum. This method, inasmuch
as it implies only two lines of cord along each edge common to
four cubes, is vastly simpler than the original method described
above as in my lecture at Baltimore, which required eight lines of
cord along each common edge. The kinematic result with its
dynamic consequences, is the same in the two methods.

To avoid the assumption of an inextensible elastic cord, place
at each corner, common to eight cubes, six bell-cranks properly
pivoted to produce the effect of cords running, as it were, round
pulleys, which we first realised by the six cords running through
the ring of fig. 3. And instead of each straight portion of cord,
substitute an inextensible bar of rigid matter, hooked at its two
ends to the arm-ends of the proper bell-cranks; just as are the
two ends of each copper bell-wire, so well known in the nineteenth
and preceding centuries, but perhaps to be forgotten early in the

T. L.   9

Molar.   twentieth century, when children grow up who have never seen
bell-cranks and bell-wires and only know the electric bell. Remark
now that our connecting-rods, being rigid, can transmit push as
well as pull; instead of merely the pull of the flexible cord
with which we commenced. Our model now may be constructed
wholly of matter fulfilling the Poisson-Navier condition, and it
gives us a molecular structure for matter violating that condition.

Suppose now we ideally introduce repulsions in the lines of
the body diagonals between the four pairs of corners of each cube.
This will keep stretched, all the cords between rings, or connecting-rods between arm-ends of bell-cranks; and will give a
cubically isotropic elastic solid with a certain definite aeolotropic
quality. If besides we introduce mutual forces in the edges of
the cubes between each atom and its nearest neighbour, we can
give complete isotropy, or any prescribed aeolotropy consistent
with cubic isotropy. All this is for an incompressible solid. But
lastly, by substituting india-rubber elastics for the cords, or ideal
attractions or repulsions (Boscovichian) instead of the connecting-
rods between bell-crank-arm-ends, we allow for any degree of
compressibility, and produce if we please a completely isotropic
elastic solid with any prescribed values for the moduluses of
rigidity and resistance to compression, fulfilling or not fulfilling
Poisson's ratio. It is mechanically interesting to work out details
for this problem for the case suggested by cork, that is, a perfectly
isotropic elastic solid, having rigidity much greater in proportion
to resistance to compression than in the case of Poisson's ratio,
and just sufficient to produce constancy of cross-section in a
column compressed, or elongated, by forces applied to its ends.
It is also interesting to go further and produce a solid of which a
column shall shrink transversely when compressed merely by
longitudinal force.

In my lecture at Baltimore I indicated, without going into
details, how by taking a parallelepiped of unequal edges instead
of a cube and introducing different degrees of elasticity in the
portions of the cords lying along the different edges of each
parallelepiped, and by introducing also forces of attraction or
repulsion among neighbouring atoms, we can produce a model
elastic solid with the twenty-one independent moduluses of
Green's theory.

But the method by cords and pulleys or bell-cranks which we

have been considering, though highly interesting in mechanics, Molar. and dignified by its relationship to Lagrange's original method of proving his theorem of "virtual velocities" (law of work done) in mathematical dynamics, lost much of its interest for molecular physics when I found* that the restriction to Poisson's ratio in an elastic solid held only for the case of a homogeneous assemblage of *single* Boscovich point-atoms: and that, in a homogeneous assemblage of pairs of dissimilar atoms, laws of force between the similar and between the dissimilar atoms can readily be assigned, so as to give any prescribed rigidity and any prescribed modulus of resistance to compression for an isotropic elastic solid: and for an aeolotropic homogeneous solid, Green's 21 independent moduluses of elasticity, or 18 when axes of coordinates are so chosen as to reduce the number by three.]

I want now to go through a piece of mathematical work with you which, though indicated by Green†, has not hitherto, so far as I know, been given anywhere, except partially in my Article on "Elasticity" in the *Encyclopædia Britannica*. It is to find the most general possible plane wave in a homogeneous elastic solid of the most general aeolotropy possible, expressed in terms of Green's 21 independent moduluses‡. Taking Green's general formula

---

* [*Proc. R. S. E.* July 1 and 15, 1889: *Math. and Phys. Papers*, Vol. III. Art. XCVII., p. 395: also " On the Elasticity of a Crystal according to Boscovich," *Proc. R. S.*, June 15, 1893, and republished as an Appendix to present volume.]

† Green's *Mathematical Papers* (Macmillan, 1871), pp. 307, 308.

‡ [*June* 16, 1898.—Through references in Todhunter and Pearson's *Elasticity* I have recently found three very important and suggestive memoirs by Blanchet in *Liouville's Journal*, Vols. v. and vii. (1840 and 1842), in which this problem is treated on the foundation of 36 independent coefficients in the six linear equations expressing each of the six stress-components in terms of the six strain-components. In respect to the history of the doctrine of energy in abstract dynamics, it is curious to find in a Report to the French Academy of Sciences by Poisson, Coriolis, and Sturm (*Comptes Rendus*, Vol. vii., p. 1143) on the first of these memoirs (which had been presented to the Academy on August 8, 1838), the following sentence:—" Les équations différentielles auxquelles sont assujétis les déplacements d'un point quelconque du milieu écarté de sa position d'équilibre renferment 36 coefficients constants, qui dépendent de la nature du milieu, *et qu'on ne pourrait réduire à un moindre nombre sans faire des hypothèses sur la disposition des molécules et sur les lois de leurs actions mutuelles.*" (The italicising is mine.) Blanchet's second memoir, also involving essentially 36 independent coefficients, was presented to the Academy of Sciences on June 14, 1841, and was reported on by Cauchy, Liouville, and Duhamel without any protest against the 36 coefficients. In Green's memoir " On the Propagation of Light in crystallized Media," read May 20, 1839 to the Cambridge Philosophical Society, the expression for the energy of a strain as a

Molar. with the notation with which I put it before you in Lecture II. Part I. (pp. 22, 23, 24), we find

$$
\left.\begin{aligned}
P &= 11e + 12f + 13g + 14a + 15b + 16c\\
Q &= 12e + 22f + 23g + 24a + 25b + 26c\\
R &= 13e + 23f + 33g + 34a + 35b + 36c\\
S &= 14e + 24f + 34g + 44a + 45b + 46c\\
T &= 15e + 25f + 35g + 45a + 55b + 56c\\
U &= 16e + 26f + 36g + 46a + 56b + 66c
\end{aligned}\right\}\dots\dots(1).
$$

Now, considering an ideal infinitesimal parallelepiped of the solid $\delta x \delta y \delta z$, remark that, in virtue of the stress components P, Q, R, S, T, U, it experiences pairs of opposing forces parallel to OX on its three pairs of faces as follows,

$$
\left(P \pm \frac{1}{2}\frac{dP}{dx}.\,\delta x \right)\delta y\,.\,\delta z\,;
$$

$$
\left(U \pm \frac{1}{2}\frac{dU}{dy}.\,\delta y \right)\delta z\,.\,\delta x\,;
$$

$$
\left(T \pm \frac{1}{2}\frac{dT}{dz}.\,\delta z \right)\delta x\,.\,\delta y:
$$

and the total resultant component parallel to OX is therefore

$$
\left(\frac{dP}{dx} + \frac{dU}{dy} + \frac{dT}{dz} \right)\delta x \delta y \delta z \dots\dots\dots(2).
$$

Hence if we denote by $(x + \xi,\ y + \eta,\ z + \zeta)$ the coordinates at time t of a point of the solid of which $(x,\ y,\ z)$ is the equilibrium, we have, as we found in Lecture II. (p. 26 above),

$$
\left.\begin{aligned}
\rho\frac{d^2\xi}{dt^2} &= \frac{dP}{dx} + \frac{dU}{dy} + \frac{dT}{dz}\\[4pt]
\rho\frac{d^2\eta}{dt^2} &= \frac{dU}{dx} + \frac{dQ}{dy} + \frac{dS}{dz}\\[4pt]
\rho\frac{d^2\zeta}{dt^2} &= \frac{dT}{dx} + \frac{dS}{dy} + \frac{dR}{dz}
\end{aligned}\right\}\dots\dots\dots(3).
$$

Now a plane wave, or a succession of plane waves, or the motion resulting from the superposition of sets of plane waves

quadratic function of the six strain-components had been fundamentally used; and by it the fifteen equalities among the 36 coefficients in the linear equations for stress in terms of strain, reducing the number of independent coefficients to 21, had been demonstrated without hypothesis.]

travelling in the same or in contrary directions, may be defined Molar.
generally as any motion of the solid in which every infinitely thin
lamina parallel to some fixed plane experiences a motion which is
purely translational; or in other words, a motion in which ξ, η, ζ
are functions of (p, t), where p denotes the perpendicular from O
to the plane through (x, y, z) parallel to the wave-front. If l, m, n
denote the direction-cosines of this perpendicular, we have

$$p = lx + my + nz \quad\dots\dots\dots\dots\dots(4),$$

and therefore

$$\frac{d}{dx} = l\frac{d}{dp}; \quad \frac{d}{dy} = m\frac{d}{dp}; \quad \frac{d}{dz} = n\frac{d}{dp}\dots\dots\dots(5),$$

when these symbols are applied to any function of (p, t).

Hence, according to the definitions of e, f, g, a, b, c which I
gave you in Lecture II. (pp. 22, 23), we have

$$e = l\frac{d\xi}{dp}; \qquad f = m\frac{d\eta}{dp}; \qquad g = n\frac{d\zeta}{dp}$$
$$a = n\frac{d\eta}{dp} + m\frac{d\zeta}{dp}; \quad b = l\frac{d\zeta}{dp} + n\frac{d\xi}{dp}; \quad c = m\frac{d\xi}{dp} + l\frac{d\eta}{dp} \quad\dots\dots(6).$$

Hence by (3), (5), (1) and (6) we find

$$\rho\frac{d^2\xi}{dt^2} = \frac{d^2}{dp^2}(A\xi + C'\eta + B'\zeta)$$
$$\rho\frac{d^2\eta}{dt^2} = \frac{d^2}{dp^2}(C'\xi + B\eta + A'\zeta) \quad\dots\dots\dots\dots(7),$$
$$\rho\frac{d^2\zeta}{dt^2} = \frac{d^2}{dp^2}(B'\xi + A'\eta + C\zeta)$$

where

$$A = 11l^2 + 66m^2 + 55n^2 + 2 \times 16lm + 2 \times 56mn + 2 \times 15nl$$
$$B = 66l^2 + 22m^2 + 44n^2 + 2 \times 26lm + 2 \times 24mn + 2 \times 46nl$$
$$C = 55l^2 + 44m^2 + 33n^2 + 2 \times 45lm + 2 \times 34mn + 2 \times 35nl$$
$$A' = 56l^2 + 24m^2 + 34n^2 + (25+46)lm + (23+44)mn + (36+45)nl$$
$$B' = 15l^2 + 46m^2 + 35n^2 + (14+56)lm + (36+45)mn + (13+55)nl$$
$$C' = 16l^2 + 26m^2 + 45n^2 + (12+66)lm + (25+46)mn + (14+56)nl$$
$$\dots\dots\dots\dots(8).$$

Now, for our plane waves travelling in either direction, ξ, η, ζ

must each be a function of $(p \pm vt)$, where v denotes the propagational velocity of the wave: hence

$$\frac{d^2\xi}{dt^2} = v^2 \frac{d^2\xi}{dp^2}; \qquad \frac{d^2\eta}{dt^2} = v^2 \frac{d^2\eta}{dp^2}; \qquad \frac{d^2\zeta}{dt^2} = v^2 \frac{d^2\zeta}{dp^2}\ldots\ldots\ldots(9).$$

Hence if we denote the acceleration components by $\ddot{\xi}$, $\ddot{\eta}$, $\ddot{\zeta}$, equations (7) become

$$\left.\begin{array}{l} (A - \rho v^2)\,\ddot{\xi} + \qquad C'\ddot{\eta} \qquad + \qquad B'\ddot{\zeta} \qquad = 0 \\ C'\ddot{\xi} + (B - \rho v^2)\,\ddot{\eta} + \qquad A'\ddot{\zeta} \qquad = 0 \\ B'\ddot{\xi} + \qquad A'\ddot{\eta} \qquad + (C - \rho v^2)\,\ddot{\zeta} = 0 \end{array}\right\} \ldots\ldots(10).$$

The determinant of these equations equated to zero gives, for v^2, three essentially real values $v_1{}^2$, $v_2{}^2$, $v_3{}^2$; which are essentially positive if the coefficients 11, 66, &c., are of any values capable of representing the elastic properties of a stable elastic solid. And for each value of v^2, equations (10) give determinate values, λ, μ, for the ratios $\ddot{\eta}/\ddot{\xi}$, $\ddot{\zeta}/\ddot{\xi}$ which we may denote by λ_1, μ_1; λ_2, μ_2; λ_3, μ_3; so that finally we have, for the complete solution of our problem, superimposed sets of three waves expressed as follows,

$$\left.\begin{array}{l} \xi = \phi_1\,(p + v_1 t) + \phi_2\,(p + v_2 t) + \phi_3\,(p + v_3 t) \\ \quad + \psi_1\,(p - v_1 t) + \psi_2\,(p - v_2 t) + \psi_3\,(p - v_3 t) \\ \eta = \lambda_1\,(\phi_1 + \psi_1) + \lambda_2\,(\phi_2 + \psi_2) + \lambda_3\,(\phi_3 + \psi_3) \\ \zeta = \mu_1\,(\phi_1 + \psi_1) + \mu_2\,(\phi_2 + \psi_2) + \mu_3\,(\phi_3 + \psi_3) \end{array}\right\} \ldots(11),$$

where ϕ_1, ϕ_2, ϕ_3, ψ_1, ψ_2, ψ_3 denote arbitrary functions.

This solution and the relative formulas will be very useful to us to-morrow when we shall be considering the corresponding "wave-surface" in all its generality; that is to say, the surface touched by planes perpendicular to (l, m, n) and at distances from an ideal origin of disturbance equal to $v_1 t$, $v_2 t$, $v_3 t$.

LECTURE XII.

WE will look a little more at this wave problem. Our conclusion Molar. is, that if you choose arbitrarily, in any position whatever relatively to the elastic solid, a set of parallel planes for wave-fronts, there are three directions at right angles to one another (each generally oblique to the set of planes) which fulfil the condition, that the elastic force is in the direction of the displacement; and the equations we have put down express the wave-motion. Each of the three waves will be a wave in which the oscillation of the matter in its front is as I am performing it now, i.e., an oscillation to and fro in a line oblique to the plane of the wave-front, represented by this piece of cardboard which I hold in my hand. You will find the vibrations of the three waves corresponding to the three roots of the determinantal cubic, whether they are oblique or not oblique to the wave-front, are in directions at right angles to one another.

[*Thirteen and a half years interval.* Here is a very short proof. In equations (10) of Lecture XI., put

$$A' = \sqrt{\beta\gamma}; \quad B' = \sqrt{\gamma\alpha}; \quad C' = \sqrt{\alpha\beta};$$

and

$$S = \xi \sqrt{\alpha} + \eta \sqrt{\beta} + \zeta \sqrt{\gamma} \qquad \Big\} \quad \dots\dots\dots(12).$$

With this notation equations (10) give

$$\xi = \frac{S\sqrt{\alpha}}{\rho v^2 - A + \alpha}; \quad \eta = \frac{S\sqrt{\beta}}{\rho v^2 - B + \beta}; \quad \zeta = \frac{S\sqrt{\gamma}}{\rho v^2 - C + \gamma} \dots\dots(13).$$

Multiplying these by $\sqrt{\alpha}$, $\sqrt{\beta}$, $\sqrt{\gamma}$ respectively and adding; and dividing both sides of the resulting equation by S, we find

$$1 = \frac{\alpha}{\rho v^2 - A + \alpha} + \frac{\beta}{\rho v^2 - B + \beta} + \frac{\gamma}{\rho v^2 - C + \gamma} \dots\dots(14).$$

Molar.   This is a form of the determinantal cubic for the reduction of a homogeneous quadratic function of three variables, which I gave fifty-three years ago in the *Cambridge Mathematical Journal* (*Math. and Physical Papers*, Art. xv. Vol. I. p. 55).

Writing down this equation for roots v_1^2, v_2^2, and taking the difference, we find

$$0 = \rho \left(v_1^2 - v_2^2\right) \left\{ \frac{\alpha}{(\rho v_1^2 - A + \alpha)(\rho v_2^2 - A + \alpha)} \right.$$

$$+ \frac{\beta}{(\rho v_1^2 - B + \beta)(\rho v_2^2 - B + \beta)}$$

$$\left. + \frac{\gamma}{(\rho v_1^2 - C + \gamma)(\rho v_2^2 - C + \gamma)} \right\} \quad \ldots\ldots\ldots(15).$$

Hence if v_1^2, v_2^2 are equal, the second factor of this expression may have any value. If they are unequal it must be zero, and (13) gives

$$\ddot{\xi}_1 \ddot{\xi}_2 + \ddot{\eta}_1 \ddot{\eta}_2 + \ddot{\zeta}_1 \ddot{\zeta}_2 = 0 \quad \ldots\ldots\ldots\ldots\ldots(16),$$

which shows that the lines of the vibration in any two of our three waves are necessarily perpendicular to one another*, except in the case when the two propagational velocities are equal. In this case the two waves become one, and the line of vibration may be in any direction in a plane perpendicular to the line given by the third root. The case of three equal roots may also occur: in it the three waves become one, and the line of vibration may be in any direction whatever. Both in this case and in the case of two equal roots, each particle may describe a circular or elliptic orbit, or may move to and fro in a straight line. One equation among (l, m, n) gives a cone, such that, for the plane wave-front perpendicular to any one of its generating lines, two of the three wave-velocities are equal. Two equations among (l, m, n) give a line normal to a wave-front, or wave-fronts, for which the three wave-velocities are equal.]

The consideration of the three sets of plane waves with three different propagational velocities, but with their fronts all parallel to one plane, leads us to a wave-surface different, so far as I know,

---

* This important proposition does not hold for the three directions of vibration found by Blanchet (footnote above), which, for three unequal roots of his cubic, are necessarily not all at right angles to one another unless Green's fifteen equalities are all fulfilled. Compare Thomson and Tait's *Natural Philosophy*, §§ 344, 345.

from anything that has been worked out hitherto in the dynamics Molar. of elastic solids—a wave-surface in which there will be three sheets instead of only two, as in Fresnel's wave-surface : and in which there will be condensation and rarefaction at each point of each sheet, instead of the pure distortion of the ether at every point of each of the two sheets of Fresnel's wave-surface. It is a geometrical problem of no contemptible character to work out this wave-surface.

[*May* 16, 1898. Here is the problem fully worked out, except the performance of the final elimination of l, m, n. In (14) above, for v put $lx + my + nz$ and for n, wherever it occurs, put $\sqrt{1 - l^2 - m^2}$. Let $\phi\,(l, m, x, y, z)$ denote what the second member of (14) then becomes. Take the following three equations :—

$$\phi = 1 ; \quad \frac{d\phi}{dl} = 0 ; \quad \frac{d\phi}{dm} = 0 \quad \ldots\ldots\ldots\ldots(17),$$

and eliminate l, m between them. The resulting equation expresses the wave-surface; that is to say, the surface, whose tangent plane at points of it where the direction cosines are l, m, n, is at distance v from the origin. I need scarcely say that a symmetrical treatment of l, m, n may be preferred in the process of the elimination.]

The wave-surface problem, in words, is this :—Let the solid within any small volume of space round the origin of coordinates, O, be suddenly disturbed in any manner and then left to itself. It is required to find the surface at every point of which a pulse of disturbance is experienced at time t [$t = 1$ in the mathematical solution above].

[*June* 16, 1898. This problem I now find was stated very clearly and attacked with great analytical power by Blanchet in his "Mémoire sur la Propagation et la Polarisation du Mouvement dans un milieu élastique indéfini, cristallisé d'une manière quelconque," the first of the three memoirs referred to in the footnote above. At the end of this paper he sums up his conclusion as follows :

    "1°. Dans un milieu élastique, homogène, indéfini, cristallisé d'une manière quelconque, le mouvement produit par un ébranlement central se propage par une onde plus ou moins compliquée dans sa forme.

"2°.   Pour chaque nappe de l'onde, la vitesse de propagation est constante dans une même direction, variable avec la direction suivant une loi qui dépend de la forme de l'onde.

"3°.   Pour une même direction, les vitesses de vibration sont constamment parallèles entre elles dans une même nappe de l'onde pendant la durée du mouvement, et parallèles à des droites différentes pour les différentes nappes, ce qui constitue une veritable polarisation du mouvement."

In 2° and 3° of this statement "direction" must be interpreted as meaning direction of the perpendicular to the tangent plane, and to 3° it is to be added that the three "droites différentes" mentioned in it are mutually perpendicular, because of the fifteen necessary equalities not assumed by Blanchet among the thirty-six coefficients. The second and third of Blanchet's memoirs (*Liouville's Journal*, Vol. VII., 1842) are entitled "Mémoire sur la Délimitation de l'onde dans la propagation des Mouvements Vibratoires," and "Mémoire sur une circonstance remarquable de la Délimitation de l'onde." They contain some exceedingly interesting conclusions, which Blanchet on the invitation of Liouville had worked out as extensions to a crystallised body of results previously found by Poisson for an isotropic solid, regarding the space throughout which there is some movement of the elastic solid at any time after the cessation of the disturbing action within a small finite space.   Cauchy had also worked on the same subject and had given an analytical method, his "Calcul des Residus," which Blanchet used with due acknowledgment.   The two authors, working nearly simultaneously, seem to have found, each for himself, all the main results, and each to have appreciated loyally the other's work.   It is interesting also to find Poisson, Coriolis, Sturm, Cauchy, Liouville, and Duhamel reporting favourably and suggestively on Blanchet's memoirs, and Liouville helping him with advice in the course of his investigations.

As part of his conclusion regarding "délimitation" Blanchet says, "Il n'y a, en général, ni déplacement ni vitesse au-delà de la plus grande nappe des ondes"; and Cauchy on the same subject in the *Comptes Rendus*, XIV. (1842), p. 13, excluding condensational-rarefactional waves, says, "Les déplacements et par suite les vitesses des molécules s'évanouiront par tous les points situés en dehors ou en dedans des deux ondes propagées.

M. Blanchet a remarqué avec justesse qu'on ne pouvait, en Molar. général, en dire autant des points situés entre les deux ondes. Toutefois il est bon d'observer que, même en ces derniers points, les déplacements et les vitesses se réduisent à zéro quand on suppose nulle la dilatation du volume..., c'est à dire, en d'autres termes, quand les vibrations longitudinales disparaissent."*]

Green treats the subject of waves in an aeolotropic elastic solid in a peculiar and most interesting manner for the purpose of forming a dynamical theory of "the propagation of light in crystallized media." He investigates conditions† that "transverse vibrations shall always be *accurately* in the front of the wave," or, in modern language, that the wave may be purely distortional. He finds‡ 14 relations among his 21 coefficients by which this is secured for a double-sheeted wave-surface, which he finds to be identical with Fresnel's. There is necessarily a third sheet, although Green does not mention it at all. It is ellipsoidal, and corresponds essentially to a condensational-rarefactional wave with vibrations at every point perpendicular to the tangent plane. It is quite disconnected from the double-sheeted surface of the distortional wave; and a disturbing source can be so adjusted as to produce only distortional wave-motion with the double-sheeted wave-surface, or only the condensational-rarefactional wave-motion with the ellipsoidal wave-surface, or both kinds simultaneously. The three principal axes of the ellipsoidal wave-surface coincide with the three axes of symmetry found for the wave-surface of the distortional Fresnel-Green wave-motion.

This dynamics of waves in an elastic solid is a fine subject for investigation, and I am sorry now to pass from it for a time.

But if the war is to be directed to fighting down the difficulties which confront us in the undulatory theory of light it is not of the slightest use towards solving our difficulties, for us to have a medium which kindly permits distortional waves to be propagated through it, even though it be aeolotropic. It is not enough to know that though the medium be aeolotropic it can let purely distortional waves through it, and that two out of the

---

* These quotations are copied from a very interesting account of the work of Blanchet and Cauchy on this subject on pp. 627—634 of Todhunter and Pearson's *Elasticity*, Vol. I.

† Green's *Mathematical Papers* : " On Propagation of Light in Crystallized Media," p. 293 : Reprint from *Trans. Cambridge Philosophical Society*, May 20, 1839.

‡ *Ibid.* p. 309.

Molar.

three waves will be purely distortional. What we want is a medium which, when light is refracted and reflected, will under all circumstances give rise to distortional waves alone. Green's medium would fail in this respect when waves of light come to a surface of separation between two such mediums. All that Green secures is that there can be a purely distortional wave; he does not secure that there shall not be a condensational wave. There would generally be condensational waves from the source. White-hot bodies, flames as of candles or gaslights, electric light of all kinds, would produce condensational waves, whether in an aeolotropic or isotropic medium, so far as Green's conditions here spoken of, go. What we want is a medium resisting condensation sufficiently; a medium with an infinite or practically infinite bulk-modulus—so great that the amount of energy, developed in the shape of condensational waves, has not been discovered by observation.

As an essential in every reflection and refraction there may be a little loss of energy from the want of perfect polish in the surface, but as a rule, we have practically no loss of light in reflection and refraction at surfaces of glass and clear crystals. There perhaps is some but we have not discovered it. The medium that gives us the luminiferous vibrations must be such that if there is any part of the energy of the wave expended in condensational waves after refraction and reflection, the amount of it must be so small that it has not been discovered. Numerical observations have been made with great accuracy, in which, for example, Fresnel's formula for the ratio of normally incident and reflected light $\left(\dfrac{\mu-1}{\mu+1}\right)^2$ is verified within closer than one per cent, I believe. Still a half per cent or a tenth per cent of the energy may for oblique incidences be converted into condensational waves, for all we know. But if any large percentage were converted into condensational waves, there would be a great deal of energy in condensational waves going about through space, and (to use for a moment an absurd mode of speaking of these things) there would be a "*new force*" that we know nothing of. There would be some tremendous action all through the universe produced by the energy of condensational waves if the energy of these were one-tenth, or one-hundredth per cent of the energy of the distortional waves. I believe that if in oblique reflection and refraction of light at

any surface, or in case of violent action in the source, there are Molar.
condensational waves produced with anything like a thousandth
or a ten-thousandth of the energy of the light and radiant heat
which we know, we should have some prodigious effect, but which
might, perhaps, have to be discovered by some other sense than
we have.   The want of indication of any such actions is sufficient
to prove that if there are any in nature, they must be exceedingly
small.   But that there are such waves, I believe ; and I believe
that the velocity of the unknown condensational wave that we are
speaking of is the velocity of propagation of electro-static force.

I say "believe" here in a somewhat guarded manner.   I do
not mean that I believe this as a matter of religious faith, but
rather as a matter of strong scientific probability.   If this is true
of propagation of electro-static force, it is true that there is
exceedingly little energy in the waves corresponding to the
propagation of an electro-static force.   That is however going
beyond our tether of Molecular Dynamics.   What I proposed
in the introductory statement with reference to these Lectures
was to bring what principles and results of the science of molecular
dynamics I could enter upon, to bear upon the wave theory of
light.   We are sticking closely to that for the present, and we
may say that we have nothing to do with condensational waves.
Our medium is to be incompressible, and instead of Green's
fourteen equations, we have merely one condition,—that the
medium is incompressible.   It is obvious that this condition
suffices to prevent the possibility of a wave of condensation at all
and reduces our wave-surface to a surface with two sheets, like
the Fresnel surface.   But before passing away from that beautiful
dynamical speculation of Green's, if we think of what the condensational wave must be in an aeolotropic solid fulfilling Green's
condition that it can have purely distortional waves proceeding in
all directions—the condition that two of the three waves which
we investigated three-quarters of an hour ago shall be purely
distortional—I think we shall find also condensational waves, and
that the wave-surfaces for them will be a set of concentric ellipsoids.
It will be a single-sheeted surface, that is certain, because you
have only one velocity corresponding to each tangent plane at the
wave-surface.

I shall now leave this subject for the present.   We shall come
back upon it again, perhaps, and look a little more into the

142   LECTURE XII.   PART I.

Molar.   question of moduluses of elasticity. We shall work up from an isotropic solid to the most general solid; and we shall work down from the most general solid to an isotropic solid. We shall take first the most general value for the compressibility; we shall then come to this subject again of assuming incompressibility. We shall then begin with the most general solid possible, and see what conditions we must impose to make it as symmetrical as is necessary for the Fresnel wave-surface. The molecular problem will prepare your way a good deal for this.

I had intended to prepare something about the mass of the luminiferous ether. I have not had time to take it up, but I certainly shall do so before we have done with the subject. We shall go into the question of the density of the luminiferous ether, giving superior and inferior limits. We shall also consider what fraction of a gramme may be in one of these molecules and show what an enormously smaller fraction of a gramme we may suppose it to displace in the luminiferous ether. We shall try to get into the notion of this, that the molecule must be *elastic* and that there must be an enormous mass in its interior. Its outer part feels and touches the luminiferous ether. It is a very curious supposition to make, of a molecular cavity lined with a rigid spherical shell; but that something exists in the luminiferous ether and acts upon it in the manner that is faultily illustrated by our mechanical model, I absolutely believe.

Just think of the effect of a shock consisting say of a collision between that and another molecule. Instead of its being broken into bits, let us suppose an unbroken spherical sheath around it. It will bound away, vibrating. Just imagine the central nucleus vibrating in one direction while the shell vibrates in the other, and you have a molecule with two parts going in opposite directions; but differing from what I thought of the other day (Lecture IX. p. 96 above) in that one part is inside the other. The ether gets its motion from the outside part. Therefore I say that the most fundamental supposition we can make with reference to the originating source of a sequence of waves of light is that illustrated by a globe vibrating to and fro in a straight line.

We have already investigated (Lecture VIII. pp. 86—89) the solution corresponding to that. Consider the spherical waves; no vibrations for points in one certain diameter of the sphere; maximum vibrations in all points of the equatorial plane of that

diameter and perpendicular to that plane; for all points in the Molecular.
quadrant of an arc of the spherical surface extending from axis to
equator, vibrations in the plane of and tangential to the arc; and
of magnitude proportional to the cosine of the latitude or angular
distance from the equator and of intensity proportional to the
square of the cosine of the latitude. Then in a wave travelling
outwards, let the amplitude vary inversely as the distance from
the centre, and therefore the intensity inversely as the square of
the distance from the centre; and you have a correct word-painting
of the very simplest and most frequent sequence of vibrations
constituting light.

Ten minutes interval.

Let us return to the consideration of the dynamics of refraction,
ordinary dispersion, anomalous dispersion, and absorption. Beginning with our formulas as we left them yesterday (p. 120 above),
let us consider what they become for the case $\tau = \infty$; that is to
say, for static displacement of the containing shell. The inertia
of the molecules will now not be called into play, and the case
becomes simply that of the equilibrium of the set of springs whose
stiffnesses we have denoted by $c_1, c_2, \ldots, c_j, c_{j+1}$, when S, the end
of c_1 remote from m_1, is displaced through a space ξ, and F, the
end of c_{j+1} remote from m_j, is fixed.

Denoting by X the force with which S pulls and F resists,
which, as inertia does not come into play, is therefore the force
with which each spring is stretched, we have

$$X = c_{j+1}x_j = c_j(x_{j-1} - x_j) = \ldots = c_2(x_1 - x_2) = c_1(\xi - x_1)\ldots(1);$$

or, as we may write it,

$$X = \frac{x_j}{\dfrac{1}{c_{j+1}}} = \frac{x_{j-1} - x_j}{\dfrac{1}{c_j}} = \ldots = \frac{x_1 - x_2}{\dfrac{1}{c_2}} = \frac{\xi - x_1}{\dfrac{1}{c_1}}\ldots\ldots\ldots(2);$$

whence

$$\frac{x_1}{\xi} = \frac{\dfrac{1}{c_2} + \dfrac{1}{c_3} + \ldots + \dfrac{1}{c_j} + \dfrac{1}{c_{j+1}}}{\dfrac{1}{c_1} + \dfrac{1}{c_2} + \dfrac{1}{c_3} + \ldots + \dfrac{1}{c_j} + \dfrac{1}{c_{j+1}}}\ldots\ldots\ldots\ldots(3).$$

Molecular.   Hence, unless at least one of $c_1, c_2, \ldots, c_{j+1}$ is zero, $\frac{x_1}{\xi} - 1$ is

negative for $\tau = \infty$, and therefore $\frac{1}{\zeta^2}$ diminishes to $-\infty$ as τ^2 increases to $+\infty$. Hence a large enough finite value of τ makes $\frac{1}{\zeta^2} = 0$, and all larger values of τ^2 make $\frac{1}{\zeta^2}$ negative. This is a particular, and an extreme, case of a very important result with which we shall have much to do later, in following the course of our formula when the period of the light is increased from any molecular period to the next greater molecular period; and we may get quit of it for the present by assuming c_{j+1} to be zero. We may recover it again, and perceive its true physical significance, by supposing m_j to be infinitely great; and therefore we lose nothing of generality by taking $c_{j+1} = 0$. This, in our formulas of to-day, makes $X = 0$ and $\frac{x_1}{\xi} = 1$. The latter, by putting $\tau = \infty$ in our last formula of Lecture X. (p. 120 above), gives

$$1 = \frac{c_1}{4\pi^2 m_1} (\kappa^2 R + \kappa_{\prime}^2 R_{\prime} + \kappa_{\prime\prime}^2 R_{\prime\prime} + \&c.)\ldots\ldots\ldots\ldots(4).$$

Subtracting this from our last formula of Lecture X. with τ general, we find

$$\frac{x_1}{\xi} - 1 = \frac{c_1}{4\pi^2 m_1} \left(\frac{\kappa^4 R}{\tau^2 - \kappa^2} + \frac{\kappa_{\prime}^4 R_{\prime}}{\tau^2 - \kappa_{\prime}^2} + \frac{\kappa_{\prime\prime}^4 R_{\prime\prime}}{\tau^2 - \kappa_{\prime\prime}^2} + \&c. \right)\ldots\ldots(5);$$

whence, by the preceding formula of Lecture X., (p. 120)

$$\frac{1}{\zeta^2} = \frac{\rho}{n} + \frac{c_1^2 \tau^2}{16\pi^4 m_1 n} \left(\frac{\kappa^4 R}{\tau^2 - \kappa^2} + \frac{\kappa_{\prime}^4 R_{\prime}}{\tau^2 - \kappa_{\prime}^2} + \frac{\kappa_{\prime\prime}^4 R_{\prime\prime}}{\tau^2 - \kappa_{\prime\prime}^2} + \&c. \right)\ldots(6).$$

A convenient modification of this formula is got by putting in it

$$\frac{1}{\zeta^2} = \frac{\rho}{n} \mu^2 \ldots\ldots\ldots\ldots\ldots\ldots\ldots\ldots(7),$$

and

$$c_1 = \frac{4\pi^2 m_1}{\kappa_1^2} \ldots\ldots\ldots\ldots\ldots\ldots\ldots\ldots(8).$$

Thus μ denotes the refractive index of the medium; and κ_1 denotes the period which m_1 would have as a vibrator, if the shell lining the ether were held fixed, and if the elastic connection between m_1 and interior masses were temporarily annulled. With these notations (6) becomes

$$\mu^2 = 1 + \frac{m_1 \kappa^4 R}{\rho \kappa_1^4} \cdot \frac{\tau^2}{\tau^2 - \kappa^2} + \frac{m_1 \kappa_{\prime}^4 R_{\prime}}{\rho \kappa_1^4} \cdot \frac{\tau^2}{\tau^2 - \kappa_{\prime}^2} + \frac{m_1 \kappa_{\prime\prime}^4 R_{\prime\prime}}{\rho \kappa_1^4} \cdot \frac{\tau^2}{\tau^2 - \kappa_{\prime\prime}^2} + \&c. \ (9).$$

When the period of the light is very long in comparison with Molecular. the longest of the molecular periods of the embedded molecules, it is obvious that each material particle will be carried to and fro with almost exactly the same motion as the shell, in fact almost as if it were rigidly connected with the shell; and therefore the velocity of light will be sensibly the same as if the masses of the particles were distributed homogeneously through the ether without any disturbance of its rigidity.

Let us consider now how, when the period of the light is not infinitely long in comparison with any one of the molecular periods, the internal vibrations of the molecule modify the transmission of light through the medium. For simplicity at present let us suppose our molecule to have only one vibrating particle, the m_1 of our formulas, which we will now call simply m, being the sum of the masses of the vibrators per unit volume of the ether. Imagine it connected with the shell or sheath, S, surrounding it, by massless springs (as in the accompanying diagram), through which it acts on the ether surrounding the sheath. Its influence on the transmission of light through the medium can 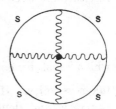 be readily understood and calculated from the diagrams of p. 147, representing the well-known elementary dynamics of a pendulum vibrating in simple harmonic motion, when freely hung from a point S, (corresponding to the sheath lining our ideal cavity in ether), which is itself compelled by applied forces to move horizontally with simple harmonic motion. Figs. 1 and 2 illustrate the cases in which the period of the point of suspension, S, is longer than the period of the suspended pendulum with S fixed; and figs. 3 and 4 cases in which it is shorter. In each diagram OM represents the length of an undisturbed simple pendulum whose period is equal to that of the motion of S, the point of support of our disturbed pendulum SM. Thus if we denote the periods by τ and κ respectively, we have $OM/SM = \tau^2/\kappa^2$. Hence if ξ and x denote respectively the simultaneous maximum displacements of S and M, we have

$$\frac{x}{\xi} = \frac{OM}{OS} = \frac{\tau^2}{\tau^2 - \kappa^2} \dots\dots\dots\dots\dots\dots(10).$$

If now for a moment we denote by w the mass of the single vibrating particle of our diagrams, the horizontal component of the

Molecular. pull exerted by the thread MS on S at the instant of maximum displacement is $w\,(2\pi/\tau)^2\,x$; and with m in place of w, we have the sum of the forces exerted by all the molecules per unit volume of the ether at the instant of maximum displacement of ether and molecules at any point. This force is subtracted from the restitutional force of the ether's elasticity, when the period of the light is greater than the molecular period (figs. 1 and 2); and is added to it, when the period of the light is less than the molecular period (figs. 3 and 4).

Now if the rigidity of the ether be taken as unity, its elastic force per unit of volume at any point at the instant of its maximum displacement is $(2\pi/l)^2$, where l is the wave-length of light in the ether with its embedded molecules. Hence the total actual restitutional force on the ether is

$$\left(\frac{2\pi}{l}\right)^2 \xi - m \left(\frac{2\pi}{\tau}\right)^2 x,$$

which must be equal to $\left(\dfrac{2\pi}{\tau}\right)^2 \xi$ if the density of the ether be taken as unity. Hence dividing both members of our equation by $(2\pi/\tau)^2\,\xi$, and denoting by ζ the velocity of light through the medium consisting of ether and embedded molecules, and by μ its refractive index, we have

$$\mu^2 = \frac{1}{\zeta^2} = \left(\frac{\tau}{l}\right)^2 = 1 + m\frac{x}{\xi} \quad \ldots\ldots\ldots\ldots\ldots(11);$$

whence by (10),

$$\mu^2 = \frac{1}{\zeta^2} = \left(\frac{\tau}{l}\right)^2 = 1 + m\frac{\tau^2}{\tau^2 - \kappa^2} \quad \ldots\ldots\ldots\ldots(12).$$

By quite analogous elementary dynamical considerations we find

$$\mu^2 = 1 + m\frac{\tau^2}{\tau^2 - \kappa^2} + m_{,}\frac{\tau^2}{\tau^2 - \kappa_{,}^2} + m_{,,}\frac{\tau^2}{\tau^2 - \kappa_{,,}^2} + \&c.\ldots\ldots(13),$$

if, instead of one set of equal and similar molecules, there are several such sets, the molecules of one set differing from those of the others. The period κ, or $\kappa_{,}$, or $\kappa_{,,}$, &c., of any one of the sets is simply the period of vibration of the interior mass, when the rigid spherical lining of the ether around it, which for brevity we shall henceforth call the sheath of the molecule, is held fixed. Instead of calling this sheath massless as hitherto we shall suppose it to have mass equal to that of the displaced

Molecular.

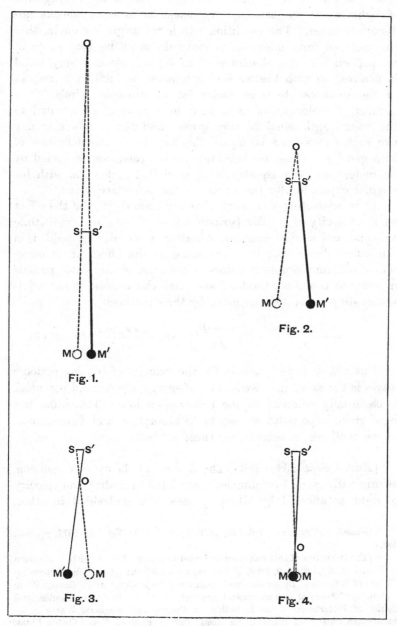

Fig. 1.

Fig. 2.

Fig. 3.

Fig. 4.

Molecular. ether; so that if there were no interior masses the propagation of light would be sensibly the same as in homogeneous undisturbed ether. The condition which we originally made, that the distance from molecule to molecule must be very great in comparison with the diameter of each molecule and very small in comparison with the wave-length, must be fulfilled in respect to the distances between molecules of different kinds. The number of molecules of each kind in a cube of edge equal to the wave-length must be very great; and the numbers in any two such cubes must be equal, this last being the definition of homogeneity. These are substantially the conditions assumed by Sellmeier*, and our equation (13) is virtually identical with his original expression for the square of the refractive index.

It is interesting to remark that our formula (9) for the effect on the velocity of regular periodic waves of light, of a multitude of equal and similar complex vibrating molecules embedded in the ether, shows that it is the same as the effect of as many sets of different simple vibrators as there are fundamental periods in our one complex vibrator; and that the masses of the equivalent simple vibrators are given by the equations

$$m = \frac{m_1 \kappa^4 R}{\rho \kappa_1^4}; \quad m_{,} = \frac{m_1 \kappa_{,}^4 R_{,}}{\rho \kappa_1^4}; \quad m_{,,} = \frac{m_1 \kappa_{,,}^4 R_{,,}}{\rho \kappa_1^4}; \quad \&c....(14).$$

But although the formula for the velocity of regular periodic waves is the same, the distribution of energy, kinetic and potential, is essentially different on the two suppositions. This difference is of great importance in respect to absorption and fluorescence, as we shall see in considering these subjects later.

[*Added Sept.* 23†, 1898. The dynamical theory of dispersion, as originally given by Sellmeier*, consisted in finding the velocity of light as affected by vibratory molecules embedded in ether,

---

* Sellmeier, *Pogg. Ann.*, Vol. 145, 1872, pp. 399, 520 ; Vol. 147, 1872, pp. 387, 525.

† [The substance of this was communicated to Sec. A of the British Association at Bristol on September 9, 1898, in two papers under the titles " The Dynamical Theory of Refraction, Dispersion, and Anomalous Dispersion," and " Continuity in Undulatory Theory of Condensational-rarefactional Waves in Gases, Liquids, and Solids, of Distortional Waves in Solids, of Electric and Magnetic Waves in all Substances capable of transmitting them, and of Radiant Heat, Visible Light, Ultra-Violet Light and Röntgen Rays."]

such as those which had been suggested by Stokes[*] to account Molecular. for the dark lines of the solar spectrum. Sellmeier's mathematical work was founded on the simplest ideal of a molecular vibrator, which may be taken as a single material particle connected by a massless spring or springs with a rigid sheath lining a small vesicle in ether. He investigated the propagation of distortional waves, and found the following expression (which I give with slightly altered notation) for the square of the refractive index of light passing through ether studded with a very large number of vibratory molecules in every volume equal to the cube of the wave-length :—

$$\mu^2 = 1 + m\,\frac{\tau^2}{\tau^2 - \kappa^2} + m_{,}\,\frac{\tau^2}{\tau^2 - \kappa_{,}^2} + m_{,,}\,\frac{\tau^2}{\tau^2 - \kappa_{,,}^2} + \&c.$$

where τ denotes the period of the light; $\kappa, \kappa_{,}, \kappa_{,,}$, &c., the vibratory periods of the embedded molecules on the supposition of their sheaths held fixed; and $m, m_{,}, m_{,,}$, &c., their masses. He showed that this formula agreed with all that was known in 1872 regarding ordinary dispersion, and that it contained what we cannot doubt is substantially the true dynamical explanation of anomalous dispersions, which had been discovered by Fox-Talbot[†] for the extraordinary ray in crystals of a chromium salt, by Leroux[‡] for iodine vapour, and by Christiansen[§] for liquid solution of fuchsin, and had been experimentally investigated with great power by Kundt[**].

Sellmeier himself somewhat marred[††] the physical value of his mathematical work by suggesting a distinction between refractive and absorptive molecules ("refractive und absorptive Theilchen"), and by seeming to confine the application of his formula to cases in which the longest of the molecular periods is small in comparison with the period of the light. But the splendid value to physical science of his non-absorptive formula has been quite wonderfully proved by Rubens (who, however, inadvertently

[*] See Kirchhoff-Stokes-Thomson, *Phil. Mag.*, March and July 1860.
[†] Fox-Talbot, *Proc. Roy. Soc. Edin.*, 1870—71.
[‡] Leroux, *Comptes rendus*, 55, 1862, pp. 126—128.
[§] Christiansen, *Ann. Phys. Chem.*, 141, 1870, pp. 479, 480; *Phil. Mag.*, 41, 1871, p. 244; *Annales de Chimie*, 25, 1872, pp. 213, 214.
[**] Kundt, *Pogg. Ann.*, Vols. 142, 143, 144, 145, 1871—72.
[††] *Pogg. Ann.*, Vol. 147, 1872, p. 525.

Molecular. quotes* it as if due to Ketteler). Fourteen years ago Langley†
had measured the refractivity of rock-salt for light and radiant
heat of wave-lengths (in air or ether) from ·43 of a mikrom‡ to
5·3 mikroms (the mikrom being 10^{-6} of a metre, or 10^{-4} of a
centimetre), and without measuring refractivities further, had
measured wave-lengths as great as 15 mikroms in radiant heat.
Within the last six years measurements of refractivity by Rubens,
Paschen, and others, agreeing in a practically perfect way with
Langley's through his range, have given us very accurate knowledge of the refractivity of rock-salt and of sylvin (chloride of
potassium) through the enormous range of from ·4 of a mikrom to
23 mikroms.

Rubens began by using empirical and partly theoretical
formulas which had been suggested by various theoretical and
experimental writers, and obtained fairly accurate representations
of the refractivities of flint-glass, quartz, fluorspar, sylvin, and
rock-salt through ranges of wave-length from ·4 to nearly 12
mikroms§. Two years later, further experiments extending the
measure of refractivities of sylvin and rock-salt for light of wave-
lengths up to 23 mikroms, showed deviations from the best of the
previous empirical formulas increasing largely with increasing

* *Wied. Ann.*, Vol. 53, 1894, p. 267. In the formula quoted by Rubens from
Ketteler, substitute for μ_∞ the value of μ found by putting $\tau = \infty$ in Sellmeier's
formula, and Ketteler's formula becomes identical with Sellmeier's. Remark that
Ketteler's "M" is Sellmeier's "$m\kappa^2$" according to my notation in the text.

† Langley, *Phil. Mag.*, 1886, 2nd half-year.

‡ [For a small unit of length Langley, fourteen years ago, used with great
advantage and convenience the word "mikron" to denote the millionth of a
metre. The letter *n* has no place in the metrical system, and I venture to suggest
a change of spelling to "mikrom" for the millionth of a metre, after the analogy
of the English usage for millionths (mikrohm, mikro-ampere, mikrovolt). For a
conveniently small corresponding unit of time I further venture to suggest
"michron" to denote the period of vibration of light whose wave-length in ether
is 1 mikrom. Thus, the velocity of light in ether being 3×10^8 metres per second,
the michron is $\frac{1}{3} \times 10^{-14}$ of a second, and the velocity of light is 1 mikrom of space
per michron of time. Thus the frequency of the highest ultra-violet light investigated by Schumann (·1 of a mikrom wave-length, *Sitzungsber. d. k. Gesellsch. d.
Wissensch. zu Wien*, CII. pp. 415 and 625, 1893) is 10 periods per michron of time.
The period of sodium light (mean of lines *D*) is ·58932 of a michron ; the periods
of the "Reststrahlen" of rock-salt and sylvin found by Rubens and Aschkinass
(*Wied. Ann.* LXV. (1898), p. 241) are 51·2 and 61·1 michrons respectively. No
practical inconvenience can ever arise from any possible confusion, or momentary
forgetfulness, in respect to the similarity of sound between michrons of time and
mikroms of space.—K.]

§ Rubens, *Wied. Ann.*, Vols. 53, 54, 1894—95.

wave-lengths.   Rubens then fell back[*] on the simple unmodified Sellmeier formula, and found by it a practically perfect expression of the refractivities of those substances from ·434 to 22·3 mikroms.

And now for the splendid and really wonderful confirmation of the dynamical theory.   One year later a paper by Rubens and Aschkinass[†] describes experiments proving that radiant heat after five successive reflections from approximately parallel surfaces of rock-salt; and again of sylvin; is of mean wave-length 51·2 and 61·1 mikroms respectively.   The two formulas which Rubens had given in February 1897, as deduced solely from refractivities measured for wave-lengths of less than 23 mikroms, made μ^2 negative for radiant heat of wave-lengths from 37 to 55 mikroms in the case of reflection from rock-salt, and of wave-lengths from 45 to 67 mikroms in the case of reflection from sylvin! (μ^2 negative means that waves incident on the substance cannot enter it, but are totally reflected).

These formulas, written with a somewhat important algebraic modification serving to identify them with Sellmeier's original expression, (13) above, and thus making their dynamical meaning clearer, are as follows:—

Rock-salt $\mu^2 = 1\cdot1875 + 1\cdot1410 \dfrac{\tau^2}{\tau^2 - \cdot01621} + 2\cdot8504 \dfrac{\tau^2}{\tau^2 - 3419\cdot3}$.

Sylvin $\mu^2 = 1\cdot5329 + \ \cdot6410 \dfrac{\tau^2}{\tau^2 - \cdot0234} + 2\cdot3792 \dfrac{\tau^2}{\tau^2 - 4517\cdot1}$.

The accompanying diagrams (pp. 152, 153) represent the squares of the refractive index of rock-salt and sylvin, calculated from these formulas through a range of periods from ·434 of a michron to 100 michrons.   In each diagram the scale both of ordinate and abscissa from 0 to 10 michrons is ten times the scale of the continuation from 10 to 100 michrons.   Additions and subtractions to keep the ordinates within bounds are indicated on the diagrams.   The circle and cross, on portion b of each curve, represent respectively the points where $\mu^2 = 1$ and $\mu^2 = 0$.   The critical periods exhibited in the formula are

for rock-salt,

·1273 and 56·116, being the square roots of ·01621 and 3149·3;

for sylvin,

·1529 and 67·209,   „   „   „   „   ·0234   „   4517·1.

    * Rubens and Nichols, *Wied. Ann.*, Vol. 60, 1896—97, p. 454.
    † Rubens and Aschkinass, *Wied. Ann.*, Vol. 65, 1898, p. 241.

Molecular.

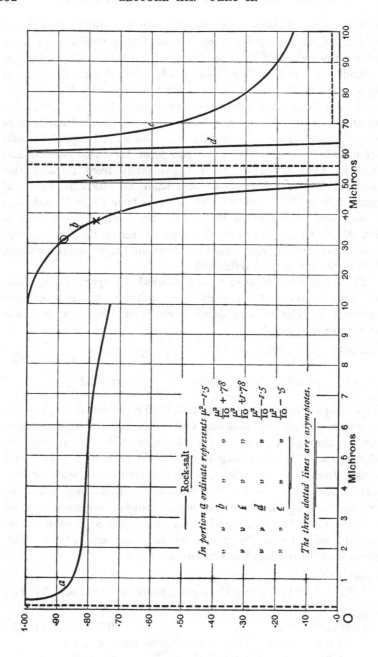

Molecular.

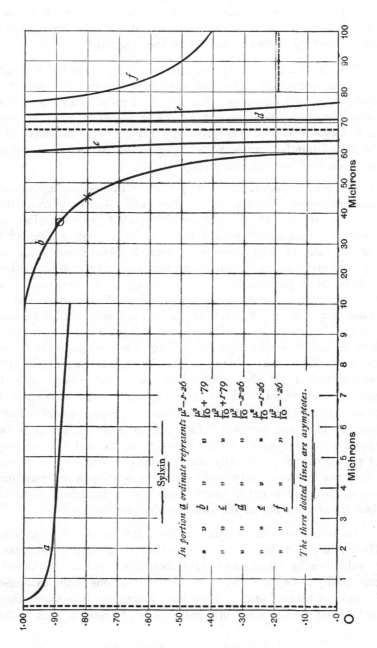

In portion *a* ordinate represents $\mu^2 - 1 \cdot 26$

"	*b*	"	$\frac{\mu^2}{10} + \cdot 79$
"	*c*	"	$\frac{\mu^2}{10} + 1 \cdot 79$
"	*d*	"	$\frac{\mu^2}{10} - 2 \cdot 26$
"	*e*	"	$\frac{\mu^2}{10} - 1 \cdot 26$
"	*f*	"	$\frac{\mu^2}{10} - \cdot 26$

The three dotted lines are asymptotes.

Sylvin

Molecular. The value of μ^2 passes from $-\infty$ to $+\infty$ as τ rises through these values, and we have accordingly asymptotes at these points, represented by dotted ordinates in the diagrams.

The agreement with observation is absolutely perfect through the whole range of Langley, Rubens, and Paschen from ·434 of a michron to 22·3 michrons. The observed refractivities are given by them to five significant figures, or four places of decimals ; and in a table of comparison given by Rubens and Nichols* the calculated numbers agree with the observed to the last place of decimals, both for rock-salt and sylvin.

While in respect to refractivity there is this perfect agreement with Sellmeier's formula through the range of periods from ·434 of a michron to fifty-one times this period (corresponding to nearly six octaves in music), it is to be remarked that with radiant heat of 22·3 michrons period, Rubens found, both for rock-salt and sylvin, so much absorption, increasing with increasing periods, as to prevent him from carrying on his measurements of refractivity to longer periods than 22 or 23 michrons, and to extinguish a large proportion of the radiant heat of 23 michrons period in its course through his prisms. Hence although Sellmeier's formula makes no allowance for the large absorptivity of the transparent medium, such as that thus proved in rock-salt and sylvin with radiant heat of periods less than half the critical period κ in each case, it is satisfactory to know that large as it is the absorptivity produces very little effect on the velocity of propagation of radiant heat through the medium. This indeed is just what is to be expected from dynamical theory, which shows that the velocity of propagation is necessarily affected but very little by the forces which produce absorption, unless the absorptivity is so great that the intensity of a ray is almost annulled in travelling three or four times its wave-length through the medium.

In an addition to a later lecture I intend to refer again to Helmholtz's introduction of resisting forces in simple proportion to velocities, by which he extended Sellmeier's formula to include true bodily absorption, and to recent modifications of the extended formula by himself and by Ketteler. Meantime it, is interesting, and it may correct some misapprehension, to remark that so far

* *Wied. Ann.*, Vol. 60, 1896–97, p. 454.

as the dynamics of Sellmeier's single vibrating masses, or of my Molecular. complex molecule, goes, there is no necessity to expect any absorption at all, even for light or radiant heat of one of the chief critical periods κ, as we shall see by the following general view of the circumstances of this and other critical periods. We shall see in fact that there are three kinds of critical period,

(1)   a period for which $\mu^2 = 1$, or velocity in the medium equal to velocity in pure undisturbed ether;

(2)   a period for which $\mu^2 = 0$, or wave-velocity infinite in the medium;

(3)   a period (any one of the κ-periods) such that if we imagine passing through it with augmenting period, μ^2 changes from $-\infty$ to $+\infty$.

The dynamical explanation of points where $\mu^2 = 1$ (marked by circles on the curves), is simpler with my complex molecule than with Sellmeier's several sets of single vibrators. With my complex molecule it means a case in which No. 1 spring, that between m_1 and the sheath, experiences no change of length. In order that this may be the case, the period of the light must be equal to some one of the free periods of the complex molecule, detached from the sheath (c_1 temporarily annulled). With Sellmeier's arrangement, it is a case in which tendency to quicken the vibrations of the ether by one set of vibrators, is counterbalanced by tendency to slow them by another set, or other sets. It is a case which could not occur with only one set of equal and similar simple vibrators.

The case we have next to consider, $\mu^2 = 0$ (or $\zeta^2 = \infty$), marked with crosses on the curve, occurs essentially just once for a single set of simple vibrators; and it occurs as many times as there are sets of vibrators, if there are two or more sets with different periods; or, just once for each of the critical periods κ, $\kappa_{\prime}$, $\kappa_{\prime\prime}$, &c. of my complex molecule. The dynamical explanation is particularly simple for a single set of Sellmeier vibrators; it is the whole ether remaining undistorted and vibrating in one direction, while the masses m all vibrate in the opposite direction; the whole system being as it were two masses E and M connected by a single massless spring, and vibrating to and fro in opposite directions in a straight line with their common centre of inertia

Molecular. at rest. Thus we see exactly the meaning of wave-velocity becoming infinite for the critical cases marked with a cross on the curves.

Increase of the period beyond this critical value makes μ^2 negative until we reach the next one of our critical values κ, $\kappa_{,}$, $\kappa_{,,}$, &c., or the one of these, if there is only one. The meaning clearly is that light cannot penetrate the medium and is totally reflected from it wherever it falls on it. Thus for light or radiant heat of all periods corresponding to the interval between a point on one of our curves marked with a cross and the next asymptotic ordinate, the medium acts as silver does to visible light; that is to say, it is impervious and gives total reflection. When we increase the period through one of our chief critical values κ, $\kappa_{,}$, $\kappa_{,,}$, &c., μ^2 passes from $-\infty$ to $+\infty$. With exactly the critical period we have infinitely small velocity of propagation of light through the medium, and still total reflection of incident light. There would be infinitely great amplitude of the molecular vibrators, if the light could get into the medium; but it cannot get in. With period just a little greater than one of these critical values, the reflection of the incident light is very nearly total; the velocity of propagation of the light which enters is very small; and the energy (kinetic and potential) of the molecular vibrators is very great in comparison with the energy (kinetic and potential) of the ether.

Lastly, if there were no greater critical period than this which we have just passed, we should now have ordinary refraction with ordinary dispersion, at first large, becoming less and less with increased period, and μ^2 diminishing asymptotically to the value $\dfrac{\rho}{M+\rho}$, where ρ denotes the density of the ether, and M the sum of all the masses within the sheaths and connected to them by springs. But when we consider that the whole mass of the ponderable matter is embedded in the ether, and we cannot conceive of merely an infinitesimal portion of it clogging the ether in its luminiferous or electric vibrations, and that therefore for glass, or water, or even for rarefied air, M must be millions of millions of times ρ, we see how utterly our dynamical theory fails to carry us with any tolerable comfort, in trying to follow and understand the nature of waves and vibrations of periods longer than the 60 michrons (or 2×10^{-13} of a second) touched by Rubens and Aschkinass*, up to 10^{-6} of a second, or Molar. further up to one thousandth of a second, or up to still longer periods.

Yet we must try somehow to find and thoroughly understand continuity in the undulatory theory of condensational-rarefactional waves in gases, liquids, and solids, of distortional waves in solids, of electric and magnetic waves in all substances capable of transmitting them, and of radiant heat, visible light, ultra-violet light and Röntgen rays.

Consider the following three analogous cases :—I. mechanical; II. electrical; III. electromagnetic.

I.   Imagine an ideally rigid globe of solid platinum of 12 centim. diameter, hung inside an ideal rigid massless spherical shell of 13 centim. internal diameter, and of any convenient thickness.   Let this shell be hung in air or under water by a very long cord, or let it be embedded in a great block of glass, or rock, or other elastic solid, electrically conductive or non-conductive, transparent or non-transparent for light.

I.   (1)   By proper application of force between the shell and the nucleus cause the shell and nucleus to vibrate in opposite directions with simple harmonic motion through a relative total range of 10^{-3} of a centimetre.   We shall first suppose the shell to be in air.   In this case, because of the small density of air compared with that of platinum, the relative total range will be practically that of the shell, and the nucleus may be considered as almost absolutely fixed.   If the period is $\frac{1}{32}$ of a second, (frequency 32 according to Lord Rayleigh's designation), a humming sound will be heard, certainly not excessively loud, but probably amply audible to an ear within a metre or half a metre of the shell.   Increase the frequency to 256, and a very loud sound of the well-known musical character (C_{256}) will be heard†.

Increase the frequency now to 32 times this, that is to 8192 periods per second, and an exceedingly loud note 5 octaves higher will be heard.   It may be too loud a shriek to be tolerable; if so,

---

* *Wied. Ann.*, Vol. 64, 1898.

† Lord Rayleigh has found that with frequency 256, periodic condensation and rarefaction of the marvellously small amount 6×10^{-9} of an atmosphere, or "addition and subtraction of densities far less than those to be found in our highest vacua," gives a perfectly audible sound.   The amplitude of the aerial vibration, on each side of zero, corresponding to this, is $1 \cdot 27 \times 10^{-7}$ of a centimetre.—*Sound*, Vol. II, p. 439 (2nd edition).

diminish the range till the sound is not too loud.   Increase the
frequency now successively according to the ratios of the diatonic
scale, and the well-known musical notes will be each clearly and
perfectly perceived through the whole of this octave.   To some
or all ears the musical notes will still be clear up to the G (24756
periods per second) of the octave above, but we do not know
from experience what kind of sound the ear would perceive
for higher frequencies than 25000.   We can scarcely believe
that it would hear nothing, if the amplitude of the motion is
suitable.

   To produce such relative motions of shell and nucleus as we
have been considering, whether the shell is embedded in air, or
water, or glass, or rock, or metal, a certain amount of work, not
extravagantly great, must be done to supply the energy for the
waves (both condensational and rarefactional), which are caused
to proceed outwards in all directions.   Suppose now, for example,
we find how much work per second is required to maintain vibration with a frequency of 1000 periods per second, through total
relative motion of 10^{-3} of a centimetre.   Keeping to the same
rate of doing work, raise the frequency to 10^4, 10^5, 10^6, 10^9, 10^{12},
500×10^{12}.   We now hear nothing; and we see nothing from any
point of view in the line of the vibration of the centre of the
shell which I shall call the axial line.   But from all points of
view, not in this line, we see a luminous point of homogeneous
polarized yellow light, as it were in the centre of the shell, with
increasing brilliance as we pass from any point of the axial line
to the equatorial plane, keeping at equal distances from the centre.
The line of vibration is everywhere in the meridional plane, and
perpendicular to the line drawn to the centre.

   When the vibrating shell is surrounded by air, or water, or
other fluid, and when the vibrations are of moderate frequency, or
of anything less than a few hundred thousand periods per second,
the waves proceeding outwards are condensational-rarefactional,
with zero of alternate condensation and rarefaction at every point
of the equatorial plane and maximum in the axial line.   When
the vibrating shell is embedded in an elastic solid extending to
vast distances in all directions from it, two sets of waves, distortional and condensational-rarefactional, according respectively to
the two descriptions which have been before us, proceed outwards
with different velocities, that of the former essentially less than

that of the latter in all known elastic solids*. Each of these Molar.
propagational velocities is certainly independent of the frequency
up to 10^4, 10^5, or 10^6, and probably up to any frequency not so
high but that the wave-length is a large multiple of the distance
from molecule to molecule of the solid. When we rise to fre-
quencies of 4×10^{12}, 400×10^{12}, 800×10^{12}, and 3000×10^{12}, corresponding to the already known range of long-period invisible
radiant heat, of visible light, and of ultra-violet light, what
becomes of the condensational-rarefactional waves which we have
been considering? How and about what range do we pass from
the propagational velocities of 3 kilometres per second for distortional waves in glass, or 5 kilometres per second for the condensational waves in glass, to the 200,000 kilometres per second for
light in glass, and, perhaps, no condensational wave? Of one
thing we may be quite sure; the transition is continuous. Is it
probable (if ether is absolutely incompressible, it is certainly
possible) that the condensational-rarefactional wave becomes less
and less with frequencies of from 10^6 to 4×10^{12}, and that there is
absolutely none of it for periodic disturbances of frequencies of
from 4×10^{12} to 3000×10^{12}? There is nothing unnatural or
fruitlessly ideal in our ideal shell, and in giving it so high a
frequency as the 500×10^{12} of yellow light. It is absolutely
certain that there is a definite dynamical theory for waves of
light, to be enriched, not abolished, by electromagnetic theory;
and it is interesting to find one certain line of transition from our
distortional waves in glass, or metal, or rock, to our still better
known waves of light.

I. (2) Here is another still simpler transition from the distortional waves in an elastic solid to waves of light. Still think
of our massless rigid spherical shell, 13 centim. internal diameter,
with our solid globe of platinum, 12 centim. diameter, hung in its
interior. Instead of as formerly applying simple forces to produce
contrary rectilinear vibrations of shell and nucleus, apply now
a proper mutual forcive between shell and nucleus to give them
oscillatory rotations in contrary directions. If the shell is hung
in air or water, we should have a propagation outwards of disturbance due to viscosity, very interesting in itself; but we should
have no motion that we know of appropriate to our present
subject until we rise to frequencies of 10^9, 10×10^{12}, 400×10^{12},

---

* *Math. and Phys. Papers*, Vol. III. Art. civ. p. 522.

Molar. 800×10^{12}, or 3000×10^{12}, when we should have radiant heat, or
visible light, or ultra-violet light proceeding from the outer surface
of the shell, as it were from a point-source of light at the centre,
with a character of polarization which we shall thoroughly consider a little later.   But now let our massless shell be embedded
far in the interior of a vast mass of glass, or metal, or rock, or of
any homogeneous elastic solid, firmly attached to it all round, so
that neither splitting away nor tangential slip shall be possible.
Purely distortional waves will spread out in all directions except
the axial.   Suppose, to fix our ideas, we begin with vibrations of
one-second period, and let the elastic solid be either glass or iron.
At distances of hundreds of kilometres (that is to say, distances
great in comparison with the wave-length and great in comparison
with the radius of the shell), the wave-length will be approximately 3 kilometres*.   Increase the frequency now to 1000
periods per second: at distances of hundreds of metres the wave-
length will be about 3 metres.   Increase now the frequency to
10^6 periods per second; the wave-length will be 3 millim., and
this not only at distances of several times the radius of the shell,
but throughout the elastic medium from close to the outer surface
of the shell; because the wave-length now is a small fraction
of the radius of the shell.   Increase the frequency further to
1000×10^6 periods per second; the wave-length will be 3×10^{-3} of
a millim., or 3 mikroms, if, as in all probability is the case, the
distance between the centres of contiguous molecules in glass and
in iron is less than a five-hundredth of a mikrom.   But it is
probable that the distance between centres of contiguous molecules
in glass and in iron is greater than 10^{-5} of a mikrom, and therefore it is probable that neither of these solids can transmit waves
of distortional motion of their own ponderable matter, of so short
a wave-length as 10^{-5} of a mikrom.   Hence it is probable that if
we increase the frequency of the rotational vibrations of our shell
to one hundred thousand times 1000×10^6, that is to say, to
100×10^{12}, no distortional wave of motion of the ponderable
matter can be transmitted outwards; but it seems quite certain
that distortional waves of radiant heat in ether will be produced
close to the boundary of the vibrating shell, although it is also
probable that if the surrounding solid is either glass or iron,
these waves will not be transmitted far outwards, but will be

* *Math. and Phys. Papers*, Vol. III. Art. civ. p. 522.

absorbed, that is to say converted into non-undulatory thermal Molar. motions, within a few mikroms of their origin.

Lastly, suppose the elastic solid around our oscillating shell to be a concentric spherical shell of homogeneous glass of a few centimetres, or a few metres, thickness and of refractive index 1·5 for D light. Let the frequency of the oscillations be increased to $5·092 \times 10^{14}$ periods per second, or its period reduced to ·58932 of a michron: homogeneous yellow light of period equal to the mean of the periods of the two sodium lines will be propagated outwards through the glass with wave-length of about $\frac{2}{3} \times$ ·58932 of a mikrom, and out from the glass into air with wave-length of ·58932 of a mikrom. The light will be of maximum intensity in the equatorial plane and zero in either direction along the axis, and its plane of polarization will be everywhere the meridional plane. It is interesting to remark that the axis of rotation of the ether for this case coincides everywhere with the line of vibration of the ether in the case first considered; that is to say, in the case in which the shell vibrated to and fro in a straight line, instead of, as in the second case, rotating through an infinitesimal angle round the same line.

A full mathematical investigation of the motion of the elastic medium at all distances from the originating shell, for each of the cases of I. (1) and I. (2), will be found later (p. 190) in these lectures.

II. An electrical analogy for I. (1) is presented by substituting for our massless shell an ideally rigid, infinitely massive shell of glass or other non-conductor of electricity, and for our massive platinum nucleus a massless non-conducting globe electrified with a given quantity of electricity. For simplicity we shall suppose our apparatus to be surrounded by air or ether. Vibrations to and fro in a straight line are to be maintained by force between shell and nucleus as in I. (1). Or, consider simply a fixed solid· non-conducting globe coated with two circular caps of metal, leaving an equatorial non-conducting zone between them, and let thin wires from a distant alternate-current dynamo, or electrostatic inductor, give periodically varying opposite electrifications to the two caps. For moderate frequencies we have a periodic variation of electrostatic force in the air or ether surrounding the apparatus, which we can readily follow in imagination, and can measure by proper electrostatic measuring apparatus.

Molar.

Its phase, with moderate frequencies, is very exactly the same as that of the electric vibrator. Now suppose the frequency of the vibrator to be raised to several hundred million million periods per second. We shall have polarized light proceeding as if from an ideal point-source at the centre of the vibrator and answering fully to the description of I. (1). Does the phase of variation of the electrostatic force in the axial line outside the apparatus remain exactly the same as that of the vibrator? An affirmative answer to this question would mean that the velocity of propagation of electrostatic force is infinite. A negative answer would mean that there is a finite velocity of propagation for electrostatic force.

III. The shell and interior electrified non-conducting massless globe being the same as in II., let now a forcive be applied between shell and nucleus to produce rotational oscillations as in I. (2). When the frequency of the oscillations is moderate, there will be no alteration of the electrostatic force and no perceptible magnetic force in the air or ether around our apparatus. Let now the frequency be raised to several hundred million million periods per second; we shall have visible polarized light proceeding as if from an ideal point-source at the centre and answering fully to the description of the light of I. (2). The same result would be obtained by taking simply a fixed solid non-conducting globe and laying on wire on its surface approximately along the circumferences of equidistant circles of latitude, and, by the use of a distant source (as in II.), sending an alternate current through this wire. In this case, while there is no manifestation of electrostatic force, there is strong alternating magnetic force, which in the space outside the globe is as if from an ideal infinitesimal magnet with alternating magnetization, placed at the centre of the globe and with its magnetic axis in our axial line.]

LECTURE XIII.

SATURDAY, *October* 11, 8 P.M.

PROF. MORLEY has already partially solved the definite dyna- Molecular. mical problem that I proposed to you last Wednesday (p. 103 above) so far as determining four of the fundamental periods; and you may be interested in knowing the result. He finds roots, κ^{-2}, $\kappa_{,}^{-2}$, &c., $= 3\cdot46$, $1\cdot005$, $\cdot298$, $\cdot087$; each root not being very different from three times the next after it. I will not go into the affair any further just now. I just wish to call your attention to what Prof. Morley has already done upon the example that I suggested for our arithmetical laboratory. I think it will be worth while also to work out the energy ratios* (p. 74). In selecting this example, I designed a case for which the arithmetic would of necessity be highly convergent. But I chose it primarily because it is something like the kind of thing that presents itself in the true molecule:—An elastic complex molecule consisting of a finite number of discontinuous masses elastically connected (with enormous masses in the central parts, that seems certain): the whole embedded in the ether and acted on by the ether in virtue of elastic connections which, unless the molecule were rigid and embedded in the ether simply like a rigid mass embedded in jelly, must consist of elastic bonds analogous to springs.

I think you will be interested in looking at this model which, by the kindness of Prof. Rowland, I am now able to show you. It is made on the same plan as a wave machine which I made many years ago for use in my Glasgow University classes, and finally modified in preparations for a lecture given to the Royal

---

* This was done by Mr Morley who kindly gave his results to the "Coefficients" on the 17th Oct. See Lecture XIX. below.

Molecular. Institution about two years ago on "The Size of Atoms*." I
think those who are interested in the illustration of dynamical

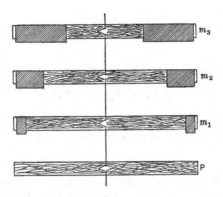

problems will find this a very nice and convenient method. If
you will look at it, you will see how the thing is done : Pianoforte

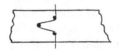

wire, bent around three pins in the way you
see here, supports each bar. These pins are
slanted in such a way as to cause the wire
to press in close to the bar so as to hold it
quite firm. The wood is slightly cut away
to prevent the wire from touching it above and below the pins,
so that there may be no impairment of elastic action due to slip of
steel on wood. The wire used is fine steel pianoforte wire; that
is the most elastic substance available, and it seems to me, indeed,
by far the most elastic of all the materials known to us [except
crystals; *Jan.* 19, 1899]. A heavy weight is hung on the lower
end of the wire to keep it tightly stretched.

Prof. Rowland is going to have another machine made, which
I think you will be pleased with—a continuous wave machine.
This of mine is not a wave machine, but a machine for illustrating the vibrations of a finite group of several elastically
connected particles. The connecting springs are represented by
the torsional springiness of the three portions of connecting wire
and the fourth portion by which the upper mass is hung. In this
case gravity contributes nothing to the effect except to stretch
the wire. If we stretch the wire between two sides of a portable

* February 3, 1883 ; *Proc. Royal Institution*, Vol. x. p. 185 ; *Popular Lectures
and Addresses*, Vol. i. p. 154.

frame, we might take our model to an ideal lecture-room at the Molecular.
centre of the earth, and it would work exactly as you see it working
now. You will understand that these upper masses correspond
to m_1, m_2, m_3. In all we have four masses here, of which the
lowest represents the spherical shell lining the ether around
our ideal cavity. I will just apply a moving force to this lower
mass, P. To realize the circumstances of our case more fully, we
should have a spring connected with a vibrator to pull P with,
and perhaps we may get that up before the next lecture. [Done
by Professor Rowland; see Lecture XIV.] I shall attempt no
more at present than to cause this first particle to move to and
fro in a period perceptibly shorter than the shortest of the three
fundamental periods which we have when the lowest bar is held
fixed. The result is scarcely sensible motion of the others.
I do not know that there would be any sensible motion at all if
I had observed to keep the greatest range of this lowest particle
to its original extent on the two sides of its mean position.

The first part of our lecture this evening I propose to be a
continuation of our conference regarding æolotropy. The second
part will be molecular dynamics. I propose to look at this question
a little, but I want to look very particularly to some of the points
connected with the conceivable circumstances by which we can
account for not merely regular refraction, but anomalous dispersion,
and both the absorption that we have in liquids and very opaque
bodies and such absorption as is demonstrated by the very fine
dark lines of the solar spectrum which are now shown more
splendidly than ever by Prof. Rowland's gratings.

I shall speak now of æolotropy. The equations by which Molar.
Green realized the condition that two of the three waves having
fronts parallel to one plane shall be distortional, are in this respect
equivalent to a very easily understood condition that I may
illustrate first by considering the more general problem. That
problem is similar to another of the very greatest simplicity,
which is the well-known problem of the displacement of a particle
subject to forces acting upon it in different directions from fixed
centres. An infinitesimal displacement in any direction being
considered, the question is, when is the return force in the
direction of the displacement? As we know, there are three
directions at right angles to one another, in which the return
force is in the direction of the displacement. The sole difference

between that very trite problem and our problem of yesterday (Lecture XII. p. 135), is that in yesterday's the question is put with reference to a whole infinite plane in an infinite homogeneous solid, which is displaced in any direction between two fixed parallel planes on its two sides, to which it is always kept parallel. Considering force per unit of area, we have the same question, when is the return force in the direction of the displacement? And the answer is, there are three directions at right angles to one another in which the return force is in the direction of the displacement. Those three directions are generally oblique to the plane; but Green found the conditions under which one will be perpendicular to the plane, and the other two in the plane.

I shall now enter upon the subject more practically in respect to the application to the wave theory of light, and begin by preparing to introduce the condition of incompressibility. Take first the well-known equations of motion for an isotropic solid and express in them the condition that the body is incompressible. The equations are:

$$\left.\begin{array}{l} \rho \dfrac{d^2\xi}{dt^2} = (k + \tfrac{1}{3}n)\dfrac{d\delta}{dx} + n\nabla^2\xi \\[2mm] \rho \dfrac{d^2\eta}{dt^2} = (k + \tfrac{1}{3}n)\dfrac{d\delta}{dy} + n\nabla^2\eta \\[2mm] \rho \dfrac{d^2\zeta}{dt^2} = (k + \tfrac{1}{3}n)\dfrac{d\delta}{dz} + n\nabla^2\zeta \end{array}\right\}.$$

Suppose now the resistance to compression is infinite, which means, make $k = \infty$ at the same time that we have $\delta = 0$. What then is to become of the first term of the second members of these equations? We simply take $(k + \tfrac{1}{3}n)\delta = p$, and write the second members $\dfrac{dp}{dx} + n\nabla^2\xi$, &c., accordingly. This requires no hypothesis whatever. We may now take $k = \infty$, $\delta = 0$, without interfering with the form of our equations. These equations, without any condition whatever as to ξ, η, ζ, with the condition $p = (k + \tfrac{1}{3}n)\delta$, are the equations necessary and sufficient for the problem. On the other hand, if $k = \infty$, the condition that that involves is

$$\frac{d\xi}{dx} + \frac{d\eta}{dy} + \frac{d\zeta}{dz} = 0,$$

which gives four equations in all for the four unknown quantities Molar.
ξ, η, ζ, p.

Precisely the same thing may be done in respect to equations (1), (2), (3) of p. 132 above for a solid with 21 independent coefficients. We will have this equation, $\delta = 0$, again for an æolotropic body, and a corresponding equality to infinity. I am not going to introduce any of these formulas at present. In the meantime, I tell you a principle that is obvious. In order to introduce the condition $\dfrac{d\xi}{dx} + \dfrac{d\eta}{dy} + \dfrac{d\zeta}{dz} = 0$ into our general equation of energy with its 21 coefficients, which involves a quadratic expression in terms of the six quantities that we have denoted by e, f, g, a, b, c, we must modify the quadratic into a form in which we have $(e + f + g)^2$ with a coefficient. That coefficient equated to infinity, and $e + f + g = 0$, leave us the general equations of equilibrium of an incompressible æolotropic elastic solid.

I want to call your attention to the kind of deviation from isotropy which is annulled by Green's equations among the coefficients expressing that two out of the three waves shall be purely distortional, and the third shall be condensational-rarefactional.

The next thing to an isotropic body is one possessing what Rankine calls cyboid asymmetry. Rankine marks an era in philology and scientific nomenclature. In England, and I believe in America also, there has been a classical reaction, or reform, according to which, instead of taking all our Greek words through the French mill changing κ (kappa) into c, and υ (upsilon) into y, we spell in English, and we pronounce Greek words, and even some Latin words, more nearly according to what we may imagine to be the actual usage of the ancients. We cannot however in the present generation get over Kuros instead of Cyrus, Kikero instead of Cicero. Rankine is a curious specimen of the very last of the French classical style. Rankine was the last writer to speak of *cinematics* instead of *kinematics*. *Cyboid*, which he uses, is a very good word, but I do not know that there is any need of introducing it instead of *Cubic*. *Cubic* is an exception to the older classical derivation in that u is not changed into y; it should be cybic, and cube should be *cybe* (I suppose $\kappa\upsilon\beta o\varsigma$ to be the Greek word). *Cyboid* obviously means

Molar. cube-like, or *cubic,* and it is taken from the Greek in Rankine's manner, now old-fashioned.

Rankine gives the equations that will leave cubic asymmetry. He afterwards makes the very apposite remark that Sir David Brewster discovered that kind of variation from isotropy in analcime. I only came to this in Rankine two or three days ago. But I remember going through the same thing myself not long ago, and I said to Stokes—(I always consult my great authority Stokes whenever I get a chance)—"Surely there may be something found in nature to exemplify this kind of asymmetry; would it not be likely to be found in crystals of the cubic class?" Stokes—he knows almost everything—instantly said, "Sir David Brewster thought he had found it in cubic crystals, but there is another explanation; it may be owing to the effect of the cleavage planes, or the separation of the crystal into several crystalline laminas"—I do not remember all that Stokes said, but he distinctly denied that Brewster's experiment showed a true instance of cubic optical asymmetry. He pointed out that an exceedingly slight deviation from cubic isotropy would show very markedly on elementary phenomena of light, and might be very readily tested by means of ordinary optical instruments. The fact that nothing of the kind has been discovered is absolute evidence that the deviation, if there is any, from optic isotropy in a crystal of the cubic class, is exceedingly small in comparison with the deviation from isotropy presented by ordinary doubly refracting crystals.

Molecular. As a matter of fact, square æolotropy is found in a pocket handkerchief or piece of square-woven cloth, supposing the warp and woof to be accurately similar, a supposition that does not hold of ordinary cloth. Take wire-cloth carefully made in squares; that is symmetrical and equal in its moduluses with reference to two axes at right angles to one another. There will be a vast difference according as you pull out one side and compress the other, or pull out one diagonal and compress the other. Take the extreme case of a cloth woven up with inextensible frictionless threads, and there is an absolute resistance to distortion in two directions at right angles to one another, and no resistance at all to distortion of the kind that is presented in changing it from square to rhombic shape. That is to say, a framework of this kind has no resistance to shearing parallel to the sides; in other words, to the distortion produced Molecular. by lengthening one diagonal and shortening the other. Now imagine cut out of this pocket hand-kerchief, a square with sides parallel to the dia-gonals, making a pattern of this kind. There is a square that has infinite resistance to shearing parallel to its sides, and zero resistance to pulling out in the direction perpendicular to either pair of sides. This is not altogether a trivial illustration. Surgeons make use of it in their bandages. A person not familiar with the theory of elastic solids might cut a strip of cloth parallel to warp or woof; but cut it obliquely and you have a conveniently pliable and longitudinally semielastic character that allows it to serve for some kinds of bandage.

Imagine an elastic solid made up in that kind of way, with that kind of deviation from isotropy; and you have clearly two different rigidities for different distortions in the same plane. I remember that Rankine in one of his early papers proved this to be impossible! He proved a proposition to the effect that the rigidity was the same for all distortions in the same plane. That was no doubt founded on some special supposition as to arrangement of molecules and may be true for the particular arrangement assumed; but it is clearly fallacious in respect to true elastic solids. Rankine in his first paper made too short work of the elastic solid in respect to possibilities of æolotropy. He soon after took it up very much on the same foundation as Green with his 21 coefficients, but still under the bondage of his old proposition that rigidity is the same for all distortions in the same plane. Yet a little later he escaped from the yoke, and took his revenge splendidly by giving a fine Greek name " cyboid asymmetry" to designate the special crystalline quality of which he had proved the impossibility!

I must read to you some of the fine words that Rankine has Molar. introduced into science in his work on the elasticity of solids. That is really the first place I know of, except in Green, in which this thing has been gone into thoroughly. It is not really satisfactory in Rankine except in the way in which he carries out the algebra of the subject, and the determinants and matrices that he goes into so very finely. But what I want to call attention to now is his grand names. I do not know

Molar.   whether Prof. Sylvester ever looked at these names; I think he
would be rather pleased with them. "Thlipsinomic transforma-
tions," "Umbral surfaces," and so on. Any one who will learn the
meaning of all these words will obtain a large mass of knowledge
with respect to an elastic solid. The simple, good words, "strain
and stress," are due to Rankine; "potential" energy also. Hear
also the grand words "Thlipsinomic, Tasinomic, Platythliptic,
Euthytatic, Metatatic, Heterotatic, Plagiotatic, Orthotatic, Pan-
tatic, Cybotatic, Goniothliptic, Euthythliptic," &c.

You may now understand what cyboid asymmetry is, or as I
prefer to call it, cubic æolotropy. Rankine had not the word
æolotropy; that came in from myself* later. Cyboid or cubic
æolotropy is the kind of æolotropy exhibited by a cubic grating;
as it were a structure built up of uniform cubic frames. *There*
[*Feb.* 14, 1899; a skeleton cube, of twelve equal wooden rods with
their ends fixed in eight india-rubber balls forming its corners]
is a thing that would be isotropic, except for its smaller rigidity
for one than for the other of the two principal distortions in each
one of the planes of symmetry.

I will go no further into that just now; but I hope that in
the next lecture, or somehow before we have ended, we may be
able to face the problem of introducing the relations among the
21 moduluses which are sufficient to do away with all obliquities
with reference to three rectangular axes. But you can do this in
a moment—equate to zero enough of the 21 coefficients to fulfil
two conditions, (1) that if you compress a cube of the body by
three balancing pairs of pressures, equal or unequal, perpendicular
to its three pairs of faces, it will remain rectangular, and (2) that
if you apply, in four planes meeting in four parallel edges, balancing tangential forces perpendicular to these edges, the angles at
these edges will be made alternately acute and obtuse, and the
angles at the other eight edges will remain right angles. [*Feb.* 14,
1899; here are the required annulments of coefficients to fulfil
those two conditions, with OX, OY, OZ taken parallel to edges of
the undisturbed cube, and with the notation of Lecture II. p. 23:—

$$42 = 0; \; 43 = 0; \; 51 = 0; \; 53 = 0; \; 61 = 0; \; 62 = 0$$
$$41 = 0; \; 52 = 0; \; 63 = 0; \; 56 = 0; \; 64 = 0; \; 45 = 0 \Big\} \dots(1).$$

* See p. 118 above.

Thus the formula for potential energy becomes reduced to

$$2E = 11e^2 + 22f^2 + 33g^2 + 2\,(23fg + 31ge + 12ef)$$
$$+ 44a^2 + 55b^2 + 66c^2 \ldots \ldots (2).$$

With these nine coefficients, 11, 22, 33; 23, 31, 12; 44, 55, 66; all independent; the elastic solid would present two different rigidities for the two distortions of each of the planes (yz), (zx), (xy), one by shearing motion parallel to either of the two other planes, the other by shearing motion parallel to either of the planes bisecting the right angles between those planes. The values of these rigidities are as follows for the cases of shearing motions parallel, and at 45°, to our principal planes:—

Distortion	Plane of Distortion	Line of Motion	Rigidity
a	(yz)	y or z	44
$f-g$		45° to y and z	$\frac{1}{2}\{\frac{1}{2}(22+33)-23\}$
b	(zx)	z or x	55
$g-e$		45° to z and x	$\frac{1}{2}\{\frac{1}{2}(33+11)-31\}$
c	(xy)	x or y	66
$e-f$		45° to x and y	$\frac{1}{2}\{\frac{1}{2}(11+22)-12\}$

From the fact that squares only of a, b, c appear in the equation of energy, we see that in equilibrium[*] these distortions are separately balanced by the tangential stresses S, T, U. And we conclude that there can be plane waves of purely longitudinal motion (condensational-rarefactional) and waves of pure distortion travelling in the directions x, y, z with their fronts perpendicular to these directions; and that their velocities[†] are as shown in the following table:—

---

[*] Lecture II. p. 24.

[†] Equations (1) above, with equations (4), (7), (8) of Lecture XI. p. 133, applied respectively to the cases
$$m=0,\ n=0,\ \xi=0\,;\quad n=0,\ l=0,\ \eta=0\,;\quad l=0,\ m=0,\ \zeta=0.$$

Molar.

Wave-front	Line of Vibration	Quality	Velocity
(yz)	x	condensational-rarefactional	$\sqrt{\dfrac{11}{\rho}}$
	y	purely distortional	$\sqrt{\dfrac{66}{\rho}}$
	z	,, ,,	$\sqrt{\dfrac{55}{\rho}}$
(zx)	y	condensational-rarefactional	$\sqrt{\dfrac{22}{\rho}}$
	z	purely distortional	$\sqrt{\dfrac{44}{\rho}}$
	x	,, ,,	$\sqrt{\dfrac{66}{\rho}}$
(xy)	z	condensational-rarefactional	$\sqrt{\dfrac{33}{\rho}}$
	x	purely distortional	$\sqrt{\dfrac{55}{\rho}}$
	y	,, ,,	$\sqrt{\dfrac{44}{\rho}}$

Selecting from these the purely distortional waves, and taking them in pairs having equal velocities, we have the following convenient table :—

Wave-front	Line of Vibration	Plane of Distortion	Velocity
(xy)	y	(yz)	$\sqrt{\dfrac{44}{\rho}}$
(xz)	z		
(yz)	z	(zx)	$\sqrt{\dfrac{55}{\rho}}$
(yx)	x		
(zx)	x	(xy)	$\sqrt{\dfrac{66}{\rho}}$
(zy)	y		

In this table we find one and the same velocity, $\sqrt{44/\rho}$, for two Molar. different waves with fronts parallel to x, and having their lines of vibration parallel to y and z respectively, and therefore each having yz for its plane of distortion. If now we apply the formulas of Lecture XI., p. 133, to investigate the velocities of other waves having the same plane of distortion—for instance, waves with fronts parallel to x and lines of vibration at 45° to y and z,—we find different propagational velocities unless

$$44 = \tfrac{1}{2}\left\{\tfrac{1}{2}\left(22 + 33\right) - 23\right\}*,$$

which makes them all equal. Thus, and by similar considerations relatively to y and z, we see that each of the three propagational velocities given in the preceding table is the same for all waves having the same plane of distortion, if the following conditions are fulfilled:—

$$\left.\begin{aligned}
44 &= \tfrac{1}{2}\left\{\tfrac{1}{2}\left(22 + 33\right) - 23\right\} \\
55 &= \tfrac{1}{2}\left\{\tfrac{1}{2}\left(33 + 11\right) - 31\right\} \\
66 &= \tfrac{1}{2}\left\{\tfrac{1}{2}\left(11 + 22\right) - 12\right\}
\end{aligned}\right\} \quad \ldots\ldots\ldots\ldots\ldots\ldots(3).$$

These three equations simply express the condition that in each of the three coordinate planes the rigidity due to a shear parallel to either of the two other coordinate planes is equal to the rigidity due to a shear parallel to either plane bisecting the right angle between them.

Green† found fourteen equations among his 21 coefficients to express that there can be a purely distortional wave with wave-front in any plane whatever. Three more equations‡ express further that the planes chosen for the coordinates are planes of symmetry. The conditions which we have considered have given us, in (1) and (3) above, fifteen equations which are identical with fifteen of Green's. His other two are

$$11 = 22 = 33 \ldots\ldots\ldots\ldots\ldots\ldots\ldots(4).$$

With all these seventeen equations among the coefficients, the equation of energy becomes reduced to

$$2E = 11\left(e + f + g\right)^2 + 44\left(a^2 - 4fg\right) + 55\left(b^2 - 4ge\right) + 66\left(c^2 - 4ef\right)\ldots(5).$$

---

* Compare with formula $n = \tfrac{1}{2}\left(\mathfrak{A} - \mathfrak{B}\right)$ in Lecture II. p. 25, which is the condition that the rigidity is the same for all distortions in any one of the three coordinate planes for the case of $11 = 22 = 33$, there considered.

† *Collected Papers*, p. 309.

‡ *Ibid.*, pp. 303, 309.

It is easy now to verify Green's result. Remember that if we denote by ξ, η, ζ infinitesimal displacements of a point of the solid from its undisturbed position (x, y, z), we have

$$e = \frac{d\xi}{dx}; \qquad f = \frac{d\eta}{dy}; \qquad g = \frac{d\zeta}{dz}$$
$$a = \frac{d\zeta}{dy} + \frac{d\eta}{dz}; \quad b = \frac{d\xi}{dz} + \frac{d\zeta}{dx}; \quad c = \frac{d\eta}{dx} + \frac{d\xi}{dy} \left.\right\} \quad \dots\dots(6).$$

Denoting now by $\iiint dx\,dy\,dz$ integration throughout a volume V, with the condition that ξ, η, ζ are each zero at every point of the boundary, we find by a well-known method of double integration by parts*,

$$\iiint dx\,dy\,dz\ \frac{d\eta}{dy}\frac{d\zeta}{dz} = \iiint dx\,dy\,dz\ \frac{d\eta}{dz}\frac{d\zeta}{dy}$$
$$\iiint dx\,dy\,dz\ \frac{d\zeta}{dz}\frac{d\xi}{dx} = \iiint dx\,dy\,dz\ \frac{d\zeta}{dx}\frac{d\xi}{dz} \left.\right\}\dots\dots(7).$$
$$\iiint dx\,dy\,dz\ \frac{d\xi}{dx}\frac{d\eta}{dy} = \iiint dx\,dy\,dz\ \frac{d\xi}{dy}\frac{d\eta}{dx}$$

And by aid of this transformation we find

$$2\iiint dx\,dy\,dz\ E = \iiint dx\,dy\,dz\ 11\left(\frac{d\xi}{dx} + \frac{d\eta}{dy} + \frac{d\zeta}{dz}\right)^2$$
$$+ \iiint dx\,dy\,dz\ \left\{44\left(\frac{d\zeta}{dy} - \frac{d\eta}{dz}\right)^2 + 55\left(\frac{d\xi}{dz} - \frac{d\zeta}{dx}\right)^2 + 66\left(\frac{d\eta}{dx} - \frac{d\xi}{dy}\right)^2\right\}$$
$$\dots\dots\dots(8).$$

Let now V be the space between two infinite planes, paralle' to the fronts of any series of purely distortional waves traversing the space between them, these planes being taken at places at which for an instant the displacement is zero. They may, for instance, be the planes, half a wave-length asunder, of two consecutive zeros of displacement in a series of simple harmonic waves. Let P be any plane of the vibrating solid parallel to them; let p be its distance from the origin of coordinates; and q, its displacement at any instant. Then $\frac{1}{2}\frac{dq}{dp}$ is the molecular

---

* Compare *Math. and Phys. Papers*, Art. xcix. Vol. iii. p. 448 ; and "On the Reflexion and Refraction of Light," *Phil. Mag.* 1888, 2nd half-year.

rotation of the solid infinitely near to this plane on each side ; and Molar. by the rectangular resolution of rotations, we have

$$l\frac{dq}{dp} = \frac{d\zeta}{dy} - \frac{d\eta}{dz}; \quad m\frac{dq}{dp} = \frac{d\xi}{dz} - \frac{d\zeta}{dx}; \quad n\frac{dq}{dp} = \frac{d\eta}{dx} - \frac{d\xi}{dy} \dots(9),$$

where l, m, n denote the direction cosines of the line in P perpendicular to the direction of q, this being the axis of the molecular rotation. Using these new values in (8), and remarking that the first chief term vanishes because the displacement is purely distortional, we have

$$2\iiint dx\,dy\,dz\, E = (44l^2 + 55m^2 + 66n^2)\iiint dx\,dy\,dz\,\left(\frac{dq}{dp}\right)^2\dots(10).$$

Hence we see that no condensation or rarefaction accompanies our plane distortional waves, and that their velocity of propagation is

$$\sqrt{\frac{44l^2 + 55m^2 + 66n^2}{\rho}} \quad\dots\dots\dots\dots\dots(11).$$

This is Fresnel's formula for the propagational velocity of a plane wave in a crystal with (l, m, n) denoting the direction of vibration ; while in Green's theory (l, m, n) is the line in the wave-front perpendicular to the direction of vibration. The wave-surface is identical with Fresnel's.]

I will read to you Green's own statement of the relative tactics of the motion in his and in Fresnel's waves. Here it is at the bottom of page 304 of Green's *Collected Papers* : " We " thus see that if we conceive a section made in the ellipsoid " to which the equation (10) belongs, by a plane passing through " its centre and parallel to the wave's front, this section, when " turned 90 degrees in its own plane, will coincide with a similar " section of the ellipsoid to which the equation (8) belongs, and " which gives the directions of the disturbance that will cause " a plane wave to propagate itself without subdivision, and the " velocity of propagation parallel to its own front. The change " of position here made in the elliptical section is evidently " equivalent to supposing the actual disturbances of the ethereal " particles to be parallel to the plane usually denominated as the " *plane of polarization.*"

Before we separate this evening, return for a few minutes to our problem of vibratory molecules embedded in an elastic

Molecular. solid; and let us consider particularly the application of this dynamical theory to the Fraunhofer double dark line D of sodium-vapour.

[*March* 1, 1899.* For a perfectly definite mechanical representation of Sellmeier's theory, imagine for each molecule of sodium-vapour a spherical hollow in ether, lined with a thin rigid spherical shell, of mass equal to the mass of homogeneous ether which would fill the hollow. This rigid lining of the hollow we shall call the sheath of the molecule, or briefly the sheath. Within this put two rigid spherical shells, one inside the other, each moveable and each repelled from the sheath with forces, or distribution of force, such that the centre of each is attracted towards the centre of the hollow with a force varying directly as the distance. These suppositions merely put two of Sellmeier's single-atom vibrators into one sheath.

Imagine now a vast number of these diatomic molecules, equal and similar in every respect, to be distributed homogeneously through all the ether which we have to consider as containing sodium-vapour. In the first place, let the density of the vapour be so small that the distance between nearest centres is great in comparison with the diameter of each molecule. And in the first place also, let us consider light whose wave-length is very large in comparison with the distance from centre to centre of nearest molecules. Subject to these conditions we have (Sellmeier's formula)

$$\left(\frac{v_e}{v_s}\right)^2 = 1 + \frac{m\tau^2}{\tau^2 - \kappa^2} + \frac{m_{\prime}\tau^2}{\tau^2 - \kappa_{\prime}^2} \dots\dots\dots\dots(1);$$

where $m, m_{\prime}$ denote the ratios of the sums of the masses of one and the other of the moveable shells of the diatomic molecules in any large volume of ether, to the mass of undisturbed ether filling the same volume; $\kappa, \kappa_{\prime}$ the periods of vibration of one and the other of the two moveable shells of one molecule, on the supposition that the sheath is held fixed; v_e the velocity of light in pure undisturbed ether; v_s the velocity of light of period τ in the sodium-vapour.

---

* This replaces the concluding portion of the Lecture as originally delivered. It was read before the Royal Society of Edinburgh on Feb. 6, 1899, and reprinted in *Phil. Mag.* for March, 1899, under the title "Application of Sellmeier's Dynamical Theory to the Dark Lines D_1, D_2 produced by Sodium-Vapour."

For sodium-vapour, according to the measurements of Rowland Molecular. and Bell*, published in 1887 and 1888 (probably the most accurate hitherto made), the periods of light corresponding to the exceedingly fine *dark* lines D_1, D_2 of the solar spectrum are ·589618 and ·589022 of a michron†. The mean of these is so nearly one thousand times their difference that we may take

$$\kappa = \tfrac{1}{2}(\kappa + \kappa_{\prime})\left(1 - \frac{1}{2000}\right); \quad \kappa_{\prime} = \tfrac{1}{2}(\kappa + \kappa_{\prime})\left(1 + \frac{1}{2000}\right) \quad\text{......} \quad (2).$$

Hence if we put

$$\tau = \tfrac{1}{2}(\kappa + \kappa_{\prime})\left(1 + \frac{x}{1000}\right) \quad\text{................}(3);$$

and if x be any numeric not exceeding 4 or 5 or 10, we have

$$\left(\frac{\kappa}{\tau}\right)^2 \doteqdot 1 - \frac{1}{1000}(2x + 1); \quad \left(\frac{\kappa_{\prime}}{\tau}\right)^2 \doteqdot 1 - \frac{1}{1000}(2x - 1) \quad\text{......}(4);$$

whence

$$\frac{\tau^2}{\tau^2 - \kappa^2} \doteqdot \frac{1000}{2x + 1}; \quad \frac{\tau^2}{\tau^2 - \kappa_{\prime}^2} \doteqdot \frac{1000}{2x - 1} \quad\text{...........}(5).$$

Using this in (1), and denoting by μ the refractive index from ether to an ideal sodium-vapour with only the two disturbing atoms m, $m_{\prime}$, we find

$$\left(\frac{v_e}{v_s}\right)^2 = \mu^2 = 1 + \frac{1000m}{2x + 1} + \frac{1000m_{\prime}}{2x - 1} \quad\text{...........} \quad (6).$$

When the period, and the corresponding value of x according to (3), is such as to make μ^2 negative, the light cannot enter the sodium-vapour. When the period is such as to make μ^2 positive, the proportion, according to Fresnel and according to the most probable dynamics, of normally incident light which enters the vapour would be

$$1 - \left(\frac{\mu - 1}{\mu + 1}\right)^2 \quad\text{.....................} \quad (7),$$

if the transition from space, where the propagational velocity is v_e, to medium in which it is v_s, were infinitely sudden.

Judging from the approximate equality in intensity of the

---

* Rowland, *Phil. Mag.* 1887, first half-year; Bell, *Phil. Mag.* 1888, first half-year.

† See footnote, p. 150.

Molecular. bright lines D_1, D_2 of incandescent sodium-vapour; and from the approximately equal strengths of the very fine dark lines D_1, D_2 of the solar spectrum; and from the approximately equal strengths, or equal breadths, of the dark lines D_1, D_2 observed in the analysis of the light of an incandescent metal, or of the electric arc, seen through sodium-vapour of insufficient density to give much broadening of either line; we see that m and $m_{,}$ cannot be very different, and we have as yet no experimental knowledge to show that either is greater than the other. I have therefore assumed them equal in the calculations and numerical illustrations described below.

At the beginning of the present year I had the great pleasure to receive from Professor Henri Becquerel, enclosed with a letter of date Dec. 31, 1898, two photographs of anomalous dispersion by prisms of sodium-vapour*, by which I was astonished and delighted to see not merely a beautiful and perfect demonstration of the "anomalous dispersion" towards infinity on each side of the zero of refractivity, but also an illustration of the characteristic nullity of absorption and finite breadth of dark lines, originally shown in Sellmeier's formula† of 1872 and now, after 27 years, first actually seen. Each photograph showed dark spaces on the high sides of the D_1, D_2 lines, very narrow on one of the photographs; on the other much broader, and the one beside the D_2 line decidedly broader than the one beside the D_1 line; just as it should be according to Sellmeier's formula, according to which also the density of the vapour in the prism must have been greater in the latter case than in the former. Guessing from the ratio of the breadths of the dark bands to the space between their D_1, D_2 borders, and from a slightly greater breadth of the one beside D_2, I judged that m must in this case have been not very different from ·0002; and I calculated accordingly from (6) the accompanying graphical representation showing the value

of $1 - \dfrac{1}{\mu}$, represented by y in fig. 1. Fig. 2 represents similarly

the value of $1 - \dfrac{1}{\mu}$ for $m = ·001$, or density of vapour five times

* A description of Professor Becquerel's experiments and results will be found in *Comptes Rendus*, Dec. 5, 1898, and Jan. 16, 1899.

† Sellmeier, *Pogg. Ann.* Vol. cxlv. (1872) pp. 399, 520; Vol. cxlvii. (1872) pp. 387, 525.

Fig. 1*. $m = \cdot 0002$.   Molecular.

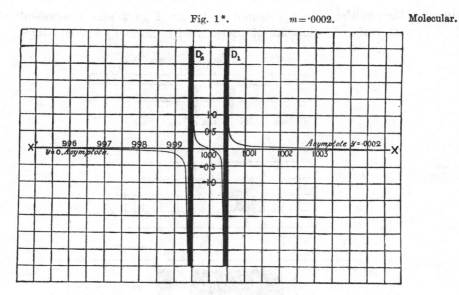

Fig. 2. $m = \cdot 001$.

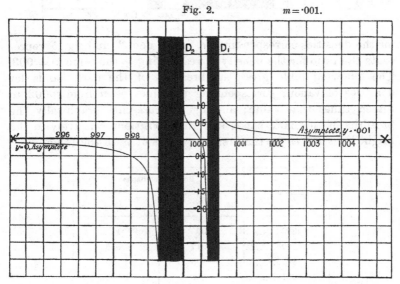

* In figs. 1 and 2 the D_1, D_2 lines are touched by curves of finite curvature at $y = +1$; and in figs. 3, 4, and 5 at $y = 0$. The left-hand side of each dark band is an asymptote to the curves of figs. 1 and 2, and a tangent at $y = 0$ to the curves of figs. 3, 4, and 5. The diagrams could not show these characteristics clearly unless on a much larger scale.

Molecular. that in the case represented by fig. 1. Figs. 3 and 4 represent the ratio of intensities of transmitted to normally incident light

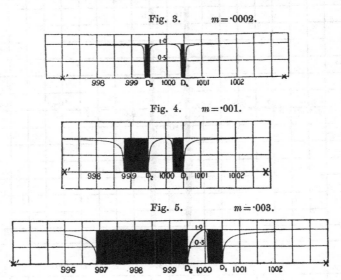

Fig. 3. $m = ·0002$.

Fig. 4. $m = ·001$.

Fig. 5. $m = ·003$.

for the densities corresponding to figs. 1 and 2, and fig. 5 represents the ratio for the density corresponding to the value $m = ·003$. The following table gives the breadths of the dark bands for densities of vapour corresponding to values of m from ·0002 to fifteen times that value; and fig. 6 represents graphically the breadths of the dark bands and their positions relatively to the bright lines D_1, D_2 for the first five values of m in the table.

Values of m	Breadths of Bands	
	D_1	D_2
·0002.........	·09	·11
·0006.........	·217	·383
·0010.........	·293	·707
·0014.........	·340	1·060
·0018.........	·371	1·429
·0022.........	·392	1·808
·0026.........	·408	2·192
·0030.........	·419	2·581

According to Sellmeier's formula the light transmitted through Molecular. a layer of sodium-vapour (or any transparent substance to which

Fig. 6.

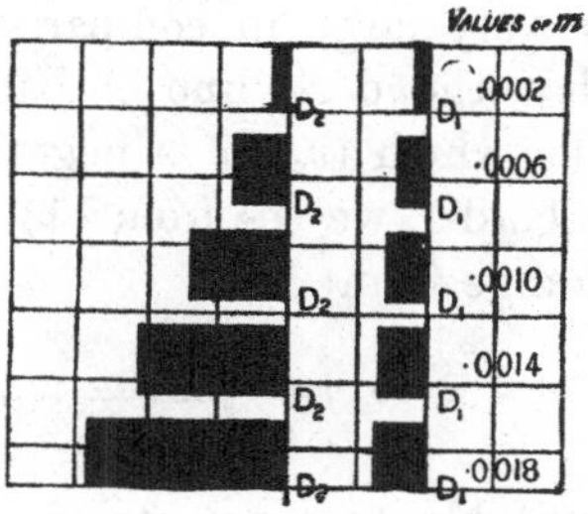

the formula is applicable) is the same whatever be the thickness of the layer (provided of course that the thickness is many times the wave-length). Thus the D_1, D_2 lines of the spectrum of solar light, which has traversed from the source a hundred kilometres of sodium-vapour in the sun's atmosphere, must be identical in breadth with those seen in a laboratory experiment in the spectrum of light transmitted through half a centimetre or a few centimetres of sodium-vapour, of the same density as the densest part of the sodium-vapour in the portion of the solar atmosphere traversed by the light analysed in any particular observation. The question of temperature cannot occur except in so far as the density of the vapour, and the clustering in groups of atoms or non-clustering (mist or vapour of sodium), are concerned.

A grand inference from the experimental foundation of Stokes' and Kirchhoff's original idea is that the periods of molecular vibration are the same to an exceedingly minute degree of accuracy through the great differences of range of vibration presented in the radiant molecules of an electric spark, electric arc, or flame, and in the molecules of a comparatively cool vapour or gas giving dark lines in the spectrum of light transmitted through it.

It is much to be desired that laboratory experiments be made, notwithstanding their extreme difficulty, to determine the density and pressure of sodium-vapour through a wide range of temperature, and the relation between density, pressure, and temperature of gaseous sodium.

Molecular.　　　Passing from the particular case of sodium, I add an application of Sellmeier's formula, (1) above, to the case of a gas or
vapour having in its constitution only a single molecular period κ.
Taking $m_{,} = 0$ in (1), we see that the square of the refractive
index for values of r very large in comparison with κ is $1 + m$.
And remembering that the dark line or band extends through
the range of values for which $(v_e/v_s)^2$ is negative, and that $(v_e/v_s)^2$
is zero at the higher border, we see from (1) that the dark band
extends through the range from

$$\tau = \kappa \ \text{ to } \ \tau = \frac{\kappa}{\sqrt{1 + m}} \quad\dots\dots\dots\dots\dots(8).$$

As an example suitable to illustrate the broadening of the
dark line by increased density of the gas, I take $m = a \times 10^{-4}$,
and take a some moderate numeric not greater than 10 or 20.
This gives for the range of the dark band from

$$\tau = \kappa \ \text{ to } \ \tau = \kappa\,(1 - \tfrac{1}{2}a \times 10^{-4})\dots\dots\dots\dots(9);$$

and for large values of τ it makes the refractive index
$1 + \tfrac{1}{2}a \times 10^{-4}$, and therefore the refractivity, $\tfrac{1}{2}a \times 10^{-4}$. If for
example we take $a = 6$, the refractivity would be ·0003, which is
nearly the same as the refractivity of common air at ordinary
atmospheric density.

Fig. 7.

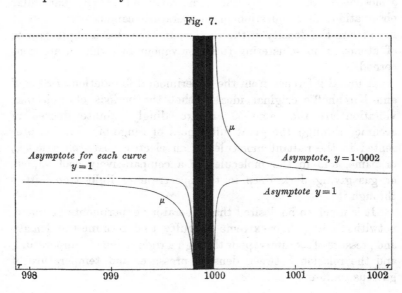

Taking $\kappa = 1000$, we have, for values of τ not differing from Molecular. 1000 by more than 10 or 20,

$$\frac{\tau^2}{\tau^2 - \kappa^2} \doteqdot \frac{1000}{2x}, \text{ where } x = \tau - 1000 \quad ....(10).$$

Thus we have

$$\mu = \sqrt{1 + \frac{a}{20x}} \quad ....(11).$$

In fig. 7 the curve marked μ represents the values of the refractive index corresponding to values of τ through a small range above and below κ, taking $a = 4$. The other curve represents the proportionate intensity of the light entering the vapour, calculated from these values of μ by (7) above.

The table on next page shows calculated values for the ordinates of the two curves; also values (essentially negative) for the formula of intensity calculated from the negative values of μ algebraically admissible from (11).

The negative values of μ have no physical interpretation for either curve; but the consideration of the algebraic prolongations of the curves through the zero of ordinates on the left-hand side of the dark band illustrates the character of their contacts. The physically interpreted part of each curve ends abruptly at this zero; which for each curve corresponds to a maximum value of x. The algebraic prolongation of the μ curve on the negative side is equal and similar to the curve shown on the positive side. But the algebraic prolongation of the intensity curve through its zero, as shown in the table, differs enormously from the curve shown on the positive side. To the degree of approximation to which we have gone, the portions of the intensity curve on the left and right hand sides of the dark band are essentially equal and similar. This proves that so far as Sellmeier's theory represents the facts, the penumbras are equal and similar on the two sides of a single dark line of the spectrum uninfluenced by others. It is also interesting to remark that according to Sellmeier as now interpreted, the broadening of a single undisturbed dark line, produced by increased density of the gas or vapour, is essentially on the high side of the finest dark line shown with the least density, and is in simple proportion to the density of the gas.]

Molecular.

x	μ	$1 - \left(\dfrac{\mu - 1}{\mu + 1}\right)^2$	
		μ positive	μ negative
$-1{\cdot}0$	$\cdot 894$		
$-\cdot 9$	$\cdot 882$		
$-\cdot 8$	$\cdot 866$		
$-\cdot 7$	$\cdot 845$	$\cdot 993$	$-141{\cdot}9$
$-\cdot 6$	$\cdot 816$	$\cdot 988$	$-82{\cdot}3$
$-\cdot 5$	$\cdot 774$	$\cdot 984$	$-61{\cdot}5$
$-\cdot 4$	$\cdot 707$	$\cdot 971$	$-33{\cdot}5$
$-\cdot 3$	$\cdot 577$	$\cdot 928$	$-12{\cdot}9$
$-\cdot 2$	0	0	0
$-\cdot 1$	imaginary	imaginary	imaginary
0	∞	0	0
$+\cdot 1$	$1{\cdot}732$	$\cdot 928$	$-12{\cdot}9$
$+\cdot 2$	$1{\cdot}414$	$\cdot 971$	$-33{\cdot}5$
$+\cdot 3$	$1{\cdot}291$	$\cdot 984$	$-61{\cdot}5$
$+\cdot 4$	$1{\cdot}247$	$\cdot 988$	$-82{\cdot}3$
$+\cdot 5$	$1{\cdot}183$	$\cdot 993$	$-141{\cdot}9$
$+\cdot 6$	$1{\cdot}155$		
$+\cdot 7$	$1{\cdot}134$		
$+\cdot 8$	$1{\cdot}118$		
$+\cdot 9$	$1{\cdot}105$		
$+1{\cdot}0$	$1{\cdot}095$		

LECTURE XIV.

AT this lecture were seen, immediately behind the model Molecular.
heretofore presented, two wires extending from the ceiling and

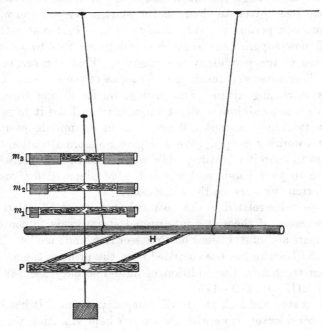

bearing a long heavy bar about three feet above the floor by
means of closely-fitting rings. By slipping these rings along the
bar, the period of vibration of this bifilar suspension could be
altered at will. Two parallel pieces of wood, jointed at each end,

served to transmit the azimuthal motion of this vibrator to the lower bar P of the model.

Let us look at this and see what it does. I have not seen it before, and it is quite new to me. Oh, see, you can vary the period; that is very nice, that is beautiful. We are going to study these vibrations a little, just as illustrations. Prof. Rowland has kindly made this arrangement for us, and I think we will all be interested in experimenting with it. We have this bar P, moved by this bifilar pendulum H, which is so massive that its period is but little affected, I suppose, by being connected with P. It takes some time before the initial free vibrations in the model are got quit of and the thing settles into simple harmonic motion corresponding to the period of the bifilar pendulum. If we keep this pendulum going long enough through nearly a constant range, the masses P, m_1, m_2, m_3 will settle into a definite simple harmonic motion, through the subsidence of any free vibrations which may have been given to them in the start. We see the whole apparatus now performing very nearly a simple harmonic motion. We will now superimpose another vibration on this by altering the period of the pendulum very slightly. That, you see, seems to have diminished very much the vibrations of the system. They are now increasing again. That will go on for a long time. I shall give this pendulum a slight impulse when I see it flagging, to keep its range constant. When it is in its middle position, I apply a working couple. We will give no more attention to it than just to keep it vibrating, while we look at these notes which you have in your hands, and which I have prepared for you so as to shorten our work on the black-board.

[These notes related to the tasinomic treatment of æolotropy. The discussion of them was interrupted at intervals to continue experiments and observations on Professor Rowland's model. That part of the Lecture has been omitted from the print as the subject has been treated in the addition of date February 14, 1899 to Lecture XIII., pp. 170—175.]

Let us stop and look at our vibrating apparatus. It has been going a considerable time with the exciter kept vibrating through a constant range, and you see but small motion transmitted to the system. That is an illustration of the most general solution of our old problem*. Our "handle" P is in firm connection with

---

* Lecture III. Pt. 2, p. 38.

the bifilar pendulum (the exciter) and is forced to agree with it. Molecular. Let us bring the system to rest. Now start the exciter and keep it going. In time the viscosity will annul the system of vibrations, representing the difference between zero and the permanent state of vibration which the three particles will acquire. If there were no loss of energy whatever, the result would be that this initial jangle, which you now see, would last for ever; consisting of a simple harmonic motion in the exciter and a compound of the three fundamental modes of these three particles viewed as a vibrating system with the bar P held fixed. Let the system with the bar P held fixed be set to vibrate in any way whatever; then its motion will be merely a compound of those three fundamental modes. But now set the exciter going, and the state of the case may be looked on as thus constituted;—the exciter and the whole system in simple harmonic motion of the same period, and, superimposed upon that, a compound of the three modes of simple harmonic vibration that the system can perform with the exciter fixed. We cannot improve on the mathematical treatment by observation; and really a model of this kind is rather a help or corrective to brain sluggishness than a means of observation or discovery. In point of fact, we can discover a great deal better by algebra. But brains are very poor after all, and this model is of some slight use in the way of making plain the meaning of the mathematics we have been working out.

The system seems to have come once more into its permanent state. Let us stop the exciter and see how long the system will hold its vibration. The reaction on the exciter is very slight, it is very nearly the same as if that massive bar H were absolutely fixed. But the motion actually communicated to it, since it is not absolutely fixed, will correspond to a considerable loss of energy. A very slight motion of H with its great length and mass has considerable energy compared even with the energy of our particle of greatest mass; so that our system will come to rest far sooner than if H were absolutely fixed. The model is at present illustrating phosphorescence. You see the particles (m_1, m_2, m_3) have gone on vibrating for a whole minute, and m_1 must have performed a couple of dozen vibrations at least. A true phosphorescence of a hundred seconds' duration is quite analogous to the giving back of vibrations which you see in our model, only instead of our two or three dozen vibrations, we have in

Molecular. phosphorescence 40,000 million-million vibrations during a hundred seconds. Now, we cannot get 1000 residual vibrations in our model because of the dissipation of energy arising from imperfect elasticity in the wire, friction between parts of the model, and the resistance of the air. That dissipation of energy is simply the conversion of energy from one state of motion, (the visible motions which we have been watching), into another (heat in the wire, heat of frictions, and heat in the air). In molecular dynamics, we have no underground way of getting quit of energy or carrying it off. We must know exactly what has been done with it when the vibration of an embedded molecule ends, even though this be not before a thousand million-million periods have been performed. [*March* 6, 1899. Imagine a homogeneous mass of rock—granite or basalt, for example—as large as the earth, or as many times larger as you please, but with no mutual gravitation between its parts to disturb it. Let there be, anywhere in it very far from a boundary, a spherical hollow of 5 cms. radius, and let a violin-string be stretched between two hooks fixed at opposite ends of a diameter of this hollow, and tuned to vibrations at the rate of 1000 per second. Let this string be set in vibration (for the present, no matter how) according to its gravest fundamental mode, through a range of one millimetre. Let the elasticity of the string and of the granite be absolutely perfect, and let there be no air in the hollow to resist the vibrations. They will not last for ever. Why not? Because two trains of waves, respectively condensational-rarefactional and purely distortional, will be caused to travel outwards, carrying away with them the energy given first to the vibrating string (see below § 28 of addition to Lecture XIV.).] We must suppose the elasticity of our matter and molecules to be perfect, and we cannot in any part of our molecular dynamics admit unaccounted for loss of energy; that is to say, we cannot admit viscous terms unless as an integral result of vibrations connected with a part of the system that is not convenient for us to look at.

In three minutes our system has come very nearly to rest. We infer therefore that in three minutes from a commencement of vibration of the exciter we shall have nearly reached the permanent state of things.

Now we vary the period of the exciter, making it as nearly as we can midway between two fundamental periods of our complex molecule. We will keep this going in an approximately constant

range for a while and look at the vibrations which it produces in Molecular.
the system.

Now you see very markedly the difference in the vibrations
of our system after it has been going for several minutes with
the exciter in a somewhat shorter period of vibration than that
with which we commenced. Here is another still shorter. In
the course of two or three minutes the superimposed vibrations
will die out. See now the tremendous difference of this case in
which the period of the exciter is approximately equal to one
of the fundamental periods of the system, or the periods for the
case in which the lowest bar is held absolutely fixed.

I had almost hoped that I would see some way of explaining
double refraction by this system of molecules, but it seems more
and more difficult. I will take you into my confidence to-morrow,
if you like, and show you the difficulties that weigh so much upon
me. I am not altogether disheartened by this, because of the
fact that such grand and complicated and highly interesting
subjects as I have named so often, absorption, dispersion and
anomalous refraction, are all not merely explained by their
means but are the inevitable results of this idea of attached
molecules.

There is one thing I want to say before we separate, and that
is, when I was speaking last of the subject, I saw what seemed
to me to be a difficulty, but on further consideration, I find it
no difficulty at all. Not very many hours after I told you it
was a difficulty, I saw that I was wrong in making it appear
to be a difficulty at all. I do not want to paint the thing any
blacker than it really is and I want to tell you that that question
I put as to the ether keeping straight with the molecules is
easily answered when there is a large number. Our assumption
was a large number of spherical cavities, lined with rigid spherical
shells and masses inside joined by springs or what not : and the
distance from cavity to cavity small in comparison with the wave-
length. It then happens that the motion of the medium relatively to the rigid shells will be exceedingly small and a portion
of the medium that will contain a large number of these shells will
all move together (see below, addition to Lec. XIV.). If the distance
from molecule to molecule is very small in comparison with the
wave-length, then you may look upon the thing as if the structure
were infinitely fine, and you may take it that the ether moves

Molecular. quite straight with them all, and not in and out among them, as I said. It is evident on the other hand, when the wave-length of the light traversing the medium is moderate in comparison with the distance between the molecules, that it must move out and in among them. But if the stiffness of the medium is such as to make the wave-length large in comparison with the distance from molecule to molecule, this stiffness is suf-ficient to keep them all together, and you may regard these rigid shells as bonds of attachment by which the molecule is pulled this way and that way, so that we may suppose our reactionary forces, of which $c_1(\xi - x_1)$ is a sample, to be absolutely the same in their effect upon the medium as if they were uniformly distributed through it.

That takes away one part of our discontent. The only diffi-culty that I see just now is that of explaining double refraction. The subject grows upon us terribly, and so does our want of time. If it is not too much for you I must have one of our double lectures to-morrow.

[Twice* in this Lecture, and indeed many times in preceding and subsequent Lectures, I felt the want of a full mathematical investigation of spherical waves originating in the application of force to an elastic solid within a limited space. I have therefore recently undertaken this work†, and I give the following statement of it as an addition to Lecture XIV.

§ 1. The complete mathematical theory of the propagation of motion through an infinite elastic solid, including the analysis of the motion into two species, equivoluminal and irrotational, was first given by Stokes in his splendid paper "On the Dynamical Theory of Diffraction‡." The object of the present communication is to investigate fully the forcive which must be applied to the boundary, S, of a hollow of any shape in the solid, in order to originate and to maintain any known motion of the surrounding solid; and to solve the inverse problem of finding the motion

---

* p. 188, and pp. 189, 190.

† Communicated to R. S. E. on May 1, 1899, and published in *Phil. Mag.* May, Aug. and Oct., 1899 under the title " On the Application of Force within a Limited Space, required to produce Spherical Solitary Waves, or Trains of Periodic Waves, of both species, Equivoluminal and Irrotational, in an Elastic Solid."

‡ Stokes, *Mathematical Papers*, Vol. II. p. 243.

when the forcive on, or the motion of, S is given, for the particular Molar. case in which S is a spherical surface kept rigid.

§ 2. Let ξ, η, ζ denote the infinitesimal displacement at any point of the solid, of which (x, y, z) is the equilibrium position. The well-known equations of motion* are

$$\left.\begin{array}{l} \rho \, \dfrac{d^2\xi}{dt^2} = (k + \tfrac{1}{3}n) \dfrac{d\delta}{dx} + n\nabla^2\xi, \\[2mm] \rho \, \dfrac{d^2\eta}{dt^2} = (k + \tfrac{1}{3}n) \dfrac{d\delta}{dy} + n\nabla^2\eta, \\[2mm] \rho \, \dfrac{d^2\zeta}{dt^2} = (k + \tfrac{1}{3}n) \dfrac{d\delta}{dz} + n\nabla^2\zeta, \end{array}\right\} \ \ldots\ldots\ldots\ldots\ldots (1),$$

where δ denotes $\dfrac{d\xi}{dx} + \dfrac{d\eta}{dy} + \dfrac{d\zeta}{dz}$. Using the notation of Thomson and Tait† for strain-components (elongations; and distortions), e, f, g; a, b, c; we have

$$\left.\begin{array}{lll} e = \dfrac{d\xi}{dx}; & f = \dfrac{d\eta}{dy}; & g = \dfrac{d\zeta}{dz}; \\[2mm] a = \dfrac{d\eta}{dz} + \dfrac{d\zeta}{dy}; & b = \dfrac{d\zeta}{dx} + \dfrac{d\xi}{dz}; & c = \dfrac{d\xi}{dy} + \dfrac{d\eta}{dx} \end{array}\right\} \ \ldots\ldots(2);$$

and with the corresponding notation P, Q, R; S, T, U, for stress-components (normal and tangential forces on the six sides of an infinitely small rectangular parallelepiped), we have

$$\left.\begin{array}{l} P = (k + \tfrac{4}{3}n)\,e + (k - \tfrac{2}{3}n)(f + g); \quad Q = (k + \tfrac{4}{3}n)f + (k - \tfrac{2}{3}n)(g + e); \\[1mm] \qquad\qquad R = (k + \tfrac{4}{3}n)\,g + (k - \tfrac{2}{3}n)\,(e + f) \\[1mm] S = na; \qquad\qquad T = nb; \qquad\qquad U = nc \end{array}\right\} \ (3).$$

Let now σ be an infinitesimal area at any point of the surface S; λ, μ, ν the direction-cosines of the normal; and $X\sigma$, $Y\sigma$, $Z\sigma$ the components of the force which must be applied from within to produce or maintain the specified motion of the matter outside. We have

$$\left.\begin{array}{l} -X = P\lambda + U\mu + T\nu \\[1mm] -Y = Q\mu + S\nu + U\lambda \\[1mm] -Z = R\nu + T\lambda + S\mu \end{array}\right\} \ \ldots\ldots\ldots\ldots\ldots\ldots(4);$$

---

* See my paper "On the Reflexion and Refraction of Solitary Plane Waves, &c." *Proc. R. S. E.* Dec. 1898, and *Phil. Mag.* Feb. 1899; reprinted below in the present volume.

† Thomson and Tait's *Natural Philosophy*, § 669, or *Elements*, § 640.

whence by (3)

$$-X = (k - \tfrac{2}{3}n)\,\lambda\delta + n\,(2\lambda e + \mu c + \nu b)$$
$$-Y = (k - \tfrac{2}{3}n)\,\mu\delta + n\,(2\mu f + \nu a + \lambda c) \Big\} \quad\ldots\ldots\ldots(5).$$
$$-Z = (k - \tfrac{2}{3}n)\,\nu\delta + n\,(2\nu g + \lambda b + \mu a)$$

These equations give an explicit answer to the question, What is the forcive? when the strain of the matter in contact with S is given. We shall consider in detail their application to the case in which S is spherical, and the motions and forces are in meridional planes through OX and symmetrical round this line. Without loss of generality we may take

$$z = 0\,;\ \text{giving}\ \nu = 0,\ a = 0,\ b = 0,\ Z = 0 \ldots\ldots\ldots\ldots(6).$$

Equations (5) therefore become

$$-X = (k - \tfrac{2}{3}n)\,\lambda\delta + n\,(2\lambda e + \mu c)$$
$$-Y = (k - \tfrac{2}{3}n)\,\mu\delta + n\,(2\mu f + \lambda c) \Big\} \quad\ldots\ldots\ldots\ldots(7).$$

§ 3. In §§ 5—26 of his paper already referred to, Stokes gives a complete solution of the problem of finding the displacement and velocity at any point of an infinite solid, which must follow from any arbitrarily given displacement and velocity at any previous time, if after that, the solid is left to itself with no force applied to any part of it. In a future communication I hope to apply this solution to the diffraction of solitary waves, plane or spherical. Meantime I confine myself to the subject stated in the title of the present communication, regarding which Stokes gives some important indications in §§ 27—29 of his paper.

§ 4. Poisson in 1819 gave a complete solution of the equation

$$\frac{d^2 w}{dt^2} = v^2 \nabla^2 w \ \ldots\ldots\ldots\ldots\ldots\ldots\ldots(8)$$

in terms of arbitrary functions of x, y, z, representing the initial values of w and $\dfrac{dw}{dt}$; and showed that for every case in which w depends only on distance (r) from a fixed point, it takes the form

$$w = \frac{1}{r}\left\{ F\left(t - \frac{r}{v}\right) + f\left(t + \frac{r}{v}\right) \right\} \ \ldots\ldots\ldots\ldots(9),$$

where F and f denote arbitrary functions. In my Baltimore Lectures of 1884 (pp. 46, 86, 87 above) I pointed out that solutions expressing spherical waves, whether equivoluminal (in which there Molar. is essentially different range of displacement in different parts of the spherical surface) or irrotational (for which the displacement may or may not be different in different parts of the spherical surface), can be very conveniently derived from (9) by differentiations with respect to x, y, z. It may indeed be proved, although I do not know that a formal proof has been anywhere published, that an absolutely general solution of (8) is expressed by the formula

$$\Sigma \left(\frac{d}{dx}\right)^h \left(\frac{d}{dy}\right)^i \left(\frac{d}{dz}\right)^j \left\{\frac{1}{r}\left[F\left(t-\frac{r}{v}\right)+f\left(t+\frac{r}{v}\right)\right]\right\};$$

$$r = \sqrt{[(x-x')^2+(y-y')^2+(z-z')^2]}\dots\dots\dots(10),$$

where Σ denotes sums for different integral values of h, i, j, and for any different values of x', y', z'.

§ 5. I propose at present to consider only the simplest of all the cases in which motion at every point $(x, 0, 0)$ and $(0, y, z)$ is parallel to $X'X$; and for all values of y and z, ξ is the same for equal positive and negative values of x. For this purpose we of course take $x' = y' = z' = 0$; and we shall find that no values of h, i, j greater than 2 can appear in our expressions for ξ, η, ζ, because we confine ourselves to the simplest case fulfilling the specified conditions. Our special subject, under the title of this paper, excludes waves travelling inwards from distant sources, and therefore annuls $f(t+r/v)$.

§ 6. In §§ 5—8 of his paper Stokes showed that any motion whatever of a homogeneous elastic solid may, throughout every part of it experiencing no applied force, be analysed into two constituents, each capable of existing without the other, in one of which the displacement is equivoluminal, and in the other it is irrotational. Hence if we denote by (ξ_1, η_1, ζ_1) the equivoluminal constituent, and by (ξ_2, η_2, ζ_2) the irrotational constituent, the complete solution of (1) may be written as follows:—

$$\xi = \xi_1 + \xi_2; \quad \eta = \eta_1 + \eta_2; \quad \zeta = \zeta_1 + \zeta_2 \dots\dots\dots(11),$$

where ξ_1, η_1, ζ_1 and ξ_2, η_2, ζ_2 fulfil the following conditions, (12) and (13), respectively:—

$$\left. \begin{array}{c} \dfrac{d\xi_1}{dx} + \dfrac{d\eta_1}{dy} + \dfrac{d\zeta_1}{dz} = 0, \\[2mm] u^2\nabla^2\xi_1 = \dfrac{d^2\xi_1}{dt^2}; \quad u^2\nabla^2\eta_1 = \dfrac{d^2\eta_1}{dt^2}; \quad u^2\nabla^2\zeta_1 = \dfrac{d^2\zeta_1}{dt^2}; \quad u^2 = \dfrac{n}{\rho} \end{array} \right\} \dots(12);$$

$$\xi_2 = \frac{dw}{dx}; \quad \eta_2 = \frac{dw}{dy}; \quad \zeta_2 = \frac{dw}{dz},$$

w being any solution of $\left.\right\}$ ....(13).

$$v^2 \nabla^2 w = \frac{d^2 w}{dt^2}; \quad v^2 = \frac{k + \frac{4}{3}n}{\rho}$$

The first equation of (12) shows that in the (ξ_1, η_1, ζ_1) constituent of the solution there is essentially no dilatation or condensation in any part of the solid; that is to say, the displacement is equivoluminal. The first three equations of (13) prove that in the (ξ_2, η_2, ζ_2) constituent the displacement is essentially irrotational.

§ 7.　We can now see that the most general irrotational solution fulfilling the conditions of § 5 is

$$\xi_2 = \frac{d^2}{dx^2} \frac{F_2}{r}; \quad \eta_2 = \frac{d^2}{dx\,dy} \frac{F_2}{r}; \quad \zeta_2 = \frac{d^2}{dx\,dz} \frac{F_2}{r} \ldots\ldots\ldots\ldots(14),$$

giving

$$\frac{d\xi_2}{dx} + \frac{d\eta_2}{dy} + \frac{d\zeta_2}{dz} = \frac{1}{v^2} \frac{d}{dx} \frac{\ddot{F}_2}{r} \ldots\ldots\ldots\ldots(14');$$

and the most general equivoluminal solution fulfilling the same conditions is

$$\xi_1 = \frac{d^2}{dx^2} \frac{F_1}{r} - \frac{\ddot{F}_1}{u^2 r}; \quad \eta_1 = \frac{d^2}{dx\,dy} \frac{F_1}{r}; \quad \zeta_1 = \frac{d^2}{dx\,dz} \frac{F_1}{r} \ldots\ldots(15),$$

giving

$$\frac{d\xi_1}{dx} + \frac{d\eta_1}{dy} + \frac{d\zeta_1}{dz} = 0 \ldots\ldots\ldots\ldots\ldots(15'),$$

where F_1 and F_2 are put for brevity to denote arbitrary functions of $\left(t - \frac{r}{u}\right)$ and $\left(t - \frac{r}{v}\right)$ respectively. Hence the most general solution fulfilling the conditions of § 5 is

$$\xi = \frac{d^2}{dx^2} \frac{\phi}{r} - \frac{\ddot{F}_1}{u^2 r}; \quad \eta = \frac{d^2}{dx\,dy} \frac{\phi}{r}; \quad \zeta = \frac{d^2}{dx\,dz} \frac{\phi}{r} \ldots\ldots\ldots(16),$$

where for brevity ϕ denotes a function of r and t, specified as follows :—

$$\phi(r, t) = F_1\left(t - \frac{r}{u}\right) + F_2\left(t - \frac{r}{v}\right) \ldots\ldots\ldots\ldots(17).$$

Denoting now by accents differential coefficients with respect to r,

and retaining the Newtonian notation of dots to signify differential $\qquad$ Molar.
coefficients with reference to t, we have

$$\phi' = -\left(\frac{\dot{F}_1}{u} + \frac{\dot{F}_2}{v}\right); \quad \phi'' = \frac{\ddot{F}_1}{u^2} + \frac{\ddot{F}_2}{v^2}; \quad \phi''' = -\left(\frac{\dddot{F}_1}{u^3} + \frac{\dddot{F}_2}{v^3}\right) \ldots\ldots\ldots(18).$$

Working out now the differentiations in (16), we find

$$\left.\begin{aligned}
\xi &= x^2\left(\frac{\phi''}{r^3} - \frac{3\phi'}{r^4} + \frac{3\phi}{r^5}\right) + \frac{\phi'}{r^2} - \frac{\phi}{r^3} - \frac{\ddot{F}_1}{u^2 r} \\[2mm]
\eta &= xy\left(\frac{\phi''}{r^3} - \frac{3\phi'}{r^4} + \frac{3\phi}{r^5}\right) \\[2mm]
\zeta &= xz\left(\frac{\phi''}{r^3} - \frac{3\phi'}{r^4} + \frac{3\phi}{r^5}\right)
\end{aligned}\right\} \ldots\ldots(19).$$

§ 8. For the determination of the force-components by (7), we shall want values of δ, e, f, and c. Using therefore (2) and going back to (16) we see that

$$c = 2\frac{d\eta}{dx} - \frac{1}{u^2}\frac{d}{dy}\frac{\ddot{F}_1}{r} \ldots\ldots\ldots\ldots\ldots(20).$$

Hence, and by (19), we find

$$c = y\left\{2\left[x^2\left(\frac{\phi'''}{r^4} - \frac{6\phi''}{r^5} + \frac{15\phi'}{r^6} - \frac{15\phi}{r^7}\right) + \frac{\phi''}{r^3} - \frac{3\phi'}{r^4} + \frac{3\phi}{r^5}\right]\right.$$
$$\left. + \frac{1}{r^2}\frac{\ddot{F}_1}{u^3} + \frac{1}{r^3}\frac{\ddot{F}_1}{u^2}\right\} \ldots\ldots\ldots\ldots(21).$$

By (16), (14′), and (15′) we find

$$\delta = \frac{1}{v^2}\frac{d}{dx}\frac{\ddot{F}_2}{r} = -x\left(\frac{1}{r^2}\frac{\ddot{F}_2}{v^3} + \frac{1}{r^3}\frac{\ddot{F}_2}{v^2}\right) \ldots\ldots\ldots\ldots(22);$$

and by (19) directly used in (2) we find

$$\left.\begin{aligned}
e &= x\left\{x^2\left(\frac{\phi'''}{r^4} - \frac{6\phi''}{r^5} + \frac{15\phi'}{r^6} - \frac{15\phi}{r^7}\right) + 3\left(\frac{\phi''}{r^3} - \frac{3\phi'}{r^4} + \frac{3\phi}{r^5}\right) + \frac{1}{r^2}\frac{\ddot{F}_1}{u^3} + \frac{1}{r^3}\frac{\ddot{F}_1}{u^2}\right\} \\[2mm]
f &= x\left\{y^2\left(\frac{\phi'''}{r^4} - \frac{6\phi''}{r^5} + \frac{15\phi'}{r^6} - \frac{15\phi}{r^7}\right) + \frac{\phi''}{r^3} - \frac{3\phi'}{r^4} + \frac{3\phi}{r^5}\right\} \\[2mm]
g &= x\left\{z^2\left(\frac{\phi'''}{r^4} - \frac{6\phi''}{r^5} + \frac{15\phi'}{r^6} - \frac{15\phi}{r^7}\right) + \frac{\phi''}{r^3} - \frac{3\phi'}{r^4} + \frac{3\phi}{r^5}\right\}
\end{aligned}\right\}$$
$$\ldots\ldots\ldots\ldots(23).$$

Remark here how by the summation of these three formulas we find for $e + f + g$ the value given for δ in (22).

§ 9. These formulas (22) and (23), used in (5), give the force-components per unit area at any point of the boundary (S of § 1) of a hollow of any shape in the solid, in order that the motion throughout the solid around it may be that expressed by (19). Supposing the hollow to be spherical, as proposed in § 2, let its radius be q. We must in (23) and (5) put

$$x = q\lambda ; \quad y = q\mu ; \quad z = q\nu \quad\dots\dots\dots\dots\dots(24);$$

and putting $\nu = 0$, we have, as in (7), the two force-components for any point of the surface in the meridian $z = 0$, expressed as follows :—

$$\left.\begin{aligned} X &= (k - \tfrac{2}{3}n)\, C_2\lambda^3 \; - n\left\{2\lambda^2 A + 2\,(2\lambda^2 + 1)\, B + (\lambda^2 + 1)\, C_1\right\} \\ Y &= (k - \tfrac{2}{3}n)\, C_2\lambda\mu - n\lambda\mu\,(2A + 4B + C_2) \end{aligned}\right\}\dots(25),$$

where

$$\left.\begin{aligned} A &= \frac{\Phi'''}{q} - \frac{6\Phi''}{q^2} + \frac{15\Phi'}{q^3} - \frac{15\Phi}{q^4} \\[1em] B &= \frac{\Phi''}{q^2} - \frac{3\Phi'}{q^3} + \frac{3\Phi}{q^4} \\[1em] C_1 &= \frac{\ddot{\mathscr{F}}_1}{qu^3} + \frac{\dddot{\mathscr{F}}_1}{q^2 u^2} \\[1em] C_2 &= \frac{\ddot{\mathscr{F}}_2}{qv^3} + \frac{\dddot{\mathscr{F}}_2}{q^2 v^2} \end{aligned}\right\}\dots\dots\dots(26),$$

Φ and $\mathscr{F}$ denoting ϕ and F with q for r.

§ 10. Returning now to (19), consider the character of the motion represented by the formulas. For brevity we shall call XX' simply the *axis*, and the plane of YY', ZZ' the *equatorial plane*. First take $y = 0$, $z = 0$, and therefore $x = r$. We find by aid of (18)

(axial) $\quad \xi = \dfrac{1}{r}\dfrac{\ddot{F}_2}{v^2} + \dfrac{2}{r^2}\left(\dfrac{\dot{F}_1}{u} + \dfrac{\dot{F}_2}{v}\right) + \dfrac{2}{r^3}\,(F_1 + F_2); \quad \eta = 0; \quad \zeta = 0\dots(27).$

Next take $x = 0$ and we find

(equatorial) $\quad \xi = -\dfrac{1}{r}\dfrac{\ddot{F}_1}{u^2} - \dfrac{1}{r^2}\left(\dfrac{\dot{F}_1}{u} + \dfrac{\dot{F}_2}{v}\right) - \dfrac{1}{r^3}\,(F_1 + F_2); \quad \eta = 0; \quad \zeta = 0$

$$\dots\dots\dots(28).$$

Hence for very small values of r we have   Molar.

$$\text{(axial)} \quad \xi \fallingdotseq \frac{2}{r^3}(F_1 + F_2) \left. \right\}$$

$$.................................(29);$$

$$\text{(equatorial)} \quad \xi \fallingdotseq -\frac{1}{r^3}(F_1 + F_2) \left. \right\}$$

and for very large values of r,

$$\text{(axial)} \quad \xi \fallingdotseq \frac{1}{r}\frac{\ddot{F}_2}{v^2} \left. \right\}$$

$$................................(30).$$

$$\text{(equatorial)} \quad \xi \fallingdotseq -\frac{1}{r}\frac{\ddot{F}_1}{u^2} \left. \right\}$$

Thus we see that for very distant places, the motion in the axis is approximately that due to the irrotational wave alone; and the motion at the equatorial plane is that due to the equivoluminal wave alone: also that with equal values of F_1 and F_2 the equivoluminal and the irrotational constituents contribute to these displacements inversely as the squares of the propagational velocities of the two waves. On the other hand, for places very near the centre, (29) shows that both in the axis and in the equatorial plane the irrotational and the equivoluminal constituents contribute equally to the displacements.

§ 11. Equations (25) and (26) give us full specification of the forcive which must be applied to the boundary of our spherical hollow to cause the motion to be precisely through all time that specified by (19), with F_1 and F_2 any arbitrary functions. Thus we may suppose $F_1(t - q/u)$ and $F_2(t - q/v)$ to be each zero for all negative values of t, and to be zero again for all values of t exceeding a certain limit τ. At any distance r from the centre, the disturbance will last during the time

from $\qquad t = \dfrac{r-q}{v}$ to $t = \dfrac{r-q}{v} + \tau,$

$$.................(31).$$

and from $\qquad t = \dfrac{r-q}{u}$ to $t = \dfrac{r-q}{u} + \tau$

Supposing $v > u$, we see that these two durations overlap by an interval equal to

$$\frac{r-q}{v} + \tau - \frac{r-q}{u} \left. \right\}$$

$$.....................(32).$$

if $\qquad\qquad r - q < \tau \left/ \left(\frac{1}{u} - \frac{1}{v}\right)\right.$

Molar.　On the other hand, at every point of space outside the radius $q + \tau/(1/u - 1/v)$ the wave of the greater propagational velocity passes away outwards before the wave of the smaller velocity reaches it, and the transit-time of each wave across it is τ. The solid is rigorously undisplaced and at rest throughout all the spaces outside the more rapid wave, between the two waves, and inside the less rapid of the two.

§ 12.　The expressions (25) and (26) for the components of the surface-forcive on the boundary of the hollow required to produce the supposed motion, involve $\dot{\mathscr{F}}_1$ and $\dot{\mathscr{F}}_2$. Hence we should have infinite values for $t = 0$ or $t = \tau$, unless $\ddot{F}_1$ and $\ddot{F}_2$ vanish for $t = 0$ and $t = \tau$, when $r = q$. Subject to this condition the simplest possible expression for each arbitrary function to represent the two solitary waves of § 11, is of the form

$$\mathscr{F} = (1 - \chi^2)^3, \quad \text{where} \quad \chi = \frac{2t}{\tau} - 1 \ldots\ldots\ldots\ldots(33).$$

Hence, by successive differentiations, with reference to t,

$$\left. \begin{aligned} \dot{\mathscr{F}} &= -\frac{12}{\tau}\chi(1 - \chi^2)^2 \\ \ddot{\mathscr{F}} &= \frac{24}{\tau^2}(-1 + 6\chi^2 - 5\chi^4) \\ \dddot{\mathscr{F}} &= \frac{48}{\tau^3}(12\chi - 20\chi^3) \end{aligned} \right\} \ldots\ldots\ldots\ldots\ldots(34).$$

The annexed diagram of four curves represents these four functions (33) and (34).

§ 13.　Take now definitively

$$\left. \begin{aligned} F_1\left(t - \frac{r}{u}\right) &= c_1 q^3 (1 - \chi_1^2)^3 \\ F_2\left(t - \frac{r}{v}\right) &= c_2 q^3 (1 - \chi_2^2)^3 \end{aligned} \right\} \ldots\ldots\ldots\ldots(35),$$

where

$$\left. \begin{aligned} \chi_1 &= \left(t - \frac{r - q}{u} - \tfrac{1}{2}\tau\right) \div \tfrac{1}{2}\tau \\ \chi_2 &= \left(t - \frac{r - q}{v} - \tfrac{1}{2}\tau\right) \div \tfrac{1}{2}\tau \end{aligned} \right\} \ldots\ldots\ldots\ldots(36).$$

Molar.

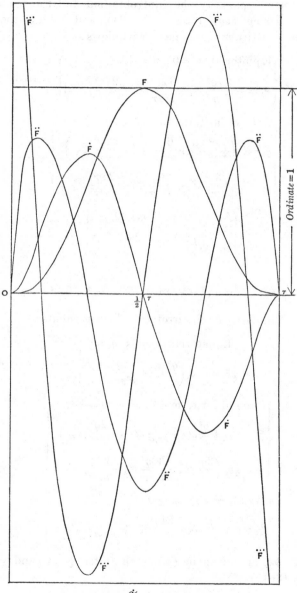

Scale of $\dot{\mathscr{F}} = \frac{1}{5}$ scale of F.

,, $\ddot{\mathscr{F}} = \frac{1}{25}$,, F.

,, $\dddot{\mathscr{F}} = \frac{1}{125}$,, F.

Molar.   Consider now separately the equivoluminal and the irrotational motions. Using (19), (18), (35), (34), and taking the equivoluminal constituents, we have as follows:—

$$\begin{cases} \text{Equatorial, } x = 0 \; ; \; \eta_1 = 0, \\ \xi_1 = -c_1 \left(\dfrac{q^3}{r^3}\mathscr{X}_1 + \dfrac{12q^3}{r^2 u \tau}\mathscr{X}_1' + \dfrac{24q^2}{r u^2 \tau^2}\mathscr{X}_1'' \right) \end{cases} \quad \dots\dots\dots(37),$$

$$\begin{cases} \text{Cone of latitude } 45°, \; x^2 = xy = \tfrac{1}{2}r^2, \\ \xi_1 = c_1 \left(\dfrac{1}{2}\dfrac{q^3}{r^3}\mathscr{X}_1 + \dfrac{6q^3}{r^2 u \tau}\mathscr{X}_1' - \dfrac{12q^3}{r u^2 \tau^2}\mathscr{X}_1'' \right), \\ \eta_1 = \tfrac{1}{2}c_1 \left(\dfrac{3q^3}{r^3}\mathscr{X}_1 + \dfrac{36q^3}{r^2 u \tau}\mathscr{X}_1' + \dfrac{24q^3}{r u^3 \tau^2}\mathscr{X}_1'' \right) \end{cases} \quad \dots\dots\dots(38),$$

$$\begin{cases} \text{Axial, } x^2 = r^2, \; \eta_1 = 0, \\ \xi_1 = c_1 \left(\dfrac{2q^3}{r^3}\mathscr{X}_1 + \dfrac{24q^3}{r^2 u \tau}\mathscr{X}_1' \right) \end{cases} \quad \dots\dots\dots\dots(39);$$

where

$$\mathscr{X}_1 = (1 - \chi_1^2)^3; \quad \mathscr{X}_1' = -(1 - \chi_1^2)^2 \chi_1; \quad \mathscr{X}_1'' = -1 + 6\chi_1^2 - 5\chi_1^4 \dots(40).$$

§ 14.   Similarly for the irrotational constituents;

$$\begin{cases} \text{Equatorial, } x = 0 \; ; \; \eta_2 = 0, \\ \xi_2 = -c_2 \left(\dfrac{q^3}{r^3}\mathscr{X}_2 + \dfrac{12q^3}{r^2 v \tau}\mathscr{X}_2' \right) \end{cases} \quad \dots\dots\dots\dots(41);$$

$$\begin{cases} \text{Cone of latitude } 45°, \; x^2 = xy = \tfrac{1}{2}r^2, \\ \xi_2 = c_2 \left(\dfrac{1}{2}\dfrac{q^3}{r^3}\mathscr{X}_2 + \dfrac{6q^3}{r^2 v \tau}\mathscr{X}_2' + \dfrac{12q^3}{r v^2 \tau^2}\mathscr{X}_2'' \right), \\ \eta_2 = \tfrac{1}{2}c_2 \left(\dfrac{3q^3}{r^3}\mathscr{X}_2 + \dfrac{36q^3}{r^2 v \tau}\mathscr{X}_2' + \dfrac{24q^3}{r v^2 \tau^2}\mathscr{X}_2'' \right) \end{cases} \quad \dots\dots\dots(42);$$

$$\begin{cases} \text{Axial, } x^2 = r^2, \; \eta_2 = 0, \\ \xi_2 = c_2 \left(\dfrac{2q^3}{r^3}\mathscr{X}_2 + \dfrac{24q^3}{r^2 v \tau}\mathscr{X}_2' + \dfrac{24q^3}{r v^2 \tau^2}\mathscr{X}_2'' \right) \end{cases} \quad \dots\dots\dots(43);$$

where $\mathscr{X}_2$ &c. are given by (40) with χ_2 for χ_1; χ_1 and χ_2 being given by (36).

§ 15.   The character of the motion throughout the solid, which is fully specified by (19), will be perfectly understood after a careful study of the details for the equatorial, conal, and axial

places, shown clearly by (37)...(43) for each constituent, the equi- Molar.
voluminal and the irrotational, separately. The curve $\mathscr{F}$ in the
diagram of § 12 shows the history of the motion that must be given
to any point of the surface S, for either constituent alone, and
therefore for the two together, in any case in which q is exceedingly small in comparison with the smaller of the two quantities
$u\tau, v\tau$, which for brevity we shall call the wave-lengths. The curve
$\ddot{\mathscr{F}}$ shows the history of the motion produced by either wave when
it is passing any point at a distance from the centre very great in
comparison with its own wave-length. But the three algebraic
functions $\mathscr{F}$, $\dot{\mathscr{F}}$, $\ddot{\mathscr{F}}$ all enter into the expression of the motion
due to either wave when the faster has advanced so far that its
rear is clear of the front of the slower, but *not* so far as to make
its wave-length (which is the constant thickness of the spherical
shell containing it) great in comparison with its inner radius.
Look at the diagram, and notice that in the origin at S, a mere
motion of each point in one direction and back, represented by $\mathscr{F}$,
causes in very distant places a motion ($\ddot{\mathscr{F}}$) to a certain displacement d, back through the zero to a displacement $1{\cdot}36 \times d$ in the
opposite direction, thence back through zero to d in the first
direction and thence back to rest at zero. Remark that the
direction of d is radial in the irrotational wave and perpendicular
to the radius in the equivoluminal wave. Remark also that the
d for every radial line varies inversely as distance from the
centre.

§ 16.　Draw any line OPK in any fixed direction through
O, the centre of the spherical surface S at which the forcive
originating the whole motion is applied. In the particular
case of §§ 12...15, and in any case in which $F_1\left(t - \dfrac{q}{u}\right)$ and
$F_2\left(t - \dfrac{q}{v}\right)$ are each assumed to be, from $t = 0$ to $t = \tau$, of the
form $t^3 (\tau - t)^3 A_i t^i$, where i denotes an integer, the time-history
of the motion of P is $B_0 + B_1 t + \ldots + B_{6+i} t^{6+i}$, and its space-history
(t constant and r variable) is $C_{-3} r^{-3} + C_{-2} r^{-2} + \ldots + C_{5+i} r^{5+i}$; the
complete formula in terms of t and r being given explicitly by (19).
The elementary algebraic character of the formula: and the *exact
nullity* of the displacement for every point of the solid for which
$r > q + vt$; and between $r = q + v(t - \tau)$ and $r = q + ut$, when

Molar. $v(t-\tau) > ut$; and between $r=q$ and $r=q+u(t-\tau)$, when $t>\tau$; these interesting characteristics of the solution of a somewhat intricate dynamical problem are secured by the particular character of the originating forcive at S, which we find according to §§ 8, 9 to be that which will produce them. But all these characteristics are lost except the first (nullity of motion through all space outside the spherical surface $r=q+vt$), if we apply an arbitrary forcive to S^*, or such a forcive as to produce an arbitrary deformation or motion of S. Let for example S be an ideal rigid spherical lining of our cavity; and let any infinitesimal arbitrary motion be given to it. We need not at present consider infinitesimal rotation of S: the spherical waves which this would produce, particularly simple in their character, were investigated in my Baltimore Lectures[†], and described in recent communications to the British Association and *Philosophical Magazine*[‡]. Neither need we consider curvilinear motion of the centre of S, because the motion being infinitesimal, independent superposition of x-, y-, z-motions produces any curvilinear motion whatever.

§ 17. Take then definitively $\mathscr{E}(t)$, or simply $\mathscr{E}$, an arbitrary function of the time, to denote excursion in the direction OX, of the centre of S from its equilibrium-position. Let $\partial^{-1}\mathscr{E}$, $\partial^{-2}\mathscr{E}$ denote $\int_0^t dt\mathscr{E}$ and $\int_0^t dt \int_0^t dt\mathscr{E}$. Our problem is, supposing the solid to be everywhere at rest and unstrained when $t=0$, to find (ξ,η,ζ) for every point of the solid $(r>q)$ at all subsequent time (t positive); with

$$\text{at } r=q, \quad \xi=\mathscr{E}(t), \quad \eta=0, \quad \zeta=0 \dots\dots\dots(44).$$

These, used in (19), give

$$0 = \frac{\ddot{\mathscr{F}}_1(t)}{u^2} + \frac{\ddot{\mathscr{F}}_2(t)}{v^2} + 3\left\{ \frac{1}{q}\left[\frac{\dot{\mathscr{F}}_1(t)}{u} + \frac{\dot{\mathscr{F}}_2(t)}{v}\right] + \frac{1}{q^2}[\mathscr{F}_1(t)+\mathscr{F}_2(t)]\right\}$$
$$\dots\dots\dots(45),$$

---

* If the space inside S is filled with solid of the same quality as outside, the solution remains algebraic, if the forcive formula is algebraic, though discontinuous. The displacement of S ends, not at time $t=\tau$ when the forcive is stopped, but at time $t=\tau+2q/u$ when the last of the inward travelling wave produced by it has travelled in to the centre, and out again to $r=q$.

† Pp. 81—83 and 159, 160 above.

‡ *B. A. Report*, 1898, p. 783; *Phil. Mag.* Nov. 1898, p. 494.

and

$$\mathscr{E}(t) = -\frac{1}{q^2}\left[\frac{\dot{\mathscr{F}_1}(t)}{u} + \frac{\dot{\mathscr{F}_2}(t)}{v}\right] - \frac{1}{q^3}[\mathscr{F}_1(t) + \mathscr{F}_2(t)] - \frac{1}{q}\frac{\ddot{\mathscr{F}_1}}{u^2}\ \ldots\ldots(46).$$

Adding $3q \times (46)$ to (45), we find

$$3q\mathscr{E}(t) = -2\frac{\ddot{\mathscr{F}_1}(t)}{u^2} + \frac{\ddot{\mathscr{F}_2}(t)}{v^2}\ \ldots\ldots\ldots(47);$$

whence

$$\frac{\mathscr{F}_2(t)}{v^2} = 2\frac{\mathscr{F}_1(t)}{u^2} + 3q\partial^{-2}\mathscr{E}(t)\ \ldots\ldots(48);$$

and by this eliminating $\mathscr{F}_2$ from (46),

$$\left[\partial^2 + \frac{1}{q}(u+2v)\partial + \frac{1}{q^2}(u^2+2v^2)\right]\mathscr{F}_1(t) = \mathscr{G}(t)\ldots\ldots(49),$$

where ∂ denotes d/dt, and

$$\mathscr{G}(t) = -qu^2\left(1 + \frac{3v}{q}\partial^{-1} + \frac{3v^2}{q^2}\partial^{-2}\right)\mathscr{E}(t)\ldots\ldots\ldots(50).$$

§ 18.   I hope later to work out this problem for the case of motion commencing from rest at $t = 0$, and $\mathscr{E}(t)$ an arbitrary function; but confining ourselves meantime to the case of S having been, and being, kept perpetually vibrating to and fro in simple harmonic motion, assume

$$\mathscr{E}(t) = h\sin\omega t\ldots\ldots\ldots\ldots\ldots(51).$$

With this, (50) gives

$$\mathscr{G}(t) = hqu^2\left[\left(\frac{3v^2}{q^2\omega^2} - 1\right)\sin\omega t + \frac{3v}{q\omega}\cos\omega t\right]\ldots\ldots\ldots(52).$$

To solve (49) in the manner most convenient for this form of $\mathscr{G}(t)$, we now have

$$\mathscr{F}_1(t) = \frac{1}{\partial^2 + \dfrac{1}{q}(u+2v)\partial + \dfrac{1}{q^2}(u^2+2v^2)}\mathscr{G}(t)$$

$$= \frac{\partial^2 + \dfrac{1}{q^2}(u^2+2v^2) - \dfrac{1}{q}(u+2v)\partial}{\left[\partial^2 + \dfrac{1}{q^2}(u^2+2v^2)\right]^2 - \left[\dfrac{1}{q}(u+2v)\partial\right]^2}\mathscr{G}(t)$$

Molar. $= hqu^2 \times$

$$\frac{\left\{\dfrac{3v^2}{q^4\omega^2}(u^2+2v^2)+\dfrac{1}{q^2}(v^2+3vu-u^2)+\omega^2\right\}\sin\omega t+\dfrac{u-v}{q\omega}\left(\dfrac{3uv}{q^2}+\omega^2\right)\cos\omega t}{\left[\dfrac{1}{q^2}(u^2+2v^2)-\omega^2\right]^2+\dfrac{\omega^2}{q^2}(u+2v)^2}$$

$$\dotfill(53).$$

With $\mathscr{F}_1$ thus determined, (48) gives $\mathscr{F}_2$ as follows,

$$\frac{\mathscr{F}_2(t)}{v^2}=\frac{2\mathscr{F}_1(t)}{u^2}-\frac{3hq}{\omega^2}\sin\omega t\dotfill(54).$$

For $\xi,\ \eta,\ \zeta$ by (19) we now have

$$\left.\begin{aligned}
\xi &= B\,(r,\,t)\,x^2-\frac{1}{r^2}\left[\frac{\dot{\mathscr{F}}_1(t_1)}{u}+\frac{\dot{\mathscr{F}}_2(t_2)}{v}\right]-\frac{1}{r^3}[\mathscr{F}_1(t_1)+\mathscr{F}_2(t_2)]-\frac{1}{r}\frac{\ddot{\mathscr{F}}_1(t_1)}{u^2}\\
\eta &= B\,(r,\,t)\,xy\\
\zeta &= B\,(r,\,t)\,xz
\end{aligned}\right\}(55),$$

where, with notation corresponding to (26) above,

$$\left.\begin{aligned}
B\,(r,\,t) &= \frac{1}{r^3}\left[\frac{\ddot{\mathscr{F}}_1(t_1)}{u^2}+\frac{\ddot{\mathscr{F}}_2(t_2)}{v^2}\right]+\frac{3}{r^4}\left[\frac{\dot{\mathscr{F}}_1(t_1)}{u}+\frac{\dot{\mathscr{F}}_2(t_2)}{v}\right]\\
&\qquad\qquad +\frac{3}{r^5}[\mathscr{F}_1(t_1)+\mathscr{F}_2(t_2)];\\
t_1 &= t-\frac{r-q}{u}; \quad t_2 = t-\frac{r-q}{v}
\end{aligned}\right\}\dotfill(56).$$

§ 19. The wave-lengths of the equivoluminal and rotational waves are respectively $\dfrac{2\pi u}{\omega}$ and $\dfrac{2\pi v}{\omega}$. For values of r very great in comparison with the greater of these, the second members of (55) become reduced approximately to the terms involving $\ddot{F}_1$ and $\ddot{F}_2$. These terms represent respectively a train of equivoluminal waves, or waves of transverse vibration, and a train of irrotational waves, or waves of longitudinal vibration; and the amplitude of each wave as it travels outwards varies inversely as r.

§ 20. For the case of an incompressible solid we have $v=\infty$, which by (53) gives

$$\frac{\ddot{\mathscr{F}}_1(t)}{u^2}=-\tfrac{3}{2}hq\sin\omega t\dotfill(57);$$

and by (55) we have, for r very great,

$$\left.\begin{array}{l} \xi = -\tfrac{3}{2}hq \sin \omega t \left(\dfrac{x^2}{r^3} - \dfrac{1}{r}\right) \\[2mm] \eta = -\tfrac{3}{2}hq \sin \omega t \dfrac{xy}{r^3} \\[2mm] \zeta = -\tfrac{3}{2}hq \sin \omega t \dfrac{xz}{r^3} \end{array}\right\} \quad \dots\dots\dots\dots(58),$$

which fully specify, for great distances from the origin, the wave-motion produced by a rigid globe of radius q, kept moving to and fro according to the formula $h \sin \omega t$.

§ 21. The strictly equivoluminal motion thus represented consists of outward-travelling waves, having direction of vibration in meridional planes, and very approximately* perpendicular to the radial direction, and amplitude of vibration equal to $\tfrac{3}{2}h\dfrac{q}{r}\sin\theta$, where θ denotes the angle between r and the axis. The gradual change from the simple motion $\xi = h \sin \omega t$ at the surface of the rigid globe, through the elastic solid at distances moderate in comparison with q, out to the greater distances where the motion is very approximately the pure wave-motion represented by (58), is a very interesting subject for detailed investigation and illustration. The formulas expressing it are found by putting $v = \infty$ in (45), and using this equation to determine $F_2(t)$ in terms of $F_1(t)$; then using (47) to determine $F_1(t)$ in terms of $\mathscr{E}(t)$ given by (51); and then using (55) and (56), for which $v = \infty$ makes $t_2 = t$, to determine ξ, η, ζ. They are as follows:—

$$\left.\begin{array}{l} \xi = B(r, t)x^2 - \left(\dfrac{q}{r}\right)^3 \dfrac{h}{2}\left[\left(1 - \dfrac{3u^2}{q^2\omega^2}\right)\sin\omega t + \dfrac{3u^2}{q^2\omega^2}\sin\omega t_1 - \dfrac{3u}{q\omega}\cos\omega t\right] \\[3mm] \qquad - \left(\dfrac{q}{r}\right)^2 \dfrac{3h}{2}\dfrac{u}{q\omega}\cos\omega t_1 + \dfrac{q}{r}\dfrac{3h}{2}\sin\omega t_1 ; \\[3mm] \eta = B(r, t)xy ; \qquad \zeta = B(r, t)xz \end{array}\right\}$$

$$\dots\dots\dots\dots(59),$$

* Rigorously so, if the wave-length and q are each infinitely small in comparison with r.

Molar.　where

$$r^2 B (r, t) = \left(\frac{q}{r}\right)^3 \frac{3h}{2} \left[\left(1 - \frac{3u^2}{q^2\omega^2}\right) \sin \omega t + \frac{3u^2}{q^2\omega^2} \sin \omega t_1 - \frac{3u}{q\omega} \cos \omega t\right]$$

$$+ \left(\frac{q}{r}\right)^2 \frac{9h}{2} \frac{u}{q\omega} \cos \omega t_1 - \frac{q}{r} \frac{3h}{2} \sin \omega t_1,$$

$$t_1 = t - \frac{r - q}{u}$$

$$\dots\dots\dots\dots(60).$$

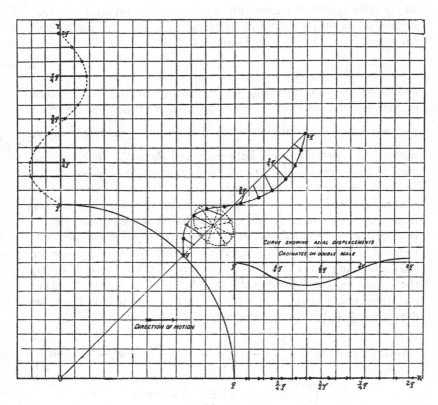

Fig. 1.

For the particular case of the wave-length equal to the radius we have

$$q = \frac{2\pi u}{\omega} \dots\dots\dots\dots\dots\dots(61),$$

which enables us to write simply $1/2\pi$ for $u/q\omega$ in equations (59) Molar. and (60). For graphic representation of this case we take $z = 0$, which makes all the displacements lie in the plane (xy).

§ 22. The accompanying drawings help us to understand thoroughly the character of the motion of the solid throughout the whole infinite space around the vibrating rigid globe. They show displacements and motions of points of which the equilibrium positions are in the equatorial plane, in the cone of 45° latitude,

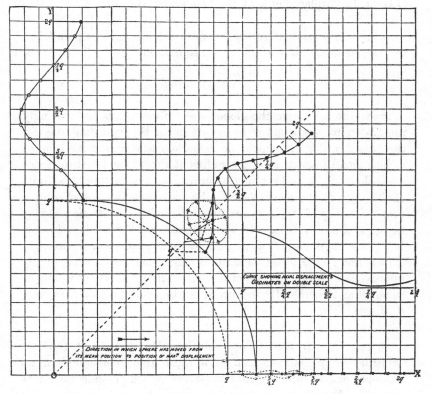

Fig. 2.

and in the axial line. Fig. 1 represents displacements at an instant when the globe is moving rightwards through its middle position. Fig. 2 shows displacements a quarter-period later, when the globe is at the end of its rightward motion. Each figure shows also the orbit of a single particle of which the equilibrium position is in the 45° cone, at a distance $\frac{5}{4}q$ from the centre of the globe. The orbital motion is in the direction of the hands of a watch. It is interesting to see illustrated in fig. 2 how the axial motion is gradually reduced from $\pm h$ at the surface of the globe to a very small range at distance q from the surface, or $2q$ from the centre, and we are helped to understand its gradual approximation to zero at greater and greater distances by the little auxiliary diagrams annexed, in which are shown by ordinates the magnitudes of the axial displacements at the two chosen times.

§ 23. The gradual transition from motion $h \sin \omega t$ parallel to the axis at the surface of the globe, to motion

$$-\frac{3}{2}\frac{q}{r}\, h \sin \theta \sin \omega t$$

at great distances from the globe in any direction, is interestingly illustrated by the conal representations in the two diagrams for the case $\theta = 45°$. It should be remarked that in reality h ought to be a small fraction of q, the radius of the globe, practically not more than $\frac{1}{100}$, in order that the strains may be within the limits of elasticity of the most elastic solid, and that the law of simple proportionality of stresses to strains (Hooke's *Ut tensio sic vis*) may be approximately true. In the diagram we have taken $h = \frac{1}{6}q$; but if we imagine every displacement reduced to $\frac{1}{20}$ of the amount shown, and in the direction actually shown, we have a true, highly approximate representation of the actual motions, which would be so small as to be barely perceptible to the eye, for a globe of 6 cms. diameter.

§ 24. Return now to our solution (53), (54), (55), (56) for arbitrary or periodic motion of a rigid globe embedded in an isotropic elastic solid of finite resistance to compression and finite rigidity. For distances from the globe very great in comparison with q, its radius, that is to say for q/r very small, (55) and (56) become

$$\xi \fallingdotseq B\,(r,\ t)\,x^2 - \frac{1}{r}\frac{\ddot{\mathscr{X}}_1\,(t_1)}{u^2}\,;$$

$$\left.\eta \fallingdotseq B\,(r,\ t)\,xy\,; \qquad \zeta \fallingdotseq B\,(r,\ t)\,xz \right\} \quad \ldots\ldots\ldots (62)\,;$$

$$r^2 B\,(r,\,t) \fallingdotseq \frac{1}{r}\left[\frac{\ddot{\mathscr{I}}_1(t_1)}{u^2} + \frac{\ddot{\mathscr{I}}_2(t_2)}{v^2}\right]$$

where $\qquad t_1 = t - \dfrac{r-q}{u}\,; \qquad t_2 = t - \dfrac{r-q}{v}$ $\qquad$ ....(63).

Putting now in the equations (62) the value of $B\,(r,\,t)$ from (63), and eliminating $\ddot{\mathscr{I}}_2(t_2)$ by (47), we find

$$\xi \fallingdotseq -\frac{r^2-x^2}{r^3}\frac{\ddot{\mathscr{I}}_1(t_1)}{u^2} + \frac{x^2}{r^3}\left[2\frac{\ddot{\mathscr{I}}_1(t_2)}{u^2} + 3q\mathscr{E}(t_2)\right]$$

$$\eta \fallingdotseq \quad\;\; \frac{xy}{r^3}\frac{\ddot{\mathscr{I}}_1(t_1)}{u^2} + \frac{xy}{r^3}\left[2\frac{\ddot{\mathscr{I}}_1(t_2)}{u^2} + 3q\mathscr{E}(t_2)\right] \Bigg\} \;...(64).$$

$$\zeta \fallingdotseq \quad\;\; \frac{xz}{r^3}\frac{\ddot{\mathscr{I}}_1(t_1)}{u^2} + \frac{xz}{r^3}\left[2\frac{\ddot{\mathscr{I}}_1(t_2)}{u^2} + 3q\mathscr{E}(t_2)\right]$$

The terms of these formulas, having t_1 and t_2 respectively for their arguments, represent two distinct systems of wave-motion, the first equivoluminal, the second irrotational, travelling outwards from the centre of disturbance with velocities u and v.

§ 25. I reserve for some future occasion the treatment of the case in which $\mathscr{E}(t)$ is discontinuous, beginning with zero when $t=0$ and ending with zero when $t=\tau$. I only remark at present in anticipation that $\mathscr{I}_1(t)$, determined by the differential equation (49), though commencing with zero at $t=0$, does not come to zero at $t=\tau$, but subsides to zero according to the logarithmic law (ϵ^{-kt}) as t goes on to infinity; and that therefore, as the same statement is proved for $\mathscr{I}_2(t)$ by (48), neither the equivoluminal nor the irrotational wave-motion is a limited solitary wave of duration τ, but on the contrary each has an infinitely long subsidential rear.

§ 26. For the general problem of the globe kept in simple harmonic motion, $h\sin\omega t$, parallel to OX, we may write (53) for brevity as follows:—

Molar.

$$\mathscr{F}_1(t) = \frac{hqu^2}{\omega^2}(K\sin\omega t + L\cos\omega t),$$

where

$$
\left.
\begin{aligned}
K &= \frac{\dfrac{3v^2}{q^4\omega^4}(u^2+2v^2)+\dfrac{1}{q^2\omega^2}(v^2+3vu-u^2)+1}{\left[\dfrac{1}{q^2\omega^2}(u^2+2v^2)-1\right]^2+\dfrac{1}{q^2\omega^2}(u+2v)^2}; \\[2em]
L &= \frac{\dfrac{u-v}{q\omega}\left(\dfrac{3uv}{q^2\omega^2}+1\right)}{\left[\dfrac{1}{q^2\omega^2}(u^2+2v^2)-1\right]^2+\dfrac{1}{q^2\omega^2}(u+2v)^2}
\end{aligned}
\right\}\ \ldots\ldots\ldots(65).
$$

In terms of this notation, (64) gives for great distances from the centre

$$
\left.
\begin{aligned}
\xi &\doteqdot \frac{hq}{r}\left\{\frac{r^2-x^2}{r^2}(K\sin\omega t_1+L\cos\omega t_1)+\frac{x^2}{r^2}[(3-2K)\sin\omega t_2-2L\cos\omega t_2]\right\} \\[1em]
\eta &\doteqdot \frac{hq}{r}\left\{-\frac{xy}{r^2}(K\sin\omega t_1+L\cos\omega t_1)+\frac{xy}{r^2}[(3-2K)\sin\omega t_2-2L\cos\omega t_2]\right\} \\[1em]
\zeta &\doteqdot \frac{hq}{r}\left\{-\frac{xz}{r^2}(K\sin\omega t_1+L\cos\omega t_1)+\frac{xz}{r^2}[(3-2K)\sin\omega t_2-2L\cos\omega t_2]\right\}
\end{aligned}
\right\}
$$

$$\ldots\ldots\ldots\ldots(66).$$

These equations represent two sets of simple harmonic waves, equivoluminal and irrotational, for which the wave-lengths are respectively $2\pi u/\omega$, $2\pi v/\omega$. The maximum displacements in the two sets at points of the cone of semi-vertical angle θ and axis OX, are respectively,

$$
\left.
\begin{aligned}
\text{(equivoluminal)}\quad & \sin\theta\,\frac{hq}{r}\,\sqrt{(K^2+L^2)}\,; \\[1.5em]
\text{(irrotational)}\quad & \cos\theta\,\frac{hq}{r}\,\sqrt{[(3-2K)^2+4L^2]}
\end{aligned}
\right\}\ \ldots\ldots\ldots(67).
$$

§ 27. The rate of transmission of energy outwards by a single set of waves of either species is equal, per period, to the sum of the kinetic and potential energies, or, which is the same, twice the whole kinetic energy, of the medium between two concentric spherical surfaces, of radii differing by a wave-length. Now the average kinetic energy throughout the wave-length in any part of the spherical shell is half the kinetic energy at the instant of maximum velocity. Hence the total energy transmitted *per period* is equal to the wave-length multiplied into the surface- Molar. integral, over the whole spherical surface, of the maximum kinetic energy at any point per unit of volume: and therefore the energy transmitted *per unit of time* is equal to the product of the propagational velocity into this surface-integral. Thus we find that the rates per unit of time of the transmission of energy by the two sets of waves, of amplitudes represented by (67), are respectively as follows:—

$$
\left.\begin{array}{ll}
\text{(equivoluminal)} & \dfrac{4\pi}{3}\,\rho h^2 q^2 \omega^2 \left(K^2 + L^2\right) u \\[2ex]
\text{(irrotational)} & \dfrac{2\pi}{3}\,\rho h^2 q^2 \omega^2 [(3 - 2K)^2 + 4L^2]\,v
\end{array}\right\} \quad \ldots\ldots(68).
$$

§ 28.   The sum of these two formulas is the whole rate of transmission of energy per unit of time, and must be equal to the average rate of doing work by the vibrating rigid globe upon the surrounding elastic solid.   Hence if w denote this rate, we must have

$$
w = \frac{2\pi\rho}{3}\, h^2 q^2 \omega^2 \left\{ 2u \left(K^2 + L^2\right) + v \left[(3 - 2K)^2 + 4L^2\right] \right\} \ldots(69).
$$

§ 29.   To verify this proposition, let us first find the resultant force, P, with which the globe presses and drags the elastic solid, and then the integral work which P does per period, and thence the average work per unit time.   Going back to § 9, we see that P is the surface-integral of X over the spherical surface of radius q.   Hence by the first of equations (25), which, in virtue of the equations

$$
k + \tfrac{4}{3}n = \rho v^2; \qquad n = \rho u^2 \cdot \ldots\ldots\ldots\ldots(70),
$$

we may write as follows:—

$$
X = \rho \left\{ C_2 \lambda^2 v^2 - [2\lambda^2 \left(A + C_2\right) + 2\left(2\lambda^2 + 1\right) B + \left(\lambda^2 + 1\right) C_1] u^2 \right\} \ldots(71),
$$

we find

$$
\begin{aligned}
P &= \frac{4\pi q^2 \rho}{3} \left\{ C_2 v^2 - 2 \left(A + 5B + 2C_1 + C_2\right) u^2 \right\} \\[2ex]
&= \frac{4\pi q^2 \rho}{3} \left\{ \frac{\dddot{\mathscr{K}}_2}{qv} + \frac{\ddot{\mathscr{K}}_2}{q^2} - 2 \left(\frac{\dddot{\mathscr{K}}_1}{qu} + \frac{\ddot{\mathscr{K}}_1}{q^2} \right) \right\} \\[2ex]
&= \frac{Q}{q} \left\{ 2 \left(v - u\right) \frac{\dddot{\mathscr{K}}_1}{qu^2} + 2 \left(v^2 - u^2\right) \frac{\ddot{\mathscr{K}}_1}{q^2 u^2} + 3v\dot{\mathscr{E}} + \frac{3v^2}{q}\, \mathscr{E} \right\} \quad \ldots\ldots(72),
\end{aligned}
$$

14—2

Molecular. where Q denotes $\frac{4}{3}\pi q^3 \rho$, being the mass which our rigid globe would have if its density were equal to that of the elastic solid. Hence by (65), for simple harmonic motion in period $2\pi/\omega$,

$$P = Q\omega^2 h \left\{ \left(2K \frac{u^2 - v^2}{q^2\omega^2} - 2L \frac{u-v}{q\omega} + 3 \frac{v^2}{q^2\omega^2} \right) \sin \omega t \right.$$
$$\left. + \left(2K \frac{u-v}{q\omega} + 2L \frac{u^2-v^2}{q^2\omega^2} + \frac{3v}{q\omega} \right) \cos \omega t \right\} \dots\dots(72');$$

or, substituting the values of K and L from (65), and denoting by D the common denominator,

$$P = \frac{Q\omega^2 h}{D}$$

$$\times \left\{ \left[\frac{9u^2v^2}{q^6\omega^6} (u^2 + 2v^2 + q^2\omega^2) - \frac{2}{q^4\omega^4} (u^2 - v^2)^2 + \frac{1}{q^2\omega^2} (4uv - v^2) \right] \sin \omega t \right.$$

$$\left. + \left[\frac{2u}{q\omega} \left(\frac{9v^4}{q^4\omega^4} + \frac{3v^2}{q^2\omega^2} + 1 \right) + \frac{v}{q\omega} \left(\frac{9u^4}{q^4\omega^4} + \frac{3u^2}{q^2\omega^2} + 1 \right) \right] \cos \omega t \right\}$$

$$\dots\dots\dots(72'').$$

This may be written for brevity

$$P = h \left(a \sin \omega t + b \cos \omega t \right) \dots\dots\dots\dots(72''').$$

Finally for w we have, denoting the period by τ,

$$w = \frac{1}{\tau} \int_0^\tau dt P \dot{\mathscr{E}}(t) = h^2 b\omega \frac{1}{\tau} \int_0^\tau dt \cos^2 \omega t = \frac{1}{2} h^2 b\omega \dots\dots\dots(73)$$

$$= \frac{Q\omega^2 h^2}{2D} \left[\frac{2u}{q} \left(\frac{9v^4}{q^4\omega^4} + \frac{3v^2}{q^2\omega^2} + 1 \right) + \frac{v}{q} \left(\frac{9u^4}{q^4\omega^4} + \frac{3u^2}{q^2\omega^2} + 1 \right) \right] \dots(73').$$

§ 30. To verify the agreement of this direct formula for the work done, with (69) which expresses the effect produced in waves travelling outwards at great distances from the centre, is a very long algebraical process, with K and L in (69) given by (65). But it becomes very simple by the aid of the following modified formulas for $\mathscr{I}_1(t)$ and $\mathscr{I}_2(t)$, which are also useful for other purposes. From (48), (49), and (50), by eliminating $\mathscr{I}_1$, we get an equation for $\mathscr{I}_2$ similar to (49), viz.,

$$\left[\partial^2 + \frac{1}{q} (u + 2v) \partial + \frac{1}{q^2} (u^2 + 2v^2) \right] \mathscr{I}_2 = \mathscr{H}(t),$$

where

$$\mathscr{H}(t) = qv^2 \left(1 + \frac{3u}{q} \partial^{-1} + \frac{3u^2}{q^2} \partial^{-2} \right) \mathscr{E}(t)$$

$$\left. \right\} \dots(74).$$

We may now write (50) in the form

$$\mathscr{G}(t) = hqu^2 G \sin (\omega t + \alpha),$$

and similarly

$$\mathscr{H}(t) = hqv^2 H \sin (\omega t - \beta) \qquad \cdots\cdots\cdots\cdots(75),$$

where the values of G, H, α, β, are given by the following equations :—

$$G \cos \alpha = \frac{3v^2}{q^2\omega^2} - 1 ; \quad G \sin \alpha = \frac{3v}{q\omega}$$

$$H \cos \beta = 1 - \frac{3u^2}{q^2\omega^2} ; \quad H \sin \beta = \frac{3u}{q\omega} \qquad \cdots\cdots\cdots(76).$$

From the second of equations (53) we therefore have

$$\mathscr{K}_1(t) = hqu^2 G \, \frac{M \sin (\omega t + \alpha) - N \cos (\omega t + \alpha)}{M^2 + N^2}$$

and

$$\mathscr{K}_2(t) = hqv^2 H \, \frac{M \sin (\omega t - \beta) - N \cos (\omega t - \beta)}{M^2 + N^2} \qquad \cdots\cdots(77),$$

where M^2 and N^2 denote the terms of the denominator of the third of equations (53). By the same method of investigation as that which gave us (69), we now find for the sum of the rates of transmission of energy

$$w = \frac{2\pi\rho}{3} \, h^2 q^2 \omega^2 \, \frac{2uG^2 + vH^2}{M^2 + N^2} \qquad \cdots\cdots\cdots\cdots(78).$$

Substituting the values of G, H, M, N, and introducing the notation Q, we obtain

$$w = \tfrac{1}{2} Q\omega^2 h^2 \, \frac{\dfrac{2u}{q}\left(\dfrac{9v^4}{q^4\omega^4} + \dfrac{3v^2}{q^2\omega^2} + 1\right) + \dfrac{v}{q}\left(\dfrac{9u^4}{q^4\omega^4} + \dfrac{3u^2}{q^2\omega^2} + 1\right)}{\left[\dfrac{1}{q^2\omega^2}(u^2 + 2v^2) - 1\right]^2 + \dfrac{1}{q^2\omega^2}(u + 2v)^2} \qquad \cdots\cdots(79).$$

This agrees with the value of w given by (73′); and thus the verification is complete.

§ 31. In (78) the numerator of the last factor shows the parts due to the equivoluminal and irrotational waves respectively. Denoting by J the ratio of the energy of the equivoluminal wave to that of the irrotational, we have

$$J = \frac{2uG^2}{vH^2} = \frac{2u\left(\dfrac{9v^4}{q^4\omega^4} + \dfrac{3v^2}{q^2\omega^2} + 1\right)}{v\left(\dfrac{9u^4}{q^4\omega^4} + \dfrac{3u^2}{q^2\omega^2} + 1\right)} \qquad \cdots\cdots\cdots\cdots(80).$$

§ 32.　Consider the following four cases :—

(a)　$q\omega$ very large in comparison with the larger of u and v.

$$J \doteqdot \frac{2u}{v}.$$

(b)　$q\omega = v$.

$$J = \frac{26}{\dfrac{v}{u} + \dfrac{3u}{v} + \dfrac{9u^3}{v^3}}.$$

(c)　$q\omega = u$.

$$J = \frac{2}{13}\left(\frac{u}{v} + \frac{3v}{u} + \frac{9v^3}{u^3}\right).$$

(d)　$q\omega$ very small in comparison with the smaller of u and v.

$$J \doteqdot \frac{2v^3}{u^3}.$$

$$\left.\begin{array}{l}\\ \\ \\ \\ \\ \\ \\ \\ \\ \\ \\ \\ \\ \\ \end{array}\right\} \dots(81).$$

If $v = \infty$, cases (a) and (b) cannot occur; and in cases (c), (d) we see by (81) that $J = \infty$; that is to say, the whole energy is carried away by the equivoluminal waves. If v is very small in comparison with u, we find that although J is infinite in cases (a) and (c), it is zero in cases (b) and (d). This to my mind utterly disproves my old hypothesis* of a very small velocity for irrotational wave-motion in the undulatory theory of light.

§ 33.　Let us now work out some examples such as that suggested in an addition of March 6, 1899, to this Lecture (p. 188), but with the simplification of assuming a rigid massless spherical lining for the cavity, which for brevity I shall call the sheath.　But first let us work out in general the problem of finding what force in simple proportion to velocity must be applied to a mass m mounted on massless springs as described in p. 145 above, to keep the sheath vibrating in simple harmonic motion $h \sin \omega t$, and therefore to do the work of sending out the two sets of waves with which we have been concerned.　Let γ be the required force per unit of velocity of m; so that $\gamma \dot{e}$ is the working force that must be applied to m, at any time when e is its displacement from its mean position.　Now the springs, which must act on the sheath with the force P of (72') above, must react with an equal force on m because they are massless;

* " On the Reflexion and Refraction of Light," *Phil. Mag.* 1888, 2nd half year.

so that the equation of motion of m is $\qquad$

$$m \frac{d^2 e}{dt^2} = -P + \gamma \frac{de}{dt} \dots\dots\dots\dots(82).$$

And, by the law of elastic action of the springs, we have

$$P = c \,(e - h \sin \omega t) \dots\dots\dots\dots(83),$$

where c denotes what I call the "stiffness" of the spring-system.

§ 34. For e, (83) and (72''') give

$$e = h \left[\left(1 + \frac{a}{c}\right) \sin \omega t + \frac{b}{c} \cos \omega t \right] \dots\dots\dots(84);$$

and with this in (82) we find

$$m\omega^2 \left[\left(1 + \frac{a}{c}\right) \sin \omega t + \frac{b}{c} \cos \omega t \right]$$

$$= \left(a + \gamma\omega \frac{b}{c}\right) \sin \omega t + \left[b - \gamma\omega \left(1 + \frac{a}{c}\right)\right] \sin \omega t,$$

which requires that

$$\left(1 + \frac{a}{c}\right) m\omega^2 = a + \frac{b}{c} \gamma\omega, \text{ and } \frac{b}{c} m\omega^2 = b - \left(1 + \frac{a}{c}\right) \gamma\omega \,;$$

by which, solved for two unknown quantities, $\gamma\omega$ and $m\omega^2$, we find

$$\gamma\omega = \frac{bc^2}{(a+c)^2 + b^2} \dots\dots\dots\dots(85),$$

and

$$m\omega^2 = \frac{c\,[a\,(a+c) + b^2]}{(a+c)^2 + b^2} \dots\dots\dots(86).$$

If we suppose ω and c known, these equations, with (72''), (72''') for a and b, tell what m/Q must be in order that the force applied to maintain the periodic motion of the sheath shall vary in simple proportion to the velocity; and they give γ, the magnitude of this force per unit velocity.

§ 35. If we denote by E the maximum kinetic energy of m, we find immediately from (84),

$$E = \tfrac{1}{2} h^2 m\omega^2 \left[\left(1 + \frac{a}{c}\right)^2 + \left(\frac{b}{c}\right)^2 \right] \dots\dots\dots(87).$$

And by (73) we have, for the work per period done on the sheath by P,

$$\tau w = \tfrac{1}{2} \tau h^2 \omega b \dots\dots\dots\dots(88).$$

Molecular. This ought to agree with the work done by $\gamma \dot{e}$ per period, being

$$\int_0^\tau dt\dot{e}\,\gamma e, \text{ which, by (84), is } \tfrac{1}{2}\tau\gamma\omega^2h^2\left[\left(1+\frac{a}{c}\right)^2+\left(\frac{b}{c}\right)^2\right]\ \ldots(89).$$

The agreement between (89) and (88) is secured by (85).

§ 36.   By (87), (89), and $\tau\omega = 2\pi$; and by (86), (85), we find

$$\frac{E}{\tau w} = \frac{m}{\tau\gamma} = \frac{m\omega^2}{2\pi\gamma\omega} = \frac{a(a+c)+b^2}{2\pi bc}\ \ldots\ldots\ldots\ldots(90),$$

which, as we shall see, is a very important result, in respect to storage of energy in vibrators for originating trains of waves.

§ 37.   Remark now that a, b, c are each of the dimensions of a "longitudinal stiffness," that is to say Force $\div$ Length, or Mass $\div$ (Time)^2; and for clearness write out the full expressions for a and b from (72″) and (72‴) as follows;

$$\left.\begin{array}{l}
a = Q\omega^2\dfrac{\dfrac{9u^2v^2}{q^4\omega^4}\left(\dfrac{u^2+2v^2}{q^2\omega^2}+1\right)-2\left(\dfrac{u^2-v^2}{q^2\omega^2}\right)^2+\dfrac{4uv-v^2}{q^2\omega^2}}{\left(\dfrac{u^2+2v^2}{q^2\omega^2}-1\right)^2+\left(\dfrac{u+2v}{q\omega}\right)^2}\\[3em]
b = Q\omega^2\dfrac{\dfrac{2u}{q\omega}\left(\dfrac{9v^4}{q^4\omega^4}+\dfrac{3v^2}{q^2\omega^2}+1\right)+\dfrac{v}{q\omega}\left(\dfrac{9u^4}{q^4\omega^4}+\dfrac{3u^2}{q^2\omega^2}+1\right)}{\left(\dfrac{u^2+2v^2}{q^2\omega^2}-1\right)^2+\left(\dfrac{u+2v}{q\omega}\right)^2}
\end{array}\right\}\ \ldots(91).$$

§ 38.   Let $q\omega$ be very large in comparison with the larger of u or v (Case I. of § 32).   We have

$$a \fallingdotseq Q\omega^2\frac{4uv-v^2}{q^2\omega^2};\quad b \fallingdotseq Q\omega^2\frac{v}{q\omega};$$

therefore $\dfrac{a}{b} \fallingdotseq 0;\quad \dfrac{E}{\tau w} \fallingdotseq \dfrac{b}{2\pi c}\ \ldots\ldots\ldots\ldots\ldots\ldots(92).$

This case is interesting in connection with the dynamics of waves in an elastic solid, but not as yet apparently so in respect to light.

§ 39.   Let $q\omega$ be very small in comparison with the smaller of u or v (Case IV. of § 32).   We have

$$\left.\begin{array}{l}
a \fallingdotseq Q\omega^2\dfrac{9u^2/q^2\omega^2}{u^2/v^2+2};\quad b \fallingdotseq Q\omega^2\dfrac{9\,(u^3/v^3+2)\,u/q\omega}{(u^2/v^2+2)^2}\\[1.5em]
\text{therefore } \dfrac{a}{b} \fallingdotseq \dfrac{(u^2/v^2+2)\,u/q\omega}{u^3/v^3+2};\quad \dfrac{E}{\tau w} \fallingdotseq \dfrac{a(a+c)}{2\pi bc};
\end{array}\right\}\ \ldots\ldots(93).$$

This case is supremely important in respect to molecular sources Molar. of light.

§ 40. Let c be very small in comparison with $a + b^2/a$. We have

$$\frac{E}{\tau w} \doteqdot \frac{a^2/b + b}{2\pi c}; \quad \gamma\omega \doteqdot \frac{bc^2}{a^2 + b^2}; \quad m\omega^2 \doteqdot c \quad\ldots\ldots(94).$$

§ 41. Let $c = \infty$. We have

$$\frac{E}{\tau w} = \frac{a}{2\pi b}; \quad \gamma\omega = b; \quad m\omega^2 = a\ldots\ldots\ldots(95).$$

This is the simple case of a rigid globe of mass m embedded firmly in the elastic solid, and no other elasticity than that of the solid around it brought into play. It is interesting in respect to Stokes' and Rayleigh's theory of the blue sky.

§ 42. Let $v = \infty$. We have

$$a = Q\omega^2\left(\frac{9u^2}{2q^2\omega^2} - \frac{1}{2}\right); \quad b = Q\omega^2\frac{9u}{2q\omega}\ldots\ldots\ldots(96).$$

This case is of supreme interest and importance in respect to the Dynamical Theory of Light.

§ 43. Take now the particular example suggested in the addition of March 6, 1899 (p. 188 above), which is specially interesting as belonging to cases intermediate between those of § 38 and § 39; a vast mass of granite with a spherical hollow of ten centimetres diameter acted on by an internal simple harmonic vibrator of $1006\frac{1}{2}$ periods per second $\left(\text{being } 1000 \sqrt{\dfrac{10}{\pi^2}}\right)$. This makes $\omega^2 = 40 \times 10^6$, $q^2\omega^2 = 10^9$, $\omega = 6324$, $q\omega = 31620$. Now the velocities of the equivoluminal and the irrotational waves in granite[*] are about 2·2, and 4 kilometres per second; so we have $u = 2\cdot2 \times 10^5$, $v = 4 \times 10^5$. Hence, and by (80) and (91),

$$\left.\begin{array}{l}
\dfrac{u}{q\omega} = 6{\cdot}957; \quad \dfrac{u^2}{q^2\omega^2} = 48{\cdot}4; \quad \dfrac{u^4}{q^4\omega^4} = 2342{\cdot}56; \\[2mm]
\dfrac{v}{q\omega} = 12{\cdot}649; \quad \dfrac{v^2}{q^2\omega^2} = 160; \quad \dfrac{v^4}{q^4\omega^4} = 25600; \\[2mm]
J = 11{\cdot}96; \quad a = 189{\cdot}1 \times Q\omega^2; \quad b = 25{\cdot}59 \times Q\omega^2; \quad \dfrac{a}{b} = 7{\cdot}390
\end{array}\right\}\ldots(97).$$

* Gray and Milne, *Phil. Mag.*, Nov. 1881.

Molecular. And by (85), (86), (90); with for brevity $c = sQ\omega^2$,

$$\left.\begin{aligned}
\gamma\omega &= \frac{25{\cdot}59 \,.\, s^2}{(189{\cdot}1+s)^2+654{\cdot}8} \times Q\omega^2 ; \quad \frac{m}{Q} = \frac{[189{\cdot}1\,(189{\cdot}1+s)+654{\cdot}8]\,s}{(189{\cdot}1+s)^2+654{\cdot}8} \\
\frac{E}{\tau w} &= \frac{7{\cdot}390\,(189{\cdot}1+s)+25{\cdot}59}{2\pi s} = 1{\cdot}176 + \frac{226}{s}
\end{aligned}\right\}$$

$$......(98).$$

§ 44. As a first sub-case take (§ 41) $c = \infty$: we find by (95), (97), $m = 189{\cdot}1Q$; and $E/\tau w = 1{\cdot}176$. These numbers show that the kinetic energy of m at each instant of transit through its mean position, supplies only $1{\cdot}176$ of the energy carried away in the period by the outward travelling waves; though its mass is as much as 189 times that of granite enough to fill the hollow. Hence we see that if the moving force $\gamma\dot{e}$ were stopped the motion of m would subside very quickly and in the course of six or seven times τ it would be nearly annulled. The not very simple law of the subsidence presents an *extremely interesting* problem which is easily enough worked out thoroughly according to the methods suitable for § 25 above. Meantime we confine ourselves to cases in which $E/\tau w$ is very large.

§ 45. Such a case we have, under §§ 39, 41, if instead of $1006\frac{1}{2}$ periods per second we have only $1{\cdot}0065$; which makes $q^2\omega^2 = 1000$; $q\omega = 10\sqrt{10} = 31{\cdot}620$; and, by (93), (95) with still our values of u and v for granite,

$$a = 1{\cdot}892 \times 10^8 \times Q\omega^2 ; \quad \frac{a}{b} = 7394 ; \quad \frac{E}{\tau w} = 1177......(99).$$

Hence the kinetic energy of m in passing through the middle of its range is nearly 1200 times the work required to maintain its vibration at the rate of $1{\cdot}0065$ periods per second: and the value found for a, used in (95), shows that m, supposed to be a rigid globe filling the hollow, must be 189 million times as dense as the surrounding granite, in order that this period of vibration can be maintained by a force in simple proportion to velocity. (See § 40.) It is now easy to see that if the maintaining force is stopped, the rigid globe will go on vibrating in very nearly the same period, but subsidentially according to the law

$$h\epsilon^{\frac{-t/\tau}{2354}} \sin\frac{2\pi t}{\tau} \quad ....(100);$$

and there will be the corresponding subsidence in the amplitudes Molecular. of the two sets of waves travelling outwards in all directions at great distances from the origin, when, according to the propagational velocities u, v, the effects of the stoppage of the maintaining force reach any particular distance.

It is quite an interesting mathematical problem, suggested at the end of § 44, to fully determine the motion in all future time, when m is left with no applied force, after any given initial conditions, with any value, large or small, of m/Q.

§ 46. Returning now to the maintenance of vibrations at the rate of $1006\frac{1}{2}$ periods per second; and u, v for granite; and $q = 5$ cm., all as in § 43; see (98) and remark that, to make $E/\tau w$ very large, s must be a very small fractional numeric; and this makes $m/Q \doteqdot s$. To take a vibrator not differing greatly in result from the violin string suggested in the addition of March 6 (p. 188), let m be a little ball of granite of $\frac{1}{2}$ cm. diameter. This makes $m = Q/8000$, and therefore (§ 40) $s = 1/8000$. Hence by (98), $E/\tau w = 1808001$, from which, with what we know of wave-motion, we infer that if m be projected with any given velocity, V, from its position of equilibrium, it will for ever after vibrate, with amplitude diminishing according to the formula

$$C\epsilon^{\frac{-t/\tau}{3616002}} \sin \frac{2\pi t}{\tau} \quad \ldots\ldots\ldots\ldots\ldots\ldots(101).$$

Thus during 3,616,002 periods the range of m will be reduced in the ratio of ϵ to 1 (say approximately $2\frac{3}{4}$ to 1), by giving away its energy to be transmitted outwards by the two species of waves, of which, according to J of (97), the equivoluminal takes twelve times as much as the irrotational.]

LECTURE XV.

Molecular.   RETURNING to our model, we shall have in a short time a state
of things not very different from simple harmonic motion, if we
get up the motion very gradually. We have now an exciting
vibration of shorter period than the shortest of the natural
periods. We must keep the vibrator going through a uniform
range. We are not to augment it, and it will be a good thing
to place something here to mark its range. [This done.] Keep it
going long enough and we shall see a state of vibration in which
each bar will be going in the opposite direction to its neighbour. If
we keep it going long enough we certainly will have the simple
harmonic motion; and if this period is smaller than the smallest
of the three natural periods, we shall, as we know, have the alternate
bars going in opposite directions. Now you see a longer-period
vibration of the largest mass superimposed on the simple harmonic motion we are waiting for. I will try to help towards
that condition of affairs by resisting the vibration of the top
particle. In fact, that particle will have exceedingly little motion
in the proper state of things (that is to say, when the motion is
simple harmonic throughout), and it will be moving, so far as it
has motion at all, in an opposite direction to the particle immediately below it. It is nearly quit of that superimposed
motion now. We cannot give a great deal of time to this, but
I think we may find it a little interesting as illustrating
dynamical principles. Prof. Mendenhall is here acting the part
of an escapement in keeping the vibrator to its constant range.
We cannot get quit of the slow vibration of the particle. A
touch upon it in the right place may do it. A very slight touch

is more than enough. I have set it the wrong way. Now we Molecular. have got quit of that vibration and you see no sensible motion of m_3 at all. These two, (m_2, m_1), are going in opposite directions, and the lower one in opposite direction to the exciter. Therefore this is a shorter vibration than the shortest natural period. Now I set it to agree with the shortest of the periods, the first critical position. If we get time in the second lecture to-day, I am going to work upon this a little to try to get a definite example illustrating a particle of sodium. Before we enter upon any hard mathematics, let us look at this a little, and help ourselves to think of the thing. What I am doing now is very gradually getting up the oscillation. I am doing to that system exactly what is done to the sodium molecule, for example, when sodium light is transmitted through sodium vapour. We may feel quite certain, however, that the energy of vibration of the sodium molecule goes on increasing during the passage through the medium of at least two-hundred thousand waves, instead of two dozen at the most perhaps that I am taking to get up this oscillation. But just note the enormous vibration we have here, and contrast it with the state of things that we had just before. The upper particle is in motion now and is performing a vibration in the same period and phase as the lower particle, only through comparatively a very small range. The second particle, I am afraid, will overstrain the wire. (By hanging up a watch, bifilarly, so that the period of bifilar suspension approximately agrees with the balance wheel, you get likewise a state of wild vibration. But if you perform such experiments with a watch, you are apt to damage it.) This, which you see now, is a most magnificent contrast to the previous state of things when the period of the exciter was very far from agreeing with any of the fundamental periods.

We will now return to our molar subject, the elastic solid. You will see a note in the paper of yesterday to which I have referred, stating that the thlipsinomic method is more convenient than the tasinomic for dealing with incompressibility, and in point of fact it is so.

I explained to you yesterday Rankine's nomenclature of thlipsinomic and tasinomic coefficients, according to which, when the six stress-components are expressed in terms of strain-components, the coefficients are called tasinomic; and when the

Molar. strain-components are expressed in terms of the stress-components, the coefficients are called thlipsinomic. [Thus, going back to Lecture XI. (p. 132 above), we see, in the six equations (1), 36 tasinomic coefficients expressing the six stress-components as linear functions of the six strain-components: and we see, in virtue of 15 equalities among the 36 coefficients, just 21 independent values, being Green's celebrated 21 coefficients. Use now those six equations to determine the six quantities e, f, g, a, b, c in terms of P, Q, R, S, T, U. Thus we find

$$
\left.
\begin{aligned}
e &= (PP)\,P + (PQ)\,Q + (PR)\,R + (PS)\,S + (PT)\,T + (PU)\,U \\
f &= (QP)\,P + (QQ)\,Q + (QR)\,R + (QS)\,S + (QT)\,T + (QU)\,U \\
g &= (RP)\,P + (RQ)\,Q + (RR)\,R + (RS)\,S + (RT)\,T + (RU)\,U \\
a &= (SP)\,P + (SQ)\,Q + (SR)\,R + (SS)\,S + (ST)\,T + (SU)\,U \\
b &= (TP)\,P + (TQ)\,Q + (TR)\,R + (TS)\,S + (TT)\,T + (TU)\,U \\
c &= (UP)\,P + (UQ)\,Q + (UR)\,R + (US)\,S + (UT)\,T + (UU)\,U
\end{aligned}
\right\}
$$

$$\dots\dots\dots\dots(1),$$

where (PP), (PQ), &c., denote algebraic functions of 11, 12, 22, &c., found by the process of elimination. This process, in virtue of the 15 equalities $12 = 21$, &c., gives $(PQ) = (QP)$, &c.; 15 equalities in all. The 21 independent coefficients $(PP), (PQ)$, &c., thus found, are what Rankine called the thlipsinomic coefficients. Taking now from Lecture II., p. 24,

$$E = \tfrac{1}{2}\,(Pe + Qf + Rg + Sa + Tb + Uc) \dots\dots (2),$$

and eliminating P, Q, R from this formula by the equations (1) of Lecture XI., (p. 132), we find the tasinomic quadratic function for the energy which we had in Lecture II. (p. 23). And eliminating e, f, g, a, b, c from it by our present six equations, we find the corresponding thlipsinomic formula for the energy with 21 independent coefficients $(PP), (PQ)$, &c.*] In a certain sense, these coefficients, both tasinomic and thlipsinomic, may be all called moduluses of elasticity, inasmuch as each of them is a definite numerical measurement of a definite elastic quality. I have, however, specially defined a modulus as a stress divided by a strain, following the analogy of Young's modulus. If we adhere to this definition, then the tasinomic coefficients are moduluses, and the thlipsinomic coefficients are reciprocals of moduluses.

---

* Compare *Thomson and Tait*, § 673, (12)—(20).

In the Lecture notes in your hands for today you see the Molar. thlipsinomic discussion of the question of compressibility or incompressibility; which is much simpler than our tasinomic discussion of the same subject in which we failed yesterday. You see that if the dilatation, $e + f + g$, be denoted by δ, you have

$$\delta = [(PP) + (QP) + (RP)]\, P + [(PQ) + (QQ) + (RQ)]\, Q$$
$$+ [(PR) + (QR) + (RR)]\, R + [(PS) + (QS) + (RS)]\, S$$
$$+ [(PT) + (QT) + (RT)]\, T + [(PU) + (QU) + (RU)]\, U \ldots \ldots (3).$$

Thus if P is the sole stress, a dilatation $[(PP) + (QP) + (RP)]\, P$ is produced; and if S is the sole stress, a dilatation

$$[(PS) + (QS) + (RS)]\, S$$

is produced. We see therefore that S, a kind of stress which in an isotropic solid would produce merely distortion, may produce condensation or rarefaction in an æolotropic solid. The coefficients of P, Q, R, S, T, U in our equation for δ may be called compressibilities. Their reciprocals are (according to my definition of a modulus) moduluses for compressibility. In an isotropic solid, each of the last three coefficients vanishes; and the reciprocal of each of the others is three times what I have denoted by k (Lecture II., p. 25), and called the compressibility-modulus or the bulk-modulus, being P/δ, where P denotes equal pull, or negative pressure, in all directions.

An æolotropic solid is incompressible if, and is compressible unless, each of the six coefficients in our formula for δ vanishes. That is to say, it is necessary and sufficient for incompressibility that

$$\left.\begin{array}{l}(PP) + (QP) + (RP) = 0 \\ (PQ) + (QQ) + (RQ) = 0 \\ (PR) + (QR) + (RR) = 0 \\ (PS) + (QS) + (RS) = 0 \\ (PT) + (QT) + (RT) = 0 \\ (PU) + (QU) + (RU) = 0\end{array}\right\} \ldots \ldots \ldots \ldots (4).$$

Thus we see that six equations among the 21 coefficients suffice to secure that there can be no condensational-rarefactional wave, or, what is the same thing, that, in every plane wave, the vibration must be rigorously in the plane of the wave-front; and therefore that Green was not right when, in proposing to confine himself

Molar.

"to the consideration of those media only in which the directions of the transverse vibrations shall always be *accurately* in the front of the wave," he said*, "This fundamental principle of Fresnel's theory gives fourteen relations between the twenty-one constants originally entering into our function." What Green really found and proved was fourteen relations ensuring, and required to ensure, that there can be plane waves with direction of vibration *accurately* in the plane of the wave, if there can also be condensational-rarefactional waves in the medium. What we have now found is that not fourteen, but only six, equations suffice to secure incompressibility and therefore to compel the direction of the vibration in any actual plane wave to be *accurately* in the plane of the wave.

[*March* 7, 1899.—I have only today found an interesting and instructive mode of dealing with those six equations of incompressibility, which I gave in this Baltimore Lecture of October 14, 1884. By the first three of them, eliminating (PP), (QQ), (RR), and by the other three, (PS), (QT), (RU) from the equation of energy, we find

$$2E = -\left[(QR)(Q-R)^2 + (RP)(R-P)^2 + (PQ)(P-Q)^2\right]$$
$$+ 2\left\{[(QU)U - (RT)T](Q-R) + [(RS)S - (PU)U](R-P)\right.$$
$$\left. + [(PT)T - (QS)S](P-Q)\right\}$$
$$+ (SS)S^2 + (TT)T^2 + (UU)U^2$$
$$+ 2\left[(TU)TU + (US)US + (ST)ST\right] \dotsc\dotsc (5).$$

In this, P, Q, R appear only in their differences, which is an interesting expression of the dynamical truth that if $P = Q = R$, they give no contribution to the potential energy.

The three differences $Q-R$, $R-P$, $P-Q$, are equivalent to only two independent variables; thus if we put $P-Q=V$ and $P-R=W$, we have $Q-R=W-V$, and the expression for E becomes a homogeneous quadratic function of the five independents V, W, S, T, U, with fifteen independent coefficients, which we may write as follows:—

$$E = \tfrac{1}{2}\{(VV)V^2 + (WW)W^2 + (SS)S^2 + (TT)T^2 + (UU)U^2\}$$
$$+ (VW)VW + (VS)VS + (VT)VT + (VU)VU$$
$$+ (WS)WS + (WT)WT + (WU)WU$$
$$+ (ST)ST + (SU)SU + (TU)TU \dotsc\dotsc(6).$$

* Green, *Collected Papers*, p. 293.

The differential coefficients of this with reference to S, T, U Molar. are of course as before, a, b, c. Denote now by h and i its differential coefficients with reference to V and W. Thus we find

$$\left.\begin{aligned}
h &= (VV)\,V + (VW)\,W + (VS)\,S + (VT)\,T + (VU)\,U \\
i &= (WV)\,V + (WW)\,W + (WS)\,S + (WT)\,T + (WU)\,U \\
a &= (SV)\,V + (SW)\,W + (SS)\,S + (ST)\,T + (SU)\,U \\
b &= (TV)\,V + (TW)\,W + (TS)\,S + (TT)\,T + (TU)\,U \\
c &= (UV)\,V + (UW)\,W + (US)\,S + (UT)\,T + (UU)\,U
\end{aligned}\right\}\ldots(7);$$

and we have

$$E = \tfrac{1}{2}(hV + iW + aS + bT + cU) \ldots\ldots\ldots (8).$$

The dynamical interpretation shows that h must represent $-f$, and i must represent $-g$, when V and W represent $P - Q$, and $P - R$.

Solving the five linear equations (7) for V, W, S, T, U, we find the tasinomic expressions for the five stress-components in terms of the five strain-components, which we may write as follows :—

$$\left.\begin{aligned}
V &= (hh)\,h + (hi)\,i + (ha)\,a + (hb)\,b + (hc)\,c \\
W &= (ih)\,h + (ii)\,i + (ia)\,a + (ib)\,b + (ic)\,c \\
S &= (ah)\,h + (ai)\,i + (aa)\,a + (ab)\,b + (ac)\,c \\
T &= (bh)\,h + (bi)\,i + (ba)\,a + (bb)\,b + (bc)\,c \\
U &= (ch)\,h + (ci)\,i + (ca)\,a + (cb)\,b + (cc)\,c
\end{aligned}\right\}\ldots\ldots (9).$$

The algebraic process shows us that

$$(hi) = (ih);\ (ha) = (ah);\ \&c.\ldots\ldots\ldots\ldots(10);$$

so that we have now found the 15 independent tasinomic coefficients from the 15 thlipsinomic. Lastly, eliminating by (9) V, W, S, T, U from (8), we find the tasinomic quadratic expressing the energy.]

As I said in the first lecture, one fundamental difficulty is quite refractory indeed. In the wave-theory of light the velocity of the wave ought to depend on the plane of distortion. If you compare the details of motion in the wave-surface* worked out for an incompressible æolotropic elastic solid, with equalities enough among the coefficients to annul all skewnesses, you will see that it agrees exactly with Fresnel's wave-surface but that instead of the direction of the line of vibration of the particles as

* Lecture XII., p. 137; Lecture XIII., p. 175.

in Fresnel's construction we have the normal to the plane of distortion as the direction on which the propagational velocity depends.

I see no way of getting over the difficulty that the return forces in an elastic solid—the forces on which the vibration depends—are dependent on the strain experienced by the solid and on that alone. I have never felt satisfied with the ingenious method by which Green got over it. Stokes quotes in his report on Double Refraction, page 265 (British Association 1862): " In " his paper on Reflection, Green had adopted the supposition of " Fresnel that the vibrations are perpendicular to the plane of " polarization. He was naturally led to examine whether the laws " of double refraction could be explained on this hypothesis. " When the medium in its undisturbed state is exposed to pressure " differing in different directions, six additional constants are intro- " duced into the function ϕ, or three in the case of the existence of " planes of symmetry to which the medium is referred. For waves " perpendicular to the principal axes, the directions of vibration " and squared velocities of propagation are as follows :—

Wave normal		x	y	z
Direction of vibration {	x	$G+A$	$N+B$	$M+C$
	y	$N+A$	$H+B$	$L+C$
	z	$M+A$	$L+B$	$I+C$

" Green assumes, in accordance with Fresnel's theory, and with " observation if the vibrations in polarized light are supposed " perpendicular to the plane of polarization, that for waves " perpendicular to any two of the principal axes, and propagated " by vibrations in the direction of the third axis, the velocity of " propagation is the same."

Let us see what this statement means before considering whether it may be verified, as Green supposes, by the introduction of " extraneous pressure." Consider waves having their fronts parallel to the sides N and W (North and West) of this box, which are perpendicular to two of the three principal axes of the crystal, and such having its vibrations in the direction of the third axis (up and down). Take first the wave that is propagated south as I hold

the box.   There is the plane of the wave (N).   The vibration up Molar.
and down with N held fixed will give a shear like that marked 1,

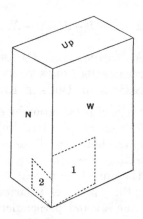

in which a square becomes a rhombic figure.   That represents the
strain in the solid corresponding to this first state of motion.   Similarly the wave propagated in the eastward direction will give rise
to a shear of this kind marked 2, the vibration still being upward.
The assumption is that one of these sets of waves is propagated
at the same speed as the other.   That is to say, the waves which
have their shear in this west plane have the same velocity as
the waves which have their shear in this north plane.   The
essence of our elastic solid is three different rigidities, one for
shearing in this plane W, one for shearing in this plane N, and
one for shearing in the other principal plane (the horizontal plane
of our box).   The incongruous assumption is that the velocities of
propagation do not depend on the planes of the shearing strain,
and do depend, simply and solely, on the direction of the vibration.

The introduction by Green (in order to accomplish this) of
what he calls "extraneous force," which gives him three other
coefficients has always seemed to me of doubtful validity.   In
the little table above, taken from Stokes, L, M, N are the three
principal rigidities, the 44, 55, 66 of our own notation. A, B, C
are the effects of extraneous pressure.   The table gives the squared
velocities of propagation and waves of different wave-normals and
directions of vibration along the axes.   The principal diagonal refers
only to condensational waves, or waves in which the direction
of vibration coincides with the wave-normal.   Green's assumption

Molar. makes, for vibrations in the x-direction, $N + B = M + C$; which, with the two corresponding equations for vibrations in the directions y, z, gives

$$A - L = B - M = C - N.$$

[*March* 7, 1899. ADDITION. ON CAUCHY'S AND GREEN'S DOCTRINE OF EXTRANEOUS FORCE TO EXPLAIN DYNAMICALLY FRESNEL'S KINEMATICS OF DOUBLE REFRACTION*.

§ 1. Green's dynamics of polarization by reflexion, and Stokes' dynamics of the diffraction of polarized light, and Stokes' and Rayleigh's dynamics of the blue sky, all agree in, as seems to me, irrefragably, demonstrating Fresnel's original conclusion, that in plane polarized light the line of vibration is perpendicular to the plane of polarization; the "plane of polarization" being defined as the plane through the ray and perpendicular to the reflecting surface, when light is polarized by reflexion.

§ 2. Now when polarized light is transmitted through a crystal, and when rays in any one of the principal planes are examined, it is found that—

(1) A ray with its plane of polarization in the principal plane travels with the same speed, whatever be its direction (whence it is called the "ordinary ray" for that principal plane); and (2) A ray whose plane of polarization is perpendicular to the principal plane, and which is called the "extraordinary ray" of that plane, is transmitted with velocity differing for different directions, and having its maximum and minimum values in two mutually perpendicular directions of the ray.

§ 3. Hence and by § 1, the velocities of all rays having their vibrations *perpendicular* to one principal plane are the same; and the velocities of rays in a principal plane which have their directions of vibration in the same principal plane, differ according to the direction of the ray, and have maximum and minimum values for directions of the ray at right angles to one another. But in the laminar shearing or distortional motion of which the wave-motion of the light consists, the "plane of the shear†" (or "plane of the distortion," as it is sometimes called) is the plane through the

---

* Reprinted, with additions, from the *Proc. R. S. E.*, Vol. xv. 1887, p. 21, and *Phil. Mag.*, Vol. xxv. 1888, p. 116.

† Thomson and Tait's *Natural Philosophy*, § 171 (or *Elements*, § 150).

direction of the ray and the direction of vibration ; and therefore Molar.
it would be the *ordinary* ray that would have its line of vibration
in the principal plane, if the ether's difference of quality in
different directions were merely the æolotropy of an unstrained
elastic solid*. Hence ether in a crystal must have something
essentially different from mere intrinsic æolotropy; something
that can give different velocities of propagation to two rays, of
one of which the line of vibration and line of propagation coincide
respectively with the line of propagation and line of vibration
of the other.

§ 4. The difficulty of imagining what this something could
possibly be, and the utter failure of dynamics to account for double
refraction without it, have been generally felt to be the greatest
imperfection of optical theory.

It is true that ever since 1839 a suggested explanation has
been before the world; given independently by Cauchy and
Green, in what Stokes has called their "Second Theories of
Double Refraction," presented on the same day, the 20th of
May of that year, to the French Academy of Sciences and the
Cambridge Philosophical Society. Stokes, in his Report on Double
Refraction†, has given a perfectly clear account of this explanation. It has been but little noticed otherwise, and somehow it
has not been found generally acceptable ; perhaps, because of a
certain appearance of artificiality and arbitrariness of assumption
which might be supposed to discredit it. But whatever may have
been the reason or reasons which have caused it to be neglected
as it has been, and though it is undoubtedly faulty, both as given
by Cauchy and by Green, it contains what seems to me, in all
probability, the true principle of the explanation, and which is,
that the ether in a doubly refracting crystal is an elastic solid,
unequally pressed or unequally pulled in different directions, by
the unmoved ponderable matter.

§ 5. Cauchy's work on the wave-theory of light is complicated
throughout, and to some degree vitiated, by admission of the

---

* The elementary dynamics of elastic solids shows that on this supposition
there might be maximum and minimum velocities of propagation for rays in
directions at 45° to one another, but that the velocities *must essentially be equal for
every two directions at* 90° *to one another*, in the principal plane, when the line of
vibration is in this plane.

† *British Association Report*, 1862.

Molar.   Navier-Poisson false doctrine* that compressibility is calculable
theoretically from rigidity; a doctrine which Green sets aside,
rightly and conveniently, by simply assuming incompressibility.
In other respects Cauchy's and Green's "Second Theories of
Double Refraction," as Stokes calls them, are almost identical.
Each supposes ether in the crystal to be an intrinsically æolotropic
elastic solid, having its æolotropy modified in virtue of internal
pressure or pull, equal or unequal in different directions, produced
by and balanced by extraneous force.   Each is faulty in leaving
intrinsic rigidity-moduluses (coefficients) unaffected by the equilibrium-pressure, and in introducing three fresh terms, with
coefficients (A, B, C in Green's notation) to represent the whole
effect of the equilibrium-pressure.   This gives for the case of an
intrinsically isotropic solid, augmentation of virtual rigidity, and
therefore of wave-velocity, by equal pull† in all directions, and
diminution by equal positive pressure in all directions; which is
obviously wrong.   Thus definitively, pull in all directions outwards
perpendicular to the bounding surface equal per unit of area to
three times the intrinsic rigidity-modulus, would give quadrupled
virtual rigidity, and therefore doubled wave-velocity!   Positive
normal pressure inwards equal to the intrinsic rigidity-modulus
would annul the rigidity and the wave-velocity—that is to say,
would make a fluid of the solid.   And, on the other hand, negative pressure, or outward pull, on an incompressible liquid, would
give it virtual rigidity, and render it capable of transmitting
laminar waves!   It is obvious that abstract dynamics can show
for pressure or pull equal in all directions, no effect on any physical
property of an incompressible solid or fluid.

§ 6.   Again, pull or pressure *unequal in different directions*, on
an isotropic incompressible solid, would, according to Green's
formula (A) in p. 303 of his collected Mathematical Papers, cause
the velocity of a laminar wave to depend simply on the wave-
front, and to have maximum, minimax, and minimum velocities

---

* See Stokes, "On the Friction of Fluids in Motion and on the Equilibrium and
Motion of Elastic Solids," *Camb. Phil. Trans.*, 1845, §§ 19, 20; reprinted in Stokes'
Mathematical and Physical Papers, Vol. I. p. 123; or Thomson and Tait's *Natural
Philosophy*, §§ 684, 685; or *Elements*, §§ 655, 656.

† So little has been done towards interpreting the formulas of either writer that
it has not been hitherto noticed that positive values of Cauchy's G, H, I, or of
Green's A, B, C, signify pulls, and negative values signify pressures.

for wave-fronts perpendicular respectively to the directions of Molecular. maximum pull, minimax pull, and minimum pull; and would make the wave-surface a simple ellipsoid! This, which would be precisely the case of foam stretched unequally in different directions, seemed to me a very interesting and important result, until (as shown in § 19 below) I found it to be not true.

§ 7. To understand fully the stress-theory of double refraction, we may help ourselves effectively by working out directly and thoroughly (as is obviously to be done by abstract dynamics) the problem of § 6, as follows:—Suppose the solid, isotropic when unstrained, to become strained by pressure so applied to its boundary as to produce, throughout the interior, homogeneous strain according to the following specification :—

The coordinates of any point M of the mass which were ξ, η, ζ when there was no strain, become in the strained solid

$$\xi \sqrt{\alpha}, \quad \eta \sqrt{\beta}, \quad \zeta \sqrt{\gamma} \dots\dots\dots\dots(1);$$

$\sqrt{\alpha}$, $\sqrt{\beta}$, $\sqrt{\gamma}$, or the "Principal Elongations*," being the same whatever point M of the solid we choose. Because of incompressibility we have

$$\alpha\beta\gamma = 1 \dots\dots\dots\dots\dots\dots(2).$$

For brevity, we shall designate as (α, β, γ) the strained condition thus defined.

§ 8. As a purely kinematic preliminary, let it be required to find the principal strain-ratios when the solid, already strained according to (1), (2), is further strained by a uniform shear, σ, specified as follows; in terms of x, y, z, the coordinates of still the same particle, M, of the solid and other notation, as explained below :—

$$\left.\begin{array}{l} x = \xi \sqrt{\alpha} + \sigma p l \\ y = \eta \sqrt{\beta} + \sigma p m \\ z = \zeta \sqrt{\gamma} + \sigma p n \end{array}\right\} \dots\dots\dots\dots\dots(3),$$

where $\qquad p = OP = \lambda\xi \sqrt{\alpha} + \mu\eta \sqrt{\beta} + \nu\zeta \sqrt{\gamma} \dots\dots\dots\dots(4)$,

with $\qquad l^2 + m^2 + n^2 = 1, \quad \lambda^2 + \mu^2 + \nu^2 = 1 \dots\dots\dots\dots(5)$,

and $\qquad l\lambda + m\mu + n\nu = 0 \dots\dots\dots\dots\dots(6)$;

* See chap. iv. of "Mathematical Theory of Elasticity" (W. Thomson), *Trans. Roy. Soc. Lond.* 1856, reprinted in Vol. III. of *Mathematical and Physical Papers*, now on the point of being published, or Thomson and Tait's *Natural Philosophy*, §§ 160, 164, or *Elements*, §§ 141, 158.

λ, μ, ν denoting the direction-cosines of OP, the normal to the shearing planes, and l, m, n the direction-cosines of shearing displacement. The principal axes of the resultant strains are the directions of OM in which it is maximum or minimum, subject to the condition

$$\xi^2 + \eta^2 + \zeta^2 = 1 \quad \dots\dots\dots\dots\dots\dots(7);$$

and its maximum, minimax, and minimum values are the three required strain-ratios. Now we have

$$OM^2 = x^2 + y^2 + z^2$$
$$= \xi^2\alpha + \eta^2\beta + \zeta^2\gamma + 2\sigma\,(l\xi\sqrt{\alpha} + m\eta\sqrt{\beta} + n\zeta\sqrt{\gamma})\,p + \sigma^2 p^2 \dots(8),$$

and to make this maximum or minimum subject to (7), we have

$$\frac{d\,(\tfrac{1}{2}OM^2)}{d\xi} = \rho\xi\,; \quad \frac{d\,(\tfrac{1}{2}OM^2)}{d\eta} = \rho\eta\,; \quad \frac{d\,(\tfrac{1}{2}OM^2)}{d\zeta} = \rho\zeta \dots(9);$$

where in virtue of (7), and because OM^2 is a homogeneous quadratic function of ξ, η, ζ,

$$\rho = OM^2.$$

The determinantal cubic, being

$$(\mathscr{A} - \rho)\,(\mathscr{B} - \rho)\,(\mathscr{C} - \rho) - a^2$$
$$(\mathscr{A} - \rho) - b^2\,(\mathscr{B} - \rho) - c^2\,(\mathscr{C} - \rho) + 2abc = 0 \quad\dots\dots(10),$$

where $\mathscr{A} = \alpha\,(1 + 2\sigma l\lambda + \sigma^2\lambda^2)$; $\quad \mathscr{B} = \beta\,(1 + 2\sigma m\mu + \sigma^2\mu^2)$;

$$\mathscr{C} = \gamma\,(1 + 2\sigma n\nu + \sigma^2\nu^2) \quad\dots\dots\dots\dots\dots(11)$$

and $a = \sqrt{(\beta\gamma)}\,[\sigma\,(m\nu + n\mu) + \sigma^2\mu\nu]$; $b = \sqrt{(\gamma\alpha)}\,[\sigma\,(n\lambda + l\nu) + \sigma^2\nu\lambda]$;

$$c = \sqrt{(\alpha\beta)}\,[\sigma\,(l\mu + m\lambda) + \sigma^2\lambda\mu] \dots\dots\dots\dots(12),$$

gives three real positive values for ρ, the square roots of which are the required principal strain-ratios.

§ 9. Entering now on the dynamics of our subject, remark that the isotropy (§ 1) implies that the work required of the extraneous pressure, to change the solid from its unstrained condition (1, 1, 1) to the strain (α, β, γ), is independent of the direction of the normal axes of the strain, and depends solely on the magnitudes of α, β, γ. Hence if E denotes its magnitude per unit of volume; or the potential energy of unit volume in the condition (α, β, γ) reckoned from zero in the condition (1, 1, 1); we have

$$E = \psi\,(\alpha, \beta, \gamma) \dots\dots\dots\dots\dots\dots(13),$$

where ψ denotes a function of which the magnitude is unaltered Molar. when the values of α, β, γ are interchanged. Consider a portion of the solid, which, in the unstrained condition, is a cube of unit side, and which in the strained condition (α, β, γ), is a rectangular parallelepiped $\sqrt{\alpha} \cdot \sqrt{\beta} \cdot \sqrt{\gamma}$. In virtue of isotropy and symmetry, we see that the pull or pressure on each of the six faces of this figure, required to keep the substance in the condition (α, β, γ), is normal to the face. Let the amounts of these forces per unit area, on the three pairs of faces respectively, be A, B, C, each reckoned as positive or negative according as the force is positive *pull*, or positive pressure. We shall take

$$A + B + C = 0 \dots\dots\dots\dots\dots\dots(14),$$

because normal pull or pressure uniform in all directions produces no effect, the solid being incompressible. The work done on any infinitesimal change from the configuration (α, β, γ) is

$$\left.\begin{array}{c} A \sqrt{(\beta\gamma)}\, d\,(\sqrt{\alpha}) + B \sqrt{(\gamma\alpha)}\, d\,(\sqrt{\beta}) + C \sqrt{(\alpha\beta)}\, d\,(\sqrt{\gamma}), \\[4pt] \text{or (because } \alpha\beta\gamma = 1) \\[4pt] \dfrac{A}{2\alpha}\, d\alpha + \dfrac{B}{2\beta}\, d\beta + \dfrac{C}{2\gamma}\, d\gamma \end{array}\right\} \dots(15).$$

§ 10. Let $\delta\alpha$, $\delta\beta$, $\delta\gamma$ be any variations of α, β, γ consistent with (2), so that we have

$$\left.\begin{array}{c} (\alpha + \delta\alpha)\,(\beta + \delta\beta)\,(\gamma + \delta\gamma) = 1 \\[4pt] \alpha\beta\gamma = 1 \end{array}\right\} \dots\dots\dots\dots(16).$$

and

Now suppose $\delta\alpha$, $\delta\beta$, $\delta\gamma$ to be so small that we may neglect their cubes and corresponding products, and all higher products. We have

$$\frac{\delta\alpha}{\alpha} + \frac{\delta\beta}{\beta} + \frac{\delta\gamma}{\gamma} + \alpha\delta\beta\delta\gamma + \beta\delta\gamma\delta\alpha + \gamma\delta\alpha\delta\beta = 0 \dots\dots(17);$$

whence

$$\left(\frac{\delta\alpha}{\alpha}\right)^2 = \left(\frac{\delta\beta}{\beta} + \frac{\delta\gamma}{\gamma}\right)^2 ;$$

whence, and by the symmetrical expressions,

$$\left.\begin{array}{c} 2\delta\beta\delta\gamma = \dfrac{1}{\alpha} \left(\dfrac{\delta\alpha^2}{\alpha^2} - \dfrac{\delta\beta^2}{\beta^2} - \dfrac{\delta\gamma^2}{\gamma^2}\right) \\[10pt] 2\delta\gamma\delta\alpha = \dfrac{1}{\beta} \left(\dfrac{\delta\beta^2}{\beta^2} - \dfrac{\delta\gamma^2}{\gamma^2} - \dfrac{\delta\alpha^2}{\alpha^2}\right) \\[10pt] 2\delta\alpha\delta\beta = \dfrac{1}{\gamma} \left(\dfrac{\delta\gamma^2}{\gamma^2} - \dfrac{\delta\alpha^2}{\alpha^2} - \dfrac{\delta\beta^2}{\beta^2}\right) \end{array}\right\} \dots\dots\dots (18).$$

§ 11. Now, if $E + \delta E$ denote the energy per unit bulk of the solid in the condition

$$(\alpha + \delta\alpha, \ \beta + \delta\beta, \ \gamma + \delta\gamma),$$

we have, by Taylor's theorem,

$$\delta E = H_1 + H_2 + H_3 + \&\text{c.},$$

where H_1, H_2, &c. denote homogeneous functions of $\delta\alpha$, $\delta\beta$, $\delta\gamma$ of the 1st degree, 2nd degree, &c. Hence, omitting cubes, &c., and eliminating the products from H_2, and taking H_1 from (15), we find

$$\delta E = \frac{1}{2}\left(\frac{A}{\alpha}\,\delta\alpha + \frac{B}{\beta}\,\delta\beta + \frac{C}{\gamma}\,\delta\gamma + G\,\frac{\delta\alpha^2}{\alpha^2} + H\,\frac{\delta\beta^2}{\beta^2} + I\,\frac{\delta\gamma^2}{\gamma^2}\right)\dots(19),$$

where G, H, I denote three coefficients depending on the nature of the function ψ (13), which expresses the energy. Thus in (19), with (14) taken into account, we have just five coefficients independently disposable, A, B, G, H, I, which is the right number because, in virtue of $\alpha\beta\gamma = 1$, E is a function of just two independent variables.

§ 12. For the case of $\alpha = 1$, $\beta = 1$, $\gamma = 1$, we have

$$A = B = C = 0 \ \text{ and } \ G = H = I = G_1, \text{ suppose};$$

which give $\delta E = \frac{1}{2}G_1\,(\delta\alpha^2 + \delta\beta^2 + \delta\gamma^2)$.

From this we see that $2G_1$ is simply the rigidity-modulus of the unstrained solid; because if we make $\delta\gamma = 0$, we have $\delta\alpha = -\,\delta\beta$, and the strain becomes an infinitesimal distortion in the plane (xy), which may be regarded in two ways as a simple shear of which the magnitude is $\delta\alpha$* (this being twice the elongation in one of the normal axes).

§ 13. Going back to (10), (11), and (12), let σ be so small that σ^3 and higher powers can be neglected. To this degree of approximation we neglect abc in (10), and see that its three roots are respectively

$$\left.\begin{array}{c} \mathscr{A} - \dfrac{b^2}{\mathscr{C} - \mathscr{A}} - \dfrac{c^2}{\mathscr{B} - \mathscr{A}}, \\[2ex] \mathscr{B} - \dfrac{c^2}{\mathscr{A} - \mathscr{B}} - \dfrac{a^2}{\mathscr{C} - \mathscr{B}}, \\[2ex] \mathscr{C} - \dfrac{a^2}{\mathscr{B} - \mathscr{C}} - \dfrac{b^2}{\mathscr{A} - \mathscr{C}} \end{array}\right\}\dots\dots\dots\dots(20),$$

---

* Thomson and Tait's *Natural Philosophy*, § 175, or *Elements*, § 154.

provided none of the differences constituting the denominators is Molar. infinitely small. The case of any of these differences infinitely small, or zero, does not, as we shall see in the conclusion, require special treatment, though special treatment would be needed to interpret for any such case each step of the process.

§ 14. Substituting now for $\mathscr{A}$, $\mathscr{B}$, $\mathscr{C}$, a, b, c in (20), their values by (11) and (12), neglecting σ^3 and higher powers, and denoting by $\delta\alpha$, $\delta\beta$, $\delta\gamma$ the excesses of the three roots above α, β, γ respectively, we find

$$
\left.
\begin{aligned}
\delta\alpha &= \alpha\left\{2\sigma l\lambda + \sigma^2\left[\lambda^2 - \frac{\gamma}{\gamma-\alpha}(n\lambda+l\nu)^2 - \frac{\beta}{\beta-\alpha}(l\mu+m\lambda)^2\right]\right\} \\
\delta\beta &= \beta\left\{2\sigma m\mu + \sigma^2\left[\mu^2 - \frac{\alpha}{\alpha-\beta}(l\mu+m\lambda)^2 - \frac{\gamma}{\gamma-\beta}(m\nu+n\mu)^2\right]\right\} \\
\delta\gamma &= \gamma\left\{2\sigma n\nu + \sigma^2\left[\nu^2 - \frac{\beta}{\beta-\gamma}(m\nu+n\mu)^2 - \frac{\alpha}{\alpha-\gamma}(n\lambda+l\nu)^2\right]\right\}
\end{aligned}
\right\} (21),
$$

and using these in (19), we find

$$
\left.
\begin{aligned}
\delta E = {}& \sigma\,(Al\lambda + Bm\mu + Cn\nu) \\
&+ \tfrac{1}{2}\sigma^2\{A\lambda^2 + B\mu^2 + C\nu^2 \\
&+ L\,(m\nu+n\mu)^2 + M\,(n\lambda+l\nu)^2 + N\,(l\mu+m\lambda)^2\} \\
&+ 2\sigma^2\,(Gl^2\lambda^2 + Hm^2\mu^2 + In^2\nu^2)
\end{aligned}
\right\} \dots (22),
$$

where

$$
L = \frac{B\gamma - C\beta}{\beta-\gamma}; \quad M = \frac{C\alpha - A\gamma}{\gamma-\alpha}; \quad N = \frac{A\beta - B\alpha}{\alpha-\beta} \dots(23).
$$

§ 15. Now from (5) and (6) we find

$$
(m\nu + n\mu)^2 = 1 - l^2 - \lambda^2 + 2\,(l^2\lambda^2 - m^2\mu^2 - n^2\nu^2)\dots\dots(24),
$$

which, with the symmetrical expressions, reduces (22) to

$$
\left.
\begin{aligned}
\delta E = {}& \sigma\,(Al\lambda + Bm\mu + Cn\nu) + \tfrac{1}{2}\sigma^2\{L+M+N \\
&+ (A-L)\lambda^2 + (B-M)\mu^2 + (C-N)\nu^2 \\
&- Ll^2 - Mm^2 - Nn^2 + 2\,[(2G+L-M-N)\,l^2\lambda^2 \\
&+ (2H+M-N-L)\,m^2\mu^2 + (2I+N-L-M)\,n^2\nu^2]\}
\end{aligned}
\right\} \dots(25).
$$

§ 16. To interpret this result statically, imagine the solid to be given in the state of homogeneous strain (α, β, γ) throughout, and let a finite plane plate of it, of thickness h, and of very large area Q, be displaced by a shearing motion according to the specification (3), (4), (5), (6) of § 8; the bounding-planes of the

Molar.   plate being unmoved, and all the solid exterior to the plate being
therefore undisturbed except by the slight distortion round the
edge of the plate produced by the displacement of its substance.
The analytical expression of this is

$$\sigma = f(p) \quad \dots\dots\dots\dots\dots\dots(26),$$

where f denotes any function of OP such that

$$\int_0^h dp f(p) = 0 \quad \dots\dots\dots\dots\dots(27).$$

If we denote by W the work required to produce the supposed
displacement, we have

$$W = Q \int_0^h dp\, \delta E + \mathcal{W} \quad \dots\dots\dots\dots(28),$$

δE being given by (25), with everything constant except σ, a
function of OP; and $\mathcal{W}$ denoting the work done on the solid
outside the boundary of the plate. In this expression the first
line of (25) disappears in virtue of (27); and we have

$$\begin{aligned}
\frac{W - \mathcal{W}}{Q} = \tfrac{1}{2} \{ & L + M + N \\
& + (A - L)\lambda^2 + (B - M)\mu^2 + (C - N)\nu^2 \\
& - Ll^2 - Mm^2 - Nn^2 + 2\,[(2G + L - M - N)\,l^2\lambda^2 \\
& + (2H + M - N - (L)\,m^2\mu^2 \\
& + (2I + N - L - M)\,n^2\nu^2]\} \int_0^h dp\sigma^2
\end{aligned} \quad \dots(29).$$

When every diameter of the plate is infinitely great in comparison
with its thickness, $\mathcal{W}/Q$ is infinitely small; and the second
member of (29) expresses the work per unit of area of the plate,
required to produce the supposed shearing motion.

§ 17. Solve now the problem of finding, subject to (5) and
(6) of § 8, the values of l, m, n which make the factor $\{\ \}$ of the
second member of (29) a maximum or minimum. This is only
the problem of finding the two principal diameters of the ellipse
in which the ellipsoid

$$\begin{aligned}
[2\,(2G + L - M - N)\lambda^2 - L]\,x^2 + [2\,(2H + M - N - L)\mu^2 - M]\,y^2 \\
+ [2\,(2I + N - L - M)\nu^2 - N]\,z^2 = \text{const.}\dots\dots(30)
\end{aligned}$$

s cut by the plane

$$\lambda x + \mu y + \nu z = 0 \dots\dots\dots\dots\dots(31).$$

If the displacement is in either of the two directions (l, m, n) thus Molar. determined, the force required to maintain it is in the direction of displacement; and the magnitude of this force per unit bulk of the material of the plate at any point within it is easily proved to be

$$\{M\} \frac{d\sigma}{d\rho} \quad \dots\dots\dots\dots\dots(32),$$

where $\{M\}$ denotes the maximum or the minimum value of the bracketed factor of (29).

§ 18.   Passing now from equilibrium to motion, we see at once that (the density being taken as unity)

$$V^2 = \{M\} \dots\dots\dots\dots\dots\dots(33),$$

where V denotes the velocity of either of two simple waves whose wave-front is perpendicular to (λ, μ, ν). Consider the case of wave-front perpendicular to one of the three principal planes; (yz) for instance: we have $\lambda = 0$; and, to make $\{\ \}$ of (29) a maximum or minimum, we see by symmetry that we must either have

(vibration *perpendicular* to principal plane)
$$l = 1, \quad m = 0, \quad n = 0 \Big\} \dots\dots(34).$$
(vibration *in* principal plane)$\dots l = 0, \quad m = -\nu, \quad n = \mu$

Hence, for the two cases, we have respectively:

Vibration *perpendicular to yz*
$$V^2 = M + N + (B - M)\mu^2 + (C - N)\nu^2 \Big\} \dots(35);$$
Vibration *in yz* $\dots V^2 = L + B\mu^2 + C\nu^2 + 4(H + I - L)\mu^2\nu^2 \dots(36).$

§ 19.   According to Fresnel's theory (35) must be constant, and the last term of (36) must vanish.   These and the corresponding conclusions relatively to the other two principal planes are satisfied if, and require that,

$$A - L = B - M = C - N \dots\dots\dots\dots(37),$$
and $\qquad H + I = L; \quad I + G = M; \quad G + H = N \dots\dots\dots(38).$

Transposing M and N in the last of equations (37), substituting for them their values by (23), and dividing each member by $\beta\gamma$, we find

$$\frac{C - A}{\beta\gamma - \alpha\beta} = \frac{A - B}{\gamma\alpha - \beta\gamma} \dots\dots\dots\dots(39);$$

Molar.

whence (sum of numerators divided by sum of denominators),

$$\frac{B-C}{\gamma\alpha - \alpha\beta} = \frac{C-A}{\alpha\beta - \beta\gamma} = \frac{A-B}{\beta\gamma - \gamma\alpha} \ldots\ldots\ldots\ldots(40).$$

The first of these equations is equivalent to the first of (37); and thus we see that the two equations (37) are equivalent to one only; and (39) is a convenient form of this one. By it, as put symmetrically in (40), and by bringing (14) into account, we find, with q taken to denote a coefficient which may be any function of (α, β, γ):

$$\left.\begin{array}{l} A = q\,(S - \beta\gamma); \quad B = q\,(S - \gamma\alpha); \quad C = q\,(S - \alpha\beta); \\ \text{where} \qquad S = \tfrac{1}{3}\,(\beta\gamma + \gamma\alpha + \alpha\beta) \end{array}\right\}\ldots(41):$$

and using this result in (23), we find

$$\left.\begin{array}{l} L = q\,[\alpha\,(\beta+\gamma) - S]; \quad M = q\,[\beta\,(\gamma+\alpha) - S]; \\ \qquad\qquad\qquad\qquad N = q\,[\gamma\,(\alpha+\beta) - S] \end{array}\right\}\ldots(42).$$

or $\quad L = q\,(2S - \beta\gamma); \quad M = q\,(2S - \gamma\alpha); \quad N = q\,(2S - \alpha\beta)$

By (2) we may put (41) and (42) into forms more convenient for some purposes as follows:—

$$A = q\left(S - \frac{1}{\alpha}\right); \quad B = q\left(S - \frac{1}{\beta}\right); \quad C = q\left(S - \frac{1}{\gamma}\right) \ldots(43),$$

$$L = q\left(2S - \frac{1}{\alpha}\right); \quad M = q\left(2S - \frac{1}{\beta}\right); \quad N = q\left(2S - \frac{1}{\gamma}\right) \ldots(44),$$

where $\qquad\qquad S = \tfrac{1}{3}\left(\dfrac{1}{\alpha} + \dfrac{1}{\beta} + \dfrac{1}{\gamma}\right) \ldots\ldots\ldots\ldots\ldots\ldots (45).$

Next, we find G, H, I; by (38), (44), and (45) we have

$$G + H + I = \tfrac{1}{2}\,(L + M + N) = \tfrac{3}{2}qS = \tfrac{1}{2}q\left(\frac{1}{\alpha} + \frac{1}{\beta} + \frac{1}{\gamma}\right)\ldots(46),$$

whence, by (38) and (44),

$$G = q\left(\frac{1}{\alpha} - \tfrac{1}{2}S\right); \quad H = q\left(\frac{1}{\beta} - \tfrac{1}{2}S\right); \quad I = q\left(\frac{1}{\gamma} - \tfrac{1}{2}S\right)\ldots(47).$$

§ 20. Using (43) and (47) in (19), we have

$$\left.\begin{array}{l} \delta E = \tfrac{1}{2}q\left\{-\dfrac{\delta\alpha}{\alpha^2} - \dfrac{\delta\beta}{\beta^2} - \dfrac{\delta\gamma}{\gamma^2} + S\left(\dfrac{\delta\alpha}{\alpha} + \dfrac{\delta\beta}{\beta} + \dfrac{\delta\gamma}{\gamma}\right)\right. \\ \left. \qquad + \dfrac{\delta\alpha^2}{\alpha^3} + \dfrac{\delta\beta^2}{\beta^3} + \dfrac{\delta\gamma^2}{\gamma^3} - \tfrac{1}{2}S\left(\dfrac{\delta\alpha^2}{\alpha^2} + \dfrac{\delta\beta^2}{\beta^2} + \dfrac{\delta\gamma^2}{\gamma^2}\right)\right\} \end{array}\right\}\ldots\ldots(48).$$

Now we have, by (2), $\log(\alpha\beta\gamma) = 0$. Hence, taking the variation Molecular. of this as far as terms of the second order,

$$\frac{\delta\alpha}{\alpha} + \frac{\delta\beta}{\beta} + \frac{\delta\gamma}{\gamma} - \tfrac{1}{2}\left(\frac{\delta\alpha^2}{\alpha^2} + \frac{\delta\beta^2}{\beta^2} + \frac{\delta\gamma^2}{\gamma^2}\right) = 0 \quad \ldots\ldots(49);$$

which reduces (48) to

$$\delta E = \tfrac{1}{2}q\left(-\frac{\delta\alpha}{\alpha^2} - \frac{\delta\beta}{\beta^2} - \frac{\delta\gamma}{\gamma^2} + \frac{\delta\alpha^2}{\alpha^3} + \frac{\delta\beta^2}{\beta^3} + \frac{\delta\gamma^2}{\gamma^3}\right)\ldots\ldots(50).$$

Remembering that cubes and higher powers are to be neglected we see that (50) is equivalent to

$$\delta E = \tfrac{1}{2}q\delta\left(\frac{1}{\alpha} + \frac{1}{\beta} + \frac{1}{\gamma}\right) \ldots\ldots\ldots\ldots\ldots\ldots(51).$$

Hence if we take q constant, we have

$$E = \tfrac{1}{2}q\left(\frac{1}{\alpha} + \frac{1}{\beta} + \frac{1}{\gamma} - 3\right)\ldots\ldots\ldots\ldots\ldots(52).$$

It is clear that q must be stationary (that is to say, $\delta q = 0$) for any particular values of α, β, γ for which (52) holds; and if (52) holds for all values, q must be constant for all values of α, β, γ.

§ 21. Going back to (29), taking Q great enough to allow $\mathscr{W}/Q$ to be neglected, and simplifying by (46), (43), and (44), we find

$$\frac{W}{Q} = q\left(\frac{l^2}{\alpha} + \frac{m^2}{\beta} + \frac{n^2}{\gamma}\right)\int_0^h dp\sigma^2 \ldots\ldots\ldots\ldots(53);$$

and the problem (§ 17) of determining l, m, n, subject to (5) and (6), to make $l^2/\alpha + m^2/\beta + n^2/\gamma$ a maximum or minimum for given values of λ, μ, ν, yields the equations

$$\omega\lambda - \omega'l + \frac{l}{\alpha} = 0; \quad \omega\mu - \omega'm + \frac{m}{\beta} = 0; \quad \omega\nu - \omega'n + \frac{n}{\gamma} = 0\ldots\ldots(54),$$

ω, ω' denoting indeterminate multipliers; whence

$$\omega' = \frac{l^2}{\alpha} + \frac{m^2}{\beta} + \frac{n^2}{\gamma} \ldots\ldots\ldots\ldots\ldots\ldots\ldots\ldots\ldots\ldots(55),$$

$$\omega^2 = l^2\left(\omega' - \frac{1}{\alpha}\right)^2 + m^2\left(\omega' - \frac{1}{\beta}\right)^2 + n^2\left(\omega' - \frac{1}{\gamma}\right)^2 \ldots\ldots(56),$$

$$\left.\begin{array}{l} \omega\lambda = l\left(-\dfrac{1 - l^2}{\alpha} + \dfrac{m^2}{\beta} + \dfrac{n^2}{\gamma}\right), \\[2ex] \omega\mu = m\left(\dfrac{l^2}{\alpha} - \dfrac{1 - m^2}{\beta} + \dfrac{n^2}{\gamma}\right), \\[2ex] \omega\nu = n\left(\dfrac{l^2}{\alpha} + \dfrac{m^2}{\beta} - \dfrac{1 - n^2}{\gamma}\right) \end{array}\right\} \ldots\ldots\ldots\ldots\ldots(57).$$

These formulas are not directly convenient for finding l, m, n from λ, μ, ν, cf. § 33 (the ordinary formulas for doing so need not be written here); but they give λ, μ, ν explicitly in terms of l, m, n supposed known; that is to say, they solve the problem of finding the wave-front of the simple laminar wave whose direction of vibration is (l, m, n). The velocity is given by

$$v^2 = q\left(\frac{l^2}{\alpha} + \frac{m^2}{\beta} + \frac{n^2}{\gamma}\right) \quad\ldots\ldots\ldots\ldots\ldots\ldots(58).$$

It is interesting to notice that this depends solely on the direction of the line of vibration; and that (except in special cases, of partial or complete isotropy) there is just one wave-front for any given line of vibration. These are precisely in every detail the conditions of Fresnel's Kinematics of Double Refraction.

§ 22.   Going back to (35) and (36), let us see if we can fit them to double refraction with line of vibration *in* the plane of polarization. This would require (36) to be the ordinary ray, and therefore requires the fulfilment of (38), as did the other supposition; but instead of (37) we now have [in order to make (36) constant]

$$A = B = C \ldots\ldots\ldots\ldots\ldots\ldots\ldots\ldots(59),$$

and therefore each, in virtue of (14), zero; and

$$\alpha = \beta = \gamma = 1;$$

so that we are driven to complete isotropy. Hence our present form (§ 7) of the stress-theory of double refraction *cannot* be fitted to give line of vibration *in* the plane of polarization. We have seen (§ 21) that it *does* give line of vibration *perpendicular to the plane of polarization with exactly Fresnel's form* of wave-surface, when fitted for the purpose, by the simple assumption that the potential energy of the strained solid is expressed by (52) with q constant! It is important to remark that q is the rigidity-modulus of the unstrained isotropic solid.

§ 23.   From (58) we see that the velocities of the waves corresponding to the three cases, $l = 1$, $m = 1$, $n = 1$, respectively are $\sqrt{(q/\alpha)}$, $\sqrt{(q/\beta)}$, $\sqrt{(q/\gamma)}$. Hence the velocity of any wave whose vibrations are parallel to any one of the three principal elongations, multiplied by this elongation, is equal to the velocity of a wave in the unstrained isotropic solid.

[§§ 24...34, added *March* 1899.]

§ 24. To fix and clear our understanding of the ideal solid, introduced in § 7 and defined in § 20 (52) and in § 22, take a bar of it of length l when unstrained, and of cross-sectional area A. For our present purpose the cross-section may be of any shape, provided it is uniform from end to end. Apply opposing forces P to the two ends; and, when there is equilibrium under this stress, let the length and cross-section be x and A'. We have $A' = lA/x$, because the solid is incompressible. The proportionate shortenings of all diameters of the cross-section will be equal, as there is no lateral constraint. Hence if the α, β, γ of (52) refer to the length of our bar and two directions perpendicular to it, we have

$$\alpha = \frac{x^2}{l^2}; \quad \beta = \gamma = \frac{l}{x} \quad\dots\dots\dots\dots\dots(60);$$

and therefore by (52)

$$E = \tfrac{1}{2}q \left(\frac{l^2}{x^2} + 2\frac{x}{l} - 3\right) lA \dots\dots\dots\dots(61),$$

where E denotes the whole work required to make the change from l to x in the length of our bar, of which the bulk is lA. Hence, as $P = dE/dx$, we find

$$P = q\left(-\frac{l^2}{x^3} + \frac{1}{l}\right) lA, \text{ and } \delta P = 3q\frac{l^3 A}{x^4}\,\delta x \dots\dots\dots(62).$$

Hence, denoting by M the Young's modulus for values of x differing not infinitely little from l, and defining it by the formula $\dfrac{\delta P \div A'}{\delta x \div x}$, we find

$$M = 3q\,\frac{l^2}{x^2} \quad\dots\dots\dots\dots\dots\dots(63).$$

Hence augmentation of length from l to x diminishes the Young's modulus, as defined above, in the ratio of l^2 to x^2.

§ 25. The property of our ideal solid expressed by the constancy of q, to which we have been forced by the assumption of Fresnel's laws for light traversing a crystal, is interesting in respect to the extension of theory and ideas regarding the elasticity of solids from infinitesimal strains, for which alone the dynamics of the mathematical theory has hitherto been developed, to strains not infinitesimal. To contrast the law of relation between force and elongation for a rod of our ideal substance, as expressed in (62) above, and for a piece of indiarubber cord or

Molar.   indiarubber band, as shown by experiment, is quite an instructive short lesson in the elements of the extended theory.

§ 26. Ten years ago, because of pressure of other avocations, I reluctantly left the stress theory of Fresnel's laws of double refraction without continuing my work far enough to find complete expressions for the equilibrium and motion of the elastic solid for any infinitesimal deviation whatever from the condition (α, β, γ). Here is the continuation which I felt wanting at that time. Let (x_0, y_0, z_0) be the coordinates of a point of the solid in its unstrained condition $(\alpha = 1, \beta = 1, \gamma = 1)$; and (x, y, z), the co-ordinates of the same point in the strained condition (α, β, γ). If the axes of coordinates are the three lines of maximum, mini-max, and minimum elongation, (which are essentially* at right angles to one another), we have

$$x = x_0 \sqrt{\alpha} ; \quad y = y_0 \sqrt{\beta} ; \quad z = z_0 \sqrt{\gamma} \quad \ldots\ldots\ldots\ldots(64).$$

Let the matter at the point (x, y, z) be displaced to $(x + \xi, y + \eta, z + \zeta)$; ξ, η, ζ having each, subject only to the condition of no change of bulk, any arbitrary value for every point (x, y, z) within some finite space S, outside of which the medium remains in the condition (α, β, γ) undisturbed.

Let now the variation of the displacement (ξ, η, ζ) from point to point of the solid be so gradual that each one of the nine ratios

$$\frac{d\xi}{dx}, \frac{d\xi}{dy}, \frac{d\xi}{dz}; \quad \frac{d\eta}{dx}, \frac{d\eta}{dy}, \frac{d\eta}{dz}; \quad \frac{d\zeta}{dx}, \frac{d\zeta}{dy}, \frac{d\zeta}{dz} \ldots\ldots(65)$$

is an infinitely small numeric. Consider the system of bodily forces, which must be applied to the solid within the disturbed region S, to produce the specified displacement (ξ, η, ζ). Let $X\Omega, Y\Omega, Z\Omega$ denote components of the force which must be applied to any infinitesimal volume, Ω, of matter around the point (x, y, z). Our directly solvable problem is to determine X, Y, Z for any point, ξ, η, ζ being given for every point.

§ 27. Our supposition as to infinitesimals, which is that the infinitesimal strain corresponding to the displacement (ξ, η, ζ), is superimposed upon the finite strain (α, β, γ), implies that if instead

* See Thomson and Tait, § 164.

of (ξ, η, ζ), the displacement be $(c\xi, c\eta, c\zeta)$, where c is any numeric Molar. less than unity having the same value for every point of the disturbed region, the force required to hold the medium in this condition is such that instead of X, Y, Z, we have cX, cY, cZ. Hence if δE denote the total work required to produce the displacement (ξ, η, ζ), we have

$$\delta E = \tfrac{1}{2} \iiint dxdydz \,(X\xi + Y\eta + Z\zeta) \quad \ldots\ldots\ldots(66),$$

where $\iiint dxdydz$ denotes integration throughout S.

§ 28.   Now according to (52) we have

$$\left.\begin{aligned} \delta E &= \tfrac{1}{2}q \iiint dxdydz \left(\frac{1}{\alpha'} + \frac{1}{\beta'} + \frac{1}{\gamma'} - \frac{1}{\alpha} - \frac{1}{\beta} - \frac{1}{\gamma}\right) \\ &= \tfrac{1}{2}q \iiint dxdydz \,(\beta'\gamma' + \gamma'\alpha' + \alpha'\beta' - \beta\gamma - \gamma\alpha - \alpha\beta) \end{aligned}\right\} \ldots(67),$$

where α', β', γ' denote the squares of the principal elongations (§ 7 above) as altered from α, β, γ, in virtue of the infinitesimal displacement (ξ, η, ζ).

To find α', β', γ', draw, parallel to our primary lines of reference OX, OY, OZ, temporary lines of reference through the point $(x+\xi, y+\eta, z+\zeta)$, which for brevity we shall call Q. Let P be a point of the solid infinitely little distant from Q, and let (f, g, h) be its coordinates relatively to these temporary reference lines. Relatively to the same lines, let (f_0, g_0, h_0) be the coordinates of the position, P_0, which P would have if the whole solid were unstrained.   We have

$$\left.\begin{aligned} f &= \left(1 + \frac{d\xi}{dx}\right) f_0 \sqrt{\alpha} + \frac{d\xi}{dy} g_0 \sqrt{\beta} + \frac{d\xi}{dz} h_0 \sqrt{\gamma} \\ g &= \frac{d\eta}{dx} f_0 \sqrt{\alpha} + \left(1 + \frac{d\eta}{dy}\right) g_0 \sqrt{\beta} + \frac{d\eta}{dz} h_0 \sqrt{\gamma} \\ h &= \frac{d\zeta}{dx} f_0 \sqrt{\alpha} + \frac{d\zeta}{dy} g_0 \sqrt{\beta} + \left(1 + \frac{d\zeta}{dz}\right) h_0 \sqrt{\gamma} \\ PQ^2 &= f^2 + g^2 + h^2; \quad P_0Q^2 = f_0^2 + g_0^2 + h_0^2 \end{aligned}\right\} \ldots\ldots(68).$$

From these we find

$$\left(\frac{PQ}{P_0Q}\right)^2 = Al^2 + Bm^2 + Cn^2 + 2\,(amn + bnl + clm) \quad \ldots\ldots\ldots(69),$$

where l, m, n denote $\dfrac{f_0}{P_0Q}$, $\dfrac{g_0}{P_0Q}$, $\dfrac{h_0}{P_0Q}$; and

Molar.

$$A = \alpha \left\{ \left(1 + \frac{d\xi}{dx}\right)^2 + \left(\frac{d\eta}{dx}\right)^2 + \left(\frac{d\zeta}{dx}\right)^2 \right\}$$

$$B = \beta \left\{ \left(\frac{d\xi}{dy}\right)^2 + \left(1 + \frac{d\eta}{dy}\right)^2 + \left(\frac{d\zeta}{dy}\right)^2 \right\}$$

$$C = \gamma \left\{ \left(\frac{d\xi}{dz}\right)^2 + \left(\frac{d\eta}{dz}\right)^2 + \left(1 + \frac{d\zeta}{dz}\right)^2 \right\}$$

$$a = \sqrt{\beta\gamma} \left\{ \frac{d\xi}{dy}\frac{d\xi}{dz} + \left(1 + \frac{d\eta}{dy}\right)\frac{d\eta}{dz} + \frac{d\zeta}{dy}\left(1 + \frac{d\zeta}{dz}\right) \right\}$$

$$b = \sqrt{\gamma\alpha} \left\{ \left(1 + \frac{d\xi}{dx}\right)\frac{d\xi}{dz} + \frac{d\eta}{dx}\frac{d\eta}{dz} + \frac{d\zeta}{dx}\left(1 + \frac{d\zeta}{dz}\right) \right\}$$

$$c = \sqrt{\alpha\beta} \left\{ \left(1 + \frac{d\xi}{dx}\right)\frac{d\xi}{dy} + \frac{d\eta}{dx}\left(1 + \frac{d\eta}{dy}\right) + \frac{d\zeta}{dx}\frac{d\zeta}{dy} \right\}$$

...(70).

The values of α', β', γ' are the maximum, minimax, and minimum values of (69), subject to the condition $l^2 + m^2 + n^2 = 1$; and therefore according to the well-known solution of this problem, they are the three roots, essentially real, of the cubic

$$(A - \rho)(B - \rho)(C - \rho)$$
$$- a^2(A - \rho) - b^2(B - \rho) - c^2(C - \rho) + 2abc = 0 \ldots\ldots(71).$$

Hence, by taking the coefficient of ρ in the expansion of this, we find

$$\beta'\gamma' + \gamma'\alpha' + \alpha'\beta' = BC + CA + AB - a^2 - b^2 - c^2$$
$$= (1 + F)\beta\gamma + (1 + G)\gamma\alpha + (1 + H)\alpha\beta \ldots (72),$$

where F, G, H are symmetrical algebraic functions of the fourth degree of the nine ratios (65). Omitting all terms above the second degree, we have

$$F = 2\left(\frac{d\eta}{dy} + \frac{d\zeta}{dz}\right) + \left(\frac{d\xi}{dy}\right)^2 + \left(\frac{d\xi}{dz}\right)^2 + \left(\frac{d\eta}{dy}\right)^2 + \left(\frac{d\zeta}{dz}\right)^2$$
$$+ 4\frac{d\eta}{dy}\frac{d\zeta}{dz} - 2\frac{d\eta}{dz}\frac{d\zeta}{dy}\ldots\ldots\ldots(73),$$

and symmetrical expressions for G and H. Going back now to (67) and remembering that $\alpha\beta\gamma = 1$ and $\alpha'\beta'\gamma' = 1$, and that α, β, γ are constant throughout the solid, we find, by (67),

$$\delta E = \tfrac{1}{2}q\left(\frac{1}{\alpha}\iiint dxdydz F + \frac{1}{\beta}\iiint dxdydz G + \frac{1}{\gamma}\iiint dxdydz H\right)\ldots(74).$$

To find $\iiint dxdydz F$ by (73), remark first that by simple in- Molar.
tegration with respect to y and with respect to z we obtain

$$\iiint dxdydz \frac{d\eta}{dy} = 0 ; \quad \iiint dxdydz \frac{d\zeta}{dz} = 0 \ldots\ldots(75),$$

because η and ζ each vanish through all the space outside the space S.   For the same reason we find by the well-known integration by parts [Lecture XIII., equations (7) above]

$$\iiint dxdydz \frac{d\eta}{dz}\frac{d\zeta}{dy} = \iiint dxdydz \frac{d\eta}{dy}\frac{d\zeta}{dz} \ldots\ldots(76).$$

We thus have

$$\iiint dxdydz\, F = \iiint dxdydz \left\{ \left(\frac{d\xi}{dy}\right)^2 + \left(\frac{d\xi}{dz}\right)^2 + \left(\frac{d\eta}{dy} + \frac{d\zeta}{dz}\right)^2 \right\} \ldots(77).$$

§ 29.   Now the condition $\alpha'\beta'\gamma' = 1$ gives

$$\frac{d\xi}{dx} + \frac{d\eta}{dy} + \frac{d\zeta}{dz} = 0 + \text{terms involving powers,}$$

$$\text{and products of the ratios}\ldots\ldots(78).$$

Therefore, neglecting higher powers than squares, we get

$$\left(\frac{d\eta}{dy} + \frac{d\zeta}{dz}\right)^2 = \left(\frac{d\xi}{dx}\right)^2 \ldots\ldots\ldots\ldots(79),$$

and (77) becomes

$$\iiint dxdydz\,.\,F = \iiint dxdydz \left\{ \left(\frac{d\xi}{dx}\right)^2 + \left(\frac{d\xi}{dy}\right)^2 + \left(\frac{d\xi}{dz}\right)^2 \right\} \ldots(80);$$

and we have corresponding symmetrical equations for G and H. With these used in (74), it becomes

$$\delta E = \tfrac{1}{2}q \iiint dxdydz \left[\frac{1}{\alpha}\left\{ \left(\frac{d\xi}{dx}\right)^2 + \left(\frac{d\xi}{dy}\right)^2 + \left(\frac{d\xi}{dz}\right)^2 \right\} \right.$$

$$\left. + \frac{1}{\beta}\left\{ \left(\frac{d\eta}{dx}\right)^2 + \left(\frac{d\eta}{dy}\right)^2 + \left(\frac{d\eta}{dz}\right)^2 \right\} + \frac{1}{\gamma}\left\{ \left(\frac{d\zeta}{dx}\right)^2 + \left(\frac{d\zeta}{dy}\right)^2 + \left(\frac{d\zeta}{dz}\right)^2 \right\} \right] \ldots(81).$$

Taking any one of the nine terms of (81), the third for example, and applying to it the process of double integration by parts, we find

$$\iiint dxdydz \left(\frac{d\xi}{dz}\right)^2 = - \iiint dxdydz\, \xi \frac{d^2\xi}{dz^2} \ldots\ldots(82).$$

Hence treating similarly all the other terms, we have

$$\delta E = -\tfrac{1}{2}q \iiint dxdydz \left(\xi \frac{\nabla^2 \xi}{\alpha} + \eta \frac{\nabla^2 \eta}{\beta} + \zeta \frac{\nabla^2 \zeta}{\gamma} \right) \dots\dots(83).$$

§ 30. This expression for δE must be equal to that of (66) above, for every possible value of ξ, η, ζ. If all values were possible, we should therefore have the coefficients of ξ, η, ζ in (66) equal respectively to the coefficients of ξ, η, ζ in (83). But in reality only values of ξ, η, ζ are possible which fulfil the condition of bulk everywhere unchanged; and Lagrange's method of indeterminate multipliers adds to the second member of (83), the following

$$\tfrac{1}{2} \iiint dxdydz \; \lambda \left(\frac{d\xi}{dx} + \frac{d\eta}{dy} + \frac{d\zeta}{dz} \right)^2 \dots\dots\dots\dots (84).$$

This, treated by the method of double integration by parts, becomes

$$\tfrac{1}{2} \iiint dxdydz \left(\xi \frac{d\varpi}{dx} + \eta \frac{d\varpi}{dy} + \zeta \frac{d\varpi}{dz} \right) \dots\dots\dots(85),$$

where
$$\varpi = -\lambda \left(\frac{d\xi}{dx} + \frac{d\eta}{dy} + \frac{d\zeta}{dz} \right) \dots\dots\dots\dots(86).$$

§ 31. We must now, according to Lagrange's splendidly powerful method, equate separately the coefficients of ξ, η, ζ in (66) to their coefficients in (83) with (85) added to it. Thus we find

$$X = -\frac{q}{\alpha} \nabla^2 \xi + \frac{d\varpi}{dx}; \quad Y = -\frac{q}{\beta} \nabla^2 \eta + \frac{d\varpi}{dy}; \quad Z = -\frac{q}{\gamma} \nabla^2 \zeta + \frac{d\varpi}{dz} \dots(87),$$

as the equations of equilibrium of our solid with every point of its boundary fixed, and its interior disturbed from the finitely strained condition (α, β, γ) by forces X, Y, Z, producing infinitesimal displacements (ξ, η, ζ); subject only to the condition

$$\frac{d\xi}{dx} + \frac{d\eta}{dy} + \frac{d\zeta}{dz} = 0 \dots\dots\dots\dots\dots(88),$$

to provide against change of bulk in any part. If this equation were not fulfilled, the equations (87), with ϖ given by (86), would be the solution of a certain definite problem regarding a compressible homogeneous solid, having certain definite quality of perfect elasticity, in respect to changes of shape and bulk, defined by the coefficients q and λ. From this problem to our actual problem there is continuous transition by making λ everywhere infinitely great, Molar. and therefore the first member of (88), being the dilatation, zero ; and leaving ϖ as a quantity to be determined to fulfil the conditions of any proposed problem.   Thus in (87), (88), supposing X, Y, Z given, we have four equations for determining the four unknown quantities ϖ, ξ, η, ζ.

§ 32.   Suppose now that the solid, after having been infinitesimally disturbed by applied forces from the $(\alpha,\ \beta,\ \gamma)$ condition, is left to itself.   According to D'Alembert's principle, the motion is determined by what the equations of equilibrium, (87), become with $-\rho\ddot{\xi}$, $-\rho\ddot{\eta}$, $-\rho\ddot{\zeta}$ substituted for X, Y, Z ; ρ denoting the density of the solid.   Thus we find for the equations of motion

$$\rho\frac{d^2\xi}{dt^2}=\frac{q}{\alpha}\nabla^2\xi-\frac{d\varpi}{dx}\ ; \qquad \rho\frac{d^2\eta}{dt^2}=\frac{q}{\beta}\nabla^2\eta-\frac{d\varpi}{dy}\ ; \qquad \rho\frac{d^2\zeta}{dt^2}=\frac{q}{\gamma}\nabla^2\zeta-\frac{d\varpi}{dz}$$

$$\dotfill(89).$$

§ 33.   We are now enabled by these equations to work out the problem of wave-motion by a more direct and synthetical process than that by which we were led to the solution in § 21 above.   The simplest mathematical expression defining plane waves in an elastic solid in terms of the notation of (89) is

$$\left. \begin{aligned} &p=\lambda x+\mu y+\nu z\\ &\xi=lf(p-vt)\ ; \quad \eta=mf(p-vt)\ ; \quad \zeta=nf(p-vt)\\ &\varpi=F(p-vt) \end{aligned} \right\} \ ...(90)\ ;$$

where λ, μ, ν are the direction-cosines of the wave-normal, and l, m, n those of the line of vibration.   Our constraint to incompressibility gives

$$l\lambda+m\mu+n\nu=0 \dotfill(91).$$

Eliminating ξ, η, ζ, ϖ from (89) by (90), we find

$$l\,(\rho v^2-q\alpha^{-1})=\lambda q\omega\ ; \quad m\,(\rho v^2-q\beta^{-1})=\mu q\omega\ ; \quad n\,(\rho v^2-q\gamma^{-1})=\nu q\omega$$

$$\dotfill(92),$$

where

$$\omega=-\frac{1}{q}\frac{dF}{dp}\Big/\frac{d^2f}{dp^2} \dotfill(93).$$

These equations agree with (54) if for ρv^2 we take $q\omega'$.

Molar.

§ 34. Multiplying equations (92) by
$$\lambda/(\rho v^2 - q\alpha^{-1}), \quad \mu/(\rho v^2 - q\beta^{-1}), \quad \nu/(\rho v^2 - q\gamma^{-1})$$
respectively, adding, and using (91), we find

$$\frac{\lambda^2}{\rho v^2 - q\alpha^{-1}} + \frac{\mu^2}{\rho v^2 - q\beta^{-1}} + \frac{\nu^2}{\rho v^2 - q\gamma^{-1}} = 0 \ldots\ldots\ldots(94).$$

This is a quadratic for the determination of v^2, with its two roots essentially real and positive. Again, by the equation
$$l^2 + m^2 + n^2 = 1,$$
we find from (92)

$$\frac{1}{\omega^2} = \left\{ \left(\frac{\lambda}{\rho v^2 - q\alpha^{-1}} \right)^2 + \left(\frac{\mu}{\rho v^2 - q\beta^{-1}} \right)^2 + \left(\frac{\nu}{\rho v^2 - q\gamma^{-1}} \right)^2 \right\} \ldots(95).$$

Equations (94) and (95) give us v^2 and ω; and (92) now gives explicitly l, m, n, when λ, μ, ν are given. Thus we complete the determination of (l, m, n), the direction of vibration, in terms of (λ, μ, ν), the wave-normal, and the constant coefficients

$$q\alpha^{-1}, \quad q\beta^{-1}, \quad q\gamma^{-1}.$$

Our solution is identical with Fresnel's, and implies exactly the same shape of wave-surface.

[§§ 35...47, added *April* 1901.]

§ 35. It will be convenient henceforth to take $\mathfrak{a}$, $\mathfrak{b}$, $\mathfrak{c}$ instead of $\alpha^{-\frac{1}{2}}$, $\beta^{-\frac{1}{2}}$, $\gamma^{-\frac{1}{2}}$. Thus, according to § 7 (1), if x_0, y_0, z_0 and x, y, z denote respectively the coordinates of one and the same particle in the unstrained and in the strained solid, we have*,

$$x = \frac{x_0}{\mathfrak{a}}; \quad y = \frac{y_0}{\mathfrak{b}}; \quad z = \frac{z_0}{\mathfrak{c}} \ldots\ldots\ldots\ldots\ldots(96),$$

or
$$e = \frac{1 - \mathfrak{a}}{\mathfrak{a}}; \quad f = \frac{1 - \mathfrak{b}}{\mathfrak{b}}; \quad g = \frac{1 - \mathfrak{c}}{\mathfrak{c}} \ldots\ldots\ldots\ldots(97);$$

where
$$e = \frac{x - x_0}{x_0}; \quad f = \frac{y - y_0}{y_0}; \quad g = \frac{z - z_0}{z_0} \ldots\ldots\ldots\ldots(98).$$

From (97) we get
$$\mathfrak{a} = \frac{1}{1 + e}; \quad \mathfrak{b} = \frac{1}{1 + f}; \quad \mathfrak{c} = \frac{1}{1 + g} \ldots\ldots\ldots\ldots(99),$$

whence
$$(1 + e)(1 + f)(1 + g) = \frac{1}{\mathfrak{abc}} = 1 \ldots\ldots\ldots\ldots(100).$$

---

* It would have been better from the beginning to have taken single letters instead of $\alpha^{-\frac{1}{2}}$, $\beta^{-\frac{1}{2}}$, $\gamma^{-\frac{1}{2}}$.

The principal elongations to pass from the unstrained to the Molar. strained solid are $\mathfrak{a}^{-1}$, $\mathfrak{b}^{-1}$, $\mathfrak{c}^{-1}$; and the principal ratios of elongation from the strained to the unstrained solid are $\mathfrak{a}$, $\mathfrak{b}$, $\mathfrak{c}$. And if E denote the work per unit volume required to bring the solid from the unstrained to the strained condition, we have by (52)

$$E = \tfrac{1}{2}q\,(\mathfrak{a}^2 + \mathfrak{b}^2 + \mathfrak{c}^2 - 3)\ldots\ldots\ldots\ldots\ldots(101).$$

From this, remembering that $\mathfrak{abc} = 1$, we find

$$P = -(q\mathfrak{a}^2 + \varpi);\quad Q = -(q\mathfrak{b}^2 + \varpi);\quad R = -(q\mathfrak{c}^2 + \varpi)\ \ldots.(102);$$

or

$$
\left.
\begin{aligned}
P &= eq\,\frac{2+e}{(1+e)^2} - \varpi - q\,;\\[1em]
Q &= fq\,\frac{2+f}{(1+f)^2} - \varpi - q\,;\\[1em]
R &= gq\,\frac{2+g}{(1+g)^2} - \varpi - q
\end{aligned}
\right\}\ldots\ldots\ldots\ldots(103);
$$

where P, Q, R denote the normal components of force per unit area (pulling outward when positive) on the three pairs of faces of a rectangular parallelepiped required to keep it in the state of strain $\mathfrak{a}$, $\mathfrak{b}$, $\mathfrak{c}$, with principal elongations perpendicular to the pairs of faces; and ϖ denotes an arbitrary pressure uniform in all directions. The proof is as follows:—Consider a cube of unit edges in the unstrained solid. In the strained condition the lengths of its edges are $1/\mathfrak{a}$, $1/\mathfrak{b}$, $1/\mathfrak{c}$, and the areas of its faces are $\mathfrak{a}$, $\mathfrak{b}$, $\mathfrak{c}$. Hence the equation of work done to augmentation of energy produced in changing $\mathfrak{a}$, $\mathfrak{b}$, $\mathfrak{c}$, to $\mathfrak{a} + \delta\mathfrak{a}$, $\mathfrak{b} + \delta\mathfrak{b}$, $\mathfrak{c} + \delta\mathfrak{c}$, is

$$P\mathfrak{a}\delta\frac{1}{\mathfrak{a}} + Q\mathfrak{b}\delta\frac{1}{\mathfrak{b}} + R\mathfrak{c}\delta\frac{1}{\mathfrak{c}} = q\delta E = q\,(\mathfrak{a}\delta\mathfrak{a} + \mathfrak{b}\delta\mathfrak{b} + \mathfrak{c}\delta\mathfrak{c})\ldots\ldots(104);$$

and by $\mathfrak{abc} = 1$ (constancy of volume) we have

$$\frac{\delta\mathfrak{a}}{\mathfrak{a}} + \frac{\delta\mathfrak{b}}{\mathfrak{b}} + \frac{\delta\mathfrak{c}}{\mathfrak{c}} = 0\ \ldots\ldots\ldots\ldots\ldots\ldots(105).$$

Hence by Lagrange's method

$$
\begin{aligned}
-\frac{P}{\mathfrak{a}}\,\delta\mathfrak{a} &- \frac{Q}{\mathfrak{b}}\,\delta\mathfrak{b} - \frac{R}{\mathfrak{c}}\,\delta\mathfrak{c}\\[0.5em]
&= \left(q\mathfrak{a} + \frac{\varpi}{\mathfrak{a}}\right)\delta\mathfrak{a} + \left(q\mathfrak{b} + \frac{\varpi}{\mathfrak{b}}\right)\delta\mathfrak{b} + \left(q\mathfrak{c} + \frac{\varpi}{\mathfrak{c}}\right)\delta\mathfrak{c}\ldots\ldots(106),
\end{aligned}
$$

where ϖ denotes an "indeterminate multiplier"; and we have definitively

$$-P = q\mathfrak{a}^2 + \varpi;\quad -Q = q\mathfrak{b}^2 + \varpi;\quad -R = q\mathfrak{c}^2 + \varpi\ldots\ldots(107),$$

which prove (102). The meaning of ϖ here, an arbitrary magnitude, is a pressure uniform in all directions, which, as the solid is incompressible, may be arbitrarily applied to the boundary of any portion of it without altering any of the effective conditions.

§ 36. By § 33 (92), we see that the propagational velocities of waves whose lines of vibration are parallel to OX, OY, OZ are respectively

$$\mathfrak{a}\sqrt{\frac{q}{\rho}}, \quad \mathfrak{b}\sqrt{\frac{q}{\rho}}, \quad \mathfrak{c}\sqrt{\frac{q}{\rho}}\dots\dots\dots\dots\dots(108).$$

Thus the propagational velocity of a wave whose front is parallel to the plane YOZ ($\lambda = 1$, $\mu = 0$, $\nu = 0$) is $\mathfrak{b}\sqrt{\frac{q}{\rho}}$ if its line of vibration is parallel to OY ($l = 0$, $m = 1$, $n = 0$), and is $\mathfrak{c}\sqrt{\frac{q}{\rho}}$ if its line of vibration is parallel to OZ ($l = 0$, $m = 0$, $n = 1$); and similarly in respect to waves whose fronts are parallel to ZX and XY.

For brevity we shall call $\mathfrak{a}\sqrt{\frac{q}{\rho}}$, $\mathfrak{b}\sqrt{\frac{q}{\rho}}$, $\mathfrak{c}\sqrt{\frac{q}{\rho}}$ the principal velocities of light in the crystal, and OX, OY, OZ its three principal lines of symmetry. We are precluded from calling these lines optic axes, by the ordinary usage of the word axis in respect to uni-axial and bi-axial crystals.

§ 37. To help us to thoroughly understand the dynamics of the stress theory of double refraction, consider as an example aragonite, a bi-axial crystal of which the three principal refractive indices are 1·5301, 1·6816, 1·6859. If, as according to our stress-theory, optic æolotropy is due to unequal extension and contraction in different directions of the ether within the crystal with volume unchanged, the principal elongations being in simple proportion to the three principal velocities of light within it, annulment of the extension and contraction would give isotropy with refractive index 1·6309, being the cube root of the product of the three principal indices. Hence, if we call V the propagational velocity corresponding to this mean index, the three principal velocities in the actual crystal (being inversely as the refractive indices) are 1·0659 V, ·9687 V, ·9679 V, of which the product is V^3. Hence according to our notation of § 7 and § 35 Molar. above we have

$$\mathfrak{a} = 1\cdot0659 ; \quad \mathfrak{b} = \cdot9687 ; \quad \mathfrak{c} = \cdot9679.$$

§ 38. Let the dotted ellipse in the diagram represent a cross section of an elliptic cylindric portion of aragonite having its axis of figure in the direction of maximum elongation of the ether. Let the diameter $A'A$ be the direction of maximum

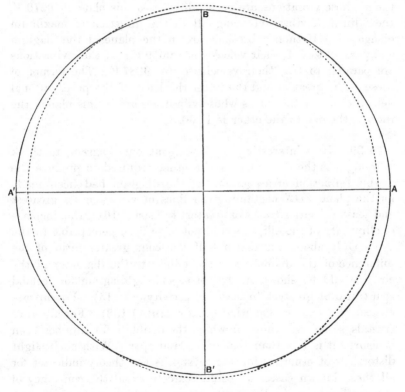

FIG. 1.

contraction, and $B'B$ that of minimax (elongation-contraction), being in fact in this case a line along which there is elongation. The circle in the diagram shows the undisturbed positions of the particles of ether, which in the crystal are forced to the positions shown by the dotted ellipse. The axes of the ellipse are equal respectively to 1/1·0659 and 1/·9687 of the diameter of the circle.

First, consider waves whose fronts are parallel to the plane of the diagram. If their vibrations are parallel to *OA*, the direction of maximum contraction, their velocities of propagation are 1·0659 *V*. If their vibrations are parallel to *OB* their velocities are ·9687 *V*.

Secondly, consider waves whose fronts are perpendicular to the plane of the diagram. The propagational velocity of all of them whose vibrations are perpendicular to this plane is ·9679 *V*, their lines of vibration being all in the direction of maximum elongation. If their vibrations are in the plane of the diagram and parallel to *OA*, their velocity is 1·0659 *V*. If their vibrations are parallel to *OB*, their velocities are ·9687 *V*. The former of these is the greatest and the latter the least of the propagational velocities of all the waves whose vibrations are in this plane: the ratio of the one to the other is 1·100.

§ 39. It is interesting, on looking at our diagram, to think how slight is the distortion of the ether required to produce the double refraction of aragonite. If the diagram had been made for the plane *ZOX* containing the lines of vibration for greatest and least velocities (ratio of greatest to least 1·101), the increase of ellipticity of the ellipse would not have been perceptible to the eye. Only about one and a half per cent greater ratio of the difference of the diameters of the ellipse to the diameter of the circle would be shown in the corresponding diagram for Iceland spar (ratio of greatest to least refractivity = 1·115). For nitrate of soda the ratio is somewhat greater still (1·188). For all other crystals, so far as I know, of which the double refraction has been measured it is less than that of Iceland spar. With such slight distortions of ether within the crystal as the theory indicates for all these known cases, we could scarcely avoid the constancy of our coefficient *q*, to the assumption of which we were forced, §§ 20, 25 above, in order to fit the theory to Fresnel's laws for light traversing a crystal, irrespectively of smallness or greatness of the amount of double refraction. Hence we see that, in fact, without any arbitrary assumption of a new property or new properties of ether, we have arrived at what would be a perfect explanation of the main phenomena of double refraction, if we

could but see how the molecules of matter could so act upon Molecular.
ether as to give a stress capable of producing the strain with
which hitherto we have been dealing.   Inability to see this has
prevented me, and still prevents me, from being convinced that
the stress-theory gives the true explanation of double refraction.

§ 40.   Now, quite recently, it has occurred to me that the
difficulty might possibly be overcome if, as seems to me necessary
on other grounds, we adopt a hypothesis regarding the motion of
ponderable matter through ether, which I suggested a year ago in
a Friday Evening Lecture, April 27th, 1900, to the Royal Institution (reproduced in the present volume as Appendix B) and
with somewhat full detail in a communication* of last July to the
Royal Society of Edinburgh, and to the Congrès Internationale
de Physique† in Paris last August (Appendix A, below).   According to this hypothesis ether is a structureless continuous elastic
solid pervading all space, and occupying space jointly with the
atoms of ponderable matter wherever ponderable matter exists;
and the action between ponderable matter and ether consists of
attractions and repulsions throughout the volume of space occupied
by each atom.   These attractions and repulsions would be essentially ineffective if ether were infinitely resistant against forces
tending to condense or dilate it, that is to say, if ether were
absolutely incompressible.   Hence, while acknowledging that ether
resists forces tending to condense it or to dilate it, sufficiently to
account for light and radiant heat by waves of purely transverse
vibration (equi-voluminal waves as I have called them), it must,
by contraction or dilatation of bulk, yield to compressing or
dilating forces sufficiently to account for known facts dependent
on mutual forces between ether and ponderable matter.   I have
suggested‡ that there may be oppositely electric atoms which
have the properties respectively of condensing and rarefying the
ether within them.   But for the present, to simplify our suppositions to the utmost, I shall assume the law of force between
the atom and the ether within it to be such that the average
density of the ether within the atom is equal to the density of

* *Proc. Roy. Soc. Edin.* July, 1900 ; *Phil. Mag.* Aug. 1900.
† *Reports,* Vol. ii. page 1.
‡ *Congrès Internationale de Physique, Reports,* Vol. ii. p. 19 ; also *Phil. Mag.*
Sept. 1900.

Molecular. the undisturbed ether outside, and that concentric spherical surfaces within the atom are surfaces of equal density. The forces
between ether and atoms we can easily believe to be enormous in
comparison with those called into play outside the atoms, in virtue
of undulatory or other motion of ether and elasticity of ether;
as for instance in interstices between atoms in a solid body, or
in the space traversed by the molecules of a gas according to
the kinetic theory of gases, or in the vacuum attainable in our
laboratories, or in interstellar space.

§ 41. To fix our ideas let ether experience condensation in
the central part of the atom and rarefaction in the outer part
according to the law explained generally in the first part of § 5,
Appendix A, and represented particularly by the formulas (9) and
(11) of that section, and fully described with numerical and
graphical illustrations for a particular case in §§ 5—8. Looking

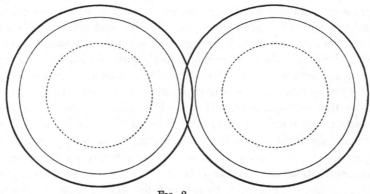

Fig. 2.

to cols. 3 and 4 of Table I, Appendix A, we see that, at distance
$r = \cdot56$ from the centre of the atom, the density of the ether is
equal to the undisturbed density outside the atom; and that from
$r = \cdot56$ to $r = 1$ the density decreases to a minimum, $\cdot35$, at
$r = \cdot865$, and augments thence to the undisturbed density, 1, at
the boundary of the atom $(r = 1)$. In each of the two atoms
represented in fig. 2, the spherical surface of undisturbed density
is indicated by a dotted circle, that of minimum density by a fine
circle, and the boundary of the atom by a heavy circle. Because
the ethereal density decreases uninterruptedly from the centre to

the surface of minimum density, the force exerted by the atom on Molecular. the ether must be towards the centre throughout this spherical space; and because the ethereal density increases uninterruptedly outwards from the surface of minimum density to the boundary of the atom, the force exerted by the atom on the ether must be repulsive in every part of the shell outside that surface.

§ 42.   Suppose now two atoms to be somehow held together in some such position as that represented in fig. 2, overlapping one another throughout a lens-shaped space lying outside the surface of minimum ethereal density in each atom.   The rarefaction of the ether in this lens-shaped space is, by the combined action of the two atoms, greater than at equal distances from the centre in non-overlapping portions of both the atoms.   Hence, remembering that each atom while attracting the ether in its central parts repels the ether in every part of it outside the spherical surface of minimum density, we see that the repulsion of each atom on the ether in the lens-shaped volume of overlap is less than its repulsion in the contrary direction on an equal and similar portion of the ether within it on the other side of its centre.   Hence the re-actions of the ether on the atoms are forces tending to bring them together; that is to say apparent attractions.   These apparent attractions are balanced by repulsions between the atoms them-selves if two atoms rest stably as indicated in fig. 2.   It seems not improbable that these are the forces concerned in the equilibrium of the two atoms of the known diatomic simple gases N_2, O_2, H_2. I assume that the law of force between ether and atoms and the law of elasticity of the ether under this force to be such that no part of the ether outside the atoms experiences any displacement in consequence of the displacements actually produced within the space occupied separately and jointly by the two atoms.   This implies that the ether drawn away from the lenticular space of overlap by the extra rarefaction there, is taken in to the central regions of the two atoms in virtue of the attractions of the atoms on the ether in those regions.   In an addition to Lecture XVIII. on the reflection and refraction of light, I shall have occasion to give explanation and justification of this assumption.

§ 43.   As a representation of an optically isotropic crystal, consider a homogeneous assemblage of atoms for simplicity taken

Molecular. in cubic order as shown in fig. 3.   Each one of the outermost
atoms experiences resultant force inwards from the ether within
it, and this force is balanced by repulsion exerted upon it by the
atom next it inside.   For every atom except those lying in the
outer faces, the forces which it experiences in different directions
from the ether within it balance one another : and so do the forces
which it experiences from the atoms around it.   But each of the
outermost atoms experiences a resultant repulsion from the other

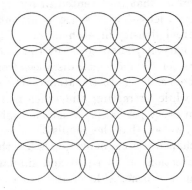

Fig. 3.

atoms in contact with it, and this repulsion outwards is balanced
by a contrary attraction of the ether on the atom.   Hence the
outermost atoms all round a cube of the crystal exert an outward
pull upon the ether within the containing cube.   In accordance
with the assumption stated at the end of § 42, the position and
shape of every particle of ether outside the atoms is undisturbed
by the forces exerted by the atoms in the spaces occupied by them
separately and jointly; and it is only in these spaces that the
ether is disturbed by the action of the atoms.

§ 44.   Suppose now that by forces applied to the atoms as
indicated by the arrow-heads in fig. 4, the distances between the
centres of contiguous atoms are increased and diminished as shown
in the diagram.   With this configuration of the atoms the ether
is pulled outwards by the atoms with stronger forces in the
direction parallel to AC, BD than in directions parallel to AB
and CD.   Hence as (§§ 42, 43 above) the ether is unstressed and
unstrained in the interstices between the atoms, the ether within
the atoms experiences an excess of outward pulling stress in the

direction parallel to AC and BD above outward pulling stress in Molecular. the direction perpendicular to these lines: and therefore, *if there were no œolotropy of inertia,* and *if* the stress theory which we have worked out (§§ 35, 36) for homogeneous ether is applicable

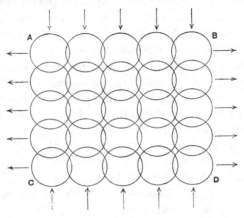

FIG. 4.

to average action of the whole ether with its great inequalities of density in the space occupied by the assemblage of atoms, the propagational velocity of distortional waves would be greater when the direction of vibration is parallel to AC and BD than when it is parallel to AB and CD. But alas! this is exactly the reverse of what, thirteen years ago, Kerr, by experimental research of a rigorously testing character, found for the bi-refringent action of strained glass*, the existence of which had been discovered by Sir David Brewster seventy years previously. We are forced to admit that one or both of our two "*if*"s must be denied.

§ 45. It seems to me now worthy of consideration whether the true explanation of the double refraction of a natural crystal or of a piece of strained glass may possibly be that given by Glazebrook in another paper† in the same volume of the *Philosophical Magazine* as Kerr's already referred to. Glazebrook made

* "Experiments on the Birefringent Action of Strained Glass," *Phil. Mag.* Oct. 1888, p. 339.

† "On the Application of Sir William Thomson's Theory of a Contractile Ether to Double Refraction, Dispersion, Metallic Reflexion, and other Optical Problems." *Phil. Mag.* Dec. 1888, p. 524.

T. L.   **17**

Molar. the remarkable discovery expressed in his equation (14) (p. 525) that, when the propagational velocity of the condensational-rarefactional wave is zero, the two propagational velocities due to æolotropic inertia in a distortional wave with same wave-front* are the same as those given by my equation (94), and originally by Fresnel; from which it follows that the wave-surface is exactly Fresnel's. It is certainly a most interesting result that the wave surface should be exactly Fresnel's, whether the optic æolotropy is due to difference of stress in different directions in an incompressible elastic solid, or to æolotropy of inertia in an ideal elastic solid endowed with a negative compressibility modulus of just such value as to make the velocity of the condensational-rarefactional wave zero. In Glazebrook's theory the direction of vibration is perpendicular to the line of the ray, and in the vibratory motion the solid experiences a slight degree of change of bulk combined with pure distortion. In the stress theory of §§ 7...34 the line of vibration is exactly in the wave-front or perpendicular to the wave normal and there is no change of bulk.

§ 46. In Lecture XVIII, when we are occupied with the reflection and refraction of light, we shall see that Fresnel's formula for the three rays, incident, reflected, and refracted, when the line of vibration is in their plane, is a strict dynamical consequence of the assumption of zero velocity for the condensational-rarefactional wave in both the mediums, or in one of the mediums only while the other medium is incompressible. But the difficulties of accepting zero velocity for condensational-rarefactional wave, whether in undisturbed ether through space or among the atoms of ponderable matter, are not overcome.

§ 47. It will be seen, (§§ 3, 4, 7, 21, 22, 24, 25, 33, 35, 37, 38, 39, 40, 45 above,) that after earnest and hopeful consideration of the stress theory of double refraction during fourteen years, I am unable to see how it can give the true explanation either of the double refraction of natural crystals, or of double refraction induced in isotropic solids by the application of unequal pressures in different directions. Nevertheless the mathematical investigations of §§ 7...21 and §§ 24...34, interesting as they are

---

* l, m, n are the direction-cosines of the normal to the wave-front in Glazebrook's paper, and λ, μ, ν, in my §§ 18, 21, 33, 34 above.

in the abstract dynamics of a homogeneous incompressible elastic Molar. solid, have an important application in respect to the influence of ponderable matter on ether.   They prove in fact the truth of the assumption at the end of § 42, however the forces within the atoms are to be explained ; because any distortion of ether in the space around a crystal, even so slight as that illustrated in fig. 1, § 38 above, would produce double refraction in air or vacuum outside the crystal, not quite as intense outside as inside ; not vastly less, close to the outside ; and diminishing with distance outside but still quite perceptible, to distances of several diameters of the mass.

LECTURE XVI.

Wednesday, *October* 15, 5 p.m.

[This was a double lecture; but as the substance of the first part, with amplification partly founded on experimental discoveries by many workers since it was delivered, has been already reproduced in dated additions on pp. 148—157 and 176—184 above, only the second part is here given.]

I want now to go somewhat into detail as to absolute magnitudes of masses and energies, in order that there may be nothing indefinite in our ideas upon this part of our subject; and I commence by reading and commenting on an old article of mine relating to the energy of sunlight and the density of ether.

[*Nov.* 20, 1899...*March* 28, 1901. From now, henceforth till the end of the Lectures, sections will be numbered continuously.]

Note on the Possible Density of the Luminiferous Medium and on the Mechanical Value of a Cubic Mile*† of Sunlight.

[From *Edin. Royal Soc. Trans.*, Vol. xxi. Part i. May, 1854; *Phil. Mag.* ix. 1854; *Comptes Rendus*, xxxix. Sept. 1854; Art. lxvii. of *Math. and Phys. Papers.*]

Molar. § 1. That there must be a medium forming a continuous material communication throughout space to the remotest visible body is a fundamental assumption in the undulatory Theory of Light. Whether or not this medium is (as appears‡ to

---

* [Note of Dec. 22, 1892. The brain-wasting perversity of the insular inertia which still condemns British Engineers to reckonings of miles and yards and feet and inches and grains and pounds and ounces and acres is curiously illustrated by the title and numerical results of this Article as originally published.]

† [Oct. 13, 1899. In the present reproduction, as part of my Lec. XVI. of Baltimore, 1884, I suggest cubic kilometre instead of "cubic mile" in the title and use the French metrical system exclusively in the article.]

‡ [Oct. 13, 1899.—Not so now. I did not in 1854 know the kinetic theory of gases.]

me most probable) a continuation of our own atmosphere, Molar. its existence is a fact that cannot be questioned, when the overwhelming evidence in favour of the undulatory theory is considered; and the investigation of its properties in every possible way becomes an object of the greatest interest. A first question would naturally occur, What is the absolute density of the luminiferous ether in any part of space? I am not aware of any attempt having hitherto been made to answer this question, and the present state of science does not in fact afford sufficient data. It has, however, occurred to me that we may assign an inferior limit to the density of the luminiferous medium in inter-planetary space by considering the mechanical value of sunlight as deduced in preceding communications to the Royal Society [*Trans. R. S. E.*; Mechanical Energies of the Solar System; re-published as Art. LXVI. of *Math. and Phys. Papers*] from Pouillet's data on solar radiation, and Joule's mechanical equivalent of the thermal unit. Thus the value of solar radiation per second per square centimetre at the earth's distance from the sun, estimated at 1235 cm.-grams, is the same as the mechanical value of sunlight in the luminiferous medium through a space of as many cubic centimetres as the number of linear centimetres of propagation of light per second. Hence the mechanical value of the whole energy, kinetic and potential, of the disturbance kept up in the space of a cubic centimetre at the earth's distance from the sun*, is $\dfrac{1235}{3 \times 10^{10}}$, or $\dfrac{412}{10^{10}}$ of a cm.-gram.

§ 2. The mechanical value of a cubic kilometre of sunlight is consequently 412 metre-kilograms, equivalent to the work of one horse-power for 5·4 seconds. This result may give some idea of the actual amount of mechanical energy of the luminiferous motions and forces within our own atmosphere. Merely to commence the illumination of eleven cubic kilometres, requires an amount of work equal to that of a horse-power for a minute; the same amount of energy exists in that space as long as light continues to traverse it; and, if the source of light be

---

* The mechanical value of sunlight in any space near the sun's surface must be greater than in an equal space at the earth's distance, in the ratio of the square of the earth's distance to the square of the sun's radius, that is, in the ratio of 46,000 to 1 nearly. The mechanical value of a cubic centimetre of sunlight near the sun must, therefore, be $\dfrac{1235 \times 46000}{3 \times 10^{10}}$ or about ·0019 of a cm.-gram.

Molecular. suddenly stopped, must pass from it before the illumination ceases*. The matter which possesses this energy is the luminiferous medium. If, then, we knew the velocities of the vibratory motions, we might ascertain the density of the luminiferous medium; or, conversely, if we knew the density of the medium, we might determine the average velocity of the moving particles.

§ 3. Without any such definite knowledge, we may assign a superior limit to the velocities, and deduce an inferior limit to the quantity of matter, by considering the nature of the motions which constitute waves of light. For it appears certain that the amplitudes of the vibrations constituting radiant heat and light must be but small fractions of the wave-lengths, and that the greatest velocities of the vibrating particles must be very small in comparison with the velocity of propagation of the waves.

§ 4. Let us consider, for instance, homogeneous plane polarized light, and let the greatest velocity of vibration be denoted by v; the distance to which a particle vibrates on each side of its position of equilibrium, by A; and the wave-length, by λ. Then, if V denote the velocity of propagation of light or radiant heat, we have

$$\frac{v}{V} = 2\pi \frac{A}{\lambda};$$

and therefore if A be a small fraction of λ, v must also be a small fraction (2π times as great) of V. The same relation holds for circularly polarized light, since in the time during which a particle revolves once round in a circle of radius A, the wave has been propagated over a space equal to λ. Now the whole mechanical value of homogeneous plane polarized light in an infinitely small space containing only particles sensibly in the same phase of vibration, which consists entirely of potential energy at the instants when the particles are at rest at the extremities of their excursions, partly of potential and partly of kinetic energy when they are moving to or from their positions of equilibrium, and wholly of kinetic energy when they are passing through these positions, is of constant amount, and must therefore be at every instant equal to half the mass multiplied by the square of the velocity which the particles have in the last-mentioned case. But the velocity of any

* Similarly we find 4140 horse-power for a minute as the amount of work required to generate the energy existing in a cubic kilometre of light near the sun.

particle passing through its position of equilibrium is the greatest Molecular. velocity of vibration. This we have denoted by v; and, therefore, if ρ denote the quantity of vibrating matter contained in a certain space, a space of unit volume for instance, the whole mechanical value of all the energy, both kinetic and potential, of the disturbance within that space at any time is $\frac{1}{2}\rho v^2$. The mechanical energy of circularly polarized light at every instant is (as has been pointed out to me by Professor Stokes) half kinetic energy of the revolving particles and half potential energy of the distortion kept up in the luminiferous medium; and, therefore, v being now taken to denote the constant velocity of motion of each particle, double the preceding expression gives the mechanical value of the whole disturbance in a unit of volume in the present case.

§ 5.   Hence it is clear, that for any elliptically polarized light the mechanical value of the disturbance in a unit of volume will be between $\frac{1}{2}\rho v^2$ and ρv^2, if v still denote the greatest velocity of the vibrating particles. The mechanical value of the disturbance kept up by a number of coexisting series of waves of different periods, polarized in the same plane, is the sum of the mechanical values due to each homogeneous series separately, and the greatest velocity that can possibly be acquired by any vibrating particle is the sum of the separate velocities due to the different series. Exactly the same remark applies to coexistent series of circularly polarized waves of different periods. Hence the mechanical value is certainly less than *half* the mass multiplied into the square of the greatest velocity acquired by a particle, when the disturbance consists in the superposition of different series of plane polarized waves; and we may conclude, for every kind of radiation of light or heat except a series of homogeneous circularly polarized waves, that *the mechanical value of the disturbance kept up in any space is less than the product of the mass into the square of the greatest velocity acquired by a vibrating particle in the varying phases of its motion.* How much less in such a complex radiation as that of sunlight and heat we cannot tell, because we do not know how much the velocity of a particle may mount up, perhaps even to a considerable value in comparison with the velocity of propagation, at some instant by the superposition of different motions chancing to agree; but we may be sure that the product of the mass into the square of an ordinary maximum velocity, or of the mean of a great many successive maximum velocities of a vibrating particle, cannot exceed in any great ratio the true mechanical value of the disturbance.

§ 6. Recurring, however, to the definite expression for the mechanical value of the disturbance in the case of homogeneous circularly polarized light, the only case in which the velocities of all particles are constant and the same, we may define the mean velocity of vibration in any case as such a velocity that the product of its square into the mass of the vibrating particles is equal to the whole mechanical value, in kinetic and potential energy, of the disturbance in a certain space traversed by it; and from all we know of the mechanical theory of undulations, it seems certain that this velocity must be a very small fraction of the velocity of propagation in the most intense light or radiant heat which is propagated according to known laws. Denoting this velocity for the case of sunlight at the earth's distance from the sun by v, and calling W the mass in grammes of any volume of the luminiferous ether, we have for the mechanical value of the disturbance in the same space, in terms of terrestrial gravitation units,

$$\frac{W}{g} v^2,$$

where g is the number 981, measuring in (C.G.S.) absolute units of force, the force of gravity on a gramme. Now, from Pouillet's observation, we found in the last footnote on § 1 above, $\frac{1235 \times 46000}{V}$ for the mechanical value, in centimetre-grams, of a cubic centimetre of sunlight in the neighbourhood of the sun; and therefore the mass, in grammes, of a cubic centimetre of the ether, must be given by the equation,

$$W = \frac{981 \times 1235 \times 46000}{v^2 V}.$$

If we assume $v = \frac{1}{n} V$, this becomes

$$W = \frac{981 \times 1235 \times 46000}{V^3} \times n^2 = \frac{981 \times 1235 \times 46000}{(3 \times 10^{10})^3} \times n^2$$

$$= \frac{20 \cdot 64}{10^{22}} \times n^2 \text{ gm.};$$

and for the mass, in grammes, of a cubic kilometre we have

$$\frac{20\cdot64}{10^7} \times n^2.$$

§ 7.   It is quite impossible to fix a definite limit to the ratio which v may bear to V; but it appears improbable that it could be more, for instance, than $\frac{1}{50}$, for any kind of light following the observed laws.   We may conclude that probably a cubic centimetre of the luminiferous medium in the space near the sun contains not less than 516×10^{-20} of a gramme of matter; and a cubic kilometre not less than 516×10^{-5} of a gramme.

§ 8.   [*Nov.* 16, 1899.   We have strong reason to believe that the density of ether is constant throughout interplanetary and interstellar space.   Hence, taking the density of water as unity according to the convenient French metrical system, the preceding statements are equivalent to saying that the density of ether in vacuum or space devoid of ponderable matter is everywhere probably not less than 5×10^{-18}.

Hence the rigidity, (being equal to the density multiplied by the square of the velocity of light), must be not less than 4500 dynes* per square centimetre.   With this enormous value as an inferior limit to the rigidity of the ether, we shall see in an addition to Lecture XIX. that it is impossible to arrange for a radiant molecule *moving through ether and displacing ether* by its translatory as well as by its vibratory motions, consistently with any probable suppositions as to magnitudes of molecules and ruptural rigidity-modulus of ether; and that it is also impossible to explain the known smallness of ethereal resistance against the motions of planets and comets, or of smaller ponderable bodies, such as those we can handle and experiment upon in our abode on the earth's surface, if the ether must be pushed aside to make way for the body moving through it.   We shall find ourselves forced to consider the necessity of some hypothesis for the free motion of ponderable bodies through ether, disturbing it only by condensations and rarefactions, with no incompatibility in respect to joint occupation of the same space by the two substances.]   See *Phil. Mag.* Aug. 1900, pp. 181—198.

* See *Math. and Phys. Papers*, Vol. III. p. 522; and in last line of Table 4, for "$\rho > 10^{-22}$" substitute "$\rho < 10^{-22}$."

§ 9. I wish to make a short calculation to show how much compressing force is exerted upon the luminiferous ether by the sun's attraction. We are accustomed to call ether imponderable. How do we know it is imponderable ? If we had never dealt with air except by our senses, air would be imponderable to us; but we know by experiment that a vacuous glass globe shows an increase of weight when air is allowed to flow into it. We have not the slightest reason to believe the luminiferous ether to be imponderable. [*Nov.* 17, 1899. I now see that we have the strongest possible reason to believe that ether is imponderable.] It is just as likely to be attracted to the sun as air is. At all events the onus of proof rests with those who assert that it is imponderable. I think we shall have to modify our ideas of what gravitation is, if we have a mass spreading through space with mutual gravitations between its parts without being attracted by other bodies. [*Nov.* 17, 1899. But is there any gravitational attraction between different portions of ether? Certainly not, unless either it is infinitely resistant against condensation, or there is only a finite volume of space occupied by it. Suppose that ether is given uniformly spread through space to infinite distances in all directions. Any large enough spherical portion of it, if held with its surface absolutely fixed, would by the mutual gravitation of its parts become heterogeneous; and this tendency could certainly not be counteracted by doing away with the supposed rigidity of its boundary and by the attraction of ether extending to infinity outside it. The pressure at the centre of a spherical portion of homogeneous gravitational matter is proportional to the square of the radius, and therefore, by taking the globe large enough, may be made as large as we please, whatever be the density. In fact, if there were mutual gravitation between its parts, homogeneous ether extending through all space would be essentially unstable, unless infinitely resistant against compressing or dilating forces. If we admit that ether is to some degree condensible and extensible, and believe that it extends through all space, then we must conclude that there is no mutual gravitation between its parts, and cannot believe that it is gravitationally attracted by the sun or the earth or any ponderable matter; that is to say, we must believe ether to be a substance outside the law of universal gravitation.]

§ 10. In the meantime, it is an interesting and definite question Molar. to think of what the weight of a column of luminiferous ether of infinite height resting on the sun, would be, supposing the sun cold and quiet, and supposing for the moment ether to be gravitationally attracted by the sun as if it were ponderable matter of density 5×10^{-18}. You all know the theorem for mean gravity due to attraction inversely as the square of the distance from a point. It shows that the heaviness of a uniform vertical column AB, of mass w per unit length, and having its length in a line through the centre of force C, is

$$\frac{mw}{CA} - \frac{mw}{CB}; \quad \text{or} \quad \frac{mw}{CA} \quad \text{if } CB = \infty,$$

where m denotes the attraction on unit of mass at unit distance. Hence writing for mw/CA, $mw\,CA/CA^2$, we see that the attraction on an infinite column under the influence of a force decreasing according to inverse square of distance, is equal to the attraction on a column equal in length to the distance of its near end from the centre, and attracted by a uniform force equal to that of gravity on the near end. The sun's radius is 697×10^8 cms. and gravity at his surface is 27 times* terrestrial gravity, or say 27000 dynes per gramme of mass. Hence the sun's attraction on a column of ether of a square centimetre section, if of density 5×10^{-18} and extending from his surface to infinity, would be $9\cdot4 \times 10^{-3}$ of a dyne, if ether were ponderable.

§ 11. Considerations similar to those of November 1899 inserted in § 9 above lead to decisive proof that the mean density of ponderable matter through any very large spherical volume of space is smaller, the greater the radius; and is infinitely small for an infinitely great radius. If it were not so a majority of the bodies in the universe would each experience infinitely great gravitational force. This is a short statement of the essence of the following demonstration.

§ 12. Let V be any volume of space bounded by a closed surface, S, outside of which and within which there are ponderable bodies; M the sum of the masses of all these bodies within S;

---

* This is founded on the following values for the sun's mass and radius and the earth's radius :—sun's mass = 324000 earth's mass; sun's radius = 697000 kilometres; earth's radius = 6371 kilometres.

Molar. and ρ the mean density of the whole matter in the volume V. We have

$$M = \rho V \dots\dots\dots\dots\dots(1).$$

Let Q denote the mean value of the normal component of the gravitational force at all points of S. We have

$$QS = 4\pi M = 4\pi\rho V \dots\dots\dots (2),$$

by a general theorem discovered by Green seventy-three years ago regarding force at a surface of any shape, due to matter (gravitational, or ideal electric, or ideal magnetic) acting according to the Newtonian law of the inverse square of the distance. It is interesting to remark, that the surface-integral of the normal component force due to matter outside any closed surface is zero for the whole surface. If normal component force acting inwards is reckoned positive, force outwards must of course be reckoned negative. In equation (2) the normal component force may be outwards at some points of the surface S, if in some places the tangent plane is cut by the surface. But if the surface is wholly convex, the normal component force must be everywhere inwards.

§ 13. Let now the surface be spherical of radius r. We have

$$S = 4\pi r^2; \quad V = \frac{4\pi}{3} r^3; \quad V = \frac{1}{3} rS \dots\dots\dots(3).$$

Hence, for a spherical surface, (2) gives

$$Q = \frac{4\pi}{3} r\rho = \frac{M}{r^2} \dots\dots\dots\dots (4).$$

This shows that the average normal component force over the surface S is infinitely great, if ρ is finite and r is infinitely great, which suffices to prove § 11.

§ 14. For example, let

$$r = 150 \cdot 10^6 \cdot 206 \cdot 10^6 = 3{\cdot}09 \cdot 10^{16} \text{ kilometres}\dots\dots(5).$$

This is the distance at which a star must be to have parallax one one-thousandth of a second; because the mean distance of the earth from the sun is one-hundred-and-fifty-million kilometres, and there are two-hundred-and-six-thousand seconds of angle in the radian. Let us try whether there can be as much matter as a thousand-million times the sun's mass, or, as we shall say for Molar. brevity, a thousand-million suns, within a spherical surface of that radius (5). The sun's mass is 324,000 times the earth's mass; and therefore our quantity of matter on trial is $3·24 . 10^{14}$ times the earth's mass. Hence if we denote by g terrestrial gravity at the earth's surface, we have by (4)

$$Q = 3·24 . 10^{14} \left(\frac{6·37 . 10^3}{3·09 . 10^{16}}\right)^2 g = 1·37 . 10^{-11} . g \ldots\ldots (6).$$

Hence if the radial force were equal over the whole spherical surface, its amount would be $1·37 . 10^{-11}$ of terrestrial surface-gravity; and every body on or near that surface would experience an acceleration toward the centre equal to

$$1·37 . 10^{-13} \text{ kilometres per second per second} \ldots\ldots (7),$$

because g is approximately 1000 centimetres per second per second, or ·01 kilometre per second per second. If the normal force is not uniform, bodies on or near the spherical surface will experience centreward acceleration, some at more than that rate, some less. At exactly that rate, the velocity acquired per year (thirty-one and a half million seconds) would be $4·32 . 10^{-6}$ kilometres per second. With the same rate of acceleration through five million years the velocity would amount to $21·6$ kilometres per second, if the body started from rest at our spherical surface; and the space moved through in five million years would be $·17 . 10^{16}$ kilometres, which is only ·055 of r (5). This is so small that the force would vary very little, unless through the accident of near approach to some other body. With the same acceleration constant through twenty-five million years the velocity would amount to 108 kilometres per second; but the space moved through in twenty-five million years would be $4·25 . 10^{16}$ kilometres, or more than the radius r, which shows that the rate of acceleration could not be approximately constant for nearly as long a time as twenty-five million years. It would, in fact, have many chances of being much greater than 108 kilometres per second, and many chances also of being considerably less.

§ 15. Without attempting to solve the problem of finding the motions and velocities of the thousand million bodies, we can see that if they had been given at rest* twenty-five million years

---

* "The potential energy of gravitation may be in reality the ultimate created

Molar. ago distributed uniformly or non-uniformly through our sphere (5) of $3\cdot09 \cdot 10^{16}$ kilometres radius, a very large proportion of them would now have velocities not less than twenty or thirty kilometres per second, while many would have velocities less than that; and certainly some would have velocities greater than 108 kilometres per second; or if thousands of millions of years ago they had been given at rest, at distances from one another very great in comparison with r (5), so distributed that they should temporarily now be equably spaced throughout a spherical surface of radius r (5), their mean velocity (reckoned as the square root of the mean of the squares of their actual velocities) would now be $50\cdot4$ kilo-metres per second*. This is not very unlike what we know of the stars visible to us. Thus it is quite possible, perhaps pro-bable, that there may be as much matter as a thousand million suns within the distance corresponding to parallax one one-thousandth of a second ($3\cdot09 \cdot 10^{16}$ kilometres). But it seems perfectly certain that there cannot be within this distance as much matter as ten thousand million suns; because if there were, we should find much greater velocities of visible stars than observation shows; according to the following tables of results, and statements, from the most recent scientific authorities on the subject.

"antecedent of all the motion, heat, and light at present in the universe." See *Mechanical Antecedents of Motion, Heat, and Light*, Art. LXIX. of my *Collected Math. and Phys. Papers*, Vol. II.

* To prove this, remark that the exhaustion of gravitational energy

$$(E = \frac{1}{8\pi} \int_{-\infty}^{+\infty} \int_{-\infty}^{+\infty} \int_{-\infty}^{+\infty} R^2 dx\, dy\, dz,$$

Thomson and Tait's *Natural Philosophy*, Part II. § 549) when a vast number, N, of equal masses come from rest at infinite distances from one another to an equably spaced distribution through a sphere of radius r is easily found to be $3/10\ Fr$, where F denotes the resultant force of the attraction of all of them on a material point, of mass equal to the sum of their masses, placed at the spherical surface. Now this exhaustion of gravitational energy is spent wholly in the generation of kinetic energy; and therefore we have $\Sigma \frac{1}{2} mv^2 = \frac{3}{10} Fr$, and by (7) $F = 1\cdot37 \cdot 10^{-13} \Sigma m$; whence

$$\frac{\Sigma mv^2}{\Sigma m} = \frac{3}{5} 1\cdot37 \cdot 10^{-13} \cdot r$$

which, for the case of equal masses, gives, with (5) for the value of r,

$$\sqrt{\frac{\Sigma v^2}{N}} = \sqrt{(\tfrac{3}{5} 1\cdot37 \cdot 10^{-13} \cdot 3\cdot09 \cdot 10^{16})} = 50\cdot4 \text{ kilometres per second.}$$

From the *Annuaire du Bureau des Longitudes* (*Paris*, 1901). Molar.

Magnitude	Name of Star	Distance from earth, in million million Kilometres	Annual proper motions	Parallax	Velocities perpendicular to line of sight, in kilometres per second
0·7	α Centauri	43	3″·62	0″·72	23·9
6·8	21185 Lalande........	64	4·75	0·48	47·1
5·1	61 Cygni	70	5·17	0·44	55·7
−1·4	Sirius	83	1·32	0·37	17·0
8·2	18609 Arg.-Œltzen ...	88	2·30	0·35	31·3
7·9	34 Groombridge	99	2·83	0·31	43·5
7·5	9352 Lacaille	110	6·97	0·28	118·5
0·5	Procyon	110	1·26	0·27	22·2
9·0	11677 Arg.-Œltzen ...	119	3·05	0·26	55·7
6·5	1643 Fedorenko	123	1·43	0·25	27·2
8·5	21258 Lalande........	128	4·40	0·24	87·1
4·7	σ Draconis	128	1·84	0·24	36·5
3·6	η Cassiopeiæ	147	1·19	0·21	27·0
0·2	α Aurigæ	147	0·43	0·21	9·8
9·0	17415 Arg.-Œltzen...	154	1·27	0·20	30·2
0·9	α Aquilæ	154	0·64	0·20	15·2
5·2	ε "Indien"	154	4·60	0·20	109·5
4·5	o² Éridani..............	181	4·05	0·17	113·2
2·4	β Cassiopeiæ	193	0·57	0·16	16·9
1·0	α Tauri.................	206	0·19	0·15	6·0
7·0	1831 Fedorenko	206	0·42	0·15	13·3
4·1	p′ Ophiuchi	206	1·13	0·15	35·8
0·2	Vega	206	0·36	0·15	11·4
2·2	α Urs. Min. (Polaris)	440	0·05	0·07	3·4

Stars which have largest of observed Velocities in the Line of Sight. (Extract by the Astronomer Royal from an Article in the *Astrophysical Journal* for 1901, January, by W. W. Campbell, Director of Lick Observatory.)

Magnitudes	Star	R. A.	Dec.	Velocity
		h. m.	° ′	
4·6	ε Andromedæ	0 33	+28 46	− 84 km. per sec.
	μ Cassiopeiæ	1 0	+54 20	− 97 ,, ,,
	δ Leporis	5 47	− 20 54	+95 ,, ,,
4·2	θ Canis Majoris	6 50	− 11 55	+96 ,, ,,
	ι Pegasi.................	21 17	+19 23	− 76 ,, ,,
4·1	μ Sagittarii	18 8	− 21 1	− 76 ,, ,,

The + sign denotes recession, the − sign approach.

Molar. Motions of Stars in the Line of Sight determined at Potsdam
Observatory, 1889–1891. (Communicated by Professor Becker,
University Observatory, Glasgow.)

Star	Magnitude	Velocity relative to the Sun	Star	Magnitude	Velocity relative to the Sun
		km.			km.
α Andromedæ ...	2·0	+ 4·5	γ Leonis............	2·0	− 38·5
β Cassiopeiæ	2·1	+ 5·2	β Ursæ Majoris ...	2·3	− 29·3
α Cassiopeiæ	var.	− 15·2	α Ursæ Majoris ...	2·0	− 11·9
γ Cassiopeiæ	2·0	− 3·5	δ Leonis	2·3	− 14·4
β Andromedæ ...	2·3	+ 11·2	β Leonis............	2·0	− 12·2
α Ursæ Minoris ...	2·0	− 25·9	γ Ursæ Majoris ...	2·3	− 26·6
γ Andromedæ ...	2·4	− 12·9	ε Ursæ Majoris ...	2·0	− 30·3
α Arietis	2·0	− 14·7	α Virginis	1·0	− 14·8
β Persei	var.	− 1·5	ζ Ursæ Majoris ...	2·1	− 31·2
α Persei	2·0	− 10·3	η Ursæ Majoris ...	2·0	− 26·2
α Tauri	1·0	+ 48·5	α Bootis	1·0	− 7·7
α Aurigæ............	1·0	+ 24·5	ε Bootis	2·0	− 16·3
β Orionis	1·0	+ 16·4	β Ursæ Minoris ...	2·0	+ 14·2
γ Orionis	2·0	+ 9·2	β Libræ	2·0	− 9·6
β Tauri	2·0	+ 8·0	α Coronæ	2·0	+ 32·0
δ Orionis............	2·5	+ 0·9	α Serpentis..	2·3	+ 22·3
ε Orionis............	2·0	+ 26·5	β Herculis	2·3	− 35·3
ζ Orionis............	2·0	+ 14·8	α Ophiuchi.........	2·0	+ 19·2
α Orionis............	var.	+ 17·2	α Lyræ	1·0	− 15·3
β Aurigæ............	2·0	− 28·1	α Aquilæ............	1·3	− 36·9
γ Geminorum ...	2·3	− 16·6	γ Cygni	2·4	− 6·4
α Canis Majoris...	1·0	− 15·6	α Cygni	1·6	− 8·0
α Geminorum ...	2·3	− 29·7	ε Pegasi	2·3	+ 8·0
α Canis Minoris...	1·0	− 9·2	β Pegasi	var	+ 6·7
β Geminorum ...	1·3	+ 1·1	α Pegasi	2·0	+ 1·3
α Leonis	1·3	− 9·1			

The velocity of the sun relatively to stars in general according
to Kempf and Risteen is probably about 19 kilometres per second*.
In respect to greatest proper motions and velocities Sir Norman
Lockyer gives me the following information :—"The star with
" the greatest known proper motion (across the line of sight) is
" 243 Cordoba = 8″·7 per annum. Velocity in kilometres not
" known.

" 1830 Groombridge has a proper motion of 7″·0 per annum
" and a parallax of 0″·089, from which it results that the velocity
" across the line of sight is 370 kms. per second. Various esti-
" mates of the parallax, however, have been made and this velocity
" is somewhat uncertain. The star with the greatest known
" velocity in the line of sight is ζ Herculis, which travels at
" 70 kms. per second.

* See footnote on § 10 of Appendix B.

"The dark line component of Nova Persei wås approaching Molar. "the earth with a velocity of over 1100 kms. per second." This last-mentioned and greatest velocity is probably that of a torrent of gas due to comparatively small particles of melted and evaporating fragments shot out laterally from two great solid or liquid masses colliding with one another, which may be many times greater than the velocity of either before collision; just as we see in the trajectories of small fragments shot out nearly horizontally when a condemned mass of cast-iron is broken up by a heavy mass of iron falling upon it from a height of perhaps twenty feet in engineering works.

§ 16. Newcomb has given a most interesting speculation regarding the very great velocity of 1830 Groombridge, which he concludes as follows:—"If, then, the star in question belongs to "our stellar system, the masses or extent of that system must "be many times greater than telescopic observation and astro- "nomical research indicate. We may place the dilemma in a "concise form, as follows :—

"Either the bodies which compose our universe are vastly "more massive and numerous than telescopic examination seems "to indicate, or 1830 Groombridge is a runaway star, flying on a "boundless course through infinite space with such momentum that "the attraction of all the bodies of the universe can never stop it.

"Which of these is the more probable alternative we cannot "pretend to say. That the star can neither be stopped, nor bent "far from its course until it has passed the extreme limit to "which the telescope has ever penetrated, we may consider "reasonably certain. To do this will require two or three millions "of years. Whether it will then be acted on by attractive forces "of which science has no knowledge, and thus carried back to "where it started, or whether it will continue straightforward for "ever, it is impossible to say.

"Much the same dilemma may be applied to the past history "of this body. If the velocity of two hundred miles or more per "second with which it is moving exceeds any that could be pro- "duced by the attraction of all the other bodies in the universe, "then it must have been flying forward through space from the "beginning, and, having come from an infinite distance, must be "now passing through our system for the first and only time."

T. L. 18

§ 17. In all these views the chance of passing another star at some small distance such as one or two or three times the sun's radius has been overlooked; and that this chance is not excessively rare seems proved by the multitude of Novas (collisions and their sequels) known in astronomical history. Suppose, for example, 1830 Groombridge, moving at 370 kilometres per second, to chase a star of twenty times the sun's mass, moving nearly in the same direction with a velocity of 50 kilometres per second, and to overtake it and pass it as nearly as may be without collision. Its own direction would be nearly reversed and its velocity would be diminished by nearly 100 kilometres per second. By two or three such casualties the greater part of its kinetic energy might be given to much larger bodies previously moving with velocities of less than 100 kilometres per second. By supposing reversed, the motions of this ideal history, we see that 1830 Groombridge may have had a velocity of less than 100 kilometres per second at some remote past time, and may have had its present great velocity produced by several cases of near approach to other bodies of much larger mass than its own, previously moving in directions nearly opposite to its own, and with velocities of less than 100 kilometres per second. Still it seems to me quite possible that Newcomb's brilliant suggestion may be true, and that 1830 Groombridge is a roving star which has entered our galaxy, and is destined to travel through it in the course of perhaps two or three million years, and to pass away into space never to return to us.

§ 18. Many of our supposed thousand million stars, perhaps a great majority of them, may be dark bodies; but let us suppose for a moment each of them to be bright, and of the same size and brightness as our sun; and on this supposition and on the further suppositions that they are uniformly scattered through a sphere (5) of radius $3·09 . 10^{16}$ kilometres, and that there are no stars outside this sphere, let us find what the total amount of starlight would be in comparison with sunlight. Let n be the number per unit of volume, of an assemblage of globes of radius a scattered uniformly through a vast space. The number in a shell of radius q and thickness dq will be $n . 4\pi q^2 dq$, and the sum of their apparent areas as seen from the centre will be

$$\frac{\pi a^2}{q^2} n . 4\pi q^2 dq \text{ or } n . 4\pi^2 a^2 dq.$$

Hence by integrating from $q = 0$ to $q = r$ we find

$$n \cdot 4\pi^2 a^2 r \dots\dots\dots\dots\dots\dots\dots(8)$$

for the sum of their apparent areas. Now if N be the total number in the sphere of radius r we have

$$n = N \Big/ \left(\frac{4\pi}{3} r^3\right) \dots\dots\dots\dots\dots(9).$$

Hence (8) becomes $N \cdot 3\pi \left(\dfrac{a}{r}\right)^2$; and if we denote by α the ratio of the sum of the apparent areas of all the globes to 4π we have

$$\alpha = \frac{3N}{4} \left(\frac{a}{r}\right)^2 \dots\dots\dots\dots\dots(10).$$

$(1 - \alpha)/\alpha$, very approximately equal to $1/\alpha$, is the ratio of the apparent area not occupied by stars to the sum of the apparent areas of all their discs. Hence α is the ratio of the apparent brightness of our star-lit sky to the brightness of our sun's disc. Cases of two stars eclipsing one another wholly or partially would, with our supposed values of r and a, be so extremely rare that they would cause a merely negligible deduction from the total of (10), even if calculated according to pure geometrical optics. This negligible deduction would be almost wholly annulled by diffraction, which makes the total light from two stars of which one is eclipsed by the other, very nearly the same as if the distant one were seen clear of the nearer.

§ 19. According to our supposition of § 18, we have $N = 10^9$, $a = 7 . 10^5$ kilometres, and therefore $r/a = 4\cdot4 . 10^{10}$. Hence by (10)

$$\alpha = 3\cdot87 . 10^{-13} \dots\dots\dots\dots\dots\dots(11).$$

This exceedingly small ratio will help us to test an old and celebrated hypothesis that if we could see far enough into space the whole sky would be seen occupied with discs of stars all of perhaps the same brightness as our own sun, and that the reason why the whole of the night-sky and day-sky is not as bright as the sun's disc is that light suffers absorption in travelling through space. Remark that if we vary r keeping the density of the matter the same, N varies as the cube of r. Hence by (10) α varies simply as r; and therefore to make α even as great as $3\cdot87/100$, or, say, the sum of the apparent

Molar. areas of discs 4 per cent. of the whole sky, the radius must be $10^{11}.r$ or $3.09.10^{27}$ kilometres. Now light travels at the rate of 300,000 kilometres per second or $9.45.10^{12}$ kilometres per year. Hence it would take $3.27.10^{14}$ or about $3\frac{1}{4}.10^{14}$ years to travel from the outlying suns of our great sphere to the centre. Now we have irrefragable dynamics proving that the whole life of our sun as a luminary is a very moderate number of million years, probably less than 50 million, possibly between 50 and 100. To be very liberal, let us give each of our stars a life of a hundred million years as a luminary. Thus the time taken by light to travel from the outlying stars of our sphere to the centre would be about three and a quarter million times the life of a star. Hence, if all the stars through our vast sphere commenced shining at the same time, three and a quarter million times the life of a star would pass before the commencement of light reaching the earth from the outlying stars, and at no one instant would light be reaching the earth from more than an excessively small proportion of all the stars. To make the whole sky aglow with the light of all the stars at the same time the commencements of the different stars must be timed earlier and earlier for the more and more distant ones, so that the time of the arrival of the light of every one of them at the earth may fall within the durations of the lights at the earth of all the others! Our supposition of uniform density of distribution is, of course, quite arbitrary; and (§§ 13, 15 above) we ought, in the greater sphere of § 19, to assume the density much smaller than in the smaller sphere (5); and in fact it seems that there may not be enough of stars (bright or dark) to make a total of star-disc-area more than 10^{-12} or 10^{-11} of the whole sky. See Appendix D, " On the Clustering of Gravitational Matter in any " part of the Universe."

§ 20. To understand the sparseness of our ideal distribution of 1000 million suns, divide the total volume of the supposed sphere of radius r (5) by 10^9, and we find $123.5.10^{39}$ cubic kilometres as the volume per sun. Taking the cube root of this we find $4.98.10^{13}$ kilometres as the edge of the corresponding cube. Hence if the stars were arranged exactly in cubic order with our sun at one of the eight corners belonging to eight neighbouring cubes, his six nearest neighbours would be each

at distance $4\cdot 98.10^{13}$ kilometres; which is the distance corre- Molar.
sponding to parallax $0''\cdot 62$. Our sun seen at so great a distance
would probably be seen as a star of something between the first
and second magnitude. For a moment suppose each of our
1000 million suns, while of the same mass as our own sun, to
have just such brightness as to make it a star of the first magnitude at distance corresponding to parallax $1'''\cdot 0$. The brightness
at distance r (5) corresponding to parallax $0'''\cdot 001$ would be one
one-millionth of this, and the most distant of our assumed
stars would be visible through powerful telescopes as stars of the
sixteenth magnitude. Newcomb (*Popular Astronomy*, 1883,
p. 424) estimated between 30 and 50 million as the number of
stars visible in modern telescopes. Young (*General Astronomy*,
p. 448) goes beyond this reckoning and estimates at 100 million
the total number of stars visible through the Lick telescope.
This is only the tenth of our assumed number. It is nevertheless probable that there may be as many as 1000 million
stars within the distance r (5); but many of them may be
extinct and dark, and nine-tenths of them though not all dark
may be not bright enough to be seen by us at their actual
distances.

§ 21. I need scarcely repeat that our assumption of equable
distribution is perfectly arbitrary. How far from being like the
truth is illustrated by Herschel's view of the form of the universe
as shown in Newcomb's *Popular Astronomy*, p. 469. It is quite
certain that the real visible stars within the distance r (5) from
us are very much more crowded in some parts of the whole
sphere than in others. It is also certain that instead of being
all equally luminous as we have taken them, they differ largely
in this respect from one another. It is also certain that the
masses of some are much greater than the masses of others;
as will be seen from the following table, which has been compiled
for me by Professor Becker from André's *Traité d'Astronomie
Stellaire*, showing the sums of the masses of the components of
some double stars, and the data from which these have been
determined.

Molar.

| | Parallax | ½ Major axis | | Period, in years | $M + M'$, in units of the sun's mass |
		in seconds	in terms of semi-major axis of earth's orbit		
α Centauri	0˙75	18˙17	25	84	2˙0
61 Cygni............	0˙44	29˙48	68	783	0˙5
Sirius	0˙39	8˙31	24	52	3˙2
Procyon	0˙27	5˙84	4	40	6˙3
o² Eridani	0˙19	5˙72	28	176	0˙9
η Cassiopeiæ	0˙15	8˙20	39	190	4˙3
ρ Ophiuchi	0˙15	4˙60	30	88	3˙6
γ Virginis	0˙05 †	3˙99	79 *	194	15˙0
γ Leonis............	0˙02 †	1˙98	102 *	407	6˙5

§ 22.   There may also be a large amount of matter in many stars outside the sphere of 3.10¹⁶ kilometres radius, but however much matter there may be outside it, it seems to be made highly probable by §§ 11—21, that the total quantity of matter within it is greater than 100 million times, and less than 2000 million times, the sun's mass.

I wish, in conclusion, to express my thanks to Sir Norman Lockyer, to the Astronomer Royal Mr Christie, to Sir Robert Ball, and to Prof. Becker, for their kindness in taking much trouble to give me information in respect to astronomical data, which has proved most useful to me in §§ 11—21 above.

* From spectroscopic observations by Belopolsky of Poulcowa, combined with elements of orbit.

† Parallax calculated from dynamical determinations of ratio of semi-major axis of double-star's orbit to semi-major axis of earth's orbit.

LECTURE XVII.

THURSDAY, *October* 16, 3.30 P.M. *Altered* (1901, 1902) *to*
extension of Lec. XVI.

§ 23. HITHERTO in all our views we have seen nothing of abso- Molecular.
lute dimensions in molecular structure, and have been satisfied to
consider the distance between neighbouring molecules in gases,
or liquids, or crystals, or non-crystalline solids to be very small in
comparison with the shortest wave-length of light with which we
have been concerned. Even in respect to dispersion, that is to
say, difference of propagational velocity for different wave-lengths,
it has not been necessary for us to accept Cauchy's doctrine that
the spheres of molecular action are comparable with the wave-
length. We have seen that dispersion can be, and probably in
fact is, truly explained by the periods of our waves of light being
not infinitely great in comparison with some of the periods of
molecular vibration; and, with this view, the dimensions of
molecular structure might, so far as dispersion is concerned, be as
small as we please to imagine them, in comparison with wave-
lengths of light. Nevertheless it is exceedingly interesting and
important for intelligent study of molecular structures and the
dynamics of light, to have some well-founded understanding in
respect to probable distances between centres of neighbouring
molecules in all kinds of ponderable matter, while for the present
at all events we regard ether as utterly continuous and structureless. It may be found in some future time that ether too has a
molecular structure, perhaps much finer than any structure of
ponderable matter; but at present we neither see nor imagine
any reason for believing ether to be other than continuous and
homogeneous through infinitely small contiguous portions of
space void of other matter than ether.

§ 24. The first suggestion, so far as we now know, for estimating the dimensions of molecular structure in ordinary matter was

Molecular. given in 1805 by Thomas Young*, as derived from his own and Laplace's substantially identical theories of capillary attraction. In this purely dynamical theory he found that the range of the attractive force of cohesion is equal to $3T/K$; where T denotes the now well-known Young's tension of the free surface of a liquid, and K denotes a multiple integral which appears in Laplace's formulas and is commonly now referred to as Laplace's K, as to the meaning of which there has been much controversy in the columns of *Nature* and elsewhere. Lord Rayleigh in his article of 1890, "On the Theory of Surface Forces†," gives the following very interesting statement in respect to Young's estimate of molecular dimensions :—

§ 25. " One of the most remarkable features of Young's treatise " is his estimate of the range a of the attractive force on the basis " of the relation $T = \frac{1}{3}aK$. Never once have I seen it alluded to ; " and it is, I believe, generally supposed that the first attempt of " the kind is not more than twenty years old. Estimating K at " 23000 atmospheres, and T at 3 grains per inch, Young finds that " ' the extent of the cohesive force must be limited to about the " ' 250 millionth of an inch [10^{-8} cm.]'; and he continues, ' nor is " ' it very probable that any error in the suppositions adopted can " ' possibly have so far invalidated this result as to have made it " ' very many times greater or less than the truth'....Young con- " tinues :—' Within similar limits of uncertainty, we may obtain " ' something like a conjectural estimate of the mutual distance " ' of the particles of vapours, and even of the actual magnitude " ' of the elementary atoms of liquids, as supposed to be nearly in " ' contact with each other; for if the distance at which the force " ' of cohesion begins is constant at the same temperature, and if " ' the particles of steam are condensed when they approach within " ' this distance, it follows that at 60° of Fahrenheit the distance " ' of the particles of pure aqueous vapour is about the 250 " ' millionth of an inch; and since the density of this vapour is " ' about one sixty thousandth of that of water, the distance of the " ' particles must be about forty times as great; consequently the " ' mutual distance of the particles of water must be about the

---

* "On the Cohesion of Fluids," *Phil. Trans.* 1805; *Collected Works*, Vol. I. p. 461.

† *Phil. Mag.* Vol. xxx. 1890, p. 474.

"'ten thousand millionth of an inch* [·025 × 10⁻⁸ cm.]. It is Molecular.
"'true that the result of this calculation will differ considerably
"'according to the temperature of the substances compared....
"'This discordance does not however wholly invalidate the general
"'tenour of the conclusion...and on the whole it appears tolerably
"'safe to conclude that, whatever errors may have affected the
"'determination, the diameter or distance of the particles of
"'water is between the two thousand and the ten thousand
"'millionth of an inch' [between ·125 × 10⁻⁸ and ·025 × 10⁻⁸ of a
"cm.]. This passage, in spite of its great interest, has been so
"completely overlooked that I have ventured briefly to quote it,
"although the question of the size of atoms lies outside the scope
"of the present paper."

§ 26.   The next suggestion, so far as I know, for estimating the
dimensions of molecular structure in ordinary matter, is to be
found in an extract from a letter of my own to Joule on the
contact electricity of metals, published in the *Proceedings* of the
Manchester Literary and Philosophical Society†, Jan. 21, 1862,
which contains the following passage :—"Zinc and copper con-
"nected by a metallic arc attract one another from any distance.
"So do platinum plates coated with oxygen and hydrogen respec-
"tively.   I can now tell the amount of the force, and calculate
"how great a proportion of chemical affinity is used up electrically,
"before two such discs come within 1/1000 of an inch of one
"another, or any less distance down to a limit within which
"molecular heterogeneousness becomes sensible.   This of course
"will give a definite limit for the sizes of atoms, or rather, as I do
"not believe in atoms, for the dimensions of molecular structures."
The theory thus presented is somewhat more fully developed in a
communication to *Nature* in March 1870, on "The Size of Atoms‡,"
and in a Friday evening lecture§ to the Royal Institution on the

---

* Young here, curiously insensible to the kinetic theory of gases, supposes the
molecules of vapour of water at 60° Fahr. to be within touch (or direct mutual
action) of one another; and thus arrives at a much finer-grainedness for liquid
water than he would have found if he had given long enough free paths to molecules
of the vapour to account for its approximate fulfilment of Boyle's law.

† Reproduced as Art. 22 of my *Electrostatics and Magnetism*.

‡ Republished as Appendix (F) in Thomson and Tait's *Natural Philosophy*,
Part II. Second Edition.

§ Republished in *Popular Lectures and Addresses*, Vol. I.

Molecular. same subject on February 3, 1883; but to illustrate it, information was wanting regarding the heat of combination of copper and zinc. Experiments by Professor Roberts-Austen and by Dr A. Galt, made within the last four years, have supplied this want; and in a postscript of February 1898 to a Friday evening lecture on "Contact Electricity," which I gave at the Royal Institution on May 21, 1897, I was able to say "We cannot avoid seeing "molecular structures beginning to be perceptible at distances of "the hundred-millionth of a centimetre, and we may consider it "as highly probable that the distance from any point in a molecule "of copper or zinc to the nearest corresponding point of a neighbour-"ing molecule is less than one one-hundred-millionth, and greater "than one one-thousand-millionth of a centimetre"; and also to confirm amply the following definite statement which I had given in my *Nature* article (1870) already referred to :—"Plates "of zinc and copper of a three hundred-millionth of a centimetre "thick, placed close together alternately, form a near approxima-"tion to a chemical combination, if indeed such thin plates could "be made without splitting atoms."

§ 27. In that same article thermodynamic considerations in stretching a fluid film against surface tension led to the following result :—"The conclusion is unavoidable, that a water-film falls "off greatly in its contractile force before it is reduced to a thick-"ness of a two hundred-millionth of a centimetre. It is scarcely "possible, upon any conceivable molecular theory, that there can "be any considerable falling off in the contractile force as long as "there are several molecules in the thickness. It is therefore "probable that there are not several molecules in a thickness of a "two-hundred-millionth of a centimetre of water." More detailed consideration of the work done in stretching a water-film led me in my Royal Institution Lecture of 1883 to substitute one one-hundred-millionth of a centimetre for one two-hundred-millionth in this statement. On the other hand a consideration of the large black spots which we now all know in a soap-bubble or soap-film before it bursts, and which were described in a most interesting manner by Newton*, gave absolute demonstration that the film retains its tensile strength in the black spot "where the

---

* Newton's *Optics*, pp. 187, 191, Edition 1721, Second Book, Part I.: quoted in my Royal Institution Lecture, *Pop. Lectures and Addresses*, Vol. I. p. 175.

"thickness is clearly much less than 1/60000 of a centimetre, Molecular. "this being the thickness of the dusky white" with which the black spot is bordered. And further in 1883 Reinold and Rücker's* admirable application of optical and electrical methods of measurement proved that the thickness of the black film in Plateau's "liquide glycérique" and in ordinary soap solution is between one eight-hundred-thousandth of a centimetre and one millionth of a centimetre. Thus it was certain that the soap-film has full tensile strength at a thickness of about a millionth of a centimetre, and that between one millionth and one one-hundred-millionth the tensile strength falls off enormously.

§ 28. Extremely interesting in connection with this is the investigation, carried on independently by Röntgen † and Rayleigh‡ and published by each in 1890, of the quantity of oil spreading over water per unit area required to produce a sensible disturbance of its capillary tension. Both experimenters expressed results in terms of thickness of the film, calculated as if oil were infinitely homogeneous and therefore structureless, but with very distinct reference to the certainty that their films were molecular structures not approximately homogeneous. Rayleigh found that olive oil, spreading out rapidly all round on a previously cleaned surface of water from a little store carried by a short length of platinum wire, produced a perceptible effect on little floating fragments of camphor at places where the thickness of the oil was 10.6×10^{-8} cm., and no perceptible effect where the thickness was 8.1×10^{-8} cm. It will be highly interesting to find, if possible, other tests (optical or dynamical or electrical or chemical) for the presence of a film of oil over water, or of films of various liquids over solids such as glass or metals, demonstrating by definite effects smaller and smaller thicknesses. Röntgen, using ether instead of camphor, found analogous evidence of layers 5.6×10^{-8} cm. thick. It will be very interesting for example to make a thorough investigation of the electric conductance of a clean rod of white glass of highest insulating quality surrounded by an atmosphere containing measured quantities of vapour of

---

* "On the Limiting Thickness of Liquid Films," *Roy. Soc. Proc.* April 19, 1883; *Phil. Trans.* 1883, Part II. p. 645.

† *Wied. Ann.* Vol. XLI. 1890, p. 321.

‡ *Proc. Roy. Soc.* Vol. XLVII. 1890, p. 364.

Molecular. water.   When the glass is at any temperature above the dew-point of the vapour, it presents, so far as we know, no optical appearance to demonstrate the pressure of condensed vapour of water upon it : but enormous differences of electric conductance, according to the density of the vapour surrounding it, prove the presence of water upon the surface of the glass, or among the interstices between its molecules, of which electric conductance is the only evidence.   Rayleigh has himself expressed this view in a recent article, " Investigations on Capillarity" in the *Philosophical Magazine.** From the estimates of the sizes of molecules of argon, hydrogen, oxygen, carbonic oxide, carbonic acid, ethylene (C_2H_4), and other gases, which we shall have to consider (§ 47 below), we may judge that in all probability if we had eyes microscopic enough to see atoms and molecules, we should see in those thin films of Rayleigh and Röntgen merely molecules of oil lying at greater and less distances from one another, but at no part of the film one molecule of oil lying above another or resting on others.

§ 29.   A very important and interesting method of estimating the size of atoms, founded on the kinetic theory of gases, was first, so far as I know, thought of by Loschmidt† in Austria and Johnstone Stoney in Ireland.   Substantially the same method occurred to myself later and was described in *Nature*, March 1870, in an article‡ on the "Size of Atoms" already referred to, § 26 above, from which the quotations in §§ 29, 30 are taken.

"The kinetic theory of gases suggested a hundred years ago "by Daniel Bernoulli has, during the last quarter of a century, "been worked out by Herapath, Joule, Clausius, and Maxwell "to so great perfection that we now find in it satisfactory ex-"planations of all non-chemical" and non-electrical "properties of "gases.   However difficult it may be to even imagine what kind "of thing the molecule is, we may regard it as an established "truth of science that a gas consists of moving molecules dis-"turbed from rectilinear paths and constant velocities by collisions "or mutual influences, so rare that the mean length of nearly

---

* *Phil. Mag.* Oct. 1899, p. 337.

† Sitzungsberichte of the Vienna Academy, Oct. 12, 1865, p. 395.

‡ Reprinted as Appendix (F) in Thomson and Tait's *Natural Philosophy*, Part II. p. 499.

"rectilinear portions of the path of each molecule is many times Molecular.
"greater than the average distance from the centre of each
"molecule to the centre of the molecule nearest it at any time.
"If, for a moment, we suppose the molecules to be hard elastic
"globes all of one size, influencing one another only through
"actual contact, we have for each molecule simply a zigzag path
"composed of rectilinear portions, with abrupt changes of direc-
"tion.......But we cannot believe that the individual molecules
"of gases in general, or even of any one gas, are hard elastic
"globes. Any two of the moving particles or molecules must act
"upon one another somehow, so that when they pass very near
"one another they shall produce considerable deflexion of the
"path and change in the velocity of each. This mutual action
"(called force) is different at different distances, and must vary,
"according to variations of the distance, so as to fulfil some
"definite law. If the particles were hard elastic globes acting
"upon one another only by contact, the law of force would be
"...zero force when the distance from centre to centre exceeds
"the sum of the radii, and infinite repulsion for any distance less
"than the sum of the radii. This hypothesis, with its 'hard and
"'fast' demarcation between no force and infinite force, seems to
"require mitigation." Boscovich's theory supplies clearly the
needed mitigation.

§ 30. To fix the ideas we shall still suppose the force absolutely
zero when the distance between centres exceeds a definite limit, λ;
but when the distance is less than λ, we shall suppose the force
to begin either attractive or repulsive, and to come gradually to
a repulsion of very great magnitude, with diminution of distance
towards zero. Particles thus defined I call Boscovich atoms. We
thus call $\frac{1}{2}\lambda$ the radius of the atom, and λ its diameter. We
shall say that two atoms are in collision when the distance
between their centres is less than λ. Thus "two molecules in
"collision will exercise a mutual repulsion in virtue of which the
"distance between their centres, after being diminished to a mini-
"mum, will begin to increase as the molecules leave one another.
"This minimum distance would be equal to the sum of the radii,
"if the molecules were infinitely hard elastic spheres; but in
"reality we must suppose it to be very different in different
"collisions."

Molecular.   § 31. The essential quality of a gas is that the straight line
of uniform motion of each molecule between collisions, called
the free path, is long in comparison with distances between centres
during collision. In an ideal perfect gas the free path would
be infinitely long in comparison with distances between centres
during collision, but infinitely short in comparison with any length
directly perceptible to our senses ; a condition which requires the
number of molecules in any perceptible volume to be exceedingly
great. We shall see that in gases which at ordinary pressures
and temperatures approximate most closely, in respect to compressibility, expansion by heat, and specific heats, to the ideal
perfect gas, as, for example, hydrogen, oxygen, nitrogen, carbon-
monoxide, the free path is probably not more than about one
hundred times the distance between centres during collisions,
and is little short of 10^{-5} cm. in absolute magnitude. Although
these moderate proportions suffice for the well-known exceedingly
close agreement with the ideal gaseous laws presented by those
real gases, we shall see that large deviations from the gaseous
laws are presented with condensations sufficient to reduce the free
paths to two or three times the diameter of the molecule, or to
annul the free paths altogether.

     § 32. It is by experimental determinations of diffusivity that
the kinetic theory of gases affords its best means for estimating
the sizes of atoms or molecules and the number of molecules
in a cubic centimetre of gas at any stated density. Let us
therefore now consider carefully the kinetic theory of these
actions, and with them also, the properties of thermal conductivity
and viscosity closely related to them, as first discovered and
splendidly developed by Clausius and Clerk Maxwell.

     § 33. According to their beautiful theory, we have three
kinds of diffusion ; diffusion of molecules, diffusion of energy, and
diffusion of momentum. Even in solids, such as gold and lead,
Roberts-Austen has discovered molecular diffusion of gold into
lead and lead into gold between two pieces of the metals when
pressed together. But the rate of diffusion shown by this admirable discovery is so excessively slow that for most purposes,
scientific and practical, we may disregard wandering of any
molecule in any ordinary solid to places beyond direct influence of

its immediate neighbours. In an elastic solid we have diffusion Molecular. of momentum by wave motion, and diffusion of energy constituting the conduction of heat through it. These diffusions are effected solely by the communication of energy from molecule to molecule and are practically not helped at all by the diffusion of molecules. In liquids also, although there is thorough molecular diffusivity, it is excessively slow in comparison with the two other diffusivities, so slow that the conduction of heat and the diffusion of momentum according to viscosity are not practically helped by molecular diffusion. Thus, for example, the thermal diffusivity* of water ($\cdot$002, according to J. T. Bottomley's first investigation, or about $\cdot$0015† according to later experimenters) is several hundred times, and the diffusivity for momentum is from one to two thousand times, the diffusivity of water for common salt, and other salts such as sulphates, chlorides, bromides, and iodides.

§ 34. We may regard the two motional diffusivities of a liquid as being each almost entirely due to communication of motion from one molecule to another. This is because every molecule is always under the influence of its neighbours and has no free path. When a liquid is rarefied, either gradually as in Andrew's experiments showing the continuity of the liquid and gaseous states, or suddenly as in evaporation, the molecules become less crowded and each molecule gains more and more of freedom. When the density is so small that the straight free paths are great in comparison with the diameters of molecules, the two motional diffusivities are certainly due, one of them to carriage of energy, and the other to carriage of momentum, chiefly by the free rectilinear motion of the molecules between collisions. Interchange of energy or of momentum between two molecules during collision will undoubtedly to some degree modify the results of mere transport; and we might expect on this account the motional diffusivities to be approximately equal to, but each somewhat greater than, the molecular diffusivity. If this view were correct, it would follow that, in a homogeneous gas when the free paths are long in comparison with the diameters of molecules, the viscosity is equal to the molecular diffusivity multiplied by the

---

* *Math. and Phys. Papers*, Vol. III. p. 226. For explanation regarding diffusivity and viscosity see same volume, pp. 428—435.

† See a paper by Milner and Chattock, *Phil. Mag.* Vol. XLVIII. 1899.

Molecular. density, and the thermal conductivity is equal to the molecular diffusivity multiplied by the thermal capacity per unit bulk, pressure constant: and that whatever deviation from exactness of these equalities there may be, would be in the direction of the motional diffusivities being somewhat greater than the molecular diffusivity. But alas, we shall see, § 45 below, that hitherto experiment does not confirm these conclusions: on the contrary the laminar diffusivities (or diffusivities of momentum) of the only four gases of which molecular diffusivities have been determined by experiment, instead of being greater than, or at least equal to, the density multiplied by the molecular diffusivity, are each somewhat less than three-fourths of the amount thus calculated.

§ 35. I see no explanation of this deviation from what seems thoroughly correct theory. Accurate experimental determinations of viscosities, whether of gases or liquids, are easy by Graham's transpirational method. On the other hand even roughly approximate experimental determinations of thermal diffusivities are exceedingly difficult, and I believe none, on correct experimental principles, have really been made*; certainly none unvitiated by currents of the gas experimented upon, or accurate enough to give any good test of the theoretical relation between thermal and material diffusivities, expressed by the following equation, derived from the preceding verbal statement regarding the three diffusivities of a gas,

$$\theta = K\rho \frac{\mu}{\rho} = K\mu = kc\mu,$$

where θ denotes the thermal conductivity, μ the viscosity, ρ the density, $K\rho$ the thermal capacity per unit bulk pressure constant, K the thermal capacity per unit mass pressure constant, c the thermal capacity per unit mass volume constant, and k the ratio of the thermal capacity pressure constant to the thermal capacity volume constant. It is interesting to remark how nearly theo-

---

* So far as I know, all attempts hitherto made to determine the thermal conductivities of gases have been founded on observations of rate of communication of heat between a thermometer bulb, or a stretched metallic wire constituting an electric resistance thermometer, and the walls of the vessel enclosing it and the gas experimented upon. See Wiedemann's *Annalen*, 1888, Vol. xxxiv. p. 623, and 1891, Vol. xliv. p. 177. For other references, see O. E. Meyer, § 107.

retical investigators* have come to the relation $\theta = kc\mu$; Clausius Molecular.
gave $\theta = \frac{5}{4}c\mu$; O. E. Meyer, $\theta = 1\cdot6027c\mu$, and Maxwell, $\theta = \frac{5}{3}c\mu$.
Maxwell's in fact is $\theta = kc\mu$ for the case of a monatomic gas.

§ 36. To understand exactly what is meant by molecular
diffusivity, consider a homogeneous gas between two infinite
parallel planes, GGG and RRR, distance a apart, and let it be
initially given in equilibrium; that is to say, with equal numbers
of molecules and equal total kinetic energies in equal volumes,
and with integral of component momentum in any and every
direction, null. Let N be the number of molecules per unit
volume. Let every one of the molecules be marked either green
or red, and whenever a red molecule strikes the plane GGG, let its
marking be altered to green, and, whenever a green molecule
strikes RRR, let its marking be altered to red. These markings
are not to alter in the slightest degree the mass or shape or elastic
quality of the molecules, and they do not disturb the equilibrium
of the gas or alter the motion of any one of its particles; they
are merely to give us a means of tracing ideally the history of any
one molecule or set of molecules, moving about and colliding with
other molecules according to the kinetic nature of a gas.

§ 37. Whatever may have been the initial distribution of the
greens and reds, it is clear that ultimately there must be a regular
transition from all greens at the plane GGG and all reds at the
plane RRR, according to the law

$$g = N\frac{x}{a}; \quad r = N\frac{a-x}{a} \quad \ldots\ldots\ldots\ldots\ldots\ldots(1),$$

where g and r denote respectively the number of green molecules
and of red molecules per unit volume at distance x from the
plane RRR. In this condition of statistical equilibrium, the
total number of molecules crossing any intermediate parallel
plane from the direction GGG towards RRR will be equal to the
number crossing from RRR towards GGG in the same time; but
a larger number of green molecules will cross towards RRR than
towards GGG, and, by an equal difference, a larger number of red
molecules will cross towards GGG than towards RRR. If we
denote this difference per unit area per unit time by QN, we have

* See the last ten lines of O. E. Meyer's book.

Molecular. for what I call the material diffusivity (called by Maxwell, "co-efficient of diffusion"),

$$D = Qa \dots\dots\dots\dots\dots\dots\dots(2).$$

We may regard this equation as the definition of diffusivity. Remark that Q is of dimensions LT^{-1}, because it is a number per unit of area per unit of time (which is of dimensions $L^{-2}T^{-1}$) divided by N, a number per unit of bulk (dimensions L^{-3}). Hence the dimensions of a diffusivity are L^2T^{-1}; and practically we reckon it in square centimetres per second.

§ 38. Hitherto we have supposed the G and the R particles to be of exactly the same quality in every respect, and the diffusivity which we have denoted by D is the inter-diffusivity of the molecules of a homogeneous gas. But we may suppose G and R to be molecules of different qualities; and assemblages of G molecules and of R molecules to be two different gases. Everything described above will apply to the inter-diffusions of these two gases; except that the two differences which are equal when the red and green molecules are of the same quality are now not equal or, at all events, must not without proof be assumed to be equal. Let us therefore denote by Q_gN the excess of the number of G molecules crossing any intermediate plane towards RRR over the number crossing towards GGG, and by Q_rN the excess of the number of R molecules crossing towards GGG above that crossing towards RRR. We have now two different diffusivities of which the mean values through the whole range between the bounding planes are given by the equations

$$D_g = Q_ga; \; D_r = Q_ra;$$

one of them, D_g, the diffusivity of the green molecules, and the other, D_r, the diffusivity of the red molecules through the heterogeneous mixture in the circumstances explained in § 37. We must not now assume the gradients of density of the two gases to be uniform as expressed by (1) of § 37, because the homogeneousness on which these equations depend no longer exists.

§ 39. To explain all this practically*, let, in the diagram, the planes GGG, and RRR, be exceedingly thin plates of dry porous material such as the fine unglazed earthenware of Graham's experi-

* For a practical experiment it might be necessary to allow for the difference of the proportions of the G gas on the two sides of the RRR plate and of the R gas on the two sides of the GGG plate. This would be exceedingly difficult, though not impossible, in practice. The difficulty is analogous to that of allowing for the

ments. Instead of our green and red marked molecules of the same Molecular.
kind, let us have two gases, which we shall call G and R, supplied in
abundance at the middles of the two ends of a non-porous tube of

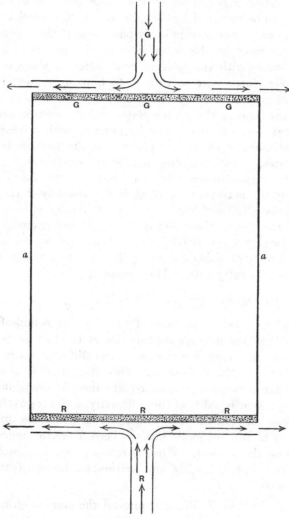

electric resistances of the connections at the ends of a stout bar of metal of which
it is desired to measure the electric resistance. But the simple and accurate
"potential method" by which the difficulty is easily and thoroughly overcome
in the electric case is not available here. I do not, however, put forward the
arrangement described in the text as an eligible plan for measuring the inter-
diffusivity of two gases. Even if there were no other difficulty, the quantities of the
two pure gases required to realize it would be impracticably great.

Molecular. glass or metal, and guided to flow away radially in contact with
the end-plates as indicated in the diagram.   If the two axial
supply-streams of the two pure gases are sufficiently abundant,
the spaces GGG, RRR, close to the inner sides of the porous
end-plates will be occupied by the gases G and R, somewhat nearly
pure.   They could not be rigorously pure even if the velocities of
the scouring gases on the outer sides of the porous end-plates
were comparable with the molecular velocities in the gases, and
if the porous plates were so thin as to have only two or three
molecules of solid matter in their thickness.   The gases in contact
with the near faces of the porous plates would, however, probably
be somewhat approximately pure in practice with a practically
realizable thinness of the porous plates, if a, the distance between
the two plates, is not less than five or six centimetres and the
scouring velocities moderately, but not impracticably, great.
According to the notation of § 37, Q_g is the quantity of the G gas
entering across GGG and leaving across RRR per sec. of time per
sq. cm. of area; Q_r is the quantity of the R gas entering across
RRR and leaving across GGG per sec. of time per sq. cm. of area;
the unit quantity of either gas being that which occupies a cubic
centimetre in its entry tube.   The equations

$$D_g = Q_g \frac{a}{(1 - g - r)}, \qquad D_r = Q_r \frac{a}{(1 - g - r)},$$

where g and r are the proportions of the G gas at R and of the R
gas at G, define the average diffusivities of the two gases in the
circumstances in which they exist in the different parts of the
length a between the end-plates.   This statement is cautiously
worded to avoid assuming either equal values of the diffusivities
of the two gases or equality of the diffusivity of either gas throughout the space between the end-plates.   So far as I know difference
of diffusivity of the two gases has not been hitherto suggested by
any writer on the subject.   What is really given by Loschmidt's
experiments, § 43 below, is the arithmetic mean of the two
diffusivities D_g and D_r.

§ 40.   In 1877 O. E. Meyer expressed the opinion on theoretical grounds, which seem to me perfectly valid, that the inter-
diffusivity of two gases varies according to the proportions of the
two gases in the mixture.   In the 1899 edition of his *Kinetic
Theory of Gases*[*] he recalls attention to this view and quotes
results of various experimenters, Loschmidt, Obermayer, Waitz,

* Baynes' translation, p. 264.

seeming to support it, but, as he says, not quite conclusively. On the other hand, Maxwell's theory (§ 41 below) gives inter-diffusivity as independent of the proportions of the two gases; and only a single expression for diffusivity, which seems to imply that the two diffusivities are equal according to his theory. The subject is of extreme difficulty and of extreme interest, theoretical and practical; and thorough experimental investigation is greatly to be desired.

§ 41. In 1873 Maxwell* gave, as a result of a theoretical investigation, the following formula which expresses the inter-diffusivity (D_{12}) of two gases independently of the proportion of the two gases in any part of the mixture: each gas being supposed to consist of spherical Boscovich atoms mutually acting according to the law, force zero for all distances exceeding the sum of the radii (denoted by s_{12}) and infinite repulsion when the distance between their centres is infinitely little less than this distance :

$$D_{12} = \frac{1}{2\sqrt{6\pi}} \frac{V}{N} \sqrt{\left(\frac{1}{w_1} + \frac{1}{w_2}\right)\frac{1}{s_{12}^2}} \dots \dots \dots \dots (1),$$

where w_1, w_2 are the masses of the molecules in the two gases in terms of that of hydrogen called unity; V is the square root of the mean of the squares of the velocities of the molecules in hydrogen at $0°$ C.; and N is the number of molecules in a cubic centimetre of a gas (the same for all gases according to Avogadro's law) at $0°$ C. and standard atmospheric pressure. I find the following simpler formula more convenient

$$D_{12} = \frac{1}{2\sqrt{6\pi}\,N s_{12}^2} \sqrt{(V_1^2 + V_2^2)} \dots \dots \dots \dots (2),$$

where V_1^2, V_2^2 are the mean squares of the molecular velocities of the two gases at $0°$ C., being the values of $3p/\rho$ for the two gases, or three times the squares of their Newtonian velocities of sound, at that temperature. For brevity, we shall call mean molecular velocity the square root of the mean of the squares of the velocities of the molecules. The same formula is, of course, applicable to the molecular diffusivity of a single gas by taking $V_1 = V_2 = V$ its mean molecular velocity, and $s_{12} = s$ the diameter of its molecules; so that we have

$$D = \frac{1}{2\sqrt{3\pi}} \frac{V}{N s^2} \dots \dots \dots \dots \dots (3).$$

* "On Loschmidt's Experiments on Diffusion in relation to the Kinetic Theory of Gases," *Nature*, Aug. 1873; *Scientific Papers*, Vol. II. pp. 343—350.

§ 42. It is impossible by any direct experiment to find the molecular diffusivity of a single gas, as we have no means of marking its particles in the manner explained in § 36 above; but Maxwell's theory gives us, in a most interesting manner, the means of calculating the diffusivity of each of three separate gases from three experiments determining the inter-diffusivities of their pairs. From the inter-diffusivity of each pair determined by experiment we find, by (2) § 41, a value of $s_{12} \sqrt{(2 \sqrt{3\pi} N)}$ for each pair, and we have $s_{12} = \frac{1}{2}(s_1 + s_2)$* whence

$$s_1 = s_{12} + s_{13} - s_{23}; \quad s_2 = s_{12} + s_{23} - s_{13}; \quad s_3 = s_{13} + s_{23} - s_{12} \ldots\ldots(1).$$

Calculating thus the three values of $s \sqrt{(2 \sqrt{3\pi} N)}$, and using them in (3) § 41, we find the molecular diffusivities of the three separate gases.

§ 43. In two communications† to the Academy of Science of Vienna in 1870, Loschmidt describes experimental determinations of the inter-diffusivities of ten pairs of gases made, by a well-devised method, with great care to secure accuracy. In each case the inter-diffusivity determined by the experiment would be, at all events, somewhat approximately the mean of the two diffusivities, § 39 above, if these are unequal. The results reduced to 0° C. and standard atmospheric pressure, and multiplied by 2·78 to reduce from Loschmidt's square metres per hour to the now usual square centimetres per second, are as follows :—

TABLE OF INTER-DIFFUSIVITIES D.

Pairs of gases	D in sq. cms. per sec.
H_2, O_2	·7214
H_2, CO	·6422
H_2, CO_2	·5558
O_2, CO	·1802
O_2, CO_2	·1409
CO, CO_2	·1406
CO_2, Air	·1423
CO_2, NO	·0984
CO_2, CH_4	·1587
SO_2, H_2	·4809

* This agrees with Maxwell's equation (4), but shows his equation (6) to be incorrect.

† "Experimental-Untersuchungen über die Diffusion von Gasen ohne poröse Scheidewände," *Sitz. d. k. Akad. d. Wissensch.*, March 10 and May 12, 1870.

In the first six of these, each of the four gases H_2, O_2, CO, CO_2 Molecular. occurs three times, and we have four sets of three inter-diffusivities giving in all three determinations of the diffusivity of each gas as follows :—

Pairs of gases	D_1	Pairs of gases	D_2
(12, 13, 23)	......1·32	(12, 13, 23)	......·193
(12, 14, 24)	......1·35	(12, 14, 24)	......·190
(13, 14, 34)	......1·26	(23, 24, 34)	......·183
	Mean 1·31		Mean ·189

Gases

H_2 ......(1)
O_2 ......(2)
CO ......(3)
CO_2......(4)

	D_3		D_4
(12, 13, 23)	......·169	(12, 14, 24)	......·106
(13, 14, 34)	......·175	(13, 14, 34)	......·111
(23, 24, 34)	......·178	(23, 24, 34)	......·109
	Mean ·174		Mean ·109

Considering the great difficulty of the experimental investigation, we may regard the agreements of the three results for each separate gas as, on the whole, very satisfactory, both in respect to the accuracy of Loschmidt's experiments and the correctness of Maxwell's theory. It certainly is a very remarkable achievement of theory and experiment to have found in the four means of the sets of three determinations, what must certainly be somewhat close approximations to the absolute values for the four gases, hydrogen, oxygen, carbon-monoxide, and carbon-dioxide, of something seemingly so much outside the range of experimental observation, as the inter-diffusivity of the molecules of a separate gas.

§ 44. Maxwell, in his theoretical writings of different dates, gave two very distinct views of the inner dynamics of viscosity in a single gas, both interesting, and each, no doubt, valid. In one[*], viscous action is shown as a subsidence from an "instantaneous rigidity of a gas." In the other[†], viscosity is shown as a diffusion of momentum: and in p. 347 of his article quoted in § 41

---

[*] *Trans. Roy. Soc.*, May 1866 ; *Scientific Papers*, Vol. II. p. 70.

[†] "Molecules," a lecture delivered before the Brit. Assoc. at Bradford, *Scientific Papers*, Vol. II. p. 378. See also O. E. Meyer's *Kinetic Theory of Gases* (Baynes' trans. 1899), §§ 74—76.

Molecular. above he gives as from "the theory," but without demonstration, a formula (5), which, taken in conjunction with (1), makes

$$\frac{\mu}{\rho} = D \quad\dots\dots\dots\dots\dots\dots\dots(1);$$

ρ denoting the density, μ the viscosity, and D the molecular diffusivity, of any single gas. On the other hand, in his 1866 paper he had given formulas making[*]

$$\frac{\mu}{\rho} = \cdot 648D \quad\dots\dots\dots\dots\dots\dots(2).$$

§ 45. Viewing viscosity as explained by diffusion of momentum we may, it has always seemed to me (§ 34 above), regard (1) as approximately true for any gas, monatomic, diatomic, or polyatomic, provided only that the mean free path is large in comparison with the sum of the durations of the collisions. Unfortunately for this view, however, comparisons of Loschmidt's excellent experimental determinations of diffusivity with undoubtedly accurate determinations of viscosity from Graham's original experiments on transpiration, and more recent experiments of Obermayer and other accurate observers, show large deviations from (1) and are much more nearly in agreement with (2). Thus taking ·0000900, ·001430, ·001234, ·001974 as the standard densities of the four gases, hydrogen, oxygen, carbon-monoxide, and carbon-dioxide, and multiplying these respectively by the diffusivities from Loschmidt's experiments and Maxwell's theory, we have the following comparison with Obermayer's viscosities at 0° C. and standard pressure, which shows the discrepance from experiment and seeming theory referred to in § 34.

Col. 1	Col. 2	Col. 3	Col. 4
Gas	Viscosity calculated by Maxwell's theory from Loschmidt's diffusivities $\mu = \rho D$	Viscosities according to Obermayer	Ratio of values in Col. 3 to those in Col. 2
H_2	·000119	·0000822	·691
O_2	·000269	·0001873	·696
CO	·000212	·0001630	·769
CO_2	·000218	·0001414	·649

[*] The formula for viscosity (*Sci. Papers*, Vol. II. p. 68) taken with the formula for molecular diffusivity of a single gas, derived from the formula of inter-diffusivity

§ 46. Leaving this discrepance unexplained, and eliminating Molecular.
D between (1) of § 44 and (3) of § 41, we find as Maxwell's latest
expression of the theoretical relation between number of molecules
per cubic centimetre, diameter of the molecules, molecular velocity,
density, and viscosity of a single gas,

$$Ns^2 = \frac{1}{2\sqrt{3\pi}} \frac{V\rho}{\mu} = \cdot 1629 \frac{V\rho}{\mu} \quad \ldots\ldots\ldots\ldots\ldots\ldots(1).$$

The number of grammes and the number of molecules in a cubic
centimetre being respectively ρ and N, ρ/N is the mass of one
molecule in grammes; and therefore, denoting this by m, we have

$$m = 2\sqrt{3\pi} \frac{\mu}{V} s^2 = 6\cdot140 \frac{\mu}{V} s^2 \quad \ldots\ldots\ldots\ldots\ldots\ldots(2).$$

In these formulas, as originally investigated by Maxwell for the
case of an ideal gas composed of hard spherical atoms, s is
definitely the diameter of the atom, and is the same at all
temperatures and densities of the gas. When we apply the
formulas to diatomic or polyatomic gases, or to a monatomic gas
consisting of spherical atoms whose spheres of action may overlap more or less in collision according to the severity of the
impact, s may be defined as the diameter which an ideal hard
spherical atom, equal in mass to the actual molecule, must have to
give the same viscosity as the real gas, at any particular temperature. This being the rigorous definition of s, we may call it
the proper mean shortest distance of inertial centres of the molecules in collision to give the true viscosity; a name or expression
which helps us to understand the thing defined.

§ 47. For the ideal gas of hard spherical atoms, remembering
that V is independent of the density and varies as $t^{\frac{1}{2}}$ (t denoting
absolute temperature), § 46 (2) proves that the viscosity is independent of the density and varies approximately as $t^{\frac{1}{2}}$. Rayleigh's
experimental determinations of the viscosity of argon at different
temperatures show that for this monatomic gas the viscosity varies
as $t^{\cdot815}$; hence § 46 (2) shows that s^2 varies as $t^{-\cdot315}$, and therefore
s varies as $t^{-\cdot16}$. Experimental determinations by Obermayer* of
viscosities and their rates of variation with temperature for carbonic acid, ethylene, ethylene-chloride, and nitrous oxide, show

of two gases of equal densities, gives $\frac{\mu}{\rho D} = \frac{A_1}{3A_2}$, which is equal to $\cdot648$ according to
the values of A_1 and A_2 shown in p. 42 of Vol. II., *Sci. Papers.*

* Obermayer, *Wien. Akad.* 1876, Mar. 16th, Vol. 73, p. 433.

Molecular. that for these the viscosity is somewhat nearly in simple proportion to the absolute temperature: hence for them s^2 varies nearly as $t^{-\cdot5}$. His determinations for the five molecularly simpler gases, air, hydrogen, carbonic oxide, nitrogen, and oxygen show that the increases of μ, and therefore of s^{-2}, with temperature are, as might be expected, considerably smaller than for the more complex of the gases on which he experimented. Taking his viscosities at 0° Cent., for carbonic acid and for the four other simple gases named above, and Rayleigh's for argon, with the known densities of all the six gases at 0° C. and standard atmospheric pressure, we have the following table of the values concerned in § 46 (1):

Col. 1	Col. 2	Col. 3	Col. 4	Col. 5	Col. 6	Col. 7	Col. 8	Col. 9
Gas	ρ in terms of grammes per cubic centimetre	μ in terms of dynes per square centimetre	V in terms of centimetres per second	Ns^2 in terms of (centimetre)$^{-1}$	Hence taking $N=10^{20}$ (§ 50) we have s at 0° Cent. in terms of centimetres	Taking $N=10^{20}$, we have m in terms of grammes	Mean free paths according to Maxwell's formula * $l=\dfrac{1}{\sqrt{2}.\pi Ns^2}$ in terms of centimetres	Ratio of volume occupied by molecules to whole volume $N\dfrac{\pi}{6}s^3$
CO_2	·001974	·0001414	39200	89200	$2\cdot99\,.\,10^{-8}$	$19\cdot74.10^{-24}$	$2\cdot52\,.\,10^{-6}$	$1\cdot390.10^{-3}$
H_2	·0000900	·0000822	184000	32800	$1\cdot81$,,	·90 ,,	$6\cdot89$,,	·311 ,,
CO	·001234	·0001630	49600	61200	$2\cdot47$,,	$12\cdot34$,,	$3\cdot68$,,	·792 ,,
N_2	·001257	·0001635	49200	61600	$2\cdot48$,,	$12\cdot57$,,	$3\cdot66$,,	·800 ,,
O_2	·00143	·0001873	46100	57300	$2\cdot39$,,	$14\cdot3$,,	$3\cdot93$,,	·719 ,,
Argon	·001781	·0002083	41300	57500	$2\cdot40$,,	$17\cdot81$,,	$3\cdot91$,,	·722 ,,

§ 48. The meaning of "s," the diameter, as defined in § 46, is simpler for the monatomic gas, argon, than for any of the others; and happily we know for argon the density, not only in the gaseous state (·001781) but also in the liquid state (1·42)†. The latter of these is 797 times the former. Now, all things considered, it seems probable that the crowd of atoms in the liquid may be slightly less dense than an assemblage of globes of diameter s just touching one another in cubic order; but, to make no hypothesis in the first place, let qs be the distance from centre to centre of a cubic arrangement of the molecules 797 times denser than the gas at 0° C. and standard atmospheric pressure; q will be greater than unity if the liquid is less dense, or less than unity

---

* Maxwell's *Collected Papers*, Vol. II. p. 348, eqn. (7). The formula as printed in this paper contains a very embarrassing mistake, $\sqrt{2\pi}$ for $\sqrt{2}\,.\,\pi$.

† See Baly and Donan, *Chem. Jour.*, July, 1902.

if the liquid is denser, than the cubic arrangement with molecules, Molecular. regarded as spherical of diameter s, just touching. We have

$$797N = 1/(qs)^3 \quad\dots\dots\dots\dots\dots\dots(3),$$

and for argon we have by § 46 (1),

$$Ns^2 = 57500 \quad\dots\dots\dots\dots\dots\dots(4).$$

Eliminating s between these equations we find

$$N = 797^2 . 57500^3 q^6 = 1\cdot21 . 10^{20} . q^6 \quad\dots\dots\dots(5).$$

If the atoms of argon were ideal hard globes, acting on one another with no force except at contact, we should almost certainly have $q \geqq 1$ (because with closer packing than that of cubic order it seems not possible that the assemblage could have sufficient relative mobility of its parts to give it fluidity) and therefore N would be $\geqq 1\cdot21 . 10^{20}$.

§ 49.   For carbonic acid, hydrogen, nitrogen, and oxygen, we have experimental determinations of their densities in the solid or liquid state; and dealing with them as we have dealt with argon, irrespectively of their not being monatomic gases, we find results for the five gases as shown in the following table:

Col. 1	Col. 2		Col. 3	Col. 4	Col. 5
Gas	Solid or liquid density		Ratio of solid or liquid density to standard gaseous density	Number of molecules per cubic centimetre of gas at standard density	Values of q (§ 48) according to $q^{-6}=1\cdot21$ for argon (liquid compared with gas at 0° and atmospheric pressure)
				N	
CO_2	Solid	1·58	800	$4\cdot55 . 10^{20} . q^6$	·777
H_2	liquid at 17° absolute	·090	1000	·352 ,,	1·191
N_2	liquid	1·047	833	1·62 ,,	·923
	solid	1·400	1114	2·90 ,,	·837
O_2	liquid at its freezing pt.	1·27	888	1·49 ,,	·936
Argon	liquid at 84° absolute	1·42	797	1·21 ,,	·969
	solid	1·396*	784	1·17 ,,	·974

In this table, q denotes the ratio to s of the distance from centre to centre of nearest molecules in an ideal cubic assemblage of the same density as the solid or liquid, as indicated in cols. 3 and 2.

* From information communicated by Prof. W. Ramsay, July 23, 1901.

Molecular.   § 50. According to Avogadro's doctrine, the number of molecules per cubic centimetre is the same for all "perfect" gases at
the same temperature and pressure; and even carbonic acid is
nearly enough a "perfect gas" for our present considerations. Hence
the actual values of q^6 are inversely proportional to the numbers
by which they are multiplied in col. 3 of the preceding table.
Now, as said in § 48, all things considered, it seems probable that
for argon, liquid at density 1·42, q may be somewhat greater, but
not much greater, than unity. If it were exactly unity, N would
be $1·21.10^{20}$; and I have chosen $q = (1·21)^{-\frac{1}{6}}$ or ·969, to make N
the round number 10^{20}. Col. 6, in the table of § 47 above, is
calculated with this value of N; but it is not improbable that the
true value of N may be considerably greater than 10^{20}*.

    § 51. As compared with the value for argon, monatomic, the
smaller values of q for the diatomic gases, nitrogen and oxygen,
and the still smaller values for carbonic acid, triatomic, are quite
as might be expected without any special consideration of law of
force at different distances between atoms. It seems that the
diatomic molecules of nitrogen and oxygen and still more so the
triatomic molecule of carbonic acid, are effectively larger when
moving freely in the gaseous condition, than when closely packed
in liquid or solid assemblage. But the largeness of q for the
diatomic hydrogen is not so easily explained: and is a most interesting subject for molecular speculation, though it or any other
truth in nature is to be explained by a proper law of force according to the Boscovichian doctrine which we all now accept (many

    * Maxwell, judging from "molecular volumes" of chemical elements estimated
by Lorentz, Meyer and Kopp, unguided by what we now know of the densities
of liquid oxygen and liquid hydrogen and of the liquid of the then undiscovered gas
argon, estimated $N = ·19.10^{20}$ (Maxwell's *Collected Papers*, Vol. II. p. 350) which is
rather less than one-fifth of my estimate 10^{20}. On the same page of his paper
is given a table of estimated diameters of molecules which are about 3·2 or 3·3 times
larger than my estimates in col. 6 of the table in § 47. In a previous part of
his paper (p. 348) Maxwell gives estimates of free paths for the same gases, from
which by his formula (7), corrected as in col. 8 of my table in § 47, I find values of
N ranging from $6·05.10^{18}$ to $6·96.10^{18}$ or about one-third of $·19.10^{20}$. His
uncorrected formula $\sqrt{2\pi}$ (instead of $\sqrt{2}.\pi$) gives values of N which are $\sqrt{\pi}$ times,
or 1·77 times as great, which are still far short of his final estimate. The discrepance
is therefore not accounted for by the error in the formula as printed, and I see no
explanation of it. The free paths as given by Maxwell are about 1·3 or 1·4 times
as large as mine.

of us without knowing that we do so) as the fundamental hypo- Molecular. thesis of physics and chemistry. I hope to return to this question as to hydrogen in a crystallographic appendix.

I am deeply indebted to Professor Dewar for information regarding the density of liquid hydrogen, and the densities of other gases, liquefied or frozen, which he has given me at various times within the last three years.

§ 52. A new method of finding an inferior limit to the number of molecules in a cubic centimetre of a gas, very different from anything previously thought of, and especially interesting to us in connection with the wave-theory of light, was given by Lord Rayleigh*, in 1899, as a deduction from the dynamical theory of the blue sky which he had given 18 years earlier. Many previous writers, Newton included, had attributed the light from the sky, whether clear blue, or hazy, or cloudy, or rainy, to fine suspended particles which divert portions of the sunlight from its regular course; but no one before Rayleigh, so far as I know, had published any idea of how to explain the blueness of the cloudless sky. Stokes, in his celebrated paper on Fluorescence †, had given the true theory of what was known regarding the polarization of the blue sky in the following "significant remark" as Rayleigh calls it : " Now this result appears to me to have no remote bearing on the " question of the directions of the vibrations in polarized light. " So long as the suspended particles are large compared with the " waves of light, reflection takes place as it would from a portion of " the surface of a large solid immersed in the fluid, and no con- " clusion can be drawn either way. But if the diameter of the " particles be small compared with the length of a wave of light, it " seems plain that the vibrations in a reflected ray cannot be per- " pendicular to the vibrations in the incident ray"; which implies that the light scattered in directions perpendicular to the exciting incident ray has everywhere its vibrations perpendicular to the plane of the incident ray and the scattered ray; provided the diameter of the molecule which causes the scattering is very small in comparison with the wave-length of the light. In conversation Stokes told me of this conclusion, and explained to me with

* Rayleigh, *Collected Papers*, Vol. i. Art. viii. p. 87.

† "On the Change of Refrangibility of Light," *Phil. Trans.* 1852, and *Collected Papers*, Vol. iii.

Molecular. perfect clearness and completeness its dynamical foundation ; and applied it to explain the polarization of the light of a cloudless sky, viewed in a direction at right angles to the direction of the sun. But he did not tell me (though I have no doubt he knew it himself) why the light of the cloudless sky seen in any direction is blue, or I should certainly have remembered it.

§ 53. Rayleigh explained this thoroughly in his first paper (1871), and gave what is now known as Rayleigh's law of the blue sky; which is, that, provided the diameters of the suspended particles are small in comparison with the wave-lengths, the proportions of scattered light to incident light for different wave-lengths are inversely as the fourth powers of the wave-lengths. Thus, while the scattered light has the same colour as the incident light when homogeneous, the proportion of scattered light to incident light is seven times as great for the violet as for the red of the visible spectrum; which explains the intensely blue or violet colour of the clearest blue sky.

§ 54. The dynamical theory shows that the part of the light of the blue sky, looked at in a direction perpendicular to the direction of the sun, which is due to sunlight incident on a single particle of diameter very small in comparison with the wave-lengths of the illuminating light, consists of vibrations perpendicular to the plane of these two directions: that is to say, is completely polarized in the plane through the sun. In his 1871 paper*, Rayleigh pointed out that each particle is illuminated, not only by the direct light of the sun, but also by light scattered from other particles, and by earth-shine, and partly also by suspended particles of dimensions not small in comparison with the wave-lengths of the actual light; and he thus explained the observed fact that the polarization of even the clearest blue sky at 90° from the sun is not absolutely complete, though it is very nearly so. There is very little of polarization in the light from white clouds seen in any direction, or even from a cloudless sky close above the horizon seen at 90° from the sun. This is partly because the particles which give it are not small in comparison with the wave-lengths, and partly because they contribute much to illuminate one another in addition to the sunlight directly incident on them.

* *Collected Papers*, Vol. i. p. 94.

§ 55. For his dynamical foundation, Rayleigh definitely Molecular.
assumed the suspended particles to act as if the ether in their
places were denser than undisturbed ether, but otherwise uninfluenced by the matter of the particles themselves. He tacitly
assumed throughout that the distance from particle to particle
is very great in comparison with the greatest diameter of each
particle. He assumed these denser portions of ether to be of the
same rigidity as undisturbed ether; but it is obvious that this
last assumption could not largely influence the result, provided
the greatest diameter of each particle is very small in comparison
with its distance from next neighbour, and with the wave-lengths
of the light: and, in fact, I have found from the investigation of
§§ 41, 42 of Lecture XIV. for *rigid* spherical molecules embedded
in ether, exactly the same result as Rayleigh's; which is as follows

$$k = \frac{8\pi^3 n}{3} \left(\frac{D'-D}{D} \frac{T}{\lambda^2} \right)^2 = 82 \cdot 67\, n \left(\frac{D'-D}{D} \frac{T}{\lambda^2} \right)^2 \quad \ldots\ldots(1);$$

where λ denotes the wave-length of the incident light supposed
homogeneous; T the volume of each suspended particle; D the
undisturbed density of the ether; D' the mean density of the
ether within the particle; n the number of particles per cubic
centimetre; and k the proportionate loss of homogeneous incident
light, due to the scattering in all directions by the suspended
particles per centimetre of air traversed. Thus

$$1 - \epsilon^{-kx} \ldots\ldots\ldots\ldots\ldots\ldots\ldots\ldots\ldots\ldots\ldots(2)$$

is the loss of light in travelling a distance x (reckoned in centimetres) through ether as disturbed by the suspended particles.

It is remarkable that D' need not be uniform throughout the
particle. It is also remarkable that the shape of the volume T
may be anything, provided only its greatest diameter is very
small in comparison with λ. The formula supposes $T(D'-D)$
the same for all the particles. We shall have to consider cases in
which differences of T and D' for different particles are essential to
the result; and to include these we shall have to use the formula

$$k = \frac{82 \cdot 67}{\lambda^4} \Sigma \left[\frac{(D'-D)\,T}{D} \right]^2 \quad \ldots\ldots\ldots\ldots\ldots(3),$$

where $\Sigma \left[\dfrac{(D'-D)\,T}{D} \right]^2$ denotes the sum of $\left[\dfrac{(D'-D)\,T}{D} \right]^2$ for all

the particles in a cubic centimetre.

Molecular.   § 56. Supposing now the number of suspended particles per cubic wave-length to be very great, and the greatest diameter of each to be small in comparison with its distance from next neighbour, we see that the virtual density of the ether vibrating among the particles is

$$D + \Sigma T (D' - D)............................(4);$$

and therefore, if u and u' be the velocities of light in pure ether, and in ether as disturbed by the suspended particles, we have (Lecture VIII. p. 80)

$$u^2 = u'^2 \left[1 + \Sigma \frac{T(D' - D)}{D} \right] ....(5).$$

Hence, if μ denote the refractive index of the disturbed ether, that of pure ether being 1, we have

$$\mu = \left[1 + \Sigma \frac{T(D'-D)}{D} \right]^{\frac{1}{2}} ....(6);$$

and therefore,

$$\mu^2 - 1 = \Sigma \frac{T(D' - D)}{D} ....(7).$$

§ 57. In taking an example to illustrate the actual transparency of our atmosphere, Rayleigh says*; "Perhaps the best "data for a comparison are those afforded by the varying bright-"ness of stars at various altitudes. Bouguer and others esti-"mate about ·8 for the transmission of light through the entire "atmosphere from a star in the zenith. This corresponds to 8·3 "kilometres (the "height of the homogeneous atmosphere" at "10° Cent.) of air at standard pressure." Hence for a medium of the transparency thus indicated we have $\epsilon^{-830000k} = ·8$; which gives $1/k = 3720000$ centimetres $= 37·2$ kilometres.

§ 58. Suppose for a moment the want of perfect transparency thus defined to be *wholly* due to the fact that the *ultimate molecules* of air are not infinitely small and infinitely numerous, so that the "suspended particles" hitherto spoken of would be merely the molecules N_2, O_2; and suppose further $(D' - D) T$ to be the same for nitrogen and oxygen. The known

* *Phil. Mag.* April, 1899, p. 382.

refractivity of air ($\mu - 1 = {\cdot}0003$), nearly enough the same for all visible light, gives by equation (7) above, with n instead of Σ, Molecular.

$$\frac{n\,(D'-D)\,T}{D} = {\cdot}0006.$$

Using this in (1) we find

$$k = \frac{29{\cdot}76}{n\lambda^4 . 10^6} \dots\dots\dots\dots\dots\dots\dots(8),$$

for what the rate of loss on direct sunlight would be, per centimetre of air traversed, if the light were all of one wave-length, λ. But we have no such simplicity in Bouguer's datum regarding transparency for the actual mixture which constitutes sunlight: because the formula makes k^{-1} proportional to the fourth power of the wave-length; and every cloudless sunset and moonset and sunrise and moonrise over the sea, and every cloudless view of sun or moon below the horizon of the eye on a high mountain, proves the transparency to be in reality much greater for red light than for the average undimmed light of either luminary, though probably not so much greater as to be proportional to the fourth power of the wave-length. We may, however, feel fairly sure that Bouguer's estimate of the loss of light in passing vertically through the whole atmosphere is approximately true for the most luminous part of the spectrum corresponding to about the D line, wave-length $5{\cdot}89 . 10^{-5}$ cm., or (a convenient round number) $6 . 10^{-5}$ as Rayleigh has taken it. With this value for λ, and $3{\cdot}72 . 10^6$ centimetres for k^{-1}, (8) gives $n = 8{\cdot}54 . 10^{18}$ for atmospheric air at $10°$ and at standard pressure. Now it is quite certain that a very large part of the loss of light estimated by Bouguer is due to suspended particles; and therefore it is certain that the number of molecules in a cubic centimetre of gas, at standard temperature and pressure, is considerably greater than $8{\cdot}54 . 10^{18}$.

§ 59. This conclusion drawn by Rayleigh from his dynamical theory of the absorption of light from direct rays through air, giving very decidedly an inferior limit to the number of molecules in a cubic centimetre of gas, is perhaps the most thoroughly well founded of all definite estimates hitherto made regarding sizes or numbers of atoms. We shall see (§§ 73...79, below) that a much larger inferior limit is found on the same principles by careful

Molecular. consideration of the loss of light due to the ultimate molecules of pure air *and to suspended matter* undoubtedly existing in all parts of our atmosphere, even where absolutely cloudless, that is to say, warmer than the dew-point, and therefore having none of the liquid spherules of water which constitute cloud or mist.

§ 60. Go now to the opposite extreme from the tentative hypothesis of § 58, and, while assuming, as we know to be true, that the observed refractivity is wholly or almost wholly due to the ultimate molecules of air, suppose the opacity estimated by Bouguer to be wholly due to suspended particles which, for brevity, we shall call dust (whether dry or moist). These particles may be supposed to be generally of very unequal magnitudes: but, for simplicity, let us take a case in which they are all equal, and their number only 1/10000 of the $8\cdot54 . 10^{18}$, which in § 59 we found to give the true refractivity of air, with Bouguer's degree of opacity for $\lambda = 6 . 10^{-5}$. With the same opacity we now find the contribution to refractivity of the particles causing it, to be only 1/100 of the known refractivity of air. The number of particles of dust which we now have is $8\cdot54 . 10^{14}$ per cubic centimetre, or 184 per cubic wave-length, which we may suppose to be almost large enough or quite large enough to allow the dynamics of § 56 for refractivity to be approximately true. But it seems to me almost certain that $8\cdot54 . 10^{14}$ is vastly greater than the greatest number of dust particles per cubic centimetre to which the well-known haziness of the clearest of cloudless air in the lower regions of our atmosphere is due; and that the true numbers, at different times and places, may probably be such as those counted by Aitken[*] at from 42500 (Hyères, 4 p.m. April 5, 1892) to 43 (Kingairloch, Argyllshire, 1 p.m. to 1.30 p.m. July 26, 1891).

§ 61. Let us, however, find how small the number of particles per cubic centimetre must be to produce Bouguer's degree of opacity, without the particles themselves being so large in comparison with the wave-length as to exclude the application of Rayleigh's theory. Try for example $T = 10^{-3} . \lambda^3$ (that is to say, the volume of the molecule 1/1000 of the cubic wave-length, or

[*] *Trans. R. S. E.* 1894, Vol. xxxvii. Part iii. pp. 675, 672.

roughly, diameter of molecule 1/10 of the wave-length) which Molecular.
seems small enough for fairly approximate application of Rayleigh's theory; and suppose, merely to make an example, D' to be
the optical density of water, D being that of ether; that is to say,
$D'/D = (1\cdot3337)^2 = 1\cdot78$. Thus we have $(D' - D)\,T/D = \cdot00078\lambda^3$:
and with $\lambda = 6.10^{-5}$, and with $k^{-1} = 3\cdot72.10^6$, (1) gives $n = 1\cdot485.10^6$,
or about one and a half million particles per cubic centimetre.
Though this is larger than the largest number counted for natural
air by Aitken, it is interesting as showing that Bouguer's degree of
opacity can be accounted for by suspended particles, few enough
to give no appreciable contribution to refractivity, and yet not too
large for Rayleigh's theory. But when we look through very
clear air by day, and see how far from azure or deep blue is the
colour of a few hundred metres, or a few kilometres of air with
the mouth of a cave or the darkest shade of mountain or forest,
for background; and when in fine sunny weather we study the
appearance of the *grayish* haze always, even on the clearest days,
notably visible over the scenery among mountains or hills; and
when by night at sea we see a lighthouse light at a distance of
45 or 50 kilometres, and perceive how little of redness it shows;
and when we see the setting sun shorn of his brilliance sufficiently
to allow us to look direct at his face, sometimes whitish, oftener
ruddy, rarely what could be called ruby red; it seems to me that
we have strong evidence for believing that the want of perfect
clearness of the lower regions of our atmosphere is in the main
due to suspended particles, too large to allow approximate fulfilment of Rayleigh's law of fourth power of wave-length.

§ 62. But even if they were small enough for Rayleigh's
theory the question would remain, Are they small enough and
numerous enough to account for the refractivity of the atmosphere? To this we shall presently see we must answer undoubtedly "No"; and much less than Bouguer's degree of opacity,
probably not as much as a quarter or a fifth of it, is due to the
ultimate molecules of air. In a paper by Mr Quirino Majorana in
the *Transactions* of the R. Accademia dei Lincei (of which a
translation is published in the *Philosophical Magazine* for May,
1901), observations by himself in Sicily, at Catania and on Mount
Etna, and by Mr Gaudenzio Sella, on Monte Rosa in Switzerland,
determining the ratio of the brightness of the sun's surface to the

Molecular. brightness of the sky seen in any direction, are described. This ratio they denote by r. One specially notable result of Mr Majorana's is that "the value of r at the crater of Etna is about five times greater than at Catania." The barometric pressures were approximately 53·6 and 76 cms. of mercury. Thus the atmosphere above Catania was only 1·42 times the atmosphere above Etna, and yet it gave five times as much scattering of light by its particles, and by the particles suspended in it. This at once proves that a great part of the scattering must be due to suspended particles; and more of them than in proportion to the density in the air below the level of Etna than in the air above it. In Majorana's observations, it was found that "except for regions "close to the horizon, the luminosity of the sky had a sensibly "constant value in all directions when viewed from the summit of "Etna." This uniformity was observed even for points in the neighbourhood of the sun, as near to it as he could make the observation without direct light from the sun getting into his instrument. I cannot but think that this apparent uniformity was only partial. It is quite certain that with sunlight shining down from above, and with light everywhere shining up from earth or sea or haze, illuminating the higher air, the intensities of the blue light seen in different directions above the crater would be largely different. This is proved by the following investigation; which is merely an application of Rayleigh's theory to the question before us. But from Majorana's narrative we may at all events assume that, as when observing from Catania, he also on Etna chose the least luminous part of the sky (*Phil. Mag.*, May 1901, p. 561), for the recorded results (p. 562) of his observations.

§ 63. The diagram, fig. 1 below, is an ideal representation of a single molecule or particle, T, with sunlight falling on it indicated by parallel lines, and so giving rise to scattered light seen by an eye at E. We suppose the molecule or particle to be so massive relatively to its bulk of ether that it is practically unmoved by the ethereal vibration; and for simplicity at present we suppose the ether to move freely through the volume T, becoming effectively denser without changing its velocity when it enters this fixed volume, and less dense when it leaves. In §§ 41, 42, of Lecture XV. above, and in Appendix A, a definite supposition,

attributing to ether no other property than elasticity as of an Molecular.
utterly homogeneous perfectly elastic solid, and the exercise of
mutual force between itself and ponderable matter occupying
the same space, is explained: according to which the ether within

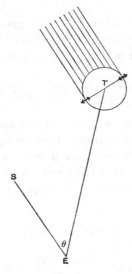

Fig. 1.

the atom will react upon moving ether outside, just as it would
if our present convenient temporary supposition of magically
augmented density within the volume of an absolutely fixed
molecule were realized in nature. For our present purpose, we
may if we please, following Rayleigh, do away altogether with
the ponderable molecule, and merely suppose T to be not fixed,
but merely a denser portion of the ether. And if its greatest
diameter is small enough relatively to a wave-length, it will
make no unnegligible difference whether we suppose the ether in
T to have the same rigidity as the surrounding free ether, or
suppose it perfectly rigid as in §§ 1—46 of Lecture XIV. dealing
with a rigid globe embedded in ether.

§ 64. Resolving the incident light into two components having
semi-ranges of vibration ϖ, ρ, in the plane of the paper and
perpendicular to it; consider first the component in the plane

having vibrations symbolically indicated by the arrow-heads, and expressed by the following formula

$$\varpi \sin \frac{2\pi u t}{\lambda},$$

where u is the velocity of light, and λ the wave-length. The greater density of the ether within T gives a reactive force on the surrounding ether outside, in the line of the primary vibration, and against the direction of its acceleration, of which the magnitude is

$$\frac{T(D'-D)\,\varpi}{D} \frac{2\pi u}{\lambda} \cos \frac{2\pi u t}{\lambda} \dots\dots\dots\dots\dots(9).$$

This alternating force produces a train of spherical waves spreading out from T in all directions, of which the displacement is, at greatest, very small in comparison with ϖ; and which at any point E at distance r from the centre of T, large in comparison with the greatest diameter of T, is given by the following expression*

$$\xi \cos \frac{2\pi}{\lambda} (ut - r),$$

with
$$\xi = \varpi \frac{\pi T(D'-D)}{r\lambda^2 D} \cos \theta \dots\dots\dots\dots(10),$$

where θ is the angle between the direction of the sun and the line TE. Formula (10), properly modified to apply it to the other component of the primary vibration, that is, the component perpendicular to the plane of the paper, gives for the displacement at E due to this component

$$\eta \cos \frac{2\pi}{\lambda} (ut - r),$$

with
$$\eta = \rho \frac{\pi T(D'-D)}{r\lambda^2 D} \dots\dots\dots\dots\dots(11).$$

Hence for the quantity of light falling from T per unit of time,

---

* This formula is readily found from §§ 41, 42 of Lecture XIV. The complexity of the formulas in §§ 8—40 is due to the inclusion in the investigation of forces and displacements at small distances from T, and to the condition imposed that T is a rigid spherical figure. The dynamics of §§ 33—36 with $c=0$, and the details of §§ 37—39 further simplified by taking $v=\infty$, lead readily to the formulas (10) and (11) in our present text.

on unit area of a plane at E, perpendicular to ET, reckoned in Molecular. convenient temporary units, we have

$$\xi^2 + \eta^2 = \left[\frac{\pi T(D'-D)}{r\lambda^2 D}\right]^2 (\varpi^2 \cos^2\theta + \rho^2)\ldots\ldots\ldots(12).$$

§ 65.   Consider now the scattered light emanating from a large horizontal plane stratum of air 1 cm. thick.   Let T of fig. 1 be one of a vast number of particles in a portion of this stratum subtending a small solid angle Ω viewed at an angular distance β from the zenith by an eye at distance r.   The volume of this portion of the stratum is $\Omega \sec\beta\, r^2$ cubic centimetres; and therefore, if Σ denotes summation for all the particles in a cubic centimetre, small enough for application of Rayleigh's theory, and q the quantity of light shed by them from the portion $\Omega \sec\beta\, r^2$ of the stratum, and incident on a square centimetre at E, perpendicular to ET, we have

$$q = \frac{\pi^2}{\lambda^4} \Sigma \left[\frac{T(D'-D)}{D}\right]^2 \Omega \sec\beta\,(\varpi^2 \cos^2\theta + \rho^2) \ldots\ldots(13).$$

Summing this expression for the contributions by all the luminous elements of the sun and taking

$$\int q = Q$$

to denote this summation, we have instead of the factor

$$\varpi^2 \cos^2\theta + \rho^2,$$

$$\cos^2\theta \int \varpi^2 + \int \rho^2 :$$

and we have
$$\int \varpi^2 = \int \rho^2 = \tfrac{1}{2}S \ldots\ldots\ldots\ldots\ldots(14),$$

where S denotes the total quantity of light from the sun falling perpendicularly on unit of area in the particular place of the atmosphere considered.   Hence the summation of (13) for all the sunlight incident on the portion $\Omega \sec\beta\, r^2$ of the stratum, gives

$$Q = \frac{\pi^2}{\lambda^4} \Sigma \left[\frac{T(D'-D)}{D}\right]^2 \Omega \sec\beta\,(\tfrac{1}{2}\cos^2\theta + \tfrac{1}{2})\,S\ldots\ldots\ldots(15).$$

§ 66.   To define the point of the sky of which the illumination is thus expressed, let ζ be the zenith distance of the sun, and ψ the azimuth, reckoned from the sun, of the place of the sky seen

Molecular. along the line ET. This place and the sun and the zenith are at the angles of a spherical triangle SZT, of which ST is equal to θ. Hence we have

$$\cos \theta = \cos \zeta \cos \beta + \sin \zeta \sin \beta \cos \psi \ldots\ldots\ldots\ldots(16).$$

Let now, as an example, the sun be vertical: we have $\zeta = 0$, $\theta = \beta$, and (15) becomes

$$Q = \frac{\pi^2}{\lambda^4} \Sigma \left[\frac{T(D'-D)}{D} \right]^2 \Omega . \tfrac{1}{2} (\cos \beta + \sec \beta) S \ldots\ldots\ldots(17).$$

This shows least luminosity of the sky around the sun at the zenith, increasing to ∞ at the horizon (easily interpreted). The law of increase is illustrated in the following table of values of $\tfrac{1}{2} (\cos \beta + \sec \beta)$ for every $10°$ of β from $0°$ to $90°$.

β	$\tfrac{1}{2}(\cos \beta + \sec \beta)$	β	$\tfrac{1}{2}(\cos \beta + \sec \beta)$
0°	1·000	50°	1·099
10°	1·000	60°	1·250
20°	1·002	70°	1·633
30°	1·010	80°	2·966
40°	1·036	90°	∞

§ 67. Instead now of considering illumination on a plane perpendicular to the line of vision, consider the illumination by light from our one-centimetre-thick great* horizontal plane stratum of air, incident on a square centimetre of horizontal plane. The quantity of this light per unit of time coming from a portion of sky subtending a small solid angle Ω at zenith distance β is $Q \cos \beta$. Taking $\Omega = \sin \beta \, d\beta \, d\psi$ and integrating, we find for the light shed by the one-centimetre-thick horizontal stratum on a horizontal square centimetre of the ground,

$$\int_0^{2\pi} d\psi \int_0^{\tfrac{1}{2}\pi} d\beta \sin \beta . Q \frac{\cos \beta}{\Omega} = \frac{4\pi^3}{3\lambda^4} \Sigma \left[\frac{T(D'-D)}{D} \right]^2 S \ldots\ldots(18).$$

Now each molecule and particle of dust sheds as much light upwards as downwards. Hence (18) doubled expresses the

---

* We are neglecting the curvature of the earth, and supposing the density and composition of the air to be the same throughout the plane horizontal stratum to distances from the zenith very great in comparison with its height above the ground.

quantity of light lost by direct rays from a vertical sun in cross- Molecular. ing the one-centimetre-thick horizontal stratum. It agrees with the expression for k in (1) of § 55, as it ought to do.

§ 68. The expression (15) is independent of the distance of the stratum above the level of the observer's eye. Hence if H denote the height above this level, of the upper boundary of an ideal homogeneous atmosphere consisting of all the ultimate molecules and all the dust of the real atmosphere scattered uniformly through it, and if s denote the whole light on unit area of a plane at E perpendicular to ET, from all the molecules and dust in the solid angle Ω of the real atmosphere due to the sun's direct light incident on them, we have

$$\frac{s}{S} = H \sec \beta \frac{\pi^2}{\lambda^4} \Sigma \left[\frac{T(D'-D)}{D} \right]^2 \Omega . \tfrac{1}{2}(\cos^2\theta + 1)\ldots\ldots\ldots(19);$$

provided we may, in the cases of application whatever they may be, neglect the diminution of the direct sunlight in its actual course through air, whether to the observer or to the portion of the air of which he observes the luminosity, and neglect the diminution of the scattered light from the air in its course through air to the observer. This proviso we shall see is practically fulfilled in Mr Majorana's observations on the crater of Etna for zenith distances of the sun not exceeding 60°, and in Mr Sella's observation on Monte Rosa in which the sun's zenith distance was 50°. But for Majorana's recorded observation on Etna at 5.50 a.m. when the sun's zenith distance was 81°·71, of which the secant is 6·928, there may have been an important diminution of the sun's light reaching the air vertically above the observer, and a considerably more important diminution of his light as seen direct by the observer. This would tend to make the sunlight reaching the observer less strong relatively to the sky-light than according to (19); and might conceivably account for the first number in col. 3 being smaller than the first number in col. 4 of the Table of § 69 below; but it seems to me more probable that the smallness of the first two numbers in col. 3, showing considerably greater luminosity of sky than according to (19), may be partly or chiefly due to dust in the air overhead, optically swelled by moisture in the early morning. Indeed the largeness of the luminosity of the sky indicated by the smallness

Molecular. of the first three numbers in col. 3, in comparison with the corresponding numbers of col. 4, is explained most probably by gray haze in the early morning.

§ 69. The results of Majorana's observations from the crater of Etna are shown in the following Table, of which the first and third columns are quoted from the *Philosophical Magazine* for May, 1901, and the second column has been kindly given to me in a letter by Mr Majorana. The values of S/s shown in col. 4 are calculated from § 68 (19), with the factor of sec β (cos² θ + 1) taken to make it equal to Majorana's r for sun's zenith distance 44°·6, on the supposition that the region of sky observed was in each case (see § 62 above) in the position of minimum luminosity as given by (21). It is obvious that this position is in a vertical

Col. 1	Col. 2	Col. 3	Col. 4	Col. 5
Time	Zenith distance of sun. ζ	Ratio of luminosity of sun's disc to luminosity of sky. r	$\dfrac{S}{s}$	Zenith distance of least luminous part of sky. β
5.50 A.M.	81°·7	2570000	4820000	5°·5
7	68·0	3125000	4610000	14·4
8	56·1	3650000	4310000	21·7
9	44·6	3930000	3930000	27·8
11	29·9	3760000	3370000	33·6

great circle through the sun, and on the opposite side of the zenith from the sun; and thus we have $\theta = \zeta + \beta$. Hence (19) becomes

$$\frac{s}{S} = H \frac{\pi^2}{\lambda^4} \Sigma \left[\frac{T(D'-D)}{D} \right]^2 \Omega \cdot \tfrac{1}{2} \sec \beta \left[\cos^2(\zeta + \beta) + 1 \right] \dots (20).$$

To make (20) a minimum we have

$$\tan \beta = \frac{2 \sin 2(\beta + \zeta)}{3 + \cos 2(\beta + \zeta)} \dots\dots\dots\dots\dots\dots (21).$$

The value of β satisfying this equation for any given value of ζ is easily found by trial and error, guided by a short preliminary table of values of β for assumed values of $\beta + \zeta$. Col. 5 shows Molecular. values of β thus found approximately enough to give the values of S/s shown in col. 4 for the several values of ζ.

§ 70. Confining our attention now to Majorana's observations at 9 a.m. when the sun's altitude was about $44°·6$; let e be the proportion of the light illuminating the air over the crater of Etna which at that hour was due to air, earth, and water below; and therefore $1 - e$ the proportion of the observed luminosity of the sky which was due to the direct rays of the sun, and expressed by § 68 (19). Thus, for $\beta = 27°·8$, $\zeta = 44°·6$, and $\theta = 72°·4$, we have $S/s = 3930000/(1 - e)$, instead of the S/s of col. 4, § 69. With this, equation (20) gives

$$\Sigma \left[\frac{T(D'-D)}{D} \right]^2 = \frac{\lambda^4 (1-e)}{H\Omega} \cdot 4·18 \cdot 10^{-8} \ldots\ldots\ldots\ldots(22).$$

Here, in order that the comparison may be between the whole light of the sun and the light from an equal apparent area of the sky, we must take

$$\Omega = \pi/214·6^{2*} = 1/14660,$$

being the apparent area of the sun's disc as seen from the earth. As to H, it is what is commonly called the "height of the homogeneous atmosphere" and, whether at the top of Etna or at sea-level, is

$$7·988 \cdot 10^5 \left(1 + \frac{t}{273} \right) \text{ centimetres};$$

where t denotes the temperature at the place above which H is reckoned. Taking this temperature as $15°$ C., we find

$$H = 8·43 \cdot 10^5 \text{ centimetres.}$$

Thus (22) becomes

$$\Sigma \left[\frac{T(D'-D)}{D} \right]^2 = \lambda^4 (1-e) \cdot ·728 \cdot 10^{-9} \ldots\ldots\ldots(23).$$

§ 71. Let us now denote by f and $1 - f$ the proportions of (23) due respectively to the ultimate molecules of air and to dust. We have

$$n \left[\frac{T(D'-D)}{D} \right]^2 = \lambda^4 f(1-e) \cdot ·728 \cdot 10^{-9} \ldots\ldots(24);$$

* The sun's distance from the earth is 214·6 times his radius.

Molecular. where n denotes the number of the ultimate molecules in a cubic centimetre of the air at the top of Etna; and $T(D'-D)/D$ relates to any one of these molecules; any difference which there may be between oxygen and nitrogen being neglected. Now assuming that the refractivity of the atmosphere is practically due to the ultimate molecules, and that no appreciable part of it is due to the dust in the air, we have by § 56 (7),

$$·0002 = n\,\frac{T(D'-D)}{2D} \quad\dots\dots\dots\dots\dots\dots(25),$$

the first number being approximately enough the refractivity of air at the crater of Etna (barometric pressure, 53·6 centimetres of mercury). Hence

$$\left[\frac{T(D'-D)}{D}\right]^2 = \frac{1}{n^2}\,1·6\,.\,10^{-7} \quad\dots\dots\dots\dots\dots(26),$$

and using this in (24) we find

$$n = \frac{220}{\lambda^4 f(1-e)} \quad\dots\dots\dots\dots\dots\dots(27).$$

Here, as in § 58 in connection with Bouguer's estimate for loss of light in transmission through air, we have an essential uncertainty in respect to the effective wave-length; and, for the same reasons as in § 58, we shall take $\lambda = 6\,.\,10^{-5}$ cm. as the proper mean for the circumstances under consideration. With this value of λ, (27) becomes

$$n = \frac{1}{f(1-e)}\,1·69\,.\,10^{19} \quad\dots\dots\dots\dots\dots(28).$$

§ 72. In Mr Sella's observations on Monte Rosa the zenith distance of the sun was 50°, and the place of the sky observed was in the zenith. He found the brightness of the sun's disc to be about 5000000 times the brightness of the sky in the zenith. Dealing with this result as in §§ 70, 71, with $\beta = 0$ in (20), and supposing the temperature of the air at the place of observation to have been 0° C., we find

$$n' = \frac{1}{f'\,(1-e')}\,2·34\,.\,10^{19}\dots\dots\dots\dots\dots(29),$$

where e', f', and n' are the values of e, f, and n, at the place of observation on Monte Rosa. Denoting now by N the number of molecules in a cubic centimetre of air at 0° C. and pressure

75 centimetres of mercury, we have, by the laws of Boyle and Molecular. Charles, on the supposition that the temperature of the air was 15° on the summit of Etna, and 0° on Monte Rosa

$$N = n \frac{75}{53\cdot6}\left(1 + \frac{15}{273}\right) = n'\frac{75}{49},$$

or
$$N = 1\cdot48n = 1\cdot53n' \dots\dots\dots\dots(30).$$

From these, with (28) and (29), we find

$$N = \frac{2\cdot50}{f(1-e)}\cdot10^{19} = \frac{3\cdot58}{f'(1-e')}\cdot10^{19}\dots\dots(31).$$

§ 73. To estimate the values of e and e' as defined in §§ 70, 72, consider the albedos* of the earth as might be seen from a balloon in the blue sky observed by Majorana and G. Sella over Etna and over Monte Rosa respectively. These might be about ·2 and ·4, the latter much the greater because of the great amount of snow contributing to illuminate the sky over Monte Rosa. With so much of guess-work in our data we need not enter on the full theory of the contribution to sky-light by earth-shine from below according to the principle of §§ 67, 68, interesting as it is; and we may take as very rough estimates ·2 and ·4 as the values of e and e'. Thus (31) becomes

$$N = \frac{3\cdot12}{f}\cdot10^{19} = \frac{5\cdot97}{f'}\cdot10^{19}\dots\dots\dots(32).$$

§ 74. Now it would only be if the whole light of the sky were due to the ultimate molecules on which the refractivity depends that f or f' could have so great a value as unity. If this were the case for the blue sky seen over Monte Rosa by G. Sella in

* Albedo is a word introduced by Lambert 150 years ago to signify the ratio of the total light emitted by a thoroughly unpolished solid or a mass of cloud to the total amount of the incident light. The albedo of an ideal perfectly white body is 1. My friend Professor Becker has kindly given me the following table of albedos from Müller's book *Die Photometrie der Gestirne* (Leipsic, 1897) as determined by observers and experimenters.

Mercury	0·14	Uranus	0·60
Venus	0·76	Neptune	0·52
Moon	0·34	Snow	0·78
Mars	0·22	White Paper	0·70
Jupiter	0·62	White Sandstone	0·24
Saturn	0·72	Damp Soil	0·08

Molecular. 1900, we should have $f' = 1$, and therefore $N = 5 \cdot 97 \cdot 10^{19}$. But it is most probable that even in the very clearest weather on the highest mountain, a considerable portion of the light of the sky is due to suspended particles much larger than the ultimate molecules N_2, O_2, of the atmosphere; and therefore the observations of the luminosity of the sky over Monte Rosa in the summer of 1900 render it probable that N is greater than $5 \cdot 97 \cdot 10^{19}$. If now we take our estimate of § 50, for the number of molecules in a cubic cm. of air at 0°, and normal pressure, $N = 10^{20}$, we have $1 - f = \cdot 688$ and $1 - f' = \cdot 403$; that is to say, according to the several assumptions we have made, $\cdot 688$ of the whole light of the portion of sky observed over Etna by Majorana was due to dust, and only $\cdot 403$ of that observed by Sella on Monte Rosa was due to dust. It is quite possible that this conclusion might be exactly true, and it is fairly probable that it is an approximation to the truth. But on the whole these observations indicate, so far as they can be trusted, the probability of at least as large a value as 10^{20} for N.

§ 75.   All the observations referred to in §§ 57—74 are vitiated by essentially involving the physiological judgment of perception of difference of strengths of two lights of different colours. In looking at two very differently tinted shadows of a pencil side by side, one of them blue or violet cast by a comparatively near candle, the other reddish-yellow cast by a distant brilliantly white incandescent lamp or by a more distant electric arc-lamp, or by the moon; when practising Rumford's method of photometry; it is quite wonderful to find how unanimous half-a-dozen laboratory students, or even less skilled observers, are in declaring This is the stronger! or, That is the stronger! or, Neither is stronger than the other! When the two shadows are declared equally strong, the declaration is that the differently tinted lights from the two shadowed places side by side on the white paper are, according to the physiological perception by the eye, equally strong. But this has no meaning in respect to any definite component parts of the two lights; and the unanimity, or the greatness of the majority, of the observers declaring it, only proves a physiological agreement in the perceptivity of healthy average eyes (from which colour-blind eyes would no doubt differ wildly). Two circular areas of white paper in Sella's observations on

Monte Rosa, a circle and a surrounding area of ground glass in Molecular. Majorana's observations with his own beautiful sky-photometer on Etna, are seen illuminated respectively by diminished sunlight of unchanged tint and by light from the blue sky. The sun-lit areas seem reddish-yellow by contrast with the sky-lit areas which are azure blue. What is meant when the two areas differing so splendidly are declared to be equally luminous? The nearest approach to an answer to this question is given at the end of § 71 above, and is eminently unsatisfactory. The same may be truly said of the dealing with Bouguer's datum in § 57, though the observers on whom Bouguer founded do not seem to have been disturbed by knowledge that there was anything indefinite in what they were trying to define or to find by observation.

§ 76. To obtain results not vitiated by the imperfection of the physiological judgment described in § 75, Newton's prismatic analysis of the light observed, or something equivalent to it, is necessary. Prismatic analysis was used by Rayleigh himself for the blue light of the sky, actually before he had worked out his dynamical theory. He compared the prismatic spectrum of light from the zenith with that of sunlight diffused through white paper; and by aid of a curve drawn from about thirty comparisons ranging over the spectrum from C to beyond F, found the following results for four different wave-lengths.

	C	D	b_3	F
Wave-length	656·2	589·2	517·3	486·2
Observed brightness . . .	1	1·64	2·84	3·60
Calculated according to λ^{-4} .	1	1·54	2·52	3·34

On these he makes the following remarks:—"It should be "noticed that the sky compared with diffused light was even "bluer than theory makes it, on the supposition that the diffused "light through the paper may be taken as similar to that whose "scattering illuminates the sky. It is possible that the paper "was slightly yellow; or the cause may lie in the yellowness "of sunlight as it reaches us compared with the colour it "possesses in the upper regions of the atmosphere. It would "be a mistake to lay any great stress on the observations in "their present incomplete form; but at any rate they show that "a colour more or less like that of the sky would result from

Molecular. " taking the elements of white light in quantities proportional
" to λ^{-4}. I do not know how it may strike others; but in-
" dividually I was not prepared for so great a difference as the
" observations show, the ratio for F being more than three times
" as great as for C." For myself I thoroughly agree with this
last sentence of Rayleigh's. There can be no doubt of the
trustworthiness of his observational results; but it seems to me
most probable, or almost certain, that the yellowness, or orange-
colour, of the sunlight seen through the paper, caused by larger
absorption of green, blue, and violet rays, explains the extreme
relative richness in green blue, and violet rays which the results
show for the zenith blue sky observed.

§ 77. An elaborate series of researches on the blue of the
sky on twenty-two days from July, 1900, to February, 1901, is
described in a very interesting paper, "Ricerche sul Bleu del
Cielo," a dissertation presented to the Royal University of Rome
by Dr Giuseppe Zettwuch, as a thesis for his degree of Doctor
in Physics. In these researches, prismatically analysed light from
the sky was compared with prismatically analysed direct sunlight
reduced by passage through a narrow slit; and the results were
therefore not vitiated by unequal absorptions of direct sunlight
in the apparatus. A translation of the author's own account
of his conclusions is published in the *Philosophical Magazine*
for August, 1902; by which it will be seen that the blueness
of the sky, even when of most serene azure, was always much
less deep than the true Rayleigh blue defined by the λ^{-4} law.
Hence, according to Rayleigh's theory (see § 53 above) much of
the light must always have come from particles not exceedingly
small in proportion to the wave-length. Thus in Zettwuch's
researches we have a large confirmation of the views expressed
in §§ 54, 58, 61, 74 above, and §§ 78, 79 below.

§ 78. Through the kindness of Professor Becker, I am now
able to supplement Bouguer's 170-year old information with the
results of an admirable extension of his investigation by Professor
Müller of the Potsdam Observatory, in which the proportion
(denoted by p in the formula below) transmitted down to sea-level
of homogeneous light entering our atmosphere vertically is found
for all wave-lengths from $4{\cdot}4 . 10^{-5}$ to $6{\cdot}8 . 10^{-5}$, by comparison of

the solar spectrum with the spectrum of a petroleum flame, for Molecular. different zenith distances of the sun. It is to be presumed, although I do not find it so stated, that only the clearest atmosphere available at Potsdam was used in these observations. For the sake of comparison with Rayleigh's theory, Professor Becker has arithmetically resolved Müller's logarithmic results into two parts; one constant, and the other varying inversely as the fourth power of the wave-length. The resulting formula*, modified to facilitate comparison with §§ 57—59 above, is as follows :

$$p = \epsilon^{-('0887 + '0772z^{-4})} = \cdot9152\epsilon^{-\cdot0772z^{-4}} \quad \ldots\ldots\ldots(33),$$

where $z = \lambda \div 6.10^{-5}$. In respect to the two factors here shown, we may say roughly that the first factor is due to suspended particles too large, and the second to particles not too large, for the application of Rayleigh's law. For the case of $\lambda = 6.10^{-5}$ ($z = 1$) this gives

$$p = \cdot9152 . \cdot9258 = \cdot847 \quad \ldots\ldots\ldots\ldots(34).$$

§ 79. Taking now the last term in the index and the last factor shown in (34) and dealing with it according to §§ 57—59 above; and still, as in § 55, using k to denote the proportionate loss of light per centimetre due to particles small enough for Rayleigh's theory, whether "suspended particles" or ultimate molecules of air, or both; we have $\epsilon^{-830000k} = \cdot9258$ which gives $k^{-1} = 10\cdot76.10^6$ cms. Hence if, as in § 58, we suppose for a moment the want of perfect transparency thus defined to be wholly due to the ultimate molecules of air, we should have, by the dynamics of refractivity, $n\frac{T(D'-D)}{D} = \cdot0006$; and thence by (1) of § 55 with $\lambda = 6.10^{-5}$ we should find for the number of molecules per cubic centimetre $n = 2\cdot47.10^{19}$. But it is quite certain that a part, and most probable that a large part, of the want of transparency produced by particles small enough for Rayleigh's theory is due to "suspended particles" larger than the ultimate molecules: and we infer that the number of ultimate molecules per cubic centimetre is greater than, and probably very much greater than, $2\cdot47.10^{19}$. Thus from the surer and more complete data of Müller regarding extinction of light of different wave-

* Müller, *Die Photometrie der Gestirne*, p. 140.

T. L. 21

Molecular. lengths traversing the air, we find an inferior limit for the number of molecules per cubic centimetre nearly three times as great as that which Rayleigh showed to be proved from Bouguer's datum.

§ 80.   Taking, somewhat arbitrarily, as the result of §§ 23—77 that the number of molecules in a cubic centimetre of a perfect gas at standard temperature and pressure is 10^{20}, we have the following interesting table of conclusions regarding the weights of atoms and the molecular dimensions of liquefied gases, of water, of ice, and of solid metals.

Substance	Mass of atom or of H_2O in grammes	Density		Number of atoms, or of molecules H_2O, in cub. cm.	Distance in cm. from centre to centre if ranged in cubic order with actual density
H	$0.45 \cdot 10^{-24}$	liquid at 17° absolute	·090	$200 \cdot 10^{21}$	$1·71 \cdot 10^{-8}$
O	7·15 ,,	,, ,, freezing point	1·27	178 ,,	1·78 ,,
H_2O	8·05 ,,	water	1·00	124 ,,	2·00 ,,
H_2O	8·05 ,,	ice	·917	114 ,,	2·06 ,,
H_2O	8·05 ,,	vapour at 0° C.	$·487 \cdot 10^{-5}$	$605 \cdot 10^{15}$	118·2 ,,
N	6·29 ,,	liquid	1·047	$166 \cdot 10^{21}$	1·82 ,,
Argon	17·81 ,,	,,	1·42	79·7 ,,	2·32 ,,
Gold	88·52 ,,	solid	19·32	218 ,,	1·66 ,,
Silver	48·47 ,,	,,	10·53	217 ,,	1·66 ,,
Copper	28·43 ,,	,,	8·95	315 ,,	1·47 ,,
Iron	25·15 ,,	,,	7·86	313 ,,	1·47 ,,
Zinc	29·30 ,,	,,	7·15	245 ,,	1·60 ,,

§ 81.   In the introductory Lecture (p. 14) we considered the question " are the vibrations of light *perpendicular to*, or are they "*in*, the plane of polarization ?—defining the plane of polarization "as the plane through the incident and reflected rays, for light "polarized by reflection." We are now able to answer *perpendicular to the plane of polarization*, with great confidence founded on two experimental proofs both given by Stokes, and each of them alone sufficient, I believe, for the conclusion.

I.   The observed fact that a large proportion of the light of the blue sky looked at in any direction perpendicular ·to the direction of the sun, and not too nearly horizontal, is polarized in the plane through the sun ; interpreted according to the dynamics of Lecture VIII., pp. 88, 89, 90, and of Lecture XVII., §§ 52, 54, 63.

II.   Observation of the change of the plane of polarization experienced by plane polarized light when diffractionally changed in direction through a large angle by passage across a Fraunhofer Molar. grating; described and interpreted in Part II. of his great paper "On the Dynamical Theory of Diffraction*." In a short Appendix, pp. 327, 328, added to his *Reprint*, Stokes notices experiments by Holtzmann †, published soon after that paper, leading to results seemingly at variance with its conclusions, and gives a probable explanation of the reason for the discrepance; and he refers to later experiments agreeing with his own, by Lorenz of Denmark.

Thus we find in II. confirmation of the conclusion, first drawn by Stokes from I., that the vibrations in plane polarized light are perpendicular to the plane of polarization.

§ 81′. Finally, we have still stronger confirmation of Stokes's original conclusion, in fact an irrefragable independent demonstration, that the vibrations of light are perpendicular to the plane of polarization, by Rayleigh; founded on a remarkable discovery, made independently by himself and Lorenz of Denmark, regarding the reflection of waves at a plane interface between two elastic solids of different rigidities but equal densities: That, when the difference of rigidities is small, and when the vibrations are in the plane of incidence, there are two angles of incidence ($\pi/8$ and $3\pi/8$), each of which gives total extinction of the reflected light. That is to say; instead of one "polarizing-angle," $\tan^{-1}\mu$; there are two "polarizing-angles" 22°·5 and 67°·5; which is utterly inconsistent with observation. The old-known fact (proved in § 123 below) that, if densities are equal and rigidities unequal, vibrations perpendicular to the plane of incidence give reflected light obeying Fresnel's "tangent" law and therefore vanishing when the angle of incidence is $\tan^{-1}\mu$, had rendered very tempting, the false supposition that light polarized by reflection at incidence $\tan^{-1}\mu$ has its vibrations in the plane of incidence; but this supposition is absolutely disproved by the two angles of extinction of the reflected ray which it implies for vibrations in the plane of incidence and reflection.

We may consider it as one of the surest doctrines through the whole range of natural philosophy, that plane-polarized light consists of vibrations of ether perpendicular to the plane of polarization.

---

* *Reprint of Mathematical and Physical Papers*, Vol. II. pp. 290—327.

† Poggendorff's *Annalen*, Vol. XCIX. (1856), p. 446, or *Philosophical Magazine*, Vol. XIII. p. 135.

LECTURE XVIII.

THURSDAY, *October* 16, 5 P.M. *Written afresh* 1902.

Reflection of Light.

Molar.   § 82. THE subject of this Lecture when originally given was " Reflection and Refraction of Light." I have recently found it convenient to omit " Refraction " from the title because (§ 130 below), if the reflecting substance is transparent, and if we know the laws of propagation of ethereal waves through it, we can calculate the amount and quality of the refracted light for every quality of incident light, and every angle of incidence, when we know the amount and quality of the reflected light. When the reflecting substance is opaque there is no " refracted " light; but, co-periodic with the motion which constitutes the incident light, there is a vibratory motion in the ether among the matter of the reflecting body, diminishing in amplitude according to the exponential law, ϵ^{-mD}, with increasing distance D from the interface. When m^{-1} is equal to 10,000 wave-lengths, say half-a-centimetre, the amplitude of the disturbance at one-half centimetre inwards from the interface would be ϵ^{-1} of the amplitude of the entering light, and the intensity would be ϵ^{-2}, or 1/7·39, of the intensity of the entering light. The substance might or might not, as we please, be called opaque; but it would be so far from being perfectly opaque that both theoretically and experimentally we might conveniently deal with the case according to the ordinary doctrine of " reflected " and " refracted " light. The index of refraction might be definitely measured by using very small prisms, not thicker than 1 mm. in the thickest part. But, on the other hand, when the opacity is so nearly perfect that m^{-1} is ten wave-lengths, or two or three wave-lengths, or less than a wave-length, we have no proper application of the ordinary law of refraction, and no modification of it can conveniently be used. Omission to perceive this negative truth has led some experimenters to waste precious time and work in making, and experimenting with, transparent metal prisms of

thickness varying from as nearly nothing as possible at the edge, Molar.
to something like the thickness of ordinary gold-leaf at distance
of two or three millimetres from the edge.

§ 83. A much more proper mode of investigating the propagation of ethereal vibrations through metals is that followed first, I believe, by Quincke, and afterwards by other able experimenters; in which an exceedingly thin uniform plate of the solid is used, and the retardation of phase (found negative for metals by Quincke!) of light transmitted through it is measured by the well-known interferential method. Some mathematical theorists have somewhat marred their work by holding on to "refraction," and giving wild sets of real numbers for refractive indices* of metals (different for different incidences!), in their treatment of light reflected from metals; after MacCullagh and Cauchy had pointed out that the main observed results regarding the reflection of light from metals can be expressed with some approach to accuracy by Fresnel's formulas with $\mu = \iota p + q$, where p and q are real numbers, and p *certainly not zero*: q very nearly zero for silver, or quite zero for what I propose to call *ideal* silver (§§ 150, 151 below). It was consideration of these circumstances which led me to drop "Refraction" from the title of Lecture XVIII.

§ 84. The theory of propagation of the ethereal motion through an opaque solid, due to any disturbance in any part of it, including that produced by light incident on its surface, is a most important subject for experimental investigation, and for work in mathematical dynamics. The theory to be tried for (§ 159) is that of the propagation of vibratory motion through ether, when under the influence of molecules of ponderable matter, causing the formulas for wave-motion to be modified by making the square of the propagational velocity a real negative quantity, or a complex. We shall see in § 159 below that a new molecular doctrine gives a thoroughly satisfactory dynamical theory of what I have called ideal silver, a substance which reflects light without loss at every incidence and in every polarizational azimuth; its quality in wave-theory being defined by a real negative quantity for the square of the propagational velocity of imaginary waves through ether within it. Liquid mercury,

* μ is essentially the ratio of the propagational velocities in the two mediums, one of which may, with good dynamical reason, be a complex imaginary. See §§ 144—152.

Molar. or quicksilver as it used to be called, may possibly, when tested with a very perfectly purified surface, be found to reflect light as well as, or nearly as well as, Sir John Conroy succeeded in getting a silver surface to do (§ 88 below); and may therefore fulfil my definition of *ideal silver*: though Rayleigh found only ·753 of the incident light, to be reflected from fairly clean mercury under air*.

§ 85. Without any scientific photometry, any one can see that the light reflected from gold, steel, copper, brass, tin, zinc, however well polished, is very much less than from silver, and hence very much less than the whole incident light. Though many excellent investigations have been made by many able experimenters on the polarizational analysis of light reflected from all these metals, few of them have given results as to the proportion of the whole reflected, to the whole incident, light. It seems indeed that, besides Rayleigh, only two observers, Potter† and Conroy‡, have directly compared the whole light reflected from metals with the whole incident light (see § 88 below). Conroy, using his most perfect silver mirror, and angle of incidence 30°, found the proportion of reflected light to incident light to be 97·3 per cent., for light vibrating perpendicular to the plane of incidence, and 99·9 per cent. for light vibrating in§ the plane of incidence; and therefore 98·6 per cent. for common unpolarized light. With speculum-metal and steel and tin mirrors, for 30° of incidence, and unpolarized light, he found the reflected light to be;—speculum-metal 66·9 per cent., steel 54·9 per cent., and tin 44·4 per cent.‖ From these measurements, as well as from ordinary non-scientific observation, we see that there is essentially much light lost in reflection from other metals than silver; and, as the light does not travel through the metal, it is quite certain that its energy must be converted into heat in an exceedingly thin surface-region of the metal (certainly not more than two or three wave-lengths). Hence the square of the imaginary velocity of the imaginary light-wave in other metals than silver cannot be a real negative quantity;

---

* *Phil. Mag.* 1892; *Collected Papers*, Vol. iv.

† *Edinburgh Journal of Science*, Vol. iii. 1830, pp. 278—288.

‡ *Proc. R. S.* Vol. xxviii. Jan. 1879, pp. 242—250; Vol. xxxi. Mar. 1881, pp. 486—500; Vol. xxxv. Feb. 1883, pp. 26—41; Vol. xxxvi. Jan. 1884, pp. 187—198; Vol. xxxvii. May, 1884, pp. 36—42.

§ I have found no other recorded case of greater reflectivity of vibrations in than of vibrations perpendicular to the plane of incidence.

‖ *Proc. R. S.* Feb. 1883, Vol. xxxv. pp. 31, 32.

because, as we readily see from the mathematical treatment, this Molar.
would imply no conversion of light-energy into heat, and would
therefore (§§ 150, 157 below) imply total reflection at every angle
of incidence. Dynamical theory is suggested in § 159, to explain
conversion of incident light into heat, among the molecules
within two or three wave-lengths of the boundary of the metallic
mirror. This is only an outlying part of the whole field of investigation required to explain all kinds of opacity in all kinds
of matter, solid, liquid, and gaseous.

§ 86.   Let us think now of the reflection of light from perfectly
polished surfaces. If the substance is infinitely fine-grained the
polish may be practically perfect; so that no light is reflected
except according to the law of equal angles of incidence and
reflection. But in reality the molecular structure of solids gives
a surface which is essentially not infinitely fine-grained: and the
nearest approach to perfect polish produced by art, or found in
nature, on the surfaces of liquids and on crystalline or fractured
surfaces of solids, is illustrated by a gravel-covered road made *as
smooth as a steam-roller can make it.* Optically, the polish would
be little less than perfect if the distances between nearest neighbours in the molecular structure are very small in comparison with
the wave-length of the incident light. In § 80 we estimated the
distances between nearest molecular neighbours in ordinary solids
and liquids at from $1{\cdot}5\,.\,10^{-8}$ to $2\,.\,10^{-8}$ cm., which is from 1/4000
to 1/3000 of the mean wave-length of visible light ($6\,.\,10^{-5}$ cm.).
The best possible polish is therefore certainly almost quite practically perfect in respect to the reflection of light. But if the work
of the steam-roller is anything less than as perfect as it can possibly be, there are irregular hollows (bowls or craters) of breadths
extending, say, to as much as 1/200 of the wave-length of mean
visible light; and the polish would probably not be optically perfect.
The want of perfectness would be shown by a very faint blue light,
scattered in all directions from a polished mirror illuminated only
by a single lamp or by the sun. The best way to look for this
blue light would be to admit sunlight into an otherwise dark room
through a round hole in a window-shutter or a metal screen, as
thoroughly blackened as possible on the side next the room; and
place the mirror to be tested in the centre of the beam of light at
a convenient distance from the hole. If the polish is optically
perfect, no light is seen from any part of the mirror by an eye

Molar. placed anywhere not in the course of the properly reflected light from sun or sky. That is to say, the whole room being dark, and the screen around the hole perfectly black, the whole mirror would seem perfectly black when looked at by an eye so placed as not to see an image (of the hole, and therefore) of any part of the sky or sun. A condition of polish *very nearly, but not quite, optically perfect*, would be shown by a faint violet-blue light from the surface of the mirror instead of absolute blackness. The tint of this light would be the true Rayleigh λ^{-4} blue, if the want of perfectness of the polish is due to craters small in comparison with the wave-length (even though large in comparison with distances between nearest molecular neighbours). When this condition is fulfilled, the blue light due to want of perfectness in the polish, seen on the surface if viewed in any direction perpendicular to the direction of the incident beam, would be found completely polarized in the plane of these two directions. Everything in § 86 is applicable to reflection from any kind of surface, whether the reflecting body be solid or liquid, or metallic, or opaque with any kind of opacity, or transparent. Experimental examination of the polish of natural faces of crystals will be interesting.

§ 87. Valuable photometric experiments with reference to reflection of light by metals have been made by Bouguer, Biot, Brewster, Potter, Jamin, Quincke, De Senarmont, De la Provostaye and Desains, and Conroy*. But much more is to be desired, not only as to direct reflection from metals, but as to reflection at all angles of incidence from metals and other opaque and transparent solids and liquids. In each case the intensity of the whole reflected light should be compared with the intensity of the whole incident light; in the first place without any artificial polarization of the incident, and without polarizational analysis of the reflected, light. It is greatly to be desired that thorough investigation of this kind should be made. It would be quite an easy kind of work† because roughly approximate photometry is

---

* Bouguer, *Traité d'Optique*, 1760, pp. 27, 131: Biot, *Ann. de Chimie*, 1815, Vol. xciv. p. 209: Brewster, *A Treatise on new Philosophical Instruments*, Edin. 1813, p. 347; *Phil. Trans.* 1830, p. 69: Potter, *Edin. Jour. of Science*, 1830: De la Provostaye and Desains, *Ann. de Chim. et de Phys.* [3], 1849, Vol. xxvii. p. 109, and 1850, Vol. xxx. pp. 159, 276: Conroy, *Proc. R. S.*; see § 84 above. See Mascart, *Traité d'Optique*, Vol. ii. 1891, pp. 441—459.

† Bouguer, *Traité d'Optique*, 1760. pp. 27, 131: Arago, *Œuvres Complètes*, Vol. x. pp. 150, 185, 216, 468. See Mascart, *Traité d'Optique*, Vol. ii. 1891, pp. 441—501.

always easy, except when rendered impossible by difference of Molar. colour in the lights to be compared.

Absolute determinations of reflected light per unit of incident light cannot be made with *great accuracy* because of the inherent difficulty, or practical impossibility, of *very accurate* photometric observations, even when, as is largely the case for reflected lights, there is no perceptible difference of tint between the lights to be compared.

§ 88. The accompanying diagram, fig. 1, shows, for angles of incidence from 10° to 80°, the reflectivities* of silver, speculum-

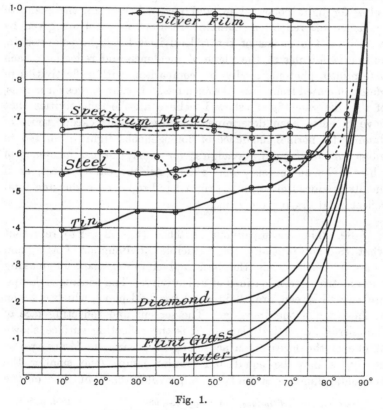

Fig. 1.

* I propose the word "reflectivity" to designate the ratio of the whole reflected light to the whole incident light, whether the incident light is unpolarized as in fig. 1; or plane polarized, either in or perpendicular to the plane of incidence, as in figs. 4—7; and whether it is ordinary white light, or homogeneous light, or light of any mixed tint.

Molar. metal, steel, and tin, according to the observations of Potter, Conroy, and Jamin; and for all incidences from 0° to 90° the reflectivities of diamond, flint-glass, and water, calculated from Fresnel's admirable formulas. These are, for almost all transparent bodies even of so high refractivity as diamond, (§§ 96, 100 below), probably very much nearer the truth than any photometric observations hitherto made, or possible to human eyes. The ordinates represent the reflected light in percentage of the incident light, at the angles of incidence represented by the abscissas. The reflectivities thus given for the three transparent bodies are the means of the reflectivities given in figs. 4, 5, 6 of § 102 below for light polarized in, and perpendicular to, the plane of incidence and reflection. Jamin's results* for steel are given by himself as the means of the reflectivities for light polarized in, and perpendicular to, the plane of incidence and reflection; each determined photometrically by comparison with reflectivities of glass calculated from Fresnel's formulas. The six curves for metals, of this diagram (fig. 1), show I believe all the reflectivities that have hitherto been determined by observation; except those of Jamin† for normal incidence of homogeneous lights from red to violet on metals, and Rayleigh's‡ for nearly normal incidence of white light on mercury and glass. All the curves meet, or if ideally produced meet, in the right-hand top corner, showing total reflection at grazing incidence.

§ 89.   The accompanying sketches (figs. 2 a, 2 b) represent what (on substantially the principle of Bouguer, Potter, and Conroy) seems to me the best and simplest plan for making photometric determinations of reflectivity. The centres of lamps, mirror, and screen, are all in one horizontal plane, which is taken as the plane of the drawings. The screens and reflecting face of the mirror are vertical planes. R is the reflecting surface, shown as one of the faces of an acute-angled prism. L, L' are two as nearly as may be equal and similar lamps (the smaller the horizontal dimensions the better §, provided the light shed on the

---

* Mascart, *Traité d'Optique*, Vol. II. pp. 534, 536.   † *Ibid.* p. 544.

‡ *Scientific Papers*, Vols. II. and IV.; *Proc. R. S.* 1886, *Phil. Mag.* 1892.

§ The best non-electric light I know for the purpose is the ordinary flat-wicked paraffin-lamp, with a screen having a narrow vertical slot placed close to the lamp-glass, with its medial line in the middle plane of the flat flame. The object of the screen is to cut off light reflected from the glass funnel, without intercepting any of the light of the flame itself; the width of the slot should therefore be very

mirror is sufficient). L' is too far from R to allow its true position Molar. to be shown on the diagram, but its direction as seen from R is indicated by a broken straight line. W is an opaque screen coated

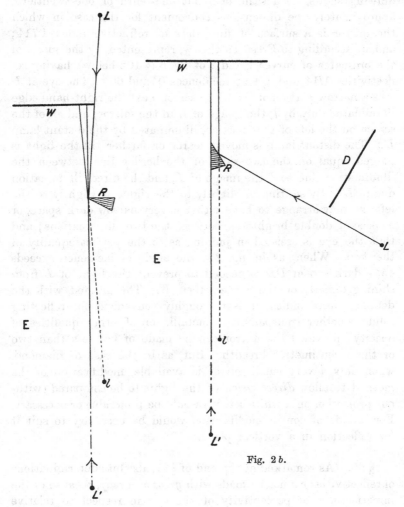

Fig. 2 b.

Fig. 2 a.

slightly greater than the thickness of the flame. This arrangement gives a much finer and steadier line of light than any unscreened candle, and vastly more light; and it gives more light than comes through a slot of the same width from a round-wicked paraffin-lamp. An electric lamp of the original Edison "hair-pin" pattern, with a slotted screen to cut off reflected light from the glass, would be more convenient than the paraffin-lamp, and give more light with finer shadows.

Molar. with white paper on the side next E, the eye of the observer. Fig. 2 a and fig. 2 b show respectively the positions for very nearly direct incidence, and for incidence about 60°. Each drawing shows, on a scale of about one-tenth or one-twentieth, approximately the dimensions convenient for the case in which the mirror is a surface of flint-glass of refractive index 1·714; and, as according to Fresnel's theory, represented by the mean of the ordinates of curves 1 and 2 of fig. 6 (§ 102 below), having reflectivities 1/14 and 1/4 at incidences 0° and 60°. The eye at E sees a narrow portion of the white screen next the right-hand edge illuminated only by l, the image of L in the mirror; and all of the screen on the left of that portion, illuminated by the distant lamp L'. The distant lamp is moved nearer or farther till the light is judged equal on the two sides of the border line between the illumination due to l, the image of L, and the direct illumination due to L'. By shifting L' slightly to the right, or slightly to the left, we may arrange to have either a very narrow dark space, or a space of double brightness, between the two illuminations; and thus the eye is assisted in judging as to the perfect equality of the two. When, as in fig. 2 b, the angle of incidence exceeds 45° a dark screen DD is needed to prevent the light of L from shining directly on the white screen, W. The method, with the details I have indicated, is thoroughly convenient for reflecting solids, whether transparent or metallic or of other qualities of opacity; provided the mirror can be made of not less than two or three centimetres breadth. But, as in the case of diamond, when only a very small mirror is available, modification of the method to allow *direct vision* of the lights to be compared (without projection on a white screen) would be preferable or necessary. For liquids, of course, modification would be necessary to suit it for reflection in a vertical plane.

§ 90. As remarked at the end of § 87, absolute determinations of reflectivities cannot be made with *great* accuracy, because of the imperfectness of perceptivity of the eye in respect to relative strengths of light, even when the tints are *exactly* the same. On the other hand, a *very high degree of accuracy* is readily attainable by the following method*, when the problem is to compare, for any or every angle of incidence, the reflected lights due to the

---

* Given originally by MacCullagh. See his *Collected Works*, p. 239; also Stokes, *Collected Works*, Vol. III. p. 199.

incidence of equal quantities of light vibrating in, and vibrating Molar. perpendicular to, the plane of incidence.

Use two Nicol's prisms, which, for brevity, I shall call N_1 and N_2, in the course of the incident and reflected light respectively. Use also a Fresnel's rhomb (F) between the reflecting surface and N_2. Set N_2, and keep it permanently set, with its two principal planes at 45° to the plane of reflection, but with facilities for turning from any one to any other of the eight positions thus defined, to secure any needful accuracy of adjustment. For brevity, I shall call the zeros of N_1 and F, positions when their principal planes are at angles of 45° to the plane of incidence and reflection. In the course of an observation N_1 and F are to be turned through varying angles, n and f, from their zeros, till perfect extinction of light coming through N_2 is obtained.

§ 91.   To begin an observation, with any chosen angle of incidence of the light on the reflecting surface, turn N_1 to a position giving as nearly as possible complete extinction of light emerging from N_2. Improve the extinction, if you can, by turning F in either direction, and get the best extinction possible by alternate turnings of N_1 and F. Absolutely complete extinction is thus obtained *at one point of the field* if homogeneous light is used, and if the Nicols and rhomb are theoretically perfect instruments. The results of the completed observation are the two angles (n, f) through which N_1 and F must be turned from their zero positions to give perfect extinction by N_2. From f thus found we calculate by two simple formulas, (7) of § 93, the ratios of the vibrational amplitudes of the two constituents defined below, $(\mathscr{G}, g)$, of the reflected vibrations in the plane of reflection, to the vibrational amplitude (C) of the reflected vibrations perpendicular to that plane. The definite constituents here referred to are $(\mathscr{G})$ vibrations in the same phase as C; and (g) vibrations in phase advanced by a quarter-period relatively to C. It is clear that if $g = 0$, complete extinction would be had without turning F from its zero position, and would be found by the same adjustment of N_1 as if there were no Fresnel's rhomb in the train.   (C corresponds to the C' of §§ 117, 123, below.)

§ 92.   Figs. 3a, 3b, are diagrams in planes respectively perpendicular to the reflected, and to the incident, light. Let O be a point in the course of the light between the reflecting surface and the

Molar. Fresnel's rhomb. Let OG be the plane of reflection of the light from the reflecting surface, and let OZ be perpendicular to that plane.

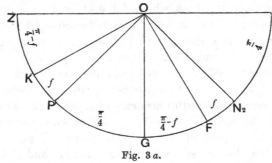

Fig. 3 a.

Let ON_2 be the vibrational plane* of the second Nicol; and OP its "plane of polarization" (being perpendicular to ON_2). Let OF be the plane through the entering ray, and perpendicular to the facial intersections† of the Fresnel's rhomb; shown in the diagram as turned through an angle $f = N_2OF$ from the zero position ON_2. Let OK be perpendicular to OF. In respect to signs we see by § 158^v (1) that in fig. 3a, for reflected, and 3b for incident, vibrations, OG is positive.

Considering the light coming from the reflecting surface and incident on the Fresnel, let $C \sin \omega t$ be the component along OZ, and $\mathscr{S} \sin \omega t - g \cos \omega t$ be the component along OG, of the displacement at time t of a particle of ether of which O is the equilibrium position. $\mathscr{S}/C$ and g/C are two functions of the angle of incidence to be determined by the observation now described. By proper resolutions and additions for vibrational components in the principal planes of the Fresnel, we find as follows:

$$\left. \begin{aligned} \sin \omega t \left[\mathscr{S} \cos \left(\tfrac{\pi}{4} - f \right) - C \sin \left(\tfrac{\pi}{4} - f \right) \right] \text{ along } OF \\ \sin \omega t \left[\mathscr{S} \sin \left(\tfrac{\pi}{4} - f \right) + C \cos \left(\tfrac{\pi}{4} - f \right) \right] \quad ,, \quad OK \end{aligned} \right\} \dots(1),$$

and

$$\left. \begin{aligned} - \cos \omega t \; g \cos \left(\tfrac{\pi}{4} - f \right) \text{ along } OF \\ - \cos \omega t \; g \sin \left(\tfrac{\pi}{4} - f \right) \quad ,, \quad OK \end{aligned} \right\} \dots\dots\dots(2).$$

---

* I use this expression for brevity to denote the plane of the vibrations of light transmitted through a Nicol.

† An expression used to denote the intersections of the traversed faces and the reflecting surfaces. See § 158^x below.

By the two total internal reflections at the oblique faces of the Molar.
Fresnel, vibrational components in the plane OF are advanced a
quarter-period relatively to the vibrational components perpendicular to OF.   Hence to find the vibrational components at time
t at a point in the course of the light emerging from the Fresnel,
we must, in the components along OF of (1) and (2), change the
$\sin \omega t$ into $\cos \omega t$, and change the $\cos \omega t$ into $-\sin \omega t$; and leave
unchanged the components along OK.   Thus we find for the
vibrational components at the chosen point in the course of the
light from the Fresnel towards the second Nicol, as follows:

$$\sin \omega t \quad g \cos \left(\frac{\pi}{4}-f\right) \qquad\qquad \text{along } OF$$
$$\sin \omega t \left[\mathscr{G} \sin \left(\frac{\pi}{4}-f\right) + C \cos \left(\frac{\pi}{4}-f\right) \right] \quad ,, \quad OK \Bigg\}\ldots(3),$$

and

$$\cos \omega t \left[\mathscr{G} \cos \left(\frac{\pi}{4}-f\right) - C \sin \left(\frac{\pi}{4}-f\right) \right] \text{along } OF$$
$$-\cos \omega t \quad g \sin \left(\frac{\pi}{4}-f\right) \qquad\qquad ,, \quad OK \Bigg\}\ldots(4).$$

§ 93.   For extinction by the second Nicol, the sum of all the
vibrational components parallel to ON_2 of the light reaching it
must be null; and therefore, by the proper resolutions and
additions, and by equating to zero separately the coefficients
of $\sin \omega t$ and $\cos \omega t$ thus found, we have

$$-\left[\mathscr{G} \sin \left(\frac{\pi}{4}-f\right) + C \cos \left(\frac{\pi}{4}-f\right) \right] \sin f + g \cos \left(\frac{\pi}{4}-f\right) \cos f = 0$$
$$\ldots\ldots\ldots\ldots(5),$$
$$\left[\mathscr{G} \cos \left(\frac{\pi}{4}-f\right) - C \sin \left(\frac{\pi}{4}-f\right) \right] \cos f + g \sin \left(\frac{\pi}{4}-f\right) \sin f = 0$$
$$\ldots\ldots\ldots\ldots(6).$$

Solving these equations for $\mathscr{G}$ and g, we find

$$\mathscr{G} = \frac{1+\cos 4f}{2+\sin 4f} C; \qquad g = \frac{2 \sin 2f}{2+\sin 4f} C \ldots\ldots\ldots(7).$$

§ 94.   Go back now to the light emerging from the first
Nicol and incident on the reflecting surface.   Let I be its

Molar. vibrational amplitude. The vibrational amplitudes of its components in, and perpendicular to, the plane of incidence are

$$I \cos\left(\frac{\pi}{4} - n\right), \text{ and } -I \sin\left(\frac{\pi}{4} - n\right),$$

when the Nicol is turned from its zero position, OQ, through the angle n, as indicated in fig. $3b$; ON_1 being the vibrational plane of the first Nicol.

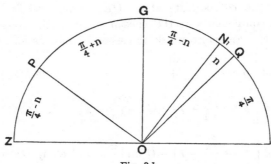

Fig. 3 b.

§ 94'. Let S be the ratio of reflected to incident vibrational amplitude for vibrational component *perpendicular* to the plane OG. Considering next the component of the incident light having vibrations *in* the plane OG; let T and E be the ratios of the vibrational amplitudes of two particular constituents of its reflected light both vibrating in the plane OG, to the vibrational amplitude of the incident component; these two constituents being respectively *in the same phase* as the component of the reflected light vibrating perpendicular to OG, and *in phase behind it by a quarter-period*[*]. We have

$$C = -I \sin\left(\frac{\pi}{4} - n\right) S, \quad \mathscr{G} = I \cos\left(\frac{\pi}{4} - n\right) T, \quad g = I \cos\left(\frac{\pi}{4} - n\right) E \ldots (8).$$

From these and (7) we have

$$\frac{T}{S} = -\frac{1 + \cos 4f}{2 + \sin 4f} \tan\left(\frac{\pi}{4} - n\right), \quad \frac{E}{S} = \frac{-2 \sin 2f}{2 + \sin 4f} \tan\left(\frac{\pi}{4} - n\right) \ldots \ldots (9).$$

* When the reflecting body is glass, or other transparent isotropic solid or liquid, Fresnel's prophecy (we cannot call it physical or dynamical theory) declares

$$S = -\frac{\sin(i - {,}i)}{\sin(i + {,}i)}, \qquad T = -\frac{\tan(i - {,}i)}{\tan(i + {,}i)}, \qquad E = 0.$$

The notation in the text is partially borrowed from Rayleigh (*Scientific Papers*, Vol. III. pp. 496—512) who used S, and T, to denote respectively the "sine-formula," and the "tangent-formula," of Fresnel. What I have denoted in the text by E is, for all transparent solids and liquids, certainly very small; and though generally

Thus our experimental result (f, n) gives the two constituents Molar. (T, E) of the reflected vibrations *in* the plane of reflection, when the single constituent (S) of the reflected vibrations perpendicular to the plane of reflection, is known.

§ 95.   Going back to § 90, note that there are four independent variables to be dealt with :—the angle of incidence, the orientations of the two Nicol's prisms, and the orientation of the Fresnel's rhomb.   Any two of these four variables may be definitely chosen for variation while the other two are kept constant; to procure, when homogeneous light is used, extinction of the light which enters the eye from the centre of the field ; that is to say, to produce perfect blackness at the central point of the field.   Theoretically there is, in general, just one point of the field (one point of absolute blackness) where the extinction is perfect ; and always before the desired adjustment is perfectly reached a black spot is conceivably to be seen, but not on the centre of the field.   By changing any one of the four independent variables the black spot is caused to move ; and generally two of them must be varied to cause it to move towards the centre of the field for the desired adjustment.   What in reality is generally perceived is, I believe, not a black spot, but a black band; and this is caused to travel till it passes through the middle of the field when the nearest attainable approach to the desired adjustment is attained.   When faint or moderate light, such as the light of a white sky, is used, the whole field may seem absolutely dark, and may continue so while any one of the four variable angles is altered by half a degree or more.   For more minutely accurate measurements, more intense light must be used; a brilliant flame; or electric arc-light; or lime-light; or, best of all, an unclouded sun as in Rayleigh's very searching investigation of light reflected from water at nearly the polarizing angle, of which the result is given in § 105 below.

§ 95′.   An easy way to see that just two independent variables are needed to obtain the desired extinction, is to confine our attention to the centre of the field, and imagine the light reaching the eye

believed to be perceptible for substances of high refrangibility such as diamond, Rayleigh has questioned its existence for any of them, and suggested that its non-nullity may be due to extraneous matter on the reflecting surface. I believe, however, that Airy, and Brewster, and Stokes who called it the adamantine property, and Fresnel himself though it was outside his theory, were right in believing it real. See § 158, below.

Molar. from it when the extinction is not perfect, to be polarizationally resolved into two components having their vibrational lines perpendicular to one another. The desired extinction requires the annulment of each of these two components, and nothing else. Proper change of the two chosen independent variables determinately secures these two annulments when what is commonly called homogeneous light, that is, light of which all the vibrations are of the same period, is used.

§ 95″. In the detailed plan of §§ 90—94 the two independent variables chosen are the orientation of the first Nicol, and the orientation of the Fresnel's rhomb. This is thoroughly convenient if N_2 is mounted on a proper mechanism to give it freedom to move in a plane perpendicular to its axis, and to keep its orientation round this axis constant. The Fresnel should be mounted so as to be free to move round an axis fixed in the direction of the light entering it: and it should carry a short tube round the light emerging from it into which N_2 fits easily. Thus, when the Fresnel is turned round the line of the light reflected from the mirror, it carries N_2 round in a circle (as it were with a hollow crank-pin), so that it is always in the proper position for carrying the emergent light to the eye of the observer.

Two other choices of independent variables, each, I believe, as well-conditioned and as convenient as that of §§ 90—94, but simpler in not wanting the special mechanism for carrying N_2, are described in § 98 below.

§ 96. Nothing in §§ 90—95 involves any hypothesis: and we have, in them, an observational method for fully, without any photometry, determining T/S and E/S; which are, for incident vibrations at 45° to the plane of incidence, the intensities of the two constituents of the reflected light vibrating in the plane of incidence, in terms of the intensity of the component of the reflected light vibrating perpendicularly to that plane. Fully carried through, it would give interesting and important information, for transparent liquids and solids, and for metals and other opaque solids, through the whole range of incidence from 0° to 90°. It can give extremely accurate values of T/S for transparent liquids and solids; and it will be interesting to find how nearly they agree with the formula $\dfrac{\cos(i+,i)}{\cos(i-,i)}$, which Fresnel's "tangent-law" and "sine-law" imply.

§ 96′. Fresnel himself used it[*] for reflection from water Molar. and from glass with incidences from 24° to 89° (not using the Fresnel's rhomb), and found values for $\tan^{-1}\dfrac{T}{S}$ differing from $\tan^{-1}\dfrac{\cos(i+,i)}{\cos(i-,i)}$ by sometimes more than 1°; but it has been supposed that these differences may be explained by the imperfection of his apparatus, and by the use of white light[†]. Similar investigation was continued by Brewster[‡], on several species of glasses and on diamond. With, for example, a glass of refractive index 1·4826, his observed results for $\tan^{-1}\dfrac{T}{S}$ differed from $\tan^{-1}\dfrac{\cos(i+,i)}{\cos(i-,i)}$ by ± 1° 4′; which he considered might be within the limits of his observational errors. For diamond, he found greater deviations which seemed systematic, and not errors of observation. With a Fresnel's rhomb used according to the method of §§ 90—94 he might probably have found the definite correction on Fresnel's formula, required to represent the polarizational analysis of reflection from diamond. Can it be that both Fresnel and Brewster underestimated the accuracy of their own experiments, and that even for water and glasses, deviations which they found from Fresnel's $\dfrac{\cos(i+,i)}{\cos(i-,i)}$ may have been real, and not errors of observation ? The subject urgently demands full investigation according to the method of §§ 90—98, with all the accuracy attainable by instruments of precision now available.

§ 97. The reader may find it interesting to follow the formulas of § 94 through the whole range of incidences from 0° to 90°. For the present, consider only the case of the angle of incidence which makes $T = 0$. This, being the incidence which, when the incident light is polarized in any plane oblique to the plane of incidence, gives 90° difference of phase for the components of the reflected light vibrating in that plane and perpendicular to it, has been called by Cauchy, and I believe by all following writers, the *principal incidence*. We shall see presently (§§ 97″, 99, 105) that, for every transparent substance, observation and dynamics show one incidence, or an odd number of incidences, fulfilling this condition.

[*] Fresnel, *Œuvres*, Vol. i. p. 646.
[†] Mascart, *Traité d'Optique*, Vol. ii. p. 466.
[‡] Brewster, *Phil. Trans.* 1830. See also Mascart, Vol. ii. p. 466.

Molar. By observation and dynamics we learn (§ 81 above), that it is fulfilled *not* by the vanishing of S, but by the vanishing of T. To make $T = 0$ we have by (9) for the case of *principal incidence*, $1 + \cos 4f = 0$, and therefore $f = \pm 45°$. This, by (9), makes (if we take $f = +45°$) for Principal Incidence,

$$E/S = \tan (n - \pi/4) = k \dots\dots\dots\dots\dots(10).$$

§ 97′. The k here introduced is Jamin's notation, adopted also by Rayleigh. It is the ratio of the vibrational amplitude of reflected vibrations in the plane of incidence to the vibrational amplitude perpendicular to it, when the incident light is polarized in a plane inclined at 45° to the plane of incidence, and when the angle of incidence is such as to make the phases of those two components of the reflected light differ by 90°. k is positive or negative according as the phase of the reflected vibration in the plane of incidence lags or leads by 90° relatively to the component perpendicular to it. It is positive when (as in every well assured case* whether of transparent or of metallic mirrors) the observation makes $n > 45°$; it would be negative if $n < 45°$.

§ 97″. The angle $n - \pi/4$ found by our observation of § 97 is called the "Principal Azimuth." See § 158[xvii] below. It has been the usage of good writers regarding the polarization of light, particularly in relation to reflection and refraction, to give the name "azimuth†" to the angle between two planes through the direction of a ray of light; for instance, the angle between the plane of incidence and the plane of vibration of rectilineally polarized light. A "Principal Azimuth," for reflection at any polished surface, I define as the angle between the vibrational plane of polarized light incident at Principal Incidence, and the plane of the incidence, to make the reflected light circularly polarized. There is one, and only one, Principal Incidence for every known mirror: except internal reflection in diamond and other substances whose refractive indices exceed 2·414; these have three Principal Incidences (§ 158′″ below). The number is essentially odd: on this is founded the theory of the polarization of light by reflection.

* See Jamin, *Cours de Physique*, Vol. II. pp. 694, 695 ; also below, §§ 105, 154, 158ᵛ, 159′″, 179, 182.

† This, when understood, is very convenient; though it is not strictly correct. Azimuth in astronomy is essentially an angle in a horizontal plane, or an angle between two vertical planes. A reader at all conversant with astronomy would naturally think this is meant when a writer on optics uses the expression "Principal Azimuth," in writing of reflection from a horizontal mirror.

§ 98.   The vibrational plane of the incident light is inclined Molar. to the plane of incidence at an angle of $n - 45°$; which, for the Principal Incidence, is such as to render the two components of the reflected light equal; and therefore to make the light circularly polarized.   However a Fresnel's rhomb is turned, circularly polarized light entering it, leaves it plane polarized.   In the observation, with the details of §§ 90—94, it is turned so that the vibrational plane of the light emerging from it is perpendicular to the *fixed* vibrational plane of N_2.   Hence it occurs to us to think that a useful modification of those details might be;—to fix the Fresnel's rhomb with its principal planes at 45° to the plane of reflection, and to mount N_2 so as to be free to turn round the line of light leaving the Fresnel's rhomb. Alternate turnings of N_2 and N_1 give the desired extinction for any angle of incidence.   Take $f = 0$ in (1), (2), (3), (4); which makes $GOF = 45°$ in fig. 3 a.   Let n_2 be the angle (clockwise in the altered fig. 3 a) from OF to ON_2; and put $n = \frac{1}{4}\pi + \alpha$ in fig. 3 b. Eliminating $\mathscr{f}$ and g from (3), (4), by (8) and proceeding as in § 93, we find

$$T/S = \tan \alpha \cos 2n_2 ; \quad E/S = - \tan \alpha \sin 2n_2 ; \quad E/T = - \tan 2n_2 \ldots (10').$$

If α is positive observation makes n_2 negative when taken acute. For principal incidence it is $- 45°$, and $E/S = \tan \alpha$.   The phasal lag of (E, T) behind (S) is $- 2n_2$.   Negative is anti-clockwise in fig. 3 a.   See § 158[xvii].

§ 99.   The following table shows, for six different metals, determinations of principal incidences and principal azimuths which have been made by Jamin and Conroy, experimenting on light from different parts of the solar spectrum.   This table expresses, I believe, practically almost all that is known from observations hitherto made as to the polarizational analysis of homogeneous light reflected from metals.   The differences between the two observers for silver are probably real, and dependent on differences of condition of the mirror-surfaces at the times of the experiments, as modified by polishing and by lapses of time. It will be seen that for each colour Jamin's k^2 is intermediate between Conroy's two values for the same plate, after polishing with rouge and with putty powder respectively.   On the other hand, each of Conroy's Principal Incidences for silver is greater

Jamin*

Colour		Silver	Speculum Metal	Bell Metal	Steel	Zinc
Extreme red...	P. i. =	75° 45'	76° 45'	75° 16'	77° 52'	75° 45'
	P. a. =	41° 37'	29° 15'	29° 25'	16° 20'	15° 50'
	k² =	·7892	·3136	·3180	·0859	·0804
Orange	P. i. =	72° 48'	74° 36'	74° 5'	76° 37'	74° 54'
	P. a. =	40° 23'	27° 15'	28° 38'	16° 33'	18° 16'
	k² =	·7235	·2652	·2981	·0883	·1090
D ray	P. i. =	72° 30'	74° 7'	73° 28'	76° 40'	74° 27'
	P. a. =	40° 9'	27° 21'	28° 24'	16° 48'	18° 45'
	k² =	·7117	·2675	·2924	·0911	·1153
E "	P. i. =	71° 30'	73° 35'	72° 20'	75° 47'	73° 28'
	P. a. =	40° 19'	25° 52'	25° 31'	17° 30'	21° 13'
	k² =	·7201	·2351	·2278	·0994	·1507
F "	P. i. =	69° 34'	73° 4'	71° 21'	75° 8'	72° 32'
	P. a. =	39° 46'	26° 15'	23° 55'	18° 29'	22° 44'
	k² =	·6926	·2431	·1966	·1118	·1756
H "	P. i. =	66° 12'	71° 36'	70° 2'	74° 32'	71° 18'
	P. a. =	39° 50'	28° 0'	23° 21'	20° 7'	25° 18'
	k² =	·6959	·2827	·1864	·1342	·2234

Conroy†

Colour	Gold polished with Chamois and — Rouge	Gold polished with Chamois and — Putty powder	Silver polished with — Rouge	Silver polished with — Putty powder	Tin
Red ...	74° 21' / 40° 31' / ·7303	73° 57' / 41° 52' / ·8032	76° 11' / 32° 43' / ·4127	76° 29' / 43° 51' / ·9228	soda light 74° 17' / 31° 26' / ·3736
Yellow	72° 26' / 40° 9' / ·7117	71° 43' / 41° 14' / ·7683	75° 27' / 35° 41' / ·5157	74° 37' / 43° 22' / ·8922	
Blue...	68° 31' / 33° 54' / ·4516	67° 10' / 35° 40' / ·5151	72° 19' / 39° 28' / ·6780	71° 33' / 43° 0' / ·8696	

* See Mascart's Traité d'Optique, Vol. II. p. 589.   † Proc. R. S. Vol. xxxi. 1880–81, pp. 490, 495; and Vol. xxxv. 1883, p. 33.

than Jamin's; and by greater differences for the yellow and blue Molar. light than for the red.

Experimental determinations of T/S and E/S, (9) or (10′), through the whole range of incidence below and above the Principal Incidence are still wanting.

§ 100.   As for transparent solids and liquids, we may consider it certain that Fresnel's laws, giving

$$-S = \frac{\sin(i-,i)}{\sin(i+,i)}, \quad -T = \frac{\tan(i-,i)}{\tan(i+,i)}, \quad E = 0 \ldots\ldots(11),$$

are very approximately true through the whole range of incidence from 0° to 90°; but, as said in § 95, it is still much to be desired that experimental determinations of T/S should be made through the whole range; in order either to prove that it differs much less from $\dfrac{\cos(i+,i)}{\cos(i-,i)}$ than found experimentally by Fresnel himself and Brewster; or, if it differs discoverably from this formula, to determine the differences.   It is certain, however, that at the Principal Incidence the agreement with Fresnel's formula (implying $E = 0$ in the notation of § 94′) is exceedingly close; but the very small deviations from it found experimentally by Jamin and Rayleigh and represented by the values of k shown in the table of § 105 below, are probably real.   An exceedingly minute scrutiny as to the agreement of the Principal Incidence with $\tan^{-1}\mu$, Brewster's estimate of it;—a scrutiny such as Rayleigh made relatively to the approach to nullity of k for purified water surfaces; is still wanted; and, so far as I know, has not hitherto been attempted for water or any transparent body.   See §§ 180, 182 below.

Hitherto, except in §§ 81, 84, 86, we have dealt exclusively with what may be called the natural history of the subject, and have taken no notice of the dynamical theory; to the consideration of which we now proceed.

§ 101.   Green's doctrine* of incompressible elastic solid with equal rigidity, but unequal density, on the two sides of an interface, to account for the reflexion and refraction of light, brings out for vibrations *perpendicular to* the plane of incidence (§ 123 below) exactly the sine-law which Fresnel gave for *light polarized in the plane of incidence*.   On the other hand, for vibrations *in* the

* *Camb. Phil. Soc.*, Dec. 1837; Green's *Collected Papers*, pp. 246, 258, 267, 268.

Molar. plane of incidence it gives a formula (§§ 104, 105, 146 below) which, only when the refractive index differs infinitely little from unity, agrees with the tangent-law given by Fresnel for *light polarized perpendicular to the plane of incidence*;—but differing enormously from Fresnel, and from the results of observation, in all cases in which the refractive index differs sufficiently from unity to have become subject of observation or measurement.

§ 102. The accompanying diagrams, figs. 4, 5, 6, illustrate, each by a single curve (Curve 1), the perfect agreement between Green and Fresnel for the law of reflection at different incidences when the vibrations are perpendicular to the plane of incidence; and by two other curves the large disagreement when the vibrations are in the plane of incidence.

Curve 1 in each diagram shows for vibrations perpendicular to the plane of incidence the ratio of the reflected to the incident light according to Fresnel's sine-law

$$\left[\frac{\sin (i - {}_{,}i)}{\sin (i + {}_{,}i)} \right]^2$$

dynamically demonstrated by Green on the hypothesis of equal rigidities and unequal densities of the two mediums.

Curve 2 shows, for vibrations in the plane of incidence, the ratio of the reflected to the incident light according to Fresnel's tangent-law,

$$\left[\frac{\tan (i - {}_{,}i)}{\tan (i + {}_{,}i)} \right]^2.$$

Curve 3 shows, for vibrations in the plane of incidence, the ratio of reflected to incident activity (rate of doing work) per unit area of wave-plane, rigorously demonstrated by Green (§ 146 below) for plane waves incident on a plane interface between elastic solids of different densities but the same rigidity; on the supposition that each solid is absolutely incompressible, and that the two are in slipless contact at the interface.

In each diagram abscissas from 0° to 90° represent "angles of incidence," that is to say, angles between wave-normals and the line-normal to the interface, or angles between the wave-planes and the interface.

§ 103. Curves 2 and 3 of fig. 4 show for water a seemingly fair agreement between Fresnel and Green for vibrations in the plane of reflection. But the scale of the diagram is too small to show Molar. important differences for incidences less than 60° or 65°, especially in the neighbourhood of the polarizing angle, 53°·1 : this want is remedied by the larger scale diagram fig. 7 showing Curves 2 and 3 of fig. 4; on a scale of ordinates 48·5 times as large, in which, for vibrations in the plane of incidence, the unit for intensity of light is the *reflected* light at zero angle of incidence, instead

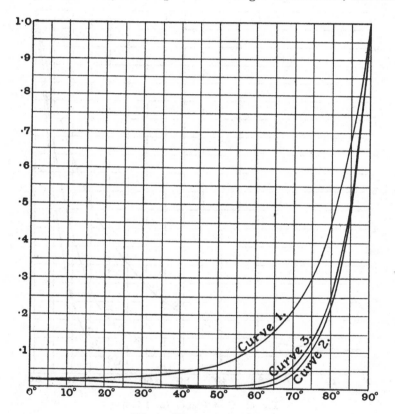

Fig. 4. Water. ($\mu = 1\cdot334$)

of the *incident* light at incidence i as in the other diagrams. Curve 2 in figs. 4 and 7 shows, for water, the absolute extinction at angle of incidence $\tan^{-1}\mu$ given by Fresnel's formula. Curve 3 (Green's formula) shows, for a slightly smaller angle of incidence (50°·0 instead of 53°·1), a minimum intensity equal to ·295 of that of directly reflected light; that is to say Green's formula

Molar. makes the directly reflected light from water only 3⅓ times
as strong as the light reflected at the angle which gives least
of it.

§ 104.   To test whether Fresnel or Green is more nearly right,
take a black japanned tray with water poured into it enough to
cover its bottom, and look through a Nicol's prism at the image

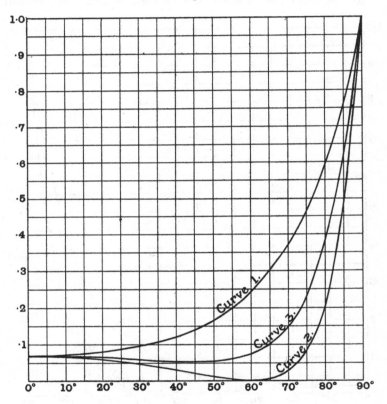

Fig. 5.   Flint Glass.   ($\mu = 1\cdot714$)

of a candle in the water-surface. You will readily find in half-a-
minute's trial a proper inclination of the light and orientation of
the Nicol to give what seems to you extinction of the light. To
test the approach to completeness of the extinction let an assistant
raise and lower alternately a piece of black cloth between the candle
and the water surface, taking care that it is lowered sufficiently
to eclipse the image of the candle when it is not extinguished by

the Nicol. By holding the Nicol very steadily in your hand, and Molar. turning to give the best extinction you can produce by it, you will find no difference in what you see whether the screen is down or up, which proves a very close approach to perfect extinction. Judging by the brilliance of the image of the candle when viewed through the Nicol at nearly normal incidence, and distances at

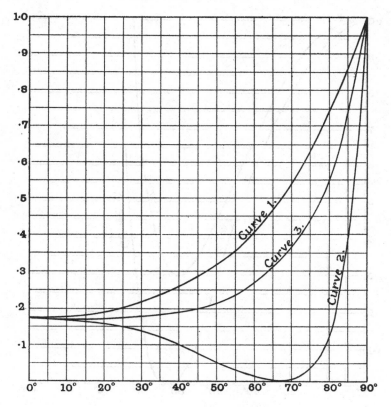

Fig. 6. Diamond. ($\mu = 2{\cdot}434$)

which this image can be seen, I think we may safely guess that the light reflected from the water at normal incidence was at least 500 (instead of Green's $3\frac{1}{3}$) times as strong as the imperceptible light of the nearest approach to extinction which the Nicol gave at the polarizing angle of incidence. And from Rayleigh's accurate experiments (§ 105 below) we know that it is 25,000,000 times, when the water surface is uncontaminated by oil or scum or impurity of any

Molar. kind. It would be only 3⅓ times if Green's dynamics were applicable
(without change of hypotheses) to the physical problem.   Hence,
looking at fig. 7, we see that, for water, Green's theory (Curve 3)
differs enormously from the truth, while Fresnel's formula
(Curve 2) shows perfect agreement with the truth, at the angle of
incidence 53°·1.   Looking at fig. 4 we see still greater differences

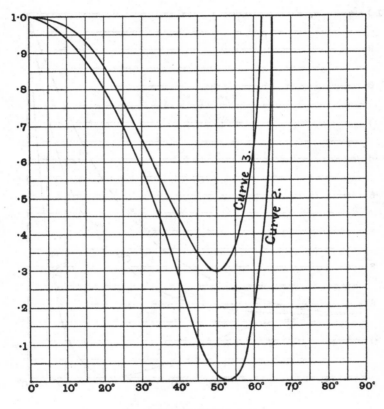

Fig. 7.   Water.

between Curves 2 and 3 even at as high angles of incidence as
75°, though there is essentially a perfect concurrence at 90°.

§ 105.   Looking at Curve 3 in figs. 5 and 6 we see that for flint
glass Green's theory gives *scarcely any diminution* of reflected
light, while for diamond it actually gives *increase*, when the angle
of incidence is increased from zero to $\tan^{-1}\mu$, Brewster's polarizing

angle. Yet even a hasty observation with no other apparatus Molar. than a single Nicol's prism shows, both for flint glass and diamond, diminution from the brightness of directly reflected light to what seems almost absolute blackness at incidence $\tan^{-1}\mu$, when the reflection is viewed through the 'Nicol with its plane of polarization perpendicular to the plane of reflection (that is to say, its transmitted light having vibration in the plane of reflection).

One readily and easily observed phenomenon relating to polarization by reflection is:—with a Nicol turned so as to transmit the vibrations in the plane of reflection, diminution of reflected light to nearly zero with angle of incidence increasing to $\tan^{-1}\mu$; and after that increase of the reflected light to totality when the angle of incidence is farther increased to 90°.

Another is:—for light reflected at a constant angle of incidence, diminution from maximum to minimum when the Nicol is turned 90° round its axis so as to bring the direction of vibration of the light transmitted by it from being perpendicular to the plane of reflection to being in this plane. This diminution from maximum to minimum is the difference between the ordinates of two curves representing, respectively for vibrations perpendicular to the plane of incidence and vibrations in the plane of incidence, the intensity of the reflected light at all angles of incidence. These two curves, if drawn with absolute accuracy, would in all probability agree *almost perfectly* with Fresnel's sine-law and Fresnel's tangent-law (Curves 1 and 2 of the diagrams) for all transparent substances. We have, however, little or no accurately measured observational comparisons except for light incident at very nearly the polarizational angle. By a very different mode of experimenting from that indicated in § 95 above, Jamin* found for eight substances having refractive indices from 2·454 to 1·334, results shown in the following table; to which is added a measurement made by Rayleigh for water with its surface carefully purified of oil or scum or any other substance than water and air.

---

* *Ann. de Chimie et de Physique*, 1850, Vol. xxix., p. 303, and 1851, Vol. xxxi., p. 179 ; and p. 174 corrected by pp. 180, 181. (Confusion between Cauchy's ϵ and Jamin's k is most bewildering in Jamin's papers. After much time sadly spent in trying to find what was intended in mutually contradictory Tables of results, I have given what seems to me probably a correct statement in the Table belonging to § 105.)

Substance	μ	Angle of Incidence $\tan^{-1}\mu$	Ratios of vibrational amplitudes, and of strengths, of reflected lights due to equal incident lights vibrating in and perpendicular to the plane of incidence	
			k	k^2
"Sulfure d'arsenic transparent" (Realgar) ...	2·454	67°·8	+ ·0850	1/138
"Blende transparente" (zinc sulphide)	2·371	67°·1	+ ·0420	1/567
Diamond	2·434	67°·7	+ ·0190	1/2770
Flint Glass	1·714	59°·7	+ ·0180	1/3086
"Verre"	1·487	56°·1	+ ·0060	1/27778
Fluorine	1·441	55°·2	− ·0084	1/14172
Absolute Alcohol	1·366	53°·8	+ ·00208	1/231160
Water (Jamin).............	1·334	53°·1	− ·00577	1/30030
Water, with specially purified surface (Rayleigh)	1·334	53°·1	+ ·0002	1/25,000,000

Molar. The greatest numeric in the last column, 1/138, would be barely perceptible on the diagram, fig. 7 ; and none of those for the other seven substances would be perceptible at all without a large magnification of the scale of ordinates.

§ 106. Although these results are related only to the ratio of the ordinates of Curve 1 to those of Curve 2 for one angle of incidence in each case, and do not touch the absolute values of the reflection due to unit quantities of incident light, we may infer as almost absolutely certain, or at all events (§§ 100, 125) highly probable, that Curve 1 (Fresnel's sine-law) and Curve 2 (Fresnel's tangent-law) are each of them about as nearly correct at all other incidences as at the critical incidences for which the observations were made. We shall see in fact (§§ 125, 133 below) in the dynamical theory to which we proceed, that the sine-law is absolutely accurate for vibrations perpendicular to the plane of incidence, on the supposition that the rigidities of the two mediums are equal and their densities unequal; and that the only correction of Green's dynamical postulates which can procure approximate annulment, for the reflected light, of vibrations in the plane of incidence at the angle $\tan^{-1}\mu$, gives correspondingly close agreement with Fresnel's tangent-law throughout the whole range of incidences from 0° to 90°!

(The following, §§ 107...111, is quoted from a paper written Molar. by myself in September—October, 1888, and published in the *Philosophical Magazine* for 1888, second half year.)

§ 107. "Since the first publication of Cauchy's work on the " subject in 1830, and of Green's in 1837, many attempts have " been made by many workers to find a dynamical foundation " for Fresnel's laws of reflection and refraction of light, but all " hitherto ineffectually. On resuming my own efforts since the " meeting of the British Association at Bath, I first ascertained " that an inviscid fluid permeating among pores of an incom- " pressible, but otherwise sponge-like, solid does not diminish, but " on the contrary augments, the deviation from Fresnel's law of " reflection for vibrations *in* the plane of incidence. Having " thus, after a great variety of previous efforts which had been " commenced in connexion with preparations for my Baltimore " Lectures of this time four years ago, seemingly exhausted " possibilities in respect to incompressible elastic solid, without " losing faith either in light or in dynamics, and knowing that " the condensational-rarefactional wave disqualifies* any solid of " positive compressibility, I saw that nothing was left but a solid " of such negative compressibility as should make the velocity of " the condensational-rarefactional wave, zero or small. So I tried " it and immediately found that, with other suppositions unaltered " from Green's, it exactly fulfils Fresnel's ' tangent-law ' for vibra- " tions *in* the plane of incidence, and his ' sine-law ' for vibrations " *perpendicular to* the plane of incidence. I then noticed that " homogeneous air-less foam, held from collapse by adhesion to " a containing vessel, which may be infinitely distant all round, " exactly fulfils the condition of zero velocity for the condensational- " rarefactional wave ; while it has a definite rigidity and elasticity " of form, and a definite velocity of distortional wave, which " can be easily calculated with a fair approximation to absolute " accuracy.

§ 108. "Green, in his original paper ' On the Reflexion and " Refraction of Light ' had pointed out that the condensational- " rarefactional wave might be got quit of in two ways, (1) by its " velocity being infinitely small, (2) by its velocity being infinitely

---

* Green's *Collected Papers*, p. 246.

Molar. "great. But he curtly dismissed the former and adopted the "latter, in the following statement: 'And it is not difficult to "'prove that the equilibrium of our medium would be unstable "'unless $A/B > 4/3$. We are therefore compelled to adopt the "'latter value of $A/B*$,' (∞) 'and thus to admit that in the "'luminiferous ether, the velocity of transmission of waves pro- "'pagated by normal vibrations is very great compared with that "'of ordinary light.' Thus originated the 'jelly-' theory of ether "which has held the field for fifty years against all dynamical "assailants, and yet has hitherto failed to make good its own "foundation.

§ 109. "But let us scrutinize Green's remark about instability. "Every possible infinitesimal motion of the medium is, in the "elementary dynamics of the subject, proved to be resolvable into "coexistent equi-voluminal wave-motions, and condensational- "rarefactional wave-motions. Surely, then, if there is a real "finite propagational velocity for each of the two kinds of wave- "motion, the equilibrium *must* be stable! And so I find Green's "own formula† proves it to be *provided we either suppose the* "*medium to extend all through boundless space, or give it a fixed* "*containing vessel as its boundary.* A finite portion of Green's "homogeneous medium left to itself in space will have the same "kind of stability or instability according as $A/B > 4/3$, or "$A/B < 4/3$. In fact $A - \frac{4}{3}B$, in Green's notation, is what I have "called the 'bulk-modulus'‡ of elasticity, and denoted by k "(being infinitesimal change of pressure divided by infinitesimal "change from unit volume produced by it: or the reciprocal of "what is commonly called 'the compressibility'). B is what "I have called the 'rigidity,' as an abbreviation for 'rigidity- "modulus,' and which we must regard as essentially positive. "Thus Green's limit $A/B > 4/3$ simply means positive compres- "sibility, or positive bulk-modulus: and the kind of instability "that deterred him from admitting any supposition of $A/B < 4/3$, "is the spontaneous shrinkage of a finite portion if left to itself

* A and B are the squares of velocities of the condensational and distortional waves respectively; supposing for a moment the density of the medium unity.

† *Collected Papers*, p. 253; formula (C).

‡ *Encyclopaedia Britannica*, Article "Elasticity": reproduced in Vol. III. of my *Collected Papers*.

"in a volume infinitesimally less, or spontaneous expansion if Molar.
"left to itself in a volume infinitesimally greater, than the
"volume for equilibrium. This instability is, in virtue of the
"rigidity of the medium, converted into stability by attaching
"the bounding surface of the medium to a rigid containing vessel.
"How much smaller than 4/3 may A/B be, we now proceed to
"investigate, and we shall find, as we have anticipated, that for
"stability it is only necessary that A be positive.

§ 110.   "Taking Green's formula (C); but to make clearer the
"energy-principle which it expresses (he had not even the words
"'energy,' or 'work'!), let W denote the quantity of work re-
"quired per unit volume of the substance, to bring it from its
"unstressed equilibrium to a condition of equilibrium in which
"the matter which was at (x, y, z) is at $(x + \xi, y + \eta, z + \zeta)$;
"ξ, η, ζ being functions of x, y, z such that each of the nine
"differential coefficients $d\xi/dx$, $d\xi/dy$, ... $d\eta/dx$ ... etc. is an in-
"finitely small numeric; we have

$$
W = \frac{1}{2} \left\{ A \left(\frac{d\xi}{dx} + \frac{d\eta}{dy} + \frac{d\zeta}{dz} \right)^2 \right.
$$
$$
+ B \left[\left(\frac{d\zeta}{dy} + \frac{d\eta}{dz} \right)^2 + \left(\frac{d\xi}{dz} + \frac{d\zeta}{dx} \right)^2 + \left(\frac{d\eta}{dx} + \frac{d\xi}{dy} \right)^2 \right]
$$
$$
\left. - 4B \left(\frac{d\eta}{dy} \frac{d\zeta}{dz} + \frac{d\zeta}{dz} \frac{d\xi}{dx} + \frac{d\xi}{dx} \frac{d\eta}{dy} \right) \right\} \dots\dots\dots\dots\dots(1).
$$

"This, except difference of notation, is the same as the formula
"for energy given in Thomson and Tait's *Natural Philosophy*,
"§ 693 (7).

§ 111.   "To find the total work required to alter the given
"portion of solid from unstrained equilibrium to the strained
"condition (ξ, η, ζ) we must take $\iiint dx\,dy\,dz\,W$ throughout the
"rigid containing vessel. Taking first the last line of (1);
"integrating the three terms each twice successively by parts
"in the well-known manner, subject to the condition $\xi = 0$, $\eta = 0$,
"$\zeta = 0$ at the boundary; we transform the factor within the last
"vinculum to

$$
\iiint dx\,dy\,dz \left(\frac{d\eta}{dz} \frac{d\zeta}{dy} + \frac{d\zeta}{dx} \frac{d\xi}{dz} + \frac{d\xi}{dy} \frac{d\eta}{dx} \right).
$$

Molar. "Adding this with its factor $-4B$ to the other terms of (1) "under $\iiint dx\,dy\,dz$, we find finally

$$\iiint dx\,dy\,dz\,W = \frac{1}{2}\iiint dx\,dy\,dz\left\{A\left(\frac{d\xi}{dx}+\frac{d\eta}{dy}+\frac{d\zeta}{dz}\right)^2\right.$$

$$\left.+ B\left[\left(\frac{d\zeta}{dy}-\frac{d\eta}{dz}\right)^2+\left(\frac{d\xi}{dz}-\frac{d\zeta}{dx}\right)^2+\left(\frac{d\eta}{dx}-\frac{d\xi}{dy}\right)^2\right]\right\}\;\ldots\ldots(2).$$

"This shows that positive work is needed to bring the solid to "the condition $(\xi,\ \eta,\ \zeta)$ from its unstrained equilibrium, and "therefore its unstrained equilibrium is stable, if A and B are "both positive, however small be either of them."

§ 112.   The equations of motion of the general elastic solid taken direct from the equations of equilibrium, with ρ to denote density, are, as we found in Lecture II. pp. 25, 26

$$\left.\begin{aligned}\rho\frac{d^2\xi}{dt^2} &= \frac{dP}{dx}+\frac{dU}{dy}+\frac{dT}{dz}\\[4pt]\rho\frac{d^2\eta}{dt^2} &= \frac{dU}{dx}+\frac{dQ}{dy}+\frac{dS}{dz}\\[4pt]\rho\frac{d^2\zeta}{dt^2} &= \frac{dT}{dx}+\frac{dS}{dy}+\frac{dR}{dz}\end{aligned}\right\}\;\ldots\ldots\ldots\ldots(3)\,;$$

where $\xi,\ \eta,\ \zeta$ denote (as above from Green) displacements; P, Q, R, normal components of pull (per unit area) on interfaces respectively perpendicular to $x,\ y,\ z$; and $S,\ T,\ U$ respectively the tangential components of pull as follows :—

$$\left.\begin{aligned}S&\begin{cases}= \text{pull parallel to } y \text{ on face perpendicular to } z\\ =\quad\text{''}\quad\text{''}\quad z\quad\text{''}\quad\text{''}\quad\text{''}\quad\text{''}\quad y\end{cases}\\[4pt]T&\begin{cases}=\quad\text{''}\quad\text{''}\quad z\quad\text{''}\quad\text{''}\quad\text{''}\quad\text{''}\quad x\\ =\quad\text{''}\quad\text{''}\quad x\quad\text{''}\quad\text{''}\quad\text{''}\quad\text{''}\quad z\end{cases}\\[4pt]U&\begin{cases}=\quad\text{''}\quad\text{''}\quad x\quad\text{''}\quad\text{''}\quad\text{''}\quad\text{''}\quad y\\ =\quad\text{''}\quad\text{''}\quad y\quad\text{''}\quad\text{''}\quad\text{''}\quad\text{''}\quad x\end{cases}\end{aligned}\right\}\;\ldots(4).$$

§ 113.   For an isotropic solid we had in Lecture XIV. (p. 191, above),

$$S = n\left(\frac{d\zeta}{dy}+\frac{d\eta}{dz}\right);\quad T = n\left(\frac{d\xi}{dz}+\frac{d\zeta}{dx}\right);\quad U = n\left(\frac{d\eta}{dx}+\frac{d\xi}{dy}\right)\;\ldots(5):$$

Molar.

$$P=(k-\tfrac{2}{3}n)\,\delta+2n\frac{d\xi}{dx};\quad Q=(k-\tfrac{2}{3}n)\delta+2n\frac{d\eta}{dy};$$

$$R=(k-\tfrac{2}{3}n)\,\delta+2n\frac{d\zeta}{dz}$$

$$\qquad\qquad\ldots(6):$$

where
$$\delta=\frac{d\xi}{dx}+\frac{d\eta}{dy}+\frac{d\zeta}{dz}\ldots\ldots\ldots\ldots\ldots(7).$$

Using these values of S, T, U, P, Q, R in (3) we find

$$\rho\frac{d^2\xi}{dt^2}=(k+\tfrac{1}{3}n)\frac{d\delta}{dx}+n\nabla^2\xi$$

$$\rho\frac{d^2\eta}{dt^2}=(k+\tfrac{1}{3}n)\frac{d\delta}{dy}+n\nabla^2\eta$$

$$\rho\frac{d^2\zeta}{dt^2}=(k+\tfrac{1}{3}n)\frac{d\delta}{dz}+n\nabla^2\zeta$$

$$\qquad\ldots\ldots\ldots\ldots(8).$$

§ 114. Taking d/dx of the first of equations (8), d/dy of the second, and d/dz of the third, and adding we find

$$\rho\frac{d^2\delta}{dt^2}=A\nabla^2\delta\ldots\ldots\ldots\ldots\ldots(9),$$

where
$$A=k+\tfrac{4}{3}n\ldots\ldots\ldots\ldots\ldots(10),$$

this being Green's "A" as used in §§ 108, 111 above.

Put now

$$\xi_1=\xi-\frac{d}{dx}\nabla^{-2}\delta;\quad\eta_1=\eta-\frac{d}{dy}\nabla^{-2}\delta;\quad\zeta_1=\zeta-\frac{d}{dz}\nabla^{-2}\delta\ldots(11),$$

which implies
$$\frac{d\xi_1}{dx}+\frac{d\eta_1}{dy}+\frac{d\zeta_1}{dz}=0\ldots\ldots\ldots\ldots(12):$$

and we find, by (8),

$$\rho\frac{d^2\xi_1}{dt^2}=n\nabla^2\xi_1;\quad\rho\frac{d^2\eta_1}{dt^2}=n\nabla^2\eta_1;\quad\rho\frac{d^2\zeta_1}{dt^2}=n\nabla^2\zeta_1\ldots\ldots(13).$$

Equations (9), (12), and (13) prove that any infinitesimal disturbance whatever is composed of specimens of the condensational-rarefactional wave (9), and specimens of the distortional wave (13), coexisting; and they prove that the displacement in the condensational-rarefactional wave is irrotational, because we see by (11) that an absolutely general expression for its components, $\xi-\xi_1$, $\eta-\eta_1$, $\zeta-\zeta_1$, if denoted by ξ_2, η_2, ζ_2, is

$$\xi_2=\frac{d\Psi}{dx},\quad\eta_2=\frac{d\Psi}{dy},\quad\zeta_2=\frac{d\Psi}{dz}\ldots\ldots\ldots(14),$$

Molar. where, when δ is known, Ψ is determined by

$$\nabla^2\Psi = \delta *\ldots\ldots\ldots\ldots\ldots\ldots\ldots(15).$$

Hence, as δ satisfies (9), we have

$$\rho\,\frac{d^2\Psi}{dt^2} = A\nabla^2\Psi \ \ldots\ldots\ldots\ldots\ldots(16);$$

and we see, finally, that the most general solution of the equations of infinitesimal motions is given by

$$\xi = \xi_1 + \xi_2, \quad \eta = \eta_1 + \eta_2, \quad \zeta = \zeta_1 + \zeta_2 \ldots\ldots\ldots(17):$$

provided ξ_1, η_1, ζ_1 satisfy (12) and (13); and ξ_2, η_2, ζ_2 satisfy (14) and (16).

§ 115. The general solutions of (11) and (12) for plane equivoluminal waves, and of (14) and (16) for plane condensational-rarefactional waves are as follows (easily proved by differentiations):—

$$\left.\begin{array}{c}\dfrac{\xi_1}{A} = \dfrac{\eta_1}{B} = \dfrac{\zeta_1}{C} = f\left(t - \dfrac{\alpha_1 x + \beta_1 y + \gamma_1 z}{u}\right) \\[2mm] u = \sqrt{\dfrac{n}{\rho}}, \text{ and } \alpha_1 A + \beta_1 B + \gamma_1 C = 0\end{array}\right\} \ \ldots\ldots(18),$$

with

$$\left.\begin{array}{c}\dfrac{\xi_2}{\alpha_2} = \dfrac{\eta_2}{\beta_2} = \dfrac{\zeta_2}{\gamma_2} = Hf\left(t - \dfrac{\alpha_2 x + \beta_2 y + \gamma_2 z}{v}\right) \\[2mm] v = \sqrt{\dfrac{k + \tfrac{4}{3}n}{\rho}}\end{array}\right\} \ \ldots\ldots(19);$$

with

where H is a constant, equal to the displacement in the condensational-rarefactional wave when $f = 1$; A, B, C, are constants equal to the x-, y-, z-components of the displacement, due to the equivoluminal wave when $f = 0$; $(\alpha_1,\ \beta_1,\ \gamma_1)$, $(\alpha_2,\ \beta_2,\ \gamma_2)$ are the direction-cosines of the normals to the wave-planes† of the

---

* Poisson's well-known fundamental theorem, in the elementary mathematics of force varying inversely as the square of the distance, tells us that when δ is known, or given arbitrarily through all space, $\nabla^{-2}\delta$ is determinate; being the potential of an ideal distribution of matter, of which the density is equal to $\dfrac{-\delta}{4\pi}$.

† By "wave-plane" of a plane wave I mean any plane passing through particles all in one phase of motion. For example, in sinusoidal plane waves the "wave-plane" may be taken as one of the planes containing particles having no displacement but maximum velocity, or it may be taken as one of the planes having maximum displacement and no velocity. For an arbitrary impulsive wave (as expressed in the text with f an arbitrary function through a finite range, and zero for all values of the argument on either side of that range) the "wave-plane" may

two waves; and u, v are the propagational velocities of the equi- Molar. voluminal and condensational-rarefactional waves respectively. In the condensational-rarefactional wave in an isotropic medium, the displacement or line of vibration is in the direction (α_2, β_2, γ_2), normal to the wave-plane.

§ 116. For the problem of reflection and refraction at an interface between two mediums, which (following Green) we shall call the upper medium and the lower medium respectively, let the interface be XOZ, and let this plane be horizontal. Let the wave-planes be perpendicular to XOY. This makes $\gamma_1 = 0$, $\gamma_2 = 0$ in (18) and (19). For brevity we shall frequently denote by P, the plane of incidence and reflection.

§ 117. Beginning now with vibrations perpendicular to P, we have $A = 0$, $B = 0$; and (18) becomes, for an incident wave as represented in fig. 8,

$$\zeta = Cf(t - ax + by) \quad \dots\dots\dots\dots\dots\dots(20),$$

where

$$a = \sin i/u, \text{ and } b = \cos i/u, \text{ with } u = \sqrt{\frac{n}{\rho}}\dots\dots(21).$$

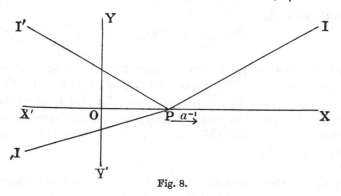

Fig. 8.

OZ is perpendicular to the diagram towards the eye.

The wave-planes of incident (I), reflected (I'), and refracted $(,I)$, waves are shown in fig. 8 for the particular case of incidence at 30°, and refractive-index for flint-glass $= 1\cdot724$; which makes $,i = 16°\cdot9$.

be taken as any plane through particles all in the same phase of motion. In this case we have a wave-front and a wave-rear; in the sinusoidal wave we have no front and no rear. I have therefore introduced the word "wave-plane" in preference to the generally used word "wave-front."

Molar.   For the reflected and refracted waves we may take respectively
$180° - i$, and $,i$ instead of i, and C', $,C$ instead of C. We have

$$\frac{\sin ,i}{,u} = \frac{\sin i}{u}, \text{ with } ,u = \sqrt{\frac{,n}{,\rho}} \dots\dots\dots(22),$$

and C' and $,C$ by equations (28) below. Thus for the displacements in the two mediums, due to the three waves, we have
$$\zeta + \zeta' = Cf(t - ax + by) + C'f(t - ax - by) \text{ in upper medium}\dots(23),$$
and
$$,\zeta = ,Cf(t - ax + ,by) \qquad\qquad \text{in lower medium}\dots(24),$$

where $,b = \cos ,i/,u,$ with $,u = \sqrt{\dfrac{,n}{,\rho}}\dots\dots\dots(25);$

$,b$ and $,u$ being both positive. Remember that y is negative in
the lower medium. Remark that, by (21), (22), (25), we have
$$b^2 = u^{-2} - a^2, \quad ,b^2 = ,u^{-2} - a^2 \dots\dots\dots(25)'.$$
The sole geometrical condition to be fulfilled at the interface is
$$\zeta + \zeta' = ,\zeta \text{ when } y = 0, \text{ which gives } C + C' = ,C \dots(26).$$
The sole dynamical condition is found by looking to § 113 (5). It is
$$S = ,S \text{ when } y = 0; \text{ giving } nb(C - C') = ,n,b,C \dots(27).$$
From these we find
$$C' = \frac{bn - ,b,n}{,b,n + bn} C, \quad ,C = \frac{2bn}{,b,n + bn} C \dots\dots(28).$$

§ 118. The interpretation of these formulas is obvious when
the quantities denoted by the several symbols are all real. But in
an important and highly interesting case of a real incident wave,
expressed by the first term of (23) with all the symbols real,
imaginaries enter into (24) and (25) by $,b$ being imaginary;
which it is when
$$a^{-1} < ,u \dots\dots\dots\dots(29);$$
or, in words, when the velocity of the trace of the wave-planes on
the interface is less than the velocity of the wave in the lower
medium. In this case $a^2 - ,u^{-2}$ is positive: denoting its value by q^2
we may put $,b = -\iota q^*$, where q is real; positive to suit notations
in § 119. Thus (28) becomes

$$\left.\begin{array}{l} C' = \dfrac{bn + \iota q,n}{bn - \iota q,n} C = \dfrac{b^2n^2 - q^2,n^2 + 2\iota qbn,n}{b^2n^2 + q^2,n^2} C \\[3mm] ,C = \dfrac{2bn}{bn - \iota q,n} C = \dfrac{2bn(bn + \iota q,n)}{b^2n^2 + q^2,n^2} C \end{array}\right\}\dots\dots(30).$$

* See foot-note on § 158″ below.

We have also in (24) an imaginary, $,b = -2q$, in the argument of f Molar. for the lower medium.

§ 119. To get real results we must choose f conveniently to make $f(t - ax - \iota qy) = F + \iota G$, where F and G are real. We may do this readily in two ways, (31), or (32), by taking τ, an arbitrary length of time, and putting

$$f(t - ax + by) = \frac{1}{t - ax + by + \iota\tau} = \frac{t - ax + by - \iota\tau}{(t - ax + by)^2 + \tau^2}.$$

This makes

$$f(t - ax + ,by) = \frac{1}{t - ax - 2qy + 2\tau} = \frac{t - ax - \iota(\tau - qy)}{(t - ax)^2 + (\tau - qy)^2}$$

$$\left.\right\} \dots(31),$$

or

$$f(t - ax + by) = \epsilon^{\iota\omega(t-ax+by)} = \cos\omega(t - ax + by) + \iota\sin\omega(t - ax + by)$$
$$f(t - ax + ,by) = \epsilon^{\iota\omega(t-ax-\iota qy)} = \epsilon^{\omega qy}[\cos\omega(t - ax) + \iota\sin\omega(t - ax)]$$

$$\left.\right\} \dots\dots\dots(32).$$

The latter of these is the proper method to show the results following the incidence of a train of sinusoidal waves; the former, which we shall take first, is convenient for the results following the incidence of a single pulse.

§ 120. With (23), (24), and (30), it gives for the incident wave in the upper medium

$$\zeta = \frac{t - ax + by - \iota\tau}{(t - ax + by)^2 + \tau^2} C \dots\dots\dots\dots(33),$$

and for the reflected wave in the upper medium

$$\zeta' = \frac{[(b^2n^2 - q^2,n^2)(t - ax - by) + 2qbn,n\tau] + \iota[2qbn,n(t - ax - by) - (b^2n^2 - q^2,n^2)\tau]}{(b^2n^2 + q^2,n^2)[(t - ax - by)^2 + \tau^2]} C$$

$$\dots\dots\dots(34),$$

and for the disturbance in the lower medium

$$,\zeta = \frac{2\{[b^2n^2(t - ax) + qbn,n(\tau - qy)] + \iota[qbn,n(t - ax) - b^2n^2(\tau - qy)]\}}{(b^2n^2 + q^2,n^2)[(t - ax)^2 + (\tau - qy)^2]} C$$

$$\dots\dots\dots(35).$$

The real parts of these three formulas represent a certain form of arbitrarily given incident wave: and the consequent reflected wave in the upper medium, and disturbance (a surface-wave,

Molar. analogous to a forced sea-wave) in the lower medium. The imaginary parts with ι removed, represent another form of incident wave and its consequences in the upper and lower mediums. In neither case does any wave travel into the lower medium away from the interface, and therefore the whole activity of the incident wave is in each case carried on by the reflected wave in the upper medium; that is to say, we have total reflection. It is interesting to see that in this total reflection, the reflected wave in each case differs in character from the incident wave, except for direct incidence; and it differs by being compounded of two constituents, one of the same character as the incident wave for that case, and the other of the same character as the incident wave for the other case. This corresponds to the change of phase (§§ 152 and 158^{iv}, below) by total internal reflection of waves of vibration perpendicular to P.

§ 121. The accompanying diagram (fig. 9) shows the characters of the two forms of incident wave, and of two constituents of the forced surface-wave in the lower medium, referred to in § 120. The two curves represent, from $t = -\infty$ to $t = +\infty$, the part of the displacement of any particle in the upper medium, due to one or other alone of the two forms of incident wave. The abscissa in each curve is t. The ordinates of the two curves are as follows:—

$$\text{Curve 1, } \frac{t}{t^2 + \tau^2}, \text{ and Curve 2, } \frac{\tau}{t^2 + \tau^2}.$$

The unit of ordinates in each case is τ^{-1}.

The disturbance in the lower medium is a forced wave, of character represented by a combination of these two curves, travelling under the interface at speed a^{-1}.

§ 122. All the words of § 120 apply also to the total reflection of sinusoidal waves, with this qualification, that the two characters of incident wave are expressed respectively by a cosine and a sine, and the "difference" becomes simply a difference of phase. Thus having taken the real parts of the formulas we get nothing new by taking the imaginary parts. The real parts of (23) and (24), with (32) for f, and with (30) for C' and $,C$, give us

Incident wave, $C \cos \omega \, (t - ax + by) \dots\dots\dots\dots(36);$

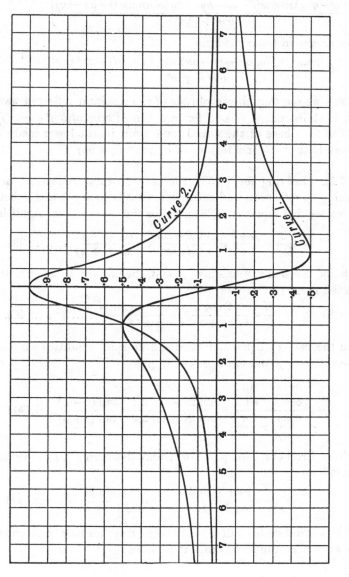

Fig. 9.

Molar. $\Bigg\{$ Reflected wave,

$$C\frac{(b^2n^2-q^2{,}n^2)\cos\omega(t-ax-by)-2qbn{,}n\sin\omega(t-ax-by)}{b^2n^2+q^2{,}n^2}\Bigg\}\dots(37);$$

$\Bigg\{$ Forced wave in lower medium,

$$C\frac{2bn\,\epsilon^{-\omega q y}\,[bn\cos\omega\,(t-ax)-q{,}n\sin\omega\,(t-ax)]}{b^2n^2+q^2{,}n^2}\Bigg\}\dots\dots(38).$$

Thus we see that the amplitude of the resultant reflected wave is C; that its phase is put forward $\tan^{-1}[2qbn{,}n/(b^2n^2-q^2{,}n^2)]$; and that the phase of the forced wave-train in the lower medium is before that of the incident wave by $\tan^{-1}(q{,}n/bn)$.

§ 123. Leaving for § 158′, the case of total reflection, and returning to reflection and refraction, that is to say, b and $,b$ both real, we see that equations (28) give for the case of equal rigidities

$$\frac{-C'}{C}=\frac{{,}b-b}{{,}b+b}=\frac{\cot{,}i-\cot i}{\cot{,}i+\cot i}=\frac{\sin(i-{,}i)}{\sin(i+{,}i)}\dots\dots(39),$$

which is Fresnel's "sine-law"; and, belonging to it for the refracted ray,

$$\frac{{,}C}{C}=\frac{2\cot i}{\cot{,}i+\cot i}=\frac{2\cos i\sin{,}i}{\sin({,}i+i)}\dots\dots(40).$$

In the case of equal densities and unequal rigidities we have $n/{,}n=\sin^2 i/\sin^2{,}i$ and equations (28) give

$$\frac{C'}{C}=\frac{\dfrac{\cos i}{\cos{,}i}-\dfrac{\sin{,}i}{\sin i}}{\dfrac{\cos i}{\cos{,}i}+\dfrac{\sin{,}i}{\sin i}}=\frac{\sin 2i-\sin 2{,}i}{\sin 2i+\sin 2{,}i}=\frac{\tan(i-{,}i)}{\tan(i+{,}i)}\dots\dots(41),$$

which is Fresnel's "tangent-law"; and, belonging to it for the refracted ray,

$$\frac{{,}C}{C}=\frac{2\sin i\cos i}{\sin{,}i\cos{,}i+\sin i\cos i}=\frac{\sin 2i}{\sin 2{,}i+\sin 2i}\dots\dots(42).$$

The third member of (41) is in some respects more convenient than Fresnel's beautiful "tangent-formula."

§ 124. These formulas, valid for pulses as well as for trains of sinusoidal waves, show, without any hypothesis in respect to compressibility or non-compressibility of ether, that, if the densities of the mediums are equal on the two sides of the interface, Molar. Fresnel's "tangent-law" is fulfilled by the reflected waves, if the vibrations are parallel to the interface; and therefore the reflected light vanishes when the angle of refraction is the complement of the angle of incidence. Hence non-polarized light incident at the angle which fulfils this condition would give reflected light consisting of *vibrations in the plane of the incident and reflected rays.* Now (§ 81) we have seen from Stokes' dynamical theory of the scattering of light from particles small in comparison with the wave-length, as in the blue sky, and also from his grating experiments and their theory, and (§ 81') confirmation by Rayleigh and by Lorenz of Denmark, that, in light polarized by reflection, *the vibrations are perpendicular to the plane of polarization defined as the plane of the incident and reflected rays*; that is to say the vibrations in the reflected ray are parallel to the interface. Hence it is certain that the densities of the mediums are *not equal* on the two sides of the interface; and that the densities and rigidities must be such as not to give evanescent reflected ray for any angle of incidence, when, as in § 123, the vibrations are parallel to the interface.

§ 125. On the other hand, formulas (39), (40) show that the supposition of equal rigidities not only does not give evanescent ray for any angle of incidence, but actually fulfils Fresnel's "sine-law" of reflection for all angles of incidence when the vibrations are parallel to the interface. Looking back to (28) and (39), we see that Fresnel's "sine-law" is expressed algebraically by

$$\frac{,b,n - bn}{,b,n + bn} = \frac{,b - b}{,b + b} \quad\dots\dots\dots\dots\dots(43).$$

If this equation is true for any one value of $b/,b$ other than 0 or ∞, we must have $n/,n = 1$. Hence if, with vibrations parallel to the interface, Fresnel's "sine-law" is exactly true for any one angle of incidence other than 90°, the rigidities of the mediums on the two sides are exactly equal, and the "sine-law" is exactly true for all angles of incidence; *a very important theorem.*

§ 126. Going back now to the more difficult case of vibrations in the plane of incidence, let us still take the interface as XOZ and horizontal; and the wave-planes perpendicular to XOY. Instead of the single displacement-component, ζ, and the single

Molar. surface-pull-component, S, of §§ 117—125, we have now two displacement-components, ξ, η; and two surface-pull-components, Q, perpendicular to the interface, and U parallel to the direction X in the interface. The interfacial conditions are that each of these four quantities has equal values on the two sides of the interface. We have now essentially the further complication of two sets of waves, equivoluminal and condensational-rarefactional, which are essentially to be dealt with. We might suppose the incident waves in the upper medium to be simply equivoluminal or simply condensational-rarefactional; but incidence on the interface between two mediums of different densities or different rigidities would give rise to reflected waves of both classes in the upper medium, and to refracted waves of both classes in the lower medium. It will therefore be convenient to begin with incident waves of both classes in the upper medium.

§ 127. Going back now to § 115, according to the details chosen in § 126, and taking j for the angle of incidence of condensational-rarefactional waves, we have $\gamma_1 = 0$, $C = 0$, $\gamma_2 = 0$; and we may put

$$\left.\begin{array}{lll} \alpha_1 = \sin i, & A = G \cos i, & \alpha_2 = \sin j, \\ \beta_1 = -\cos i; & B = G \sin i; & \beta_2 = -\cos j. \end{array}\right\} \quad \ldots\ldots(44).$$

Thus instead of (18) and (19) we now have

$$\left.\begin{array}{c} \dfrac{\xi_1}{\cos i} = \dfrac{\eta_1}{\sin i} = Gf\left(t - \dfrac{x \sin i - y \cos i}{u}\right), \\[2mm] u = \sqrt{\dfrac{n}{\rho}} \end{array}\right\} \quad \ldots\ldots(45),$$

with

$$\left.\begin{array}{c} \dfrac{\xi_2}{\sin j} = -\dfrac{\eta_2}{\cos j} = Hf\left(t - \dfrac{x \sin j - y \cos j}{v}\right), \\[2mm] v = \sqrt{\dfrac{k + \frac{4}{3}n}{\rho}} \end{array}\right\} \quad \ldots\ldots(46),$$

with

as the two constituents for the incident waves. Let now $_,u$, $_,G$, $_,i$, and $_,v$, $_,H$, $_,j$ be the values of the constants for the waves in the lower medium.

For the reflected waves in the upper medium we have $-\cos i$, $-\cos j$, instead of $\cos i$, $\cos j$; while $\sin i$, $\sin j$ are the same as for the incident waves. Let $-G'$, $+H'$, be the constants for the magnitudes of the reflected rays corresponding to G, H, for the incident rays. Thus for the total displacement-components due Molar. to the four waves in the upper medium we have

$$
\begin{aligned}
\xi = \cos i\, Gf &\left(t - \frac{x \sin i - y \cos i}{u} \right) + \sin j\, Hf \left(t - \frac{x \sin j - y \cos j}{v} \right) \\
+ \cos i\, G'f &\left(t - \frac{x \sin i + y \cos i}{u} \right) + \sin j\, H'f \left(t - \frac{x \sin j + y \cos j}{v} \right) \\
\eta = \sin i\, Gf &\left(t - \frac{x \sin i - y \cos i}{u} \right) - \cos j\, Hf \left(t - \frac{x \sin j - y \cos j}{v} \right) \\
- \sin i\, G'f &\left(t - \frac{x \sin i + y \cos i}{u} \right) + \cos j\, H'f \left(t - \frac{x \sin j + y \cos j}{v} \right)
\end{aligned}
$$

$$\dots\dots\dots(47),$$

and for total displacement-components in the lower medium the same formulas with $_{,}i$, $_{,}j$, $_{,}u$, $_{,}v$, $_{,}G$, $_{,}H$, 0, 0, in place of i, j, u, v, G, H, G', H' respectively.

§ 128.　Taking from these formulas the resultants of the (ξ, η) components of the displacements in the several waves, we have as follows :—

Equivoluminal wave
$$
\begin{cases}
\text{Incident,} & Gf \left(t - \frac{x \sin i - y \cos i}{u} \right) \\[2mm]
\text{Reflected,} & G'f \left(t - \frac{x \sin i + y \cos i}{u} \right) \\[2mm]
\text{Refracted,} & {}_{,}Gf \left(t - \frac{x \sin {}_{,}i - y \cos {}_{,}i}{{}_{,}u} \right)
\end{cases}
$$

Condensational-rare-
factional wave
$$
\begin{cases}
\text{Incident,} & Hf \left(t - \frac{x \sin j - y \cos j}{v} \right) \\[2mm]
\text{Reflected,} & H'f \left(t - \frac{x \sin j + y \cos j}{v} \right) \\[2mm]
\text{Refracted,} & {}_{,}Hf \left(t - \frac{x \sin {}_{,}j - y \cos {}_{,}j}{{}_{,}v} \right).
\end{cases}
$$

§ 129.　Putting now $y = 0$ to find displacement-components at the interface, and equating values on the two sides, we see in the first place that the arguments of f with y zero must be all equal; and that the coefficient of x is the reciprocal of the velocity of the trace of each of the four waves on the interface; and if we denote this velocity by a^{-1} we have

$$
a = \frac{\sin i}{u} = \frac{\sin {}_{,}i}{{}_{,}u} = \frac{\sin j}{v} = \frac{\sin {}_{,}j}{{}_{,}v} \quad \dots\dots\dots\dots(48).
$$

Molar. The last three of these equations express the laws of refraction of waves of either class in the lower medium consequent on waves of either class in the upper medium. They show that the sines of the angles of incidence and refraction are inversely as the propagational velocities in the two mediums, whether the two waves considered are of the same species or of the two different species. For brevity in what follows we shall put

$$\frac{\cos i}{u} = b, \quad \frac{\cos {}_{,}i}{{}_{,}u} = {}_{,}b, \quad \frac{\cos j}{v} = c, \quad \frac{\cos {}_{,}j}{{}_{,}v} = {}_{,}c \quad \ldots\ldots\ldots(49),$$

from which we find

$$\left.\begin{array}{ll} a^2 + b^2 = u^{-2} = \dfrac{\rho}{n}, & a^2 + c^2 = v^{-2} = \dfrac{\rho}{k + \frac{4}{3}n}, \\[3mm] a^2 + {}_{,}b^2 = {}_{,}u^{-2} = \dfrac{{}_{,}\rho}{{}_{,}n}; & a^2 + {}_{,}c^2 = {}_{,}v^{-2} = \dfrac{{}_{,}\rho}{{}_{,}k + \frac{4}{3}{}_{,}n}. \end{array}\right\} \quad \ldots(50).$$

§ 130. Putting now $f(t - ax) = \psi$, we find by (47) with $y = 0$

$$\left.\begin{array}{l} \xi = [bu\,(G + G') + av\,(H + H')]\,\psi \\[2mm] \eta = [au\,(G - G') + cv\,(-H + H')]\,\psi \end{array}\right\} \quad \ldots\ldots\ldots(51),$$

as the displacement-components of the upper medium at the interface.

Using now (47) to find Q and U by (6) and (5) of § 113, and putting $y = 0$ we find, as the components of surface-pull of the upper medium at the interface

$$\left.\begin{array}{l} Q = n\,[2abu\,(G + G') - hv\,(H + H')]\,\dot{\psi} \\[2mm] U = n\,[gu\,(G - G') + 2acv\,(H - H')]\,\dot{\psi} \end{array}\right\} \quad \ldots\ldots(52);$$

where

$$\left.\begin{array}{l} g = b^2 - a^2, \\[2mm] nh = (k - \tfrac{2}{3}n)\,v^{-2} + 2nc^2 = \rho + 2n\,(c^2 - v^{-2}) = \rho - 2na^2 \end{array}\right\} \quad \ldots(53).$$

Taking the terms of (51) and (52) which contain G, and H, and in them substituting ${}_{,}G$, ${}_{,}H$, ${}_{,}b$, ${}_{,}c$, ${}_{,}g$, ${}_{,}h$, ${}_{,}v$, ${}_{,}n$, ${}_{,}k$ for G, H, b, c, g, h, v, n, k, we find for the two refracted waves in the lower medium the displacement-components, and the surface-pull-components, at the interface, as follows:—

$$\left.\begin{array}{l} {}_{,}\xi = ({}_{,}b\,{}_{,}u\,G + a\,{}_{,}v\,H)\,\psi \\[2mm] {}_{,}\eta = (a\,{}_{,}u\,G - {}_{,}c\,{}_{,}v\,H)\,\psi \\[2mm] {}_{,}Q = {}_{,}n\,[2a\,{}_{,}b\,{}_{,}u\,G - {}_{,}h\,{}_{,}v\,H]\,\dot{\psi} \\[2mm] {}_{,}U = {}_{,}n\,[{}_{,}g\,{}_{,}u\,G + 2a\,{}_{,}c\,{}_{,}v\,H]\,\dot{\psi} \end{array}\right\} \quad \ldots\ldots\ldots\ldots(54).$$

Equating each component for the two sides, as said in § 126, Molar. we find the following four equations for determining the four required quantities, $,G$, $,H$, G', H', in terms of the two given quantities G, H,

$$
\left.
\begin{aligned}
bu(G + G') \quad + av\,(H + H') &= \quad ,b,u,G + \quad a,v,H \\
au(G - G') \quad - cv\,(H - H') &= \quad a,u,G - \quad ,c,v,H \\
n\,[2abu(G + G') \quad - hv\,(H + H')] &= ,n\,[2a,b,u,G - \quad ,h,v,H] \\
n\,[gu\,(G - G') + 2acv\,(H - H')] &= \quad ,n\,(,g,u,G + 2a,c,v,H)
\end{aligned}
\right\} \dots (55).
$$

§ 131. Considering for the present $G - G'$ and $H + H'$ as the known quantities, and $G + G'$, $H - H'$, $,G$, $,H$, as unknown; first find two values of $G + G'$ from the first and third of equations (55), and two values of $H - H'$ from the second and fourth of (55). Thus we have

$$
\left.
\begin{aligned}
G + G' &= \frac{1}{bu}(,b,u,G + a,v,H) - \frac{av}{bu}(H + H') \\
&= \frac{,n}{2nabu}(2a,b,u,G - ,h,v,H) + \frac{hv}{2abu}(H + H') \\
H - H' &= -\frac{1}{cv}(a,u,G - ,c,v,H) + \frac{au}{cv}(G - G') \\
&= \frac{,n}{2nacv}(,g,u,G + 2a,c,v,H) - \frac{gu}{2acv}(G - G')
\end{aligned}
\right\} \dots (56).
$$

The equalities of the second and third members of these two double equations may be taken as two equations for the determination of $,G$, $,H$ in terms of $G + G'$, and $H + H'$. With some simplifying reductions they become

$$
\left.
\begin{aligned}
2a,b(n - ,n),u,G + [,\rho + 2\,(n - ,n)a^2],v,H &= \rho v\,(H + H') \\
[,\rho + 2\,(n - ,n)\,a^2],u,G - \quad 2a,c\,(n - ,n),v,H &= \rho u.(G - G')
\end{aligned}
\right\} \dots (57).
$$

Finding $,G$ and $,H$ from these, and using the results in either the firsts or the seconds of the pairs of equations (56), we have $G + G'$ and $H - H'$ in terms of $G - G'$ and $H + H'$. These last two equations may be used to find G' and H' in terms of G and H. Thus we have finally G', H', $,G$, $,H$ in terms of the given quantities G, H. This, of course, might have been found directly from the four equations (55) by forming the proper determinants. The algebraic work is somewhat long either way.

Molar. §132. But the process we have followed in § 131 has the advantage of giving us the two intermediate equations (57); with the great simplification which they present in the case $n = {,}n$, for which they become

$${}_{,}\rho{,}v{,}H = \rho v\,(H + H'), \qquad {}_{,}\rho{,}u{,}G = \rho u\,(G - G')\ldots\ldots(58).$$

Let us now work this case out to the end, with the further simplification $H = 0$; because the particular problem which we wish to solve is to find the two reflected waves (G', H') and the two refracted waves (${}_{,}G, {}_{,}H$) due to a single incident equi-voluminal wave (G). Eliminating ${}_{,}G, {}_{,}H$ from the first two of (55), by (58) with $H = 0$; and then finding G' and H' from the two equations so got, we have

$$- G' = \frac{{}_{,}\rho b - \rho{,}b - L}{{}_{,}\rho b + \rho{,}b + L}\,G \quad\ldots\ldots\ldots\ldots\ldots (59),$$

$$- H' = 2\,\frac{\dfrac{{}_{,}\rho b}{({}_{,}\rho - \rho)\,a}\,L}{{}_{,}\rho b + \rho{,}b + L}\,\frac{u}{v}\,G \quad\ldots\ldots\ldots\ldots(60),$$

where

$$L = \frac{({}_{,}\rho - \rho)^2\,a^2}{{}_{,}\rho c + \rho{,}c} \quad\ldots\ldots\ldots\ldots\ldots\ldots(61).$$

Lastly (58) gives ${}_{,}G = 2\,\dfrac{\rho b}{{}_{,}\rho b + \rho{,}b + L}\,\dfrac{u}{{}_{,}u}\,G \ldots\ldots\ldots\ldots(62),$

$$- {}_{,}H = 2\,\frac{\dfrac{\rho b}{({}_{,}\rho - \rho)\,a}\,L}{{}_{,}\rho b + \rho{,}b + L}\,\frac{u}{{}_{,}v}\,G \quad\ldots\ldots\ldots\ldots(63).$$

This completes the theory of the reflection and refraction of waves at a plane interface of slipless contact between two ordinary elastic solids, with any given bulk-moduluses and rigidities.

§133. Remark now that by equations (49), (48), we have

$$\frac{{}_{,}b}{b} = \frac{u \cos{}_{,}i}{{}_{,}u \cos i} = \frac{\sin i \cos{}_{,}i}{\sin{}_{,}i \cos i} \quad\ldots\ldots\ldots\ldots\ldots(64);$$

and by (50) with $n = {,}n$, $\quad \rho/{,}\rho = {,}u^2/u^2 \quad\ldots\ldots\ldots\ldots\ldots\ldots(65);$

whence, by (64), $\qquad \dfrac{\rho{,}b}{{}_{,}\rho b} = \dfrac{\sin{}_{,}i \cos{}_{,}i}{\sin i \cos i} \quad\ldots\ldots\ldots\ldots\ldots(66).$

Hence, we see that if $L = 0$ we have

$$\frac{- G'}{G} = \frac{\sin 2i - \sin 2{,}i}{\sin 2i + \sin 2{,}i} = \frac{\tan (i - {,}i)}{\tan (i + {,}i)} \quad\ldots\ldots(67);$$

which is Fresnel's formula for reflection when the vibrations are in the plane of incidence.

Looking to (61) we see that this could only be the case for Molar. other than direct incidence by either c, or $_,c$, or both being infinitely great; because, § 129 (48), a can only be zero for direct incidence ($i = 0$). By (50) we see that to make c or $_,c$ very great, we must have v or $_,v$ very small. Returning to this suggestion in (Lect. XIX. § 167), we shall find good physical reason to leave, for undisturbed ether, $v = \infty$ as Green made it: and to let $_,v$ be small enough, to make L not absolutely zero but as small as required to give the closeness of approximation to truth which observation proves for Fresnel's formulas for the great majority of transparent liquids and solids, including optically isotropic crystals; and yet large enough to give modified formulas representing the deviations from Fresnel found in diamond, sulphide of zinc, sulphide of arsenic, etc. (see § 105 above and § 182 below).

§ 134. Meantime we shall consider some interesting and important characteristics of the general problem of § 130 without the limitation $n = _,n$: and of its solution for the case $n = _,n$ expressed by § 123 (39), (40) and § 132 (59)...(63), with the modification due to $v = \infty$ in space containing no ponderable matter; and with the special further modification regarding waves or vibrations of ether in the space occupied by metals (solid or liquid) to account for the observationally discovered facts of metallic reflection.

§ 135. Consider first direct incidence, whether of an equi-voluminal or of a condensational-rarefactional wave. For this we have in §§ 129, 130,

$$i = _,i = j = _,j = 0; \quad a = 0; \quad c = v^{-1}; \quad _,c = _,v^{-1};$$
$$u^{-2} = b^2 = g = h = \frac{\rho}{n}; \quad _,u^{-2} = _,b^2 = _,g = _,h = \frac{_,\rho}{_,n}.$$

These details reduce (55) to

$$\begin{aligned} G + G' &= \quad _,G \\ H - H' &= \quad _,H \\ \rho v (H + H') &= _,\rho _,v _,H \\ \rho u (G - G') &= _,\rho _,u _,G \end{aligned} \right\} \quad \dots\dots\dots\dots(68).$$

The first and fourth of these give G' and $_,G$ in terms of G; the

Molar. second and third give H' and $,H$ in terms of H. Thus our formulas verify what is obvious without them; that a *directly incident* wave, if equivoluminal, gives equivoluminal reflected and transmitted waves; and if condensational-rarefactional, gives condensational-rarefactional reflected and transmitted waves. For direct incidence, remark that in the equivoluminal waves the displacement is parallel to the reflecting surface, and in the x direction; in the condensational-rarefactional waves it is perpendicular to the reflecting surface, which is the y direction. For ratios of these displacements we find, as follows, from (68);—

(Equivoluminal) $\dfrac{-G'}{G} = \dfrac{{}_{,}\rho_{,}u - \rho u}{{}_{,}\rho_{,}u + \rho u}$; $\quad \dfrac{{}_{,}G}{G} = \dfrac{2\rho u}{{}_{,}\rho_{,}u + \rho u}$ ....(69);

(Condensational-rarefactional) $\dfrac{H'}{H} = \dfrac{{}_{,}\rho_{,}v - \rho v}{{}_{,}\rho_{,}v + \rho v}$, $\quad \dfrac{{}_{,}H}{H} = \dfrac{2\rho v}{{}_{,}\rho_{,}v + \rho v}$

......(70).

In respect to the general theory of the reflection of waves at a plane interface between two elastic solids of different quality, it is interesting to see that, provided only they are *sliplessly* connected at the interface, we have, for the case of direct incidence, the same relations between the displacements of the reflected and transmitted waves and of the incident wave in terms of densities and propagational velocities, for equivoluminal waves of transverse vibration, as for condensational-rarefactional waves (vibrations in the line of transmission). If, on the other hand, the connection between the two solids were merely by normal pressure, and if the surfaces of the two solids at the interface were perfectly frictionless, and allowed perfect freedom for tangential slipping; the reflection in the case of directly incident waves of transverse vibration would be total, and there would be no transmission of waves into the other solid.

§ 136. In respect to physical optics the solution (69) is exceedingly interesting. If we eliminate ρ, ${}_{,}\rho$ from it by

$$\rho = nu^{-2}, \quad {}_{,}\rho = {}_{,}n_{,}u^{-2},$$

we find

$$\frac{-G'}{G} = \frac{{}_{,}nu - n_{,}u}{{}_{,}nu + n_{,}u}; \quad \frac{{}_{,}G}{G} = \frac{2n_{,}u}{{}_{,}nu + n_{,}u} \quad \dots\dots\dots\dots(71).$$

Hence, denoting $u/{}_{,}u$ by μ (the refractive index), we see that, for the reflected wave, in the two cases of equal rigidities and equal densities, we have respectively,

$$\frac{G'}{G} = \frac{1-\mu}{1+\mu}, \text{ and } \frac{G'}{G} = \frac{\mu-1}{\mu+1} \quad\ldots\ldots\ldots\ldots(72).$$

Molar.

Thus G'/G is equal but with opposite signs in the two cases; and therefore, as the ratio of the intensity of the reflected light to the intensity of the incident light is equal to $(G'/G)^2$, we see that it is equal to

$$\left(\frac{\mu-1}{\mu+1}\right)^2 \quad\ldots\ldots\ldots\ldots\ldots\ldots(73),$$

and is the same in the two cases of equal densities and equal rigidities; an old known and very important result in physical optics. It was, I believe, first given by Thomas Young; it is also found by making $i = 0$ in Fresnel's formulas for reflection of polarized light at any incidence. But full dynamical theory proves that, if the refractivity $\mu - 1$ is produced otherwise than by either equal rigidities or equal densities, the ratio of reflected light to incident light would not be exactly equal to (73). This dynamical truth was referred to in my introductory Lecture (pp. 15, 16, above); and I had then come to the conclusion from Professor Rood's photometric experiments that the observed amount of reflected light from glass agrees too closely with (73) to allow any deviation from either equal rigidities or equal densities, sufficient to materially improve Green's dynamical theory of the polarization of light by reflection. This conclusion is on the whole confirmed by Rayleigh's very searching investigation of the reflection of light from glasses of different kinds*; but the great differences of reflectivity which he found in the surface of the same piece of glass in different states of polish, rendered it impossible to get thoroughly satisfactory results in respect to agreement with theory; as we see by the following statement which he gives as a summing-up of his investigation.

" Altogether the evidence favours the conclusion that recently " polished glass surfaces have a reflecting power differing not more " than 1 or 2 per cent. from that given by Fresnel's formula; but

* " On the Intensity of Light Reflected from certain surfaces at nearly Perpendicular Incidence," *Proc. R. S.*, XLI. pp. 275—294, 1886; and *Scientific Papers*, II. pp. 522—542.

24—2

Molar. "that after some months or years the reflection may fall off from
"10 to 30 per cent., and that without any apparent tarnish.

"The question as to the cause of the falling off, I am not in
"a position to answer satisfactorily. Anything like a disintegration
"of the surface might be expected to reveal itself on close in-
"spection, but nothing of this kind could be detected. A super-
"ficial layer of lower index, formed under atmospheric influence,
"even though no thicker than 1/100000 inch, would explain a
"diminished reflection. Possibly a combined examination of the
"lights reflected and transmitted by glass surfaces in various
"conditions would lead to a better understanding of the matter.
"If the superficial film act by diffusion or absorption, the trans-
"mitted light may be expected to fall off. On the other hand,
"the mere interposition of a transparent layer of intermediate
"index would entail as great an *increase* in the transmitted as
"falling off in the reflected light. There is evidently room here
"for much further investigation, but I must content myself with
"making these suggestions."

§ 137. Consider next grazing incidence ($i = 90°$, $b = 0$, $a = u^{-1}$)
of an equivoluminal wave of vibrations in the plane of incidence,
and therefore very nearly perpendicular to the reflecting surface.
We see immediately that equations (55) are satisfied by

$$H = 0, \quad H' = 0, \quad {}_{,}H = 0, \quad {}_{,}G = 0, \quad G - G' = 0.$$

This shows that, whether for equal or unequal rigidities, we have
approximately total reflection, and that the phase of the reflected
light corresponds to $G' = G$. Looking now to § 136 (71), we see
that in the case of equal rigidities of the two mediums G'/G is
negative for direct incidence, while we now find it to be positive
for grazing incidence. Hence if it is real for all incidences it
must be zero for one particular incidence. *This is fundamental
in the dynamics of polarization by reflection* and of *principal
incidences.*

§ 138. On the other hand, for equal densities and unequal
rigidities, look to § 135 (69); and we see that G'/G is positive for
direct incidence: and by § 137 it is $+1$ for grazing incidence.
Hence, it cannot vanish just once or any odd number of times
when i is increased from 0° to 90°; but it can vanish twice or
an even number of times. This is essentially concerned in the

explanation of the remarkable discovery of Lorenz and Rayleigh, Molar. referred to in § 81 above.

§ 138'.   Lastly; going back to § 123, we see that our notation has secured that, for direct incidence, C'/C is negative or positive just as is G'/G, according as the rigidities or the densities of the two mediums are equal.   But at grazing incidence, C'/C, while still negative for equal rigidities, is positive for equal densities.

§ 139.   So far everything before us, dynamical and experimental, confirms Green's original assumption of equal rigidities and different densities to account for light reflected from and transmitted through transparent bodies.   Before going on in Lecture XIX. to the promised reconciliation between Fresnel and dynamics for *transparent substances*, let us, while keeping to the supposition of equal rigidity of ether throughout vacant space and throughout space occupied by ponderables of any kind, briefly consider what suppositions we must make in respect to our solution of § 132 [(59)—(63)], to explain known truths regarding the reflection of light from metals or other opaque bodies.

§ 140.   The extremely high degree of opacity presented by all metals for light of all periods, from something considerably longer than that of the extreme red of the visible spectrum ($2 \cdot 5 \cdot 10^{-15}$ secs. for A line) to something considerably less than that of the extreme violet ($1 \cdot 3 \cdot 10^{-15}$ secs. for H line), is the most definite of the visible characteristics of metals: while the great brilliance of light reflected from them, either directly or at any angle other than grazing incidence*, compared with that reflected from glass or

---

* As obliquity of incident light is increased to approach more and more nearly grazing incidence, the brilliance of the reflected light approaches more and more nearly to equality with the incident light.   At infinitely nearly grazing incidence we find theoretically (§ 137) total reflection from every polished surface; polish being defined as in § 86 above.   Indeed we find observationally in surfaces such as sooted glass, which could scarcely be called polished according to any definition, a manifest tendency towards total reflection when the angle of incidence is increased to nearly 90°.

A striking illustrative experiment may be made by placing, on a table covered with a black cotton-velvet tablecloth, two pieces of plate glass side by side, with an arrangement of light and screen as indicated in the accompanying sketch. L is a lamp which may be held by hand, and raised or lowered slightly at pleasure. OO' is an opaque screen. PP' is a screen of white paper resting on the table.   It would be startling, if we did not expect the result, to see how much light is reflected from

Molar. crystals or from the most brilliantly polished of commonly known and seen non-metallic bodies, is their most obvious and best known quality.   The opacity of thin metal plates hitherto tested for all visible lights from red to violet has been found seemingly perfect for all thicknesses exceeding 3.10^{-5} cm. (or half the wave-length of yellow light in air).   When, in the process of gold-beating, the thickness of the gold-leaf is reduced to about 2.10^{-5} cm. (or about one-third of the wave-length of yellow light) it begins to be perceptibly translucent, transmitting faint green light when illuminated by strong white light on one side.   The thinnest of ordinary gold-leaf ($\cdot 7.10^{-5}$ cm., or about one-eighth of the wave-length of yellow light) is quite startlingly translucent, giving a strong green tinge to the transmitted light. Silver foil $1\cdot5.10^{-5}$ cm. thick (considerably thinner than translucent gold-leaf) is quite opaque to the electric light so far as our eyes allow us to judge ;

the soot above the boundary of the shadow of OO'. The experiment is rendered still more striking by placing a flat plate of polished silver beside the two glass plates, and seeing how nearly both the sooted and the clean glass plates come in rivalry with the silver plate in respect to totality of reflection, when L is lowered to

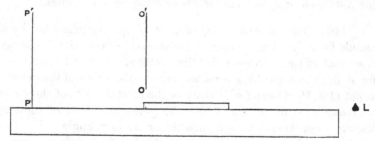

more and more nearly grazing incidence of its light. It is interesting also to take away the paper screen, and view the three plates and the lamp by an eye placed in positions to receive reflected light from the three mirrors, sooted, polished glass, and polished silver.

Another interesting experiment may be made by looking vertically downwards through a Nicol at the three surfaces, or at a clean surface of mercury or water, illuminated by light from L at nearly grazing incidence in a dark room. A surprisingly large amount of light is seen from the sooted surface, and is found to be almost wholly polarized to vibrations perpendicular to the plane through L and the line of vision. If the glass and silver are very well polished and clean, little or no light will be seen from them; unless L is very intense, when probably a faint blue light, polarized just as is the light from the soot, will be visible, indicating want of molecular perfectness in the polish, or a want of optical perfectness in the most perfect polish possible for the molecular constitution of the solid or liquid, according to the principles indicated in § 86 above.

but it is transparent to an invisible violet light through a small Molar. range of wave-length from about $3·07.10^{-5}$ to $3·32.10^{-5}$ cm.* ·(periods from $1·02.10^{-15}$ to $1·11.10^{-15}$ of a second).

§ 141. The extreme opacity of metals is quite lost for Roentgen rays (which are probably light of much shorter period than 10^{-15} of a second); sheet aluminium of thicknesses up to two or three centimetres being transparent for them. For some qualities of Roentgen rays even thick sheet lead is not perfectly opaque.

§ 142. We have no experimental knowledge in respect to the opacity of exceedingly thin metallic films for radiant heat of longer periods than that of the reddest visible light. It seems not improbable that through the whole range of periods up to 2.10^{-13} of a second, through which experiments on the refractivity and reflectivity of rock-salt and sylvin have been made by Langley, Rubens, Paschen, Rubens and Nicols, and Rubens and Aschkinaas (see above, Lecture XII. p. 150), the opacity of gold-leaf and other of the thinnest metal foils *may* be as complete, or nearly as complete, as it is for visible light.

§ 143. But, when we go to very much longer periods, we certainly find these thin metal foils quite transparent for variations of magnetic force. Thus, sheet copper of thickness 7.10^{-5} cm. (about a wave-length of orange light in air), though almost perfectly opaque to variation of magnetic force of period one-third or one-fourth of $1/(8.10^9)$ of a second, is almost perfectly transparent for periods of magnetic force of three or four times $1/(8.10^9)$ of a second, and for all longer periods.

Taking greater thicknesses, we find a copper plate two milli-metres thick, almost a perfect screen, that is, almost perfectly opaque, in respect to the transmission of the magnetic influence of a little bar-magnet rotating 8000 times per second†; somewhat opaque, but not wholly so, when the speed is 100 times per second; almost, but not perfectly, transparent, that is to say, very slightly

* See pp. 185, 186 of *Popular Lectures and Addresses*, Vol. I., Ed. 1891 (Friday Evening Royal Institution Lecture, Feb. 3, 1883), where experiments illustrating the reflectivity and the transparency of some exceedingly thin films of platinum, gold, and silver, supplied to me through the kindness of Professor Dewar, are described.

† See § 3 of Appendix K, "Variational Electric and Magnetic Screening," reprinted from *Proc. R. S.* Vol. XLIX. April 9, 1891 ; also *Math. and Phys. Papers*, Vol. III. Art. CII. "Ether, Electricity, and Ponderable Matter," § 35.

Molar. if perceptibly obstructive, when the speed is once per second; not perceptibly obstructive, when the period is 10 seconds or more. Now, according to what is without doubt really valid in the so-called electro-magnetic theory of light, we may regard as a lamp, a bar-magnet rotating about an axis perpendicular to its length, or having one pole caused to vibrate to and fro in a straight line. We may regard it as a lamp emitting light of period equal to the period of the rotation or of the vibration. For the light of this lamp, sheet copper two millimetres thick is almost perfectly transparent if the period is anything longer than one second; but it is almost perfectly opaque if the period is anything less than 1/8000 of a second down to one eight hundred million millionth of a second (the period of extreme violet light); and is probably quite opaque for still smaller periods down to those of the Roentgen rays, if we regard these rays as due to vibrators giving after each shock a sufficient number of subsiding vibrations to allow a period to be reckoned. Whatever the distinctive characteristic of the Roentgen light, sheet copper two millimetres thick is perceptibly translucent to it, and sheet aluminium much more so.

§ 144. We may reasonably look for a detailed and satisfactory investigation, mathematical and experimental intelligence acting together, by which we shall thoroughly understand the continuous relation between the reflection and translucence of metals and transparent bodies, and the phenomena of electric and magnetic vibrations in insulating matter, in non-magnetic metals, in soft iron, and in hardened steel, for all vibrational periods from those of the Roentgen rays to ten or twenty seconds or more. The investigation must, of course, include non-periodic motions of ether and atoms. It cannot but show the relation between the electric conductivity of metals and their opacity. It must involve the consideration of molecular and atomic structures. Maxwell's electro-magnetic theory of light was essentially molar*; and therefore not in touch with the dynamics of dispersion essentially

* "Suppose, however, that we leap over this difficulty [regarding electrolysis] by "simply asserting the fact of the constant value of the molecular charge, and that "we call this constant molecular charge, for convenience in description, *one* "*molecule of electricity*.

"This phrase, gross as it is, and out of harmony with the rest of this treatise, "will enable us at least to state clearly what is known about electrolysis, and to "appreciate the outstanding difficulties." Maxwell, *Electricity and Magnetism*, Vol. I. p. 312.

involved in metallic reflection and translucency;
his electro-magnetic theory, he was himself one o
leading molecularists of the nineteenth century:
kinetic theory of gases; and his estimates of the
weights of atoms; and his anticipation of the Sellmeier-h.
molecular dynamics of ordinary and anomalous dispersion.
published* Cambridge Examination question.

§ 145. First, however, without any molecular hypothesis, and
without going beyond Green's purely molar theory of infinite
resistance to compression, and equal effective rigidities of ether in
all bodies and in space void of ponderable matter, let us try how
nearly we can explain the high reflectivity and the great opacity
of metals. Either great rigidity, or great density, or both great
rigidity and great density, of ether in metals would explain these
two properties: but we have agreed not to assume differences of
rigidity, and there remains only great density. It is interesting
to remark however, that infinite rigidity would give exactly the
same law as infinite density; because each extreme hypothesis
would simply keep the ether at the interface absolutely unmoved;
and this even if we allow the ether to be compressible within
liquids and solids, as we are going to do later on.

§ 146. Go back now to § 132 (59)—(63), and, following
Green, make $v = \infty$, and $,v = \infty$. This makes, by (50), $c = ,c = -a\iota$†;
and reduces (61) and (59) to

$$L = Ka\iota \dots\dots\dots\dots\dots\dots(74),$$

and

$$\frac{-G'}{G} = \frac{,\rho b - \rho,b - Ka\iota}{,\rho b + \rho,b + Ka\iota} = \sqrt{\frac{(,\rho b - \rho,b)^2 + K^2 a^2}{(,\rho b + \rho,b)^2 + K^2 a^2}} [\cos(\phi + \psi) - \iota \sin(\phi + \psi)]$$

where $\quad K = \dfrac{(,\rho - \rho)^2}{,\rho + \rho}, \quad \phi = \tan^{-1} \dfrac{Ka}{,\rho b - \rho,b}, \quad \psi = \tan^{-1} \dfrac{Ka}{,\rho b + \rho,b}$

$$\dots\dots\dots(75).$$

---

* Rayleigh, in a footnote appended in 1899 to the end of his paper "On the
Reflexion and Refraction of Light by Intensely Opaque Matter" (*Phil. Mag.* 1872;
republished as Art. xvi. of his *Scientific Papers*, Vol. i.), writes as follows:

"I have lately discovered that Maxwell had (earlier than Sellmeier) considered
"the problem of anomalous dispersion. His results were given in the Mathematical
"Tripos Examination, Jan. 21, 1869 (see *Cambridge Calendar* for that year)."

In this examination question the *viscous term*, subsequently given by Helmholtz,
is included.

† Take $-a\iota$ (not $+a\iota$), a being positive, in order that, in § 128, H, of the upper
medium may have extinctional factor $\epsilon^{-\omega ay}$, and, H, of the lower, may have
extinctional factor $\epsilon^{\omega ay}$. See footnote on § 158″. See also § 128 and (49), and (54).

ıth Green's result*; and the square of the first final expression for $\dfrac{-G'}{G}$ is the formula by which figs. 4, 5, 6, for water, flint-glass, and diamond, in ᴏ3 above, were calculated.

ᴏ 147.   To realize this solution in the most convenient way for physical optics, put

$$f(\theta) = \epsilon^{\iota\omega\theta} = \cos \omega\theta + \iota \sin \omega\theta \dots\dots\dots\dots(76);$$

and take

for incident wave $$\theta = t - \frac{s}{u} \dots\dots\dots\dots\dots\dots(77),$$

with same continued for reflected wave, where s denotes a space in the path of the incident ray, continued in the reflected ray.

Use these in § 128, and for the real problem take the real part of each expression so found. We thus have, for the vibrational displacements,

Incident wave, $\qquad G \cos \omega \left(t - \dfrac{s}{u} \right)$

Reflected wave,

$$-G \sqrt{\frac{(_{,}\rho b - \rho_{,}b)^2 + K^2 a^2}{(_{,}\rho b + \rho_{,}b)^2 + K^2 a^2}} \cos \left[\omega \left(t - \frac{s}{u} \right) - (\phi + \psi) \right] \left.\right\} \dots(78).$$

§ 148.   Let now for a moment $_{,}\rho/\rho = \infty$. This makes $K/_{,}\rho = 1$; $\phi = \psi = i$; and gives, for vibrational displacement in the reflected wave

$$- G \sin \left[\omega \left(t - \frac{s}{u} \right) - 2i \right] = G \sin \left[\omega \left(t - \frac{s}{u} \right) + \pi - 2i \right].$$

Thus the formulas show, for vibrations in the plane of incidence, that, at every incidence, the reflection is total (which we know without the mathematical investigation, because there is no loss of energy); and that the reflected ray is advanced in phase $\pi - 2i$ relatively to the incident ray. The former proposition is proved for vibrations perpendicular to the plane of incidence by the formula (39) of § 123; but with just π for phasal change. Hence incidence at 45° instead of the 70° to 76° of metals (§ 99 above), gives 90° of phasal difference; and it is easy to see from

* Green, *Mathematical Papers*, p. 267.

the formulas that values of $_,\rho/\rho$ large enough to give the Molar. brilliance of metallic reflections could not give any approach to the elliptic and circular polarizations which observations show (§ 99 above) in all metallic reflections. Thus, though our trial hypothesis of great effective density of the ether in the substance could give reflections, and therefore general appearances, undistinguishable to the naked eye from what we see in real metals, it fails utterly to explain the qualities of the reflected light discovered by polarizational analysis. Seeing thus that no real positive effective density in the substance can explain the qualities of metallic reflection, we infer that the effective density is negative or imaginary; and thus we are led by strictly dynamical reasoning to the brilliant prevision of MacCullagh and Cauchy that metallic reflection is to be explained by an imaginary refractive index.

§ 149. In § 159 below we shall find, by a new molecular theory to which I had been led by consideration of very different subjects, a perfectly clear and simple dynamical explanation of a real negative quantity for effective density of ether traversed by light-waves of any period within certain definite limits, in a space occupied by a solid or liquid or gas. We shall also find definite molecular and dimensional conditions which may possibly give us a sure molecular foundation for an imaginary effective density; though we are still very far from a thorough working-out of the full dynamical theory.

We conclude the present Lecture with a short survey of the *quasi* wave-motion which can exist in an elastic medium having a definite negative or imaginary effective density $_,\rho$; and of the reflection of light at a plane surface of such a medium.

§ 150. Let the rigidity be n, real; and the density $_,\rho$, an unrestricted complex as follows

$$_,\rho = \varpi \, (\cos \phi - \iota \sin \phi)\ldots\ldots\ldots\ldots\ldots\ldots(79),$$

where ϖ and ϕ are real, and for simplicity may be taken both positive. This gives for propagational velocity of equivoluminal waves

$$\left.\begin{aligned}&_,u = w \, (\cos \tfrac{1}{2}\phi + \iota \sin \tfrac{1}{2}\phi); \quad _,u^{-1} = w^{-1} (\cos \tfrac{1}{2}\phi - \iota \sin \tfrac{1}{2}\phi),\\ &\text{where} \qquad w = \sqrt{\frac{n}{\varpi}}, \text{ a real velocity}\end{aligned}\right\}\ldots(80).$$

Molar. For displacement at time t, considering only equivoluminal wave-motion, let $\eta + \iota\eta'$ denote displacement in a wave-plane perpendicular to OX; and choosing for our arbitrary function

$$f(\theta) = \epsilon^{\iota\omega\theta},$$

we have $\eta + \iota\eta' = \epsilon^{\iota\omega(t - x/,u)} \ldots\ldots\ldots\ldots\ldots(81);$

whence, taking $,u^{-1}$ from (80), we find, as a real solution,

$$\eta = \epsilon^{-\omega \sin\frac{1}{2}\phi \, x/w} \cos \omega \, (t - \cos \tfrac{1}{2}\phi \, x/w)\ldots\ldots\ldots\ldots(82).$$

This shows that the wave-plane travels in the $+ x$-direction with velocity $w \sec \tfrac{1}{2}\phi$, and with vibrational amplitude diminishing according to the exponential law $\epsilon^{-\omega \sin \frac{1}{2}\phi \, x/w}$. It is interesting to remark that the propagational velocity of this subsidential wave is, in virtue of the factor $\sec \tfrac{1}{2}\phi$, essentially greater than the velocity in a medium of real density equal to the modulus, ϖ, of our imaginary density; and is infinite when $\tfrac{1}{2}\phi = 90°$. This illustrates Quincke's discovery of greater velocity of light through a thin metallic film than through air.

§ 151. Remark now that, according as the real part of the complex, $,\rho$, is positive or negative, ϕ is $< 90°$, or $> 90°$; that, in the extreme case of $,\rho$ real positive, ϕ is zero; and that, in the other extreme case ($,\rho$ a real negative quantity), ϕ is $180°$. In every case between those extremes, $\tfrac{1}{2}\phi$ is between $0°$ and $90°$, and therefore both $\cos \tfrac{1}{2}\phi$ and $\sin \tfrac{1}{2}\phi$ are positive. This implies loss of energy in the inward travelling wave: except in the second extreme case, $\tfrac{1}{2}\phi = 90°$, when its propagational velocity is infinite; and (82) becomes

$$\eta = \epsilon^{-\omega x/w} \sin \omega t \ldots\ldots\ldots\ldots\ldots\ldots(83),$$

which represents standing vibrations of ether in the substance, diminishing inwards according to the exponential law; and therefore proves no continued expenditure of energy. We conclude that, for our ideal silver, the effective quasi-density of ether in the metal is essentially *real negative*: but that for all metals of less than perfect reflectivity it must be a complex, of which the real part may be either negative or positive.

§ 152. Confining our attention for the present to ideal silver, and to merely molar results of the molecular theory promised for Lecture XIX., let us put, in §§ 123, 132,

$$- \mu^2 = \nu^2 \ldots\ldots\ldots\ldots\ldots\ldots\ldots\ldots(84),$$

where ν^2 denotes a real positive numeric; and let us find the difference of phase between the two components, given respectively by §§ 123 and 132, due to incident light polarized in any plane oblique to the plane of incidence. That difference of phase is 90° for Principal Incidence (see § 97 above). I have gone through the work with Green's supposition $v=\infty$, $,v=\infty$, in § 132; and using the formula for G'/G so given in § 147, I have found that for all values of ν^2 from 0 to ∞ it makes the Principal Incidence between 0° and 45°. Now observation gives the Principal Incidences for all colours of light and all metals, between 45° and 90° (see table of § 99, showing for all cases of metallic reflection hitherto made, so far as generally known, Principal Incidences ranging from 66° to 78°). Hence polarizational analysis of the reflected light as thoroughly disproves, for metallic reflection, Green's assumption of propagational velocity infinite for condensational-rarefactional waves within the metal, as it was disproved for transparent substances in §§ 104, 105 by mere observation of unanalysed reflectivities.

§ 153. Hence, anticipating Lecture XIX. as we did in § 133 above, let us, while still keeping $v=\infty$ in the ether outside the metal, now make $,v$ small enough inside the metal to practically annul L. This reduces (39) and (59) of § 123 and § 132, to

$$\frac{-C'}{C}=\frac{,b-b}{,b+b}; \quad \frac{-G'}{G}=\frac{,\rho b-\rho,b}{,\rho b+\rho,b} \dots\dots\dots(85).$$

Putting now in (39) and in this

$$\frac{,b}{b}=\frac{u}{,u}r; \quad \frac{,\rho}{\rho}=-\nu^2; \quad \frac{u}{,u}=-\iota\nu* \dots\dots\dots(86),$$

where
$$r=\frac{\cos,i}{\cos i}=\frac{\sqrt{(\nu^2+\sin^2 i)}}{\nu\cos i} \dots\dots\dots(87),$$

we find
$$\frac{-C'}{C}=\frac{-\iota\nu r-1}{-\iota\nu r+1}=\frac{\nu r-\iota}{\nu r+\iota} \dots\dots\dots(88),$$

and
$$\frac{-G'}{G}=\frac{\nu-\iota r}{\nu+\iota r} \dots\dots\dots(89).$$

* The sign minus is chosen here in order that $\epsilon^{p y}$ not $\epsilon^{-p y}$ may be the reducing factor of $,G$; p^{-1} being a positive length. See foot-notes on §§ 146, 158''.

Molar. Put now
$$\tan^{-1}\frac{1}{\nu r}=\theta, \quad \tan^{-1}\frac{r}{\nu}=\vartheta \dots\dots\dots\dots(90).$$

This reduces (88) and (89) to

$$\frac{-C'}{C}=\frac{\cos\theta-\iota\sin\theta}{\cos\theta+\iota\sin\theta}=\cos2\theta-\iota\sin2\theta\dots\dots\dots(91),$$

and
$$\frac{-G'}{G}=\frac{\cos\vartheta-\iota\sin\vartheta}{\cos\vartheta+\iota\sin\vartheta}=\cos2\vartheta-\iota\sin2\vartheta\dots\dots\dots(92).$$

These formulas express total reflection for the two cases respectively of vibrations perpendicular to the plane of incidence, and vibrations in this plane; and they give us 2θ and 2ϑ, which we may, for brevity, call "the phases of the vibrations." Thus, calling P the plane of incidence and reflection, we find

Phase of vibration perpendicular to P – phase of vibration in P

$$= 2\,(\vartheta-\theta)\dots\dots\dots\dots\dots\dots(93).$$

This means that the vibration perpendicular to P precedes the other by $2\,(\vartheta-\theta)$.

Whatever positive value ν^2 has, this difference is essentially zero for $i=0$; and we find that it increases through $+90°$ to $+180°$ when i is increased from $0°$ to $90°$; as illustrated in the accompanying tables for the two cases, $\nu^2=\cdot5,\ \nu^2=10$.

i	$\theta=\tan^{-1}\frac{1}{\nu r}$	$\vartheta=\tan^{-1}\frac{r}{\nu}$	$2\,(\vartheta-\theta)$
0°	54·7	54·7	0·0
10	53·5	55·9	4·8
20	50·1	59·1	18·0
30	45·0	63·4	36·8
40	38·7	68·2	59·0
50	31·7	72·9	82·4
53·2	29·3	74·3	90·0
60	24·1	77·4	106·6
70	16·2	81·7	131·0
80	8·2	85·9	155·4
90	0·0	90·0	180·0

$$\nu^2=\cdot5.$$

Molar.

i	$\theta = \tan^{-1}\dfrac{1}{\nu r}$	$\vartheta = \tan^{-1}\dfrac{r}{\nu}$	$2(\vartheta - \theta)$
$0°$	$17°\!\cdot\!5$	$17°\!\cdot\!5$	$0°\!\cdot\!0$
10	17·3	17·8	1·0
20	16·5	18·7	4·4
30	15·1	20·3	10·4
40	13·4	22·8	18·8
50	11·2	26·8	31·2
60	8·7	33·3	49·2
70	5·9	44·0	76·2
73·8	4·8	49·7	89·8
80	3·0	62·3	118·6
90	0·0	90·0	180·0

$$\nu^2 = 10.$$

§ 154. The angle of incidence ($53°\!\cdot\!2$ for $\nu^2 = \cdot 5$, $73°\!\cdot\!8$ for $\nu^2 = 10$) which gives the $90°$ advance of phase is what has been defined as the Principal Incidence (§ 97 above). The reflectivities for the two polarizational components of incident light being perfect, the azimuth which, for the "Principal Incidence," gives circular polarization for the reflected light, that is, the " Principal Azimuth," is $45°$. For all real metals observation shows the phase-difference $2(\vartheta - \theta)$ to be positive, that is to say, the vibration in the plane of reflection to lag behind the vibration perpendicular to it; as we find it for ideal silver by dynamical theory.

§ 155. The easiest mathematical method for finding the Principal Incidence, I, for any given value of ν^2 is algebraic, as follows. For Principal Incidence we have

$$2(\vartheta - \theta) = 90° \quad\dotfill\quad(94).$$

This gives $\tan 2\vartheta \tan 2\theta = -1$, or in algebra, from (90),

$$\frac{4\nu^2 r^2}{(\nu^2 r^2 - 1)(\nu^2 - r^2)} = -1 \quad\dotfill\quad(95);$$

whence,

$$r^4 - r^2(\nu^2 + 4 + \nu^{-2}) + 1 = 0 \quad\dotfill\quad(96).$$

Taking the greater root of this regarded as a quadratic for r^2, and substituting it for r^2 in the general expression

$$\tan^2 i = \nu^2(r^2 - 1)/(\nu^2 + 1) \quad\dotfill\quad(97)$$

given by (87) for i in terms of ν and r, we find $\tan^2 I$ for the

Molar. Principal Incidence. (The less root of the quadratic is rejected, because it makes $\tan^2 i$ negative.)

The following table has been thus calculated directly, to show for fourteen values of ν or ν^2, the Principal Incidence, I.

ν	ν^2	$\tan I$	I
0·00	0·00	1·000	45·0°
0·50	0·25	1·193	50·0
0·71	0·50	1·333	53·1
1·00	1·00	1·554	57·2
1·41	2·00	1·887	62·1
2·00	4·00	2·387	67·2
2·45	6·00	2·786	70·3
3·16	10·00	3·438	73·8
3·74	14·00	3·982	75·9
4·00	16·00	4·223	76·6
4·47	20·0	4·680	77·9
6·00	36·0	6·160	80·8
10·00	100·0	10·10	84·3
14·00	196·0	14·07	85·9
20·00	400·0	20·05	87·1

§ 156. The converse problem of finding ν^2 for any given Principal Incidence, by (96) and (97), yields a cubic equation for ν^2. The table of § 155 proves that this cubic equation has one, and only one, real positive root for every value of I between 45° and 90°; and no real positive root for values of I between 0° and 45°. For our present purpose it is most easily solved by trial and error, aided by the table of § 155. I have thus found for the three Principal Incidences measured by Conroy for red, yellow, and blue light respectively, incident on his silver film polished with putty powder (table of § 99 above), the following values of ν and ν^2.

	I	ν	ν^2
Red	76° 29′	3·9	15·2
Yellow......	74° 37′	3·3	10·9
Blue.........	71° 33′	2·7	7·29

§ 157. The diagram of § 88 shows that Conroy's silver film, polished as he polished it with putty powder, may be regarded as almost our ideal silver: this view is confirmed by his three

Principal Azimuths, $43°51'$, $43°52'$, $43°0'$, being each as nearly Molar. a good approximation to $45°$ as it is. (See § 159" below.) But their shortcomings of from one to two degrees below $45°$ are no doubt real, and point to the correction of the real values of ν^2 by the addition of small purely imaginary terms. Thus, to fit the formulas of § 150 to Conroy's silver, we may, keeping ϖ a positive quantity, take $\phi = 180° - \chi$; where χ, which would be zero for ideal silver, may for real silver, have some small value of a few degrees.

§ 158. It would be interesting to pursue the subject further, and include with silver, other metals for which we have the experimental data such as those shown in the table of § 99. To do this, we may conveniently in (87), (88), (89), put

$$\nu^2 = p\,(\cos\chi - \iota\sin\chi) \quad\dots\dots\dots\dots\dots(98);$$

and use the experimental data of Principal Incidence and Principal Azimuth to determine in each case the two unknown quantities $p\cos\chi$ and $p\sin\chi$: but time forbids. This would be, in fact, a working out of the theory, or empirical formula, of MacCullagh and Cauchy, to comparison with observational results regarding metallic reflection, such as has been done by Eisenlohr, Jamin, Conroy, and others*. The agreement of the theory with observation has been found somewhat approximately, but not wholly, satisfactory. Stokes, as communicator of Conroy's paper, No. III. (*Proc. R.S.*, Feb. 15, 1883), comments on this want of perfect agreement; and suggests that it is to be accounted for by the inclusion in respect to metallic reflection of what he proposes to call "the *adamantine property*" of a substance; being the property required to explain the deviation, from Fresnel's law of reflection of light by transparent bodies, discovered more than eighty years ago by Airy, in diamond. This adamantine property, as we shall see in Lecture XIX., § 173, is to be explained dynamically by assuming a small imaginary ιq, for the velocity of condensational-rarefactional waves in the substance; not small enough to utterly annul L in our formulas of § 132. Its magnitude is measured by the ratio q/u, which I propose to call *adamantinism*. Some deviation from exact equality between the effective rigidities of ether in the two mediums might also be invoked to aid in procuring agreement between dynamical theory and observation,

* Basset, *Physical Optics*, §§ 371, 380; Mascart, *Traité d'Optique*, Vol. II.

Molar. for reflection of light from all substances, transparent, or metallic or otherwise opaque.

§ 158′. Meantime, I finish this Lecture XVIII. on the Reflection of Light with an application of § 123 (39), and § 132 (59) with $L = 0$, to the theory of Fresnel's rhomb.

Let the medium in which u is the velocity of light be a transparent solid or liquid; and let the other medium be undisturbed ether. We cannot now continue Green's convenient usage adopted in § 116, and call the former "the upper" and the latter "the lower." We now have

$$,u > u;\ \rho > ,\rho;\ ,i > i;\ ,u/u = \mu;\ \rho/,\rho = \mu^2;\ \sin ,i = \mu \sin i;\ b = \cos i/u;\ ,b = \cos ,i/,u \ldots (98').$$

For convenience, to suit the case of $,i > i$, write as follows, the equations cited above,—

$$\frac{C'}{C} = \frac{b - ,b}{b + ,b};\quad \frac{G'}{G} = \frac{\rho ,b - ,\rho b}{\rho ,b + ,\rho b} \ldots\ldots\ldots\ldots(98'');$$

the sign minus being transferred from their first members because, in virtue of $,i > i$, C'/C is now positive through the whole range of non-total reflection, $[i = 0$ to $i = \sin^{-1}(1/\mu)$ giving $,i = 90°]$; and G'/G is positive through the range up to the Brewsterian angle of zero reflected ray $[i = 0$ to $i = \tan^{-1}(1/\mu)]$. From $i = \tan^{-1}(1/\mu)$ to $i = \sin^{-1}(1/\mu)$, G'/G decreases from 0 to -1; and it is imaginary-complex of modulus 1, through the whole range of total reflection $[i = \sin^{-1}(1/\mu)$ to $i = 90°]$.

§ 158″. When i is between $\sin^{-1}(1/\mu)$ and 90°, $,b^2$ is real negative; and, denoting its value by $-q^2$ (as in § 118), we have

$$,b = -\iota q^*;\ \text{where } q = b\,\frac{\sqrt{(\mu^2 \sin^2 i - 1)}}{\mu \cos i} \ldots\ldots(98''').$$

Remark that q increases from 0 to ∞ when i is increased from $\sin^{-1}(1/\mu)$ to 90°. Using (98‴) in (98″) we find,

$$\frac{C'}{C} = \frac{b + \iota q}{b - \iota q};\quad \frac{G'}{G} = \frac{\mu^2 q - \iota b}{\mu^2 q + \iota b} \ldots\ldots\ldots\ldots(98^{iv});$$

whence, by De Moivre's theorem,

$$\frac{C'}{C} = \cos 2\psi + \iota \sin 2\psi;\quad \frac{G'}{G} = \cos 2\chi - \iota \sin 2\chi \ldots\ldots(98^v),$$

where
$$\psi = \tan^{-1}\frac{q}{b};\quad \chi = \tan^{-1}\frac{b}{\mu^2 q} \ldots\ldots\ldots\ldots(98^{vi}).$$

---

* The sign *minus* is here chosen in order that $\epsilon^{q\omega y}$, not $\epsilon^{-q\omega y}$, may be the reducing factor of $,C$ and $,G$ for the disturbance in the ether outside, see § 128; y being zero at the reflecting surface, and negative outside.

By (98^{vi}) and ($98'''$) we see that ψ increases from $0°$ to $90°$ and Molar. χ decreases from $90°$ to $0°$, when i is increased from

$$\sin^{-1}(\mu^{-1}) \text{ to } 90°.$$

§ 158'''. Putting for a moment $q/b = x$, we find

$$\frac{d}{dx}(\psi + \chi) = \frac{1}{1 + x^2} - \frac{1}{\mu^2 x^2 + \mu^{-2}} \ldots\ldots\ldots(98^{vii}).$$

Equating this to zero we find, for $(\psi + \chi)$ a minimum,

$$q/b = \mu^{-1}, \text{ which makes } \psi = \chi = \tan^{-1}(\mu^{-1}) \ \ldots\ldots(98^{viii});$$

and to make $q/b = \mu^{-1}$, we have $\tan^2 i = 2/(\mu^2 - 1)$. We infer that when i is increased from $\sin^{-1}(\mu^{-1})$ to $\tan^{-1}\sqrt{[2/(\mu^2 - 1)]}$, $(\psi + \chi)$ decreases from $90°$ to $2\tan^{-1}(\mu^{-1})$ and then increases to $90°$ again, when i is increased to $90°$. This will be useful to us in § 158^{vi}.

§ 158^{iv}. Realising now, as in § 147, we find real solutions as follows, for vibrations perpendicular to, and in, the plane (P) of incidence and reflection :—

Vibrations perpendicular to P
$$\left\{ \begin{array}{l} \text{Incident wave } C \cos \omega \left(t - \dfrac{s}{u}\right) \\[2mm] \text{Reflected wave } C \cos \left[\omega\left(t - \dfrac{s}{u}\right) + 2\psi\right] \end{array} \right\} \ldots(98^{ix});$$

Vibrations in P
$$\left\{ \begin{array}{l} \text{Incident wave } G \cos \omega \left(t - \dfrac{s}{u}\right) \\[2mm] \text{Reflected wave } G \cos \left[\omega\left(t - \dfrac{s}{u}\right) - 2\chi\right] \end{array} \right\} \ldots(98^{x}).$$

Thus we see that the reflected vibrations in P are set back in phase by $2(\psi + \chi)$ relatively to the reflected vibrations perpendicular to P. By § 158''' we see that this back-set is $180°$ at the two incidences, $i = \sin^{-1}(\mu^{-1})$, and $i = 90°$ (the limits of total reflection); and that it decreases to a minimum, $4\tan^{-1}(\mu^{-1})$, at the intermediate angle of incidence $i = \tan^{-1}\sqrt{[2/(\mu^2 - 1)]}$.

The back-sets of the phase through the whole range of total reflection for four refractive indices, 1·46, 1·5, 1·51, and 1·6 are shown in the Table II. of § 158^{v} below.

Column 1 represents angles of incidence.

25—2

Molar.   Columns 2 and 3 of Table II. represent the absolute back-sets of phase by reflection respectively in, and perpendicular to, P.

Column 4 represents the back-set of phase of vibrations in P, relatively to vibrations perpendicular to P, produced by a single reflection.

The numbers in Column 4, doubled, are the total back-sets by two reflections, at the same angle of incidence i, of vibrations in, relatively to vibrations perpendicular to, P. Each of these numbers is between 180° and 360°; and therefore it indicates virtually a setting-forward of the phase of vibrations in P, relatively to the phase of those perpendicular to P, by the amount to which each number falls short of 360°. Hence the numbers in Column 5 of Table II. represent virtual *advances* of phase of the vibrations *in*, *relatively to the vibrations perpendicular to*, P, after two reflections.

§ 158ᵛ. It will be seen that all the phasal differences shown in Column 4 of Table II. are obtuse angles: with minimum values shown in Table I., which is extended to include zinc sulphide,

TOTAL INTERNAL REFLECTION.   TABLE I.

Refractive index μ	Brewsterian angle $\tan^{-1}(1/\mu)$	Limiting incidence for total internal reflection $\sin^{-1}(1/\mu)$	Incidence for minimum phasal difference $\tan^{-1}\sqrt{[2/(\mu^2-1)]}$	Minimum phasal difference $4\tan^{-1}(1/\mu)$
1·46	34° 24'·5	43° 14'	53° 3'	137° 38'
1·50	33 41·25	41 49	51 40	134 45
1·51	33 30·75	41 28	51 20	134 3
1·60	32 0·25	38 41	48 33	128 1
(Zinc Sulphide) 2·371	22 52	24 57	33 20	91 28
$\mu = \cot 22°\frac{1}{2} = 2\cdot414$	22 30	24 28	32 46	90 0
(Diamond) 2·434	22 20	24 15	32 31	89 20
(Realgar) 2·454	22 10	24 3	32 15	88 40

diamond, and realgar. By Table I. we see that the phasal differences for internal reflection in glasses, and all known transparent bodies of refractive index less than 2·414, are obtuse for all angles of incidence through the whole range of total internal reflection. This conclusion was very startling to myself, because for eighty years we have been taught that, for total internal reflection in glass, the phasal difference was an acute angle in a single reflection; and that it was 45° for each reflection in the Fresnel rhomb, instead of 135° which we now know it is. How it is so, Molar. we can easily see by the following simple considerations.

(1)   Circularly polarized light in every case of direct incidence and reflection (metallic or vitreous, external or internal), has its circular orbital motions in the same absolute directions in the direct and the reflected waves.   Hence if the orbital motion ideally seen by an eye receiving the incident light is anti-clockwise, it is clockwise to an´eye receiving the reflected light.   (Hence the arrangement in respect to signs in figs. 3 *a* and 3 *b* of § 92 above is explained thus :—First imagine the two diagrams as corresponding to normal incidence: for the incident light, fig. 3*b*, with its face down, looked up to; and for the reflected light, fig. 3*a*, with its face up, looked down on.   Then incline the two planes continuously to suit all incidences from $i = 0$, to $i = 90°$.)

(2)   Taking first, external reflection of light, of circular anti-clockwise orbits, incident on glass of negligible adamantinism* ; increase the angle of incidence from 0° to the Brewsterian $\tan^{-1}\mu$. The motion constituting the reflected light is in *clockwise* elliptic orbits, of increasing ellipticity, with long axes perpendicular to the plane of reflection, till at $\tan^{-1}\mu$ it is rectilineal.   Increase now the angle of incidence to 90° : the orbital motions are elliptic *anti-clockwise* with diminishing ellipticity (long axes still perpendicular to the plane of reflection); and become infinitely nearly circular when the incidence is infinitely nearly grazing.

(3)   Proceed now as in (2), but with internal instead of external reflection at a glass surface: the incident circularly polarized light being *anti-clockwise*.   With increasing incidences from $i = 0$, the reflected light is *clockwise* elliptic with increasing ellipticity, till it becomes rectilineal at the Brewsterian $\tan^{-1}(1/\mu)$. Increase i farther : the reflected light is now *anti-clockwise* elliptic, with diminishing ellipticity till it becomes circular at the limit of total internal reflection, $\sin^{-1}(1/\mu)$.   Increase i farther up to 90° : the reflected light is elliptic, still *anti-clockwise*, with ellipticity increasing to a maximum when $i = \sin^{-1}\sqrt{[2/(\mu^2 - 1)]}$ and diminishing till the orbit becomes again infinitely nearly circular when i is infinitely little less than 90°.

* For the change from the present statement, required by adamantinism when perceptible, see below, §§ 178—182.

Molar.   (4) With, instead of glass, an ideal substance whose refractive index is 2·414; anti-clockwise circularly polarized light, incident at angles increasing from 24° 28', gives total reflection, and anti-clockwise circular orbits, becoming elliptic, with ellipticities increasing to rectilineal vibrations when $i = 32°\ 46'$. Farther increase of i up to 90° gives elliptic orbits of ellipticities diminishing to circularity at $i = 90°$: all anti-clockwise.

(5)   Diamond and realgar, and all other substances having $\mu > 2·414$, have three principal incidences for internal reflection; one with non-total reflection at the Brewsterian angle $\tan^{-1}(1/\mu)$; and two within the range of total reflection; one less, the other greater, than the incidence which makes the phasal difference a minimum. Anti-clockwise circularly polarized light, incident at any angle between the last-mentioned two principal incidences, gives *clockwise* elliptic orbits for the reflected light; presenting a minimum deviation from circularity (minimum ellipticity we may call it) at some intermediate angle.

(6)   It is interesting to remark that in every case, according as the phasal difference is obtuse or acute, the reflected light is *anti-clockwise* or *clockwise*, when the incident light is circular anti-clockwise. This with § 158ˣᵛⁱⁱ gives a simple means for experimentally verifying the statements (1), (2), (3), (4), (5). The experimental proof thus given of the truth, that the phasal difference in every case of approximately grazing incidence is 180°, is very easy and clear; as it can be arranged to show simultaneously to the eye the incident and the reflected light; both extinguished simultaneously by the same setting of the analysing fresnel (or quarter-wave plate) and nicol.

(7)   The rule of signs referred to in (1) of the present section, and illustrated in figs. 3 a and 3 b of § 92, must be taken into account to interpret the meaning of phasal difference of two rectilineal components of reflected light, due to incident rectilineally vibrating light. Thus we see that if the vibrational plane of the incident light is turned *anti-clockwise from P*, the orbital motions of the reflected light are *anti-clockwise*, or clockwise, according as the component vibrating in, is set back or advanced, relatively to the component perpendicular to P. See § 158ˣᵛⁱⁱ below.

Total Internal Reflection. Table II.

Col. 1	Col. 2	Col. 3	Col. 4	Col. 5
i	ψ and χ calculated only to the nearest minute, except in the case of their equality		$2(\psi+\chi)$; being actual phasal back-set of vibrations in, relatively to vibrations perpendicular to, P, produced by one reflection	$360°-4(\psi+\chi)$; being virtual phasal advance of vibrations in, relatively to vibrations perpendicular to, P, produced by two reflections
	2ψ	2χ		
$\mu=1{\cdot}50$				
41° 48'·6	0° 0'	180° 0'	180° 0'	0° 0'
42	8 50	160 18	169 6	21 48
43	22 14	132 18	154 32	50 56
44	30 22	117 36	147 58	64 4
45	36 52	106 14	143 6	73 48
46	42 30	97 36	140 6	79 48
47	47 34	90 28	138 2	83 56
48	52 16	84 20	136 36	86 48
49	56 40	79 0	135 40	88 40
50	60 50	74 16	135 6	89 48
51	64 48	70 0	134 48	90 24
51 40·3	67 22·8	67 22·8	134 45·6 (min.)	90 28·8 (max.)
52	68 36	66 10	134 46	90 28
53	72 18	62 38	134 56	90 8
53 12				90 0
54	75 54	59 22	135 16	89 28
55	78 4	57 26	135 30	89 0
56	82 46	53 32	136 18	87 24
57	86 6	50 52	136 58	86 4
58	89 22	48 24	137 46	84 28
59	92 34	46 2	138 36	82 48
60	95 44	43 48	139 32	80 56
70	125 22	25 52	151 14	57 32
80	153 4	12 8	165 12	29 36
90	180 0	0 0	180 0	0 0
$\mu=1{\cdot}46$				
43 13·8	0 0	180 0	180 0	0 0
44	18 18	142 6	160 24	39 12
45	27 56	124 22	152 18	55 24
46	35 8	112 0	147 8	65 44
47	41 14	102 34	143 48	72 24
48	46 38	94 52	141 30	77 0
49	51 34	88 18	139 52	80 16
50	56 10	82 38	138 48	82 24
51	60 32	77 36	138 8	83 44
52	64 40	73 6	137 46	84 28
53	68 38	69 1	137 39	84 42
53 2·98	68 49·0	68 49·0	137 38·0 (min.)	84 44 (max.)
54	72 26	65 16	137 42	84 36
55	76 8	61 50	137 58	84 4
56	79 44	58 38	138 22	83 16
57	83 16	55 40	138 56	82 8
58	86 40	52 48	139 28	81 4
59	90 2	50 14	140 16	79 28
60	93 20	47 46	141 6	77 48
70	124 0	28 0	152 0	56 0
80	152 26	13 8	165 34	28 52
90	180 0	0 0	180 0	0 0

	Col. 1	Col. 2	Col. 3	Col. 4		Col. 5	
	i	ψ and χ calculated only to the nearest minute, except in the case of their equality		$2(\psi+\chi)$; being actual phasal back-set of vibrations in, relatively to vibrations perpendicular to, P, produced by one reflection		$360°-4(\psi+\chi)$; being virtual phasal advance of vibrations in, relatively to vibrations perpendicular to, P, produced by two reflections	
		2ψ	2χ				
$\mu=1\cdot51$	41° 28'·3	0° 0'	180° 0'	180° 0'		0° 0'	
	42	14 40	147 16	161 56		36 8	
	43	25 6	126 8	151 14		57 32	
	44	32 30	112 46	145 16		69 28	
	45	38 38	102 46	141 24		77 12	
	46	44 0	94 42	138 42		82 36	
	47	48 54	87 42	136 36		86 48	
	48	53 30	82 4	135 34		88 52	
	49	57 46	76 58	134 44		90 32	
	50	61 50	72 26	134 16		91 28	
	51	65 44	68 20	134 4		91 52	
	51 20·41	67 1·7	67 1·7	134 3·4 (min.)		91 53·2 (max.)	
	52	69 30	64 36	134 6		91 48	
	53	73 6	61 12	134 18		91 24	
	54	76 38	58 2	134 40		90 40	
	54 37					90 0	
	55	80 6	55 6	135 12		89 36	
	56	83 26	52 22	135 48		88 24	
	57	86 44	49 48	136 32		86 56	
	58	89 58	47 22	137 20		85 20	
	59	93 8	45 4	138 12		83 36	
	60	96 16	42 54	139 10		81 40	
	70	125 40	25 22	151 2		57 56	
	80	154 4	11 32	165 36		28 48	
	90	180 0	0 0	180 0		0 0	
$\mu=1\cdot60$	38 40·9	0 0	180 0	180 0		0 0	
	39	10 48	152 48	163 36		32 48	
	40	22 12	126 40	148 52		62 16	
	41	29 36	111 50	141 26		77 8	
	42	35 38	101 6	136 44		86 32	
	43	40 56	92 36	133 32		92 56	
	44	45 42	85 40	131 22		97 16	
	45	50 4	79 48	129 52		100 16	
	46	54 18	74 38	128 56		102 8	
	47	58 14	70 6	128 20		103 20	
	48	62 0	66 4	128 4		103 52	
	48 33·0	64 0·6	64 0·6	128 1·2 (min.)		103 57·6 (max.)	
	49	65 38	62 24	128 2		103 56	
	50	69 6	59 6	128 12		103 36	
	51	72 34	56 2	128 36		102 48	
	52	75 54	53 14	129 8		101 44	
	53	79 6	50 38	129 44		100 32	
	54	82 18	48 10	130 28		99 4	
	55	85 24	45 52	131 16		97 28	
	56	88 28	43 42	132 10		95 40	
	57	91 30	41 40	133 10		93 40	
	58	94 30	39 42	134 12		91 36	
	58 44					90 0	
	59	97 26	37 52	135 18		89 24	
	60	100 20	36 6	136 26		87 8	
	70	128 2	21 34	149 36		60 48	
	80	154 18	10 12	164 30		31 0	
	90	180 0	0 0	180 0		0 0	

§ 158^{vi}. That most exquisite invention, Fresnel's rhomb, Molar. produces a difference of phase of 90° between vibrations in, and vibrations perpendicular to, *P*. By § 158''' we see that this can only be with glass of refractive index greater than 1·496. Thus, the table for refractive index 1·46 shows 84° 44′ for the maximum value in Column 5. The other three tables show, for refractive indices 1·5, 1·51, and 1·6 respectively, maximum values in Column 5, 90° 28′·8, 91° 53′·2, and 103° 57′·6. In each of these cases, there are of course two incidences which make the difference of phase exactly 90°. Fresnel pointed out* that the larger of the two is to be preferred to the smaller; because, with the larger, the differences of refractivity, for the different constituents of white light, make less error from the desired 90° difference of phase. The larger of the two incidences is to be preferred, also, because small differences from it make less error on the 90° phasal differences, than equal differences from the other. For the same reason, the less the refractive index exceeds 1·496, the better. Fresnel's first rhomb was made of glass, of refractive index 1·51, from the factory of Saint Gobain: our tables show that 1·5 would reduce the phasal errors to about half those given by 1·51, for equal

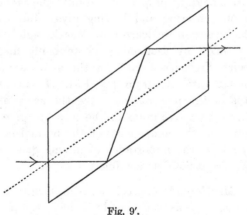

Fig. 9′.

small errors in the direction of the ray. It would thus give an appreciably better instrument than Fresnel's own. The accompanying diagram, figure 9′, represents a section perpendicular to

* *Collected Papers*, Vol. I. p. 792; original date 1823.

.raloM the traversed faces and the reflecting surfaces of the rhomb, for refractive index 1·51, as cut by Fresnel; with acute angles of 54° 37', so that the rays for its proper use may enter and leave perpendicularly to the traversed faces. If p denote the perpendicular distance between the two long sides of the parallelogram, the length of the long side is $2p \tan i$ (or $p \cdot 2·816$ with $i = 54° 37'$): the length of the short side is $p \cdot \operatorname{cosec} i$ (or $p \cdot 1·226$). The perpendicular distance between the entering and the leaving rays is $p \cdot 2 \sin i$ (or $p \cdot 1·6306$).

§ 158^{vii}. The length of the long side is chosen such that a ray, entering perpendicularly* through the centre of one face, passes out perpendicularly through the centre of the parallel face. Fresnel's diagram is far from fulfilling this condition; and so are many of the diagrams in text-books, and scientific papers on the subject. The last diagram I have seen shows only about a quarter of normally entering light to suffer two reflections; and rather more than half of it to pass straight through without any reflection at all. Airy's and Jamin's diagrams are correct and very clear. Fresnel must, in 1823, have given clear instructions to his workman (with or without a diagram); and, to this day, opticians make Fresnel's rhombs of right proportions to fulfil the proper condition in respect to the entering and leaving rays. But twenty years after Fresnel's invention we learn from MacCullagh† that Fresnel rhombs were made by Dollond (and probably also by other opticians?), with Fresnel's 54° 37' for the acute angle, but with refractive indices differing from his 1·51. I am assured that some opticians of the present day make the acute angle correct according to the refractive index of the glass, to give exactly 90° phasal difference of the components of the normal ray. I do not know if they realize the importance of having glass of refractive index as little above 1·496 as possible.

§ 158^{viii}. MacCullagh appreciated the beauty and value of Fresnel's rhomb, and as early as 1837 had begun using it for research. But he was at first much perplexed by unexpectedly large errors, until he found means of taking them into account,

---

* This, for brevity, I call a "normal ray."

† "On the Laws of Metallic Reflection and the mode of making experiments upon Elliptic Polarization," *Proc. Royal Irish Academy*, May 8, 1843. MacCullagh's *Collected Papers*, p. 240.

"and of making the rhomb itself serve to measure and to eliminate Molar.
"them." He good-naturedly adds;—"The value of the rhomb as
"an instrument of research is much increased by the circumstance
"that it can thus determine its own effect, and that it is not at all
"necessary to adapt its angle exactly to the refractive index of the
"glass." This proves a very forgiving spirit: perplexity and loss
of time in his research gratefully accepted in consideration of his
having been led to an enlarged view of the value of the Fresnel
rhomb as an accurate measuring instrument!

§ 158^{ix}. MacCullagh had two rhombs from the same maker
each "cut at an angle of 54½° as prescribed by Fresnel." He found
one of them wrong phasally by 3°, the other by 8°! He does not
say whether the errors were of excess or defect; but we see that
they must have been excess above 90°, because no refractive index
greater than 1·5, with $i = 54°\ 37'$, gives as large a defect below 90°,
as 3°. This we see by looking at the following Table, calculated

μ	i	2ψ	2χ	$2(\psi+\chi)$	$\dfrac{360°}{-4(\psi+\chi)}$	Excess above 90°
1·5	54° 37′	78° 4′	57° 28′	135° 32′	88° 56′	−1° 4′
1·51	,,	78 48	56 12	135 0	90 0	0 0
1·6	,,	84 14	46 44	130 58	98 4	+8 4
1·7	,,	88 32	39 4	127 36	104 48	14 48
1·8	,,	91 44	33 20	125 4	109 52	19 52

according to § 158^v with $i = 54°\ 37'$, for four refractive indices,
other than Fresnel's 1·51. We also see that the refractive indices
of the rhombs, supplied by Dollond to MacCullagh, must have
been between 1·51 and 1·6.

§ 158^x. Notwithstanding MacCullagh's good-natured remark,
it is important that the acute angle of a Fresnel rhomb should be
made, as accurately as possible, such that the phasal difference
shall be exactly 90°, for light of definitely specified period (sodium
light for example), when the direction of the ray is exactly perpendicular to the entering and leaving faces. But however
trustworthy may be the instrument-maker's work, MacCullagh's
principle of determining the error in the practical use of the
instrument, and eliminating it if it is perceptible, is highly important and interesting. It may be carried out, either, as he did

Molar. it´ himself, by means of observations on metallic reflection : or, as he suggested, by observations on total reflection at a separating surface of glass and air; a much simpler subject than metallic reflection. Or, as simplest method when two Fresnel's rhombs are available, we may place them with leaving face of one, and entering face of the other, parallel and close together, and mount the two rhombs so as to turn independently round a common axis perpendicular to these surfaces through their centres. The arrangement is completed by mounting two nicols so as to turn independently, one of them round the central ray entering the first rhomb, and the other round the central ray leaving the second rhomb : and providing three graduated circles, by which differences of angles turned through between the first nicol and first fresnel; between the two fresnels; and between the second fresnel and the second nicol ; may be measured. The mechanism to do this is of the simplest and easiest. The experiment consists in letting light; entering through the first nicol and traversing the two fresnels and the second nicol; be viewed by an eye seeing through the second nicol. The best approach to extinction that can be had, is to be produced by varying the three measured angles. For simplicity we may suppose that the three zeros of the three pointers, on the three circles, are set so that, using for brevity facial intersections to denote intersections of the traversed faces and reflecting surfaces of a fresnel, we have as follows :—

(1) When the first index is at zero on the first circle, the vibrational lines of the light emerging from the first nicol are perpendicular to the facial intersections of the first fresnel.

(2) When the second pointer is at zero on its circle, the facial intersections of the first fresnel are parallel to those of the second.

(3) When the third index is at the zero of the third circle, the facial intersections of the second fresnel are perpendicular to the vibrational lines of light entering the second nicol.

Thus if n_1, a, n_2 denote the readings on the three circles; n_1 is the inclination of the vibrational lines of the first nicol to the reflectional plane* of the first fresnel; a is the inclination of the facial intersections of the first fresnel to those of the second; and

* For brevity I call "reflectional plane" of a fresnel, the plane of reflection of a normal ray. It is perpendicular to the facial intersections.

n_2 is the inclination of the reflectional plane of the second fresnel to Molar. the vibrational lines of the light emerging from it. The diagram, figure 9″, shows these angles all on one circle, ideally seen by an eye looking through the second nicol. OV_1, OF_1, OF_2, OV_2, represent planes through the central emergent ray respectively parallel to the vibrational lines of the first nicol, the reflectional plane of the

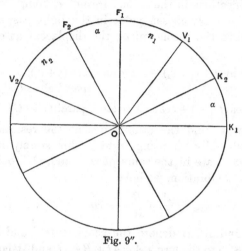

Fig. 9″.

first fresnel, the reflectional plane of the second fresnel, and the vibrational lines of the second nicol set to let pass all the light coming from the second fresnel. OK_1, OK_2, represent planes through the common axis perpendicular to OF_1 and OF_2.

§ 158^{xi}. Let now sin t denote the displacement at time t, of a point of ether in the light passing from the first nicol to the first fresnel. This implies a special unit of time, convenient for the occasion, according to which the period of the light is 2π; and it takes as unit of length, the maximum displacement of the ether between the first nicol and the first fresnel. Following the light through the apparatus;—first resolve the displacement into two components, cos n_1 sin t, sin n_1 sin t, parallel to OF_1 and OK_1. The phase of the former of these is advanced 90° + e_1 relatively to the latter by the two total reflections in the first fresnel, if e_1 denote its error, which ought to be zero. Thus, at any properly chosen point in the space traversed by the light between the two fresnels, the displacement of the ether at time t is

cos n_1 cos $(t + e_1)$ parallel to OF_1; sin n_1 sin t parallel to OK_1.

Molar. Resolving each of these into two components, we have, for the light entering the second fresnel,

$$\cos \alpha \cos n_1 \cos (t + e_1) - \sin \alpha \sin n_1 \sin t \text{ parallel to } OF_2...(98^{\text{xi}});$$

$$\sin \alpha \cos n_1 \cos (t + e_1) + \cos \alpha \sin n_1 \sin t \text{ parallel to } OK_2...(98^{\text{xii}}).$$

The former of these is advanced $90° + e_2$ relatively to the latter, by the two reflections in the second fresnel, e_2 being its error; and thus, at any properly chosen point in the light passing from the second fresnel to the second nicol, the displacement of the ether at time t is

$$- \cos \alpha \cos n_1 \sin (t + e_1 + e_2) - \sin \alpha \sin n_1 \cos (t + e_2),$$
$$\text{parallel to } OF_2...(98^{\text{xiii}});$$

$$\sin \alpha \cos n_1 \cos (t + e_1) + \cos \alpha \sin n_1 \sin t, \text{ parallel to } OK_2...(98^{\text{xiv}}).$$

We have now to find the condition that the resultant of these shall be rectilineally vibrating light. This simply implies that (98^{xiii}) and (98^{xiv}) are in the same phase; and, OV_2 being the line of the resultant vibration, we have

$$\frac{A}{A'} = \frac{B}{B'} = -\cot n_2 \ \ ....(98^{\text{xv}}),$$

where A, B and A', B' denote the coefficients found by reducing (98^{xiii}) and (98^{xiv}) to the forms $A \cos t + B \sin t$, and $A' \cos t + B' \sin t$. Thus, performing the reductions, and putting

$$\tan n_1 = h_1, \quad \tan n_2 = h_2, \quad \tan \alpha = j \ ....(98^{\text{xvi}}),$$

we find

$$\frac{-1}{h_2} = \frac{- \sin (e_1 + e_2) - h_1 j \cos e_2}{j \cos e_1} = \frac{-\cos (e_1 + e_2) + h_1 j \sin e_2}{-j \sin e_1 + h_1}$$
$$........(98^{\text{xvii}}).$$

These are two equations for finding the two unknowns, e_1, e_2; when h_1, h_2, j, are all known; one of them being chosen arbitrarily, and the other two determined by observation, worked to produce extinction of the emergent pencil (compare §§ 95, 95′, above). OV_2 in the diagram (fig. 9″) is perpendicular to the position of the vibrational line of the second nicol when set for extinction.

§ 158^{xii}. The experimenter will be guided by his mathematical judgment, or by trial and selection, to get the best conditioned values of h_1, h_2, j, for determining e_1, e_2. We must not take $h_1 = 1$ (that is to say, we must not set the entering nicol fixed with its vibrational lines at 45° to the principal planes of the first fresnel)

because in the case $e_1 = 0$ (the first fresnel true), the light entering Molar.
the second fresnel would be circularly polarized, and therefore the
turning of the second fresnel would have no effect on the quality
of the emergent light; which would be plane polarized if, and only
if, this fresnel, like the first, is true (compare § 98 above). We
must not fix $h_1 = 0$, because this would render the first fresnel
nugatory. It occurs therefore to take $n_1 = 22\frac{1}{2}°$ which makes

$$h_1^{-1} - h_1 = 2 \text{ and } h_1 = \cdot4142\ldots\ldots\ldots\ldots(98^{\text{xviii}}).$$

Probably this may be found a good selection if, for any reason, it
is thought advisable to fix h_1 and leave j variable.

If e_1, e_2 are so small that we may neglect $e_1{}^2$, $e_2{}^2$, $e_1 e_2$, (98^{xvii})
becomes

$$\frac{-1}{h_2} = -\frac{e_1 + e_2}{j} - h_1 = \frac{-1 + h_1 j e_2}{-j e_1 + h_1} \ldots\ldots(98^{\text{xix}});$$

whence we have two simple equations for e_1, e_2

$$e_1 + e_2 = j\left(\frac{1}{h_2} - h_1\right); \quad \frac{e_1}{h_2} - h_1 e_2 = \frac{1}{j}\left(\frac{h_1}{h_2} - 1\right)\ldots\ldots(98^{\text{xx}}).$$

From these we see that, as e_1 and e_2 are small, the observation must
make h_1 and h_2 each very nearly unity unless j is taken either very
small or very great. It may be convenient to fix $j = 1$, $(\alpha = 45°)$,
and to find n_1, n_2 by experiment.

§ 158^{xiii}. When both the fresnels are perfectly true $(e_1 = 0, e_2 = 0)$,
formula (98^{xx}) shows that $h_2 = h_1$, if $j = 0$; and $h_2 = 1/h_1$, if $j = \infty$:
but if j has any value between 0 and ∞, we must have $n_1 = n_2 = 45°$,
while α may have any value. That is to say, the first fresnel
produces exactly circular orbits and the second rectifies them.
When n_1 is any angle between 0 and 90°, we have elliptically
or circularly polarized light, represented by (98^{xi}) and (98^{xii}) with
$e_1 = 0$. This is light leaving the first and entering the second
fresnel. In passing through the second fresnel it becomes converted into rectilineally vibrating light, represented by (98^{xiii}) and
(98^{xiv}) with $e_1 = 0$ and $e_2 = 0$. Thus we see that §§ 158^{x}—158^{xii},
with $e_1 = 0$ and $e_2 = 0$ *passim*, expresses simply and fully the theory
of the conversion of rectilineally vibrating light into elliptically
or circularly polarized light, and *vice versâ*; by one true Fresnel's
rhomb. A diagram with rules as to directions in the use of
Fresnel's rhomb is given below in § 158^{xvii}. §§ 90—98 above

Molar. contain the application of the theory to ordinary reflection at the surfaces of transparent solids, or liquids, or metals: the first fresnel of § 158ˣ being done away with, and the reflection substituted for it. The corresponding application to total internal reflection is much simpler, because the intensities of incident and reflected components are equal. It allows the difference of phase, produced by the reflection, to be measured with great accuracy; and a careful experimental research, thus carried out, would no doubt prove the difference of phase, produced between the two components by the reflection, to agree *very accurately* with the obtuse angles calculated for different incidences according to (98ᵛⁱ) and § 158‴.

§ 158ˣⁱᵛ. About a year ago, in making some preliminary experiments by aid of a Fresnel's rhomb, to illustrate §§ 90—100, 152, 153, I interpreted the phasal difference of the rhomb according to Airy's Tracts; but found error, or confusion, in respect to phasal change by one internal reflection in glass, and by metallic reflection. I looked through all the other books of reference and scientific papers accessible to me at the time; and I have continued the inquiry to the present time, by aid of the libraries of the Royal Society of London, and the University of Glasgow; but hitherto without success, in trying to find an explicit statement as to which of the two components is advanced upon the other in the Fresnel's rhomb. I have therefore been obliged to work the problem out myself mathematically for the Fresnel's rhomb; and with the knowledge thus obtained, to find by experiment which of the two components is advanced on the other by metallic reflection. Of all the authors I have hitherto had the opportunity of studying, only Airy in respect to Fresnel's rhomb, and Jamin, and Stokes *, and Basset in respect to metallic reflection, have explicitly stated which component of the light is advanced in phase.

In Airy's Tracts, 2nd edition (1831), page 364, I find a thoroughly clear statement, agreeing with Fresnel's own, regarding the Fresnel's rhomb and total internal reflection:—"If the "light be twice reflected in the same circumstances and with the "same plane of reflection, the phase of vibrations in the plane of "incidence is more accelerated than that of the other vibrations

* *Mathematical and Physical Papers*, Vol. II. p. 360.

"by 90°." In truth, the vibrations in the plane of incidence are Molar. not advanced 90°, but set back 270°. They are therefore *virtually advanced* 90°, relatively to the vibrations perpendicular to the plane of incidence.

In Basset's *Physical Optics* (Cambridge 1892), page 339, I find, stated as a law arrived at by Jamin by experiment, with reference to metallic reflection, the following:—

"(1) *The wave which is polarized perpendicularly to the plane* "*of incidence, is more retarded than that which is polarized in the* "*plane of incidence.*"

By experiment I have verified this, working with a Fresnel's rhomb, *interpreted according to* Airy, *and* § 158^{vi} *above.* In reading Jamin's experimental paper I had felt some doubt as to his meaning because his expression " vibrations polarisées dans le plan de l'incidence " (I quote from memory) may have signified, not that the plane of polarization*, but that the line of vibration, was in the plane of incidence. That Basset's interpretation was correct is however rendered quite certain by a clear statement in Jamin's *Cours de Physique*, Vol. II. page 690, describing relative advance of phase of vibrations perpendicular to the plane of incidence of light reflected from a polished metal. This phasal relative advance he measured by a Babinet's compensator, and found it to increase from zero at normal incidence, to 90° at the principal incidence, I; and up to 180° at grazing incidence. He would have found not advance, but back-set, if he had used a Fresnel's rhomb interpreted according to the mathematical theory given (by himself as from Fresnel) in pages 783 to 787 of the same volume, with the falsified formulas (98^{xvii}).

§ 158^{xv}. The origin of the long standing mistake regarding the Fresnel's rhomb is to be found in Fresnel's original paper, " Mémoire sur la loi des modifications que la réflexion imprime " à la lumière polarisée ": reproduced in Volume I. of *Collected Papers*, Paris 1876, pages 767 to 799. In page 777 we find

---

* Considering the inevitable liability to ambiguity of this kind, I have abandoned the designation "plane of polarization"; and have resolved always to specify or describe with reference to vibrational lines. Abundant examples may be found in the earlier parts of the present volume illustrating the inconvenience of the designation "plane of polarization." In fact "polar" and "polarization" were, as is now generally admitted, in the very beginning unhappily chosen words for differences of action in different directions around a ray of light. These differences are essentially not according to what we now understand by polar quality.

Molar. Fresnel's own celebrated formulas, for reflected vibrational amplitudes and their ratio, correctly given as follows :—

$$-\frac{\tan(i-,i)}{\tan(i+,i)}, \quad -\frac{\sin(i-,i)}{\sin(i+,i)}, \quad +\frac{\cos(i-,i)}{\cos(i+,i)} \ ...(98^{\text{xxi}}).$$

It is obvious that the first two of these expressions must have the same sign, because at very nearly normal incidences the tangents are approximately equal to the sines, and at normal incidences, the two formulas mean precisely the same thing; there being, at normal incidence, no such thing as a difference between vibrations in, and vibrations perpendicular to, a plane of incidence. Yet, notwithstanding the manifest absurdity of giving different signs to the "tangent formula" and the "sine formula" of Fresnel, we find in a footnote on page 789 (by Verdet, one of Fresnel's editors), the formulas changed to

$$\frac{\tan(i-,i)}{\tan(i+,i)}, \quad -\frac{\sin(i-,i)}{\sin(i+,i)} \ ....(98^{\text{xxii}}),$$

in consequence of certain "considérations" set forth by Fresnel on pages 788, 789. I hope sometime to return to these "considérations" and to give a diagram showing the displacements of ether in a space traversed by co-existent beams of incident and reflected light, by which Fresnel's "petite difficulté" of page 787 is explained, and the erroneous change from his own originally correct formulas is obviated. The falsified formulas (98^{xxii}) have been repeated by some subsequent writers; avoided by others. But, so far as I know, *no author* has hitherto corrected the consequential error, which gives an acute angle instead of an obtuse angle for phasal difference in one total internal reflection; and gives 90° phasal difference instead of 270° in the two reflections of the Fresnel's rhomb; and gives 90° back-set, instead of the truth which is 90° virtual advance, of vibrations in the plane of incidence, relatively to vibrations perpendicular to the plane of incidence, in a Fresnel's rhomb. The serious practical error, in respect to *which of the two* components experiences phasal advance in the Fresnel's rhomb, does not occur in any published statement which I have hitherto found. Airy accidentally corrected it by another error. All the other authors limit themselves to saying that there is a phasal difference of 90° between the two components, without saying which component is in advance of the other.

§ 158^{xvi}. In Fresnel (page 782), and Airy (page 362), where Molar.
$\sqrt{-1}$ is first introduced with the factor $\sqrt{(\mu^2\sin^2 i - 1)}$, the positive
sign is taken accidentally, without reference to any other consideration. A reversal of sign, wherever $\sqrt{-1}$ occurs in the
subsequent formulas, would have given phasal advance instead
of phasal back-set, or back-set instead of advance, in each conclusion. If the authors had included the refracted wave (Airy,
page 358) in the new imaginary investigation so splendidly discovered by Fresnel, they would have found it necessary, either to
reverse the sign of $\sqrt{-1}$ throughout, or to reverse the interpretation of it in respect to phasal difference, given as purely conjectural by Fresnel, and by Airy (page 363) quoting from him:
because with this interpretation, and with the signs as they stand,
they would have found for the "refracted wave," a displacement
of the ether increasing exponentially, instead of diminishing
exponentially, with distance from the interface (see footnote on
§ 158" above). It is exceedingly interesting now to find that an
accidentally wrong choice of signs, in connection with $\sqrt{-1}$,
served to correct in the result of two reflections, the practical
error of acute instead of obtuse for the phasal difference due
to one reflection, which is entailed by the deliberate choice of
a false sign in the real formulas of (98^{xxii}): and that, thus led,
Airy gave correctly the only statement hitherto published, so far
as I know, as to which of the two components experiences phasal
advance in Fresnel's rhomb.

§ 158^{xvii}. *Note on circular polarization in metallic or vitreous
or adamantine reflection.* Referring to Basset's *Physical Optics*,
page 329, edition 1892, I find *Principal Azimuth* defined as the
angle between the *plane of polarization* of the reflected light,
and the plane of the reflection, when the incident light is *circularly
polarized light incident at the angle of principal incidence.* This
really agrees with the, at first sight, seemingly different definition
of Principal Azimuth, given in § 97" above: because it is easily
proved that when rectilineally vibrating light, is converted into
circularly polarized light by metallic or other reflection; the
azimuth of the *vibrational plane* of the incident light, is equal to
the azimuth of the *plane of polarization* of the reflected light
when circularly polarized light is converted into rectilineally
vibrating light by reflection on the same mirror. The fact that k

Molar. is positive *for every case of principal incidence* (§ 97') is included and interpreted in the following statement :—

Consider the case of plane polarized light incident on a polished metallic, or adamantine, or vitreous, reflector. For simplicity of expressions let the plane of the reflector be horizontal. To transmit the incident light let a nicol, mounted with its axis inclined to the vertical at any angle i, carry a pointer to indicate the direction of the vibrational lines of the light emerging from it, and incident on the mirror. First place the pointer upwards in the vertical plane through the axis; which is the plane of incidence of the light on the mirror. The reflected light is of rectilineal vibrations in the same plane. Now turn the pointer anti-clockwise through any angle less than 90°. The reflected light consists of elliptic, or circular, anti-clockwise orbital motions.

If $i = I$, the principal incidence; the two axes of each elliptic orbit are, one of them horizontal and perpendicular to the plane of incidence and reflection; the other in this plane.

To avoid any ambiguity in respect to "clockwise" and "anti-clockwise," the observer looks at the nicol, and at the circle in which its pointer turns, from the side towards which the light emerges after passing through it: and he looks ideally at the orbital motion of the reflected light from the side towards which the reflected light travels to his eye. See § 98 above.

In external reflection of rectilineally vibrating light by all ordinary transparent reflectors, including diamond (but not realgar), the deviation from rectilineality of the reflected light is small, except for incidences within a few degrees of the Brewsterian $\tan^{-1}\mu$. See diagram of § 178 below (figures 11 and 12), for diamond.

§ 158[xviii]. The rules for directions in elliptic and circular polarization by a Fresnel's rhomb are represented by the annexed diagram, figure 9'''. O' and O are the centres of the entering and exit faces. OK is a line parallel to the facial intersections. OF represents the plane of the reflections in the rhomb, being a plane perpendicular to the four optically effective faces.

Cases a_1, a_2, a_3. Plane polarized light enters by O'. OV_1, OV_2, OV_3, are parallels through O to the vibrational lines of the entering light; they are of equal lengths, to represent the displacements as equal in the three cases. The orbits of the exit light in the three cases are represented by (1) ellipse, (2) circle, (3) ellipse; Molar. being drawn according to the geometrical construction indicated in one quadrant of the diagram.

Cases b_1, b_2, b_3. The orbits in three cases of equally strong circularly and elliptically polarized entering light, having axes along OF and OK, shown with their centres transferred to O, are represented by (1) ellipse, (2) circle, (3) ellipse. OV_1', OV_2', OV_3', represent the displacements in the exit light, which in each of these three cases consists of rectilineal vibrations.

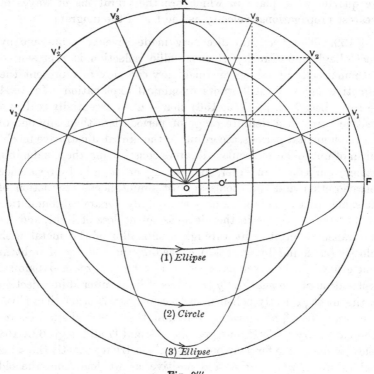

Fig. 9'''.

The direction of the orbital motions of the exit light in cases a, and of the entering light in cases b, is, in each of the six cases, anti-clockwise, as indicated by the arrowheads on the ellipses and circle.

Thus we have the following two rules for directions :—

Rule (a). When the vibrational plane of entering light must

Molar. be turned anti-clockwise to bring it into coincidence with OK, the orbital motions in the exit light are anti-clockwise.

Rule (b). When the orbital motion of circularly or elliptically polarized entering light, having axes parallel to OK and OF, is anti-clockwise, the vibrational plane of the exit light is turned anti-clockwise from OK.

These rules, and the diagram, hold with a "quarter-wave plate" substituted for the fresnel, provided homogeneous light of the corresponding wave-length is used. The principal plane of the quarter-wave plate in which are the vibrations of waves of greatest propagational velocity, is the OF of the diagram.

§ 159. The beautiful discovery made seventy years ago by MacCullagh and Cauchy, that metallic reflection is represented mathematically by taking an imaginary complex to represent the refractive index, μ, still wants dynamical explanation. In 1884 we saw (Lec. XII. pp. 155, 156) that $-\mu^2$ is essentially real and positive through a definite range of periods less than any one of the fundamental periods, according to the unreal illustrative mechanism of unbroken molecular vibrators constituting the ponderable matter; embedded in ether, and acting on it only by resistance between ether and the atoms against simultaneous occupation of the same space. This gave us a thoroughly dynamical foundation for metallic reflection in the ideal case of no loss of light, and for the transmission of light through a thin film of the metal with velocity (as found by Quincke), exceeding the velocity of light in void ether. It however gave us no leading towards a dynamical explanation of the *manifestly great* loss of light suffered in reflection at the most perfectly polished surfaces of metals other than silver or mercury (see § 88 above). But now the new realistic electro-ethereal theory set forth in Appendices A and E, and in §§ 162...168 below, while giving for non-conductors of electricity exactly the same real values of μ^2, negative and positive, as we had from the old tentative mechanism, seems to lead towards explaining loss of luminous energy both in reflection from, and in transmission through, a substance which has any electric conductivity, however small. In App. E § 30, J. J. Thomson's theory of electric conduction through gases is explained by the projection of electrions out of atoms. If this never took place, the electro-ethereal theory would, like our old mechanical vibrator, give loss of energy from transmitted light only in the exceedingly small proportion due to the size of the atoms according to Rayleigh's theory of the blue sky Molecular. (§ 58 above). On the other hand the crowds of loose electrions among under-loaded atoms, throughout the volume of a metal, by which its high electric conductivity is explained in App. E § 30, may conceivably give rise to large losses of energy from reflected light; losses spent in heating a very thin surface layer of the metal by irregular motions of the electrions. This seems to me probably the true dynamical explanation of the imaginary term in ν^2 of § 158 (98). See also § 84 above.

§ 159′.   In every case of practically complete opacity (that is to say, no perceptible translucency from the most brilliant light falling on a plate of any thickness greater than three or four wave-lengths, or about $2 \cdot 10^{-4}$ cm.), measurement of the principal azimuth, which can be performed with great accuracy by aid of two nicols and a Fresnel's rhomb (§§ 97—97″ above), gives an interesting contribution to knowledge regarding loss of luminous energy in the reflection of light.   The notation of § 94′ gives

$$S^2 \sin^2 \alpha + (T^2 + E^2) \cos^2 \alpha \dots\dots\dots\dots(98^{\text{xxi}})$$

as the reflectivity for light vibrating in azimuth α from the plane of incidence (§ 88, foot-note); so that in *every case of practical opacity*,

$$1 - [S^2 \sin^2 \alpha + (T^2 + E^2) \cos^2 \alpha]\dots\dots\dots(98^{\text{xxii}})$$

represents loss of luminous energy by conversion into heat in a thin surface stratum of less than $2 \cdot 10^{-4}$ cm. thickness.

§ 159″.   For perfect reflectivity (98^{xxii}) must be zero for every value of α, and for every incidence.   Hence, as S, T, E are independent of α, we have, at every incidence,

$$S^2 = T^2 + E^2 = 1 \dots\dots\dots\dots(98^{\text{xxiii}}).$$

Hence for the incidence making $T = 0$ (that is the *principal incidence*), $S^2 = E^2 = 1$; which makes $E/S = \pm 1$ in every case of perfect reflectivity, in total internal reflection for instance.   Therefore the *principal azimuth*, being $\tan^{-1}(E/S)$ for principal incidence, is $\pm 45°$, if the reflectivity is perfect.   (See § 157 above.)

§ 159‴.   Observation and mathematical theory agree that the principal azimuth is positive in every case: for interpretation of this see § 158^{xvii} above.   They also agree that in every case short of perfect reflectivity the principal azimuth is $< 45°$, *not* $> 45°$.

LECTURE XIX.

FRIDAY, *October* 17, 3.30 P.M., 1884. *Written afresh,* 1903.

Reconciliation between Fresnel and Green.

Molecular.   THIS Lecture or "Conference" began with the consideration of a very interesting report, presented to us by one of our twenty-one coefficients, Prof. E. W. Morley, describing a complete solution, worked out by himself, of the dynamical problem of seven mutually interacting particles, which I had proposed nine days previously (Lecture IX. p. 103 above) as an illustration of the molecular theory of dispersion with which we were occupied. His results are given in the following Table.

Solution for Fundamental Periods, Displacement, and Energy Ratios of a System of Spring-connected Particles. By EDWARD W. MORLEY, Cleveland, Ohio.

$$m = 1, \ 4, \ 16, \ 64, \ 256, \ 1024, \ 4096.$$
$$C = 1, \ 2, \ 3, \ 4, \ 5, \ 6, \ 7, \ 8.$$

Fundamental Periods corresponding to outer ends of Springs 1 and 8 held fixed.

$\tau^2 =$	·2889	·9952	3·350	11·362	39·12	137·89	680·2
$\dfrac{1}{\tau^2} =$	3·4618	1·00483	0·29849	0·0880078	0·0255607	0·0072564	0·0014701
$\dfrac{1}{\tau} =$	1·860	1·002	·546	·296	·159	·0851	·0383

Displacement Ratios, or values of $\left(\dfrac{x_i}{x_1}\right)$.   Molecular.

x_1	1	1	1	1	1	1	1
x_2	− ·231	1·000	1·351	1·456	1·487	1·496	1·499
x_3	·014	− ·341	1·047	1·589	1·761	1·813	1·829
x_4	− ·iii 27	·025	− ·431	1·129	1·787	1·997	2·066
x_5	·v 15	− ·iii 50	·033	− ·511	1·223	1·960	2·216
x_6	− ·viii 26	·v 30	− ·iii 68	·040	− ·581	1·322	2·203
x_7	·xi 13	− ·viii 51	·v 39	− ·iii 81	·045	− ·628	1·717

Energy Ratios, or values of $\dfrac{m_i x_i^2}{m_1 x_1^2}$.

$m_1 x_1^2$	1	1	1	1	1	1	1
$m_2 x_2^2$	·213	3·998	7·30	8·48	8·85	8·96	8·99
$m_3 x_3^2$	·ii 33	1·864	17·54	40·41	49·64	52·58	53·54
$m_4 x_4^2$	·v 47	·039	11·88	81·66	204·35	255·34	273·14
$m_5 x_5^2$	·ix 61	·iv 65	·28	66·73	382·71	983·10	1157·52
$m_6 x_6^2$	·xiv 7	·viii 9	·iii 47	1·62	345·60	1788·13	4968·41
$m_7 x_7^2$	·xx 7	·xii 1	·vii 63	·002	8·42	1616·99	12080·04
Sum	1·216	6·902	38·0004	199·90	1000·57	4706·10	18542·64

At present our subject is the dynamical reconciliation between Molar. Fresnel and Green, not only in respect to reflection and refraction at an interface between two isotropic transparent bodies, as promised in Lecture XVIII. §§ 133, 139, but also in respect to the propagation of light through a transparent crystal (double refraction) as promised in Lecture XV. § 45.

§ 160. In my paper on the reflection and refraction of light (*Phil. Mag.* 1888, 2nd half-year), an extract from which is quoted in §§ 107—111 of Lecture XVIII., I have shown (§§ 109—111) that a homogeneous portion of an elastic solid with its boundary

Molar. held fixed is stable if its rigidity (n) is positive; even though its bulk-modulus (k) is negative if not less than $-\frac{4}{3}n$. And in § 115 it is shown that the propagational velocity (v) of condensational-rarefactional waves in any homogeneous elastic solid (or fluid if $n = 0$) is given by the equation

$$v = \sqrt{\frac{k + \frac{4}{3}n}{\rho}} \dots\dots\dots\dots\dots\dots(99).$$

In virtue of this equation it had always been believed that the propagational velocity of condensational-rarefactional waves in an elastic solid was essentially greater than that of equivoluminal waves, which is $\sqrt{\frac{n}{\rho}}$. The Navier-Poisson doctrine, upheld by many writers long after Stokes showed it to be wrong (see Lecture XI. pp. 123, 124 above, and Appendix I below), made $k = \frac{5}{3}n$ (see Lecture VI. p. 61 above), and therefore the velocity of condensational-rarefactional waves $= \sqrt{3}$ times that of the equivoluminal wave. The deviations of real substances, such as metals, glasses, india-rubber, jelly, to which Stokes called attention, were all in the direction of making the resistance to compression greater than according to the Navier-Poisson doctrine; but it was pointed out in Thomson and Tait (§ 685) that cork deviates in the opposite direction and is, in proportion to its rigidity, much *less* resistant to compression than according to that doctrine. In truth, without violating any correct molecular theory, we may make the bulk-modulus of an elastic solid as small as we please in proportion to the rigidity provided only, for the sake of stability, we keep it positive. A zero bulk-modulus makes the velocity of condensational-rarefactional waves equal to $\sqrt{\frac{4}{3}}$ times that of equivoluminal waves. But now, to make peace between Fresnel and Green, we want for ether; if not all ether, at all events ether in the space occupied by ponderable matter, a negative bulk-modulus just a little short of $-\frac{4}{3}n$, to make the velocity (v) very small in comparison to $\sqrt{\frac{n}{\rho}}$. And now, happily (§ 167 below), a theory of atoms and electrions in ether, to which I was led by other considerations, gives us a perfectly clear and natural explanation of ether through void space practically incompressible, as Green supposed it to be; while in the interior of any ordinary solid or liquid it has a large enough negative bulk-modulus to render the propagational velocity of

condensational-rarefactional waves exceedingly small in comparison Molar.
with that of equivoluminal waves.

§ 161. In my 1888 Article I showed that if v is very small,
or infinitely small, in proportion to the propagational velocity of
equivoluminal waves in the two mediums on the two sides of a
reflecting interface, or quite zero, Fresnel's laws of reflection and
refraction are very approximately, or quite exactly, fulfilled. About
fourteen years later I found that, as said in § 46 of Lecture XV.,
it is enough for the fulfilment of Fresnel's laws that the velocity
of the condensational-rarefactional waves in *one of the two mediums*
be exceedingly small. During those fourteen years, I had been
feeling more and more the great difficulty of believing that the
compressibility-modulus of ether through all space could be negative, and so much negative as to make the propagational velocity
of condensational-rarefactional waves exceedingly small, or zero.
One chief object of the long mathematical investigation regarding
spherical waves in an elastic solid, added to Lecture XIV. (pp. 191
—219 above), was to find whether or not smallness of propagational velocity of condensational-rarefactional waves through ether
void of ponderable matter could give practical annulment of energy
carried away by this class of waves from a vibrator constituting a
source of light. I found absolute proof that the required practical
annulment was not possible; and I therefore felt forced to the
conclusion stated in § 32, p. 214: "This, to my mind, utterly
"disproves my old hypothesis of a very small velocity for irro-
"tational wave-motion in the undulatory theory of light." Now,
most happily, seeing that it is enough for the dynamical verification
of Fresnel's laws that the velocity of the condensational-rarefactional wave be exceedingly small for either one or other of the
mediums on the two sides of the interface, I can return to my old
hypothesis with a confidence I never before felt in contemplating
it. It is, I feel, now made acceptable by assuming with Green
that ether in space void of ponderable matter is practically incompressible by the forces concerned in waves proceeding from a source
of light of any kind, including radiant heat and electro-magnetic
waves; while, in the space occupied by liquids and solids, it has a
bulk-modulus largely enough negative to render the propagational
velocity of condensational-rarefactional waves exceedingly small in
comparison with that of equivoluminal waves in pure ether. We

Molar. have now no difficulty with respect to getting rid of the condensational-rarefactional waves generated by the incidence of light on a
transparent liquid or solid.   They may, with very feeble activity,
travel about through solids or liquids, experiencing internal reflections, almost total, at interfaces between solid or liquid and air
or vacuous ether; but more probably (§ 172 below) they will be
absorbed, that is, converted into non-undulatory thermal motion,
among the ponderable molecules, without ever travelling as waves
through more than an exceedingly small space containing ponderable matter (solid, or liquid, or gas).   Certain it is that neither
ethereal waves, nor any kind of dynamical action, within the body,
can give rise to condensational-rarefactional waves through void
ether if we frankly assume void ether to be incompressible.

Molecular.   § 162.   So far, we have not been supported in our faith by
any physical idea as to how ether could be practically incompressible when undisturbed by ponderable matter; and yet may be
very easily compressible, or may even have negative bulk-modulus,
in the interior of a transparent ponderable body.

Do moving atoms of ponderable matter displace ether, or do
they move through space occupied by ether without displacing it?
This is a question which cannot be evaded: when we are concerned
with definite physical speculations as to the kind of interaction
which takes place between atoms and ether; and when we
seriously endeavour to understand how it is that a transparent
body takes wave-motion just as if it were denser than the
surrounding ether outside, and were otherwise undisturbed by
the presence of the ponderable matter.   It is carefully considered
in Appendix A, and in Appendix B under the heading "Cloud I."
My answer is indicated in the long title of Appendix A, "On the
"Motion produced in an Infinite Elastic Solid by the motion
"through the space occupied by it, of a body acting on it only by
"Attraction or Repulsion." This title contradicts the old scholastic
axiom, *Two different portions of matter cannot simultaneously*
occupy the same space.   I feel it is impossible to reasonably gain-
say the contradiction.

          § 163.   Atoms move through space occupied by ether.   They
must act upon it in some way in order that motions of ponderable
matter may produce waves of light, and in order that the vibratory

motion of the waves may act with force on ponderable matter. Molecular. We know that ether does exert force on ponderable matter in producing our visual perception of light; and in photographic action; and in forcibly decomposing carbonic acid and water by sunlight in the growth of plants; and in causing expansion of bodies warmed by light or radiant heat. The title of Appendix A contains my answer to the question *What is the character of the action of atoms of matter on ether?* It is nothing else than attraction or repulsion.

§ 164. But if ether were absolutely incompressible and inextensible, an atom attracting it or repelling it would be utterly ineffectual. To render it effectual I assume that ether is capable of change of bulk; and is largely condensed and rarefied by large positive and negative pressures, due to repulsion and attraction exerted on it by an atom and its neutralising quantum of electrions. As in Appendix A, §§ 4, 5, I now for simplicity assume that an atom void of electrions *repels* the ether in it and around it with force varying directly as the distance from the centre for ether within the atom, and varying inversely as the square of the distance for ether outside. I assume that a single separate electrion *attracts* ether according to the same laws; the radius of the electrion being very small compared with the radius of any atom. I assume that all electrions are equal and similar, and exert equal forces on ether.

§ 165. While keeping in view the possibility referred to in § 6 of Appendix E, I for the present assume that the *repulsion* of a void atom* on ether outside it is equal to an integral number of times the *attraction* of one electrion on ether; same distances understood. An atom violating this equation cannot be unelectrified. I continue to make the same assumptions as in Appendix D in respect to mutual electric repulsion between void atoms and void atoms; attraction between void atoms and electrions; and repulsion between electrions and electrions. And as in Appendix A an "unelectrified atom" is an atom having its saturating quantum of electrions within it.

---

* For brevity I use the expression "void atom" to signify an atom having no electrion within it.

Molar.   § 166. I assume that the law of compressibility of ether is such as to make the mean density óf ether within any space which contains a large number of unelectrified atoms, exactly equal to the natural density of ether undisturbed by ponderable matter. If there were just one electrion within each atom this assumption would exactly annul displacement of the ether outside an atom by the repulsion of the atom and the attraction of the electrion; and would very nearly annul it when there are two or more electrions inside. Thus the ether within each atom is somewhat rarefied from the surface inwards: and farther in, it is condensed round each electrion. In a mono-electrionic atom, the spherical surface of normal density between the outer region of rarefaction and the central region of condensation, I call for brevity, *the sphere of condensation*. In a poly-electrionic atom the density of the ether decreases from enormous condensation around the centre of each electrion to exactly normal value at an enclosing surface, the space within which I shall call the electrion's *sphere of condensation*. It is very approximately spherical except when, in the course of some violent motion, two electrions come very nearly together.

§ 167. I assume that the law of compressibility of ether is such as to make the equilibrium described in § 166 stable; but so nearly unstable that the propagational velocity of condensational-rarefactional waves travelling through ether in space occupied by ponderable matter, is very small in comparison with the propagational velocity of equivoluminal waves through ether undisturbed by ponderable matter.

§ 168. Lastly I assume that the effective rigidity of ether in space occupied by ponderable matter is equal to that of pure ether undisturbed by ponderable matter. This is not an arbitrary assumption: we may regard it rather as a proposition proved by experiment, as explained in Lecture I. pp. 15, 16 and §§ 81′, 136 of Lectures XVII. and XVIII. But, as will be shown in Lec. XX., § 237 below, it is somewhat satisfactory to know that it follows directly from a natural working out of the set of assumptions of §§ 163—167.

§ 169. Return now to § 132. This is merely Green's wave-theory extended to include not only the equivoluminal waves but also the condensational-rarefactional waves resulting from the Molar. incidence of plane equivoluminal waves on a plane interface between two sliplessly connected elastic solids of equal rigidities, each capable of condensation and rarefaction. Let us suppose that in either one or other of the two mediums the propagational velocity of the condensational-rarefactional wave is very small. This makes L very small as we see by using (48) and (50) to eliminate a, c, and $,c$ from (61) which gives

$$L = \frac{(,\rho - \rho)^2 \sin^2 iu^{-2}}{,\rho \sqrt{(v^{-2} - \sin^2 iu^{-2})} + \rho \sqrt{(,v^{-2} - \sin^2 iu^{-2})}} \quad \text{...(100)}.$$

§ 170. If either the upper or the lower medium be what we commonly call vacuum, (in reality ether void of ponderable matter), the upper for example, we have $v = \infty$, and (100) becomes

$$L = \frac{(,\rho - \rho)^2 \sin^2 iu^{-2}}{,\rho\iota \sin iu^{-1} + \rho \sqrt{(,v^{-2} - \sin^2 iu^{-2})}} \quad \text{.........(101)},$$

whence when $,v$ is very small

$$L \fallingdotseq \frac{(,\rho - \rho)^2}{\rho} \sin^2 iu^{-2} ,v \quad \text{.................(102)},$$

where $\fallingdotseq$ denotes approximate equality. If v and $,v$ are each very small (100) becomes

$$L \fallingdotseq \frac{(,\rho - \rho)^2 \sin^2 iu^{-2} v ,v}{,\rho ,v + \rho v} \quad \text{..............(103)},$$

thus by (100), (102) and (103), we see not only, as said above in § 133, that Fresnel's Laws are exactly fulfilled if either v or $,v$ is zero. We see farther how nearly, to a first approximation, they are fulfilled if $v = \infty$, and $,v$ is very small; also if v and $,v$ are both very small.

Probably values of $,v$, or of v and $,v$, as small as $u/500$ or $u/1000$ may be found small enough, but $u/100$ not small enough, to give as close a fulfilment of Fresnel's Laws as is proved by observation. See § 182 below.

§ 171. Thus so far as the reflection and refraction of light are concerned, the reconciliation between Fresnel and Green is complete: that is to say we have now a thoroughly realistic dynamical foundation for those admirable laws which Fresnel's penetrating genius prophesied eighty years ago from notoriously imperfect dynamical leadings. (See §§ 106, 107 above.)

Molar.   § 172. The only hitherto known deviation from absolute rigour in Fresnel's Laws for the reflection and refraction of light at the surface of a transparent body in air or void ether, or at the interface between two transparent bodies liquid or solid, is that which was discovered by Sir George Airy for diamond in air more than eighty years ago, and many years afterwards was called by Sir George Stokes the adamantine property. (See § 158 above.) It is shown in the Table of § 105 above along with corresponding deviations discovered subsequently by Jamin in other transparent bodies solid and liquid. It now appears by (102) and (103) that the explanation of these deviations from Fresnel, which even in diamond are *exceedingly small*, is to be found by giving very small imaginary values to $_,v$, or to v and $_,v$. To suit the case of light travelling through vacuum or through air and incident on a transparent solid or liquid, we should take $v = \infty$ if the incident light travels through vacuum; or if through air, v perhaps $= \infty$, but certainly very great in comparison with $_,v$. Hence in either case the proper approximate value of L is given by (102).

       § 173. Looking to (59) and (67) of § 132 we see that if $_,v$ have a small real value (positive of course) the reflected ray, of vibrations in the plane of incidence, will vanish for an angle of incidence slightly less than the Brewsterian angle $\tan^{-1}\mu$. If we give to $_,v$ a complex value $_,p + \iota_,q$ with $_,p$ positive, we have the adamantine property, and principal incidence slightly less than with $_,p = 0$. Hitherto we have no *very* searching observations as to the perfect exactness of $\tan^{-1}\mu$ whether for the polarizing angle when the extinction is seemingly perfect or for the principal incidence in cases of perceptible adamantine property. For the present therefore we have no reason to attribute any real part to $_,v$; and we may take

$$_,v^2 = -q^2 ; \qquad _,v = \iota q \quad \dots\dots\dots\dots(104),$$

where q denotes a real velocity. Using this in the last formula of § 128 and eliminating $\sin _,j/_,v$ by (48) and $\cos _,j/_,v$ and $_,c$ by (49) and (50) we find for the displacement in the refracted condensational-rarefactional wave

$$_,Hf[t - ax - \iota \sqrt{(1 + a^2 q^2)}\, y/q] \quad \dots\dots\dots\dots(105).$$

Looking now to (63) and (102) we see that in $_,H/G$ there is an imaginary part which is exceedingly small in comparison with the real part, because, by (102), $L/,v$ is real and not small, while L is Molar. imaginary and exceedingly small. For our present approximate estimates we suppose G to be real and therefore $,H$ to be real. Taking now $f(t) = \epsilon^{\iota\omega t}$ in (105) and taking the half-sum of the two imaginary expressions as given with $\pm \iota$, we find for the real refracted condensational-rarefactional displacement the following expression; with approximate modification due to qa being a very small fraction;

$$D = ,H \cos \omega\,(t - ax)\, \epsilon^{\omega \sqrt{(1+a^2 q^2)}\, y/q} \fallingdotseq ,H \cos \omega\,(t - ax)\, \epsilon^{\omega y/q}...(106).$$

Hence q is positive. (See footnote on § 158″.) The direction of this real displacement is exceedingly nearly OY' of § 117, that is to say the direction of Y negative, because the imaginary angle $,j$ differs by an exceedingly small imaginary quantity from 90°. It would be exactly 90° if $,v$ were exactly zero; and it would be less than 90° by an exceedingly small real quantity if $,v$ were an exceedingly small real velocity.

§ 174. The motion represented by (106) is not a wave travelling into the denser medium: it is a clinging wave travelling along the interface with velocity a^{-1}. The direction of the displacement is approximately perpendicular to the interface. Its magnitude decreases inwards from the interface according to the law of proportion represented by the real exponential factor; distance inwards from the interface being $- y$. The period of the wave is $2\pi/\omega$; and the space travelled in this time with velocity q is $2\pi q/\omega$. Hence the displacement at a distance from the interface equal to this space, is $1/\epsilon^{2\pi}$ or $1/535$ of the displacement at the interface.

§ 175. Consideration of the largeness of any such distance as 2.10^{-8} cm. (§ 80 above) between centres of neighbouring atoms, and of the smallness of $,v$ if real, will probably give good dynamical reason for the assumption of $,v$ a pure imaginary. It is a very important assumption, inasmuch as it implies that there is no inward travelling condensational-rarefactional wave carrying away energy from the equivoluminal reflected and refracted rays.

§ 176. Let us now find the value of q/u, the adamantinism (§ 158); to give any observed amount of the adamantine property, as represented by the tangent of the Principal Azimuth, which is Jamin's k. In (59) of § 132 take for L the value given by (102),

and put $$\frac{(,\rho - \rho)^2}{\rho} = N ; \quad ,v = \iota q ; \quad \frac{q}{u} = \sigma............(107),$$

Molar. thus we find

$$-\frac{G'}{G} = \frac{u\,(,\!\rho b - \rho,\!b) - \iota N\sigma\sin^2 i}{u\,(,\!\rho b + \rho,\!b) + \iota N\sigma\sin^2 i} = \frac{P - \iota Q}{u^2(,\!\rho b + \rho,\!b)^2 + (N\sigma\sin^2 i)^2}$$
$$\dots\dots\dots(108)^*,$$

where
$$\left.\begin{aligned} P &= u^2\,[(,\!\rho b)^2 - (\rho,\!b)^2] - (N\sigma\sin^2 i)^2 \\ Q &= 2u,\!\rho b\, N\sigma\sin^2 i \end{aligned}\right\}\dots\dots(109).$$

Hence taking in § 128, $f(t) = \epsilon^{\iota\omega t}$; and realising by taking half-sum of solutions for $\pm\,\iota$, we find displacement

in incident wave $=$ $\quad G\cos\omega\left(t - \dfrac{s}{u}\right)$ ....(110),

in reflected wave $= -\,G\,\sqrt{(T^2 + E^2)}\cos\left[\omega\left(t - \dfrac{s}{u}\right) - \phi\right]$...(111),

$$= -\,G\,\sqrt{\frac{u^2\,(,\!\rho b - \rho,\!b)^2 + (N\sigma\sin^2 i)^2}{u^2\,(,\!\rho b + \rho,\!b)^2 + (N\sigma\sin^2 i)^2}}\cos\left[\omega\left(t - \frac{s}{u}\right) - \phi\right]\dots(111')\dagger,$$

where
$$\left.\begin{aligned} \phi &= \tan^{-1}\frac{Q}{P} = \tan^{-1}\frac{E}{T} \\ &= \tan^{-1}\frac{2\mu^2\,(\mu^2 - 1)\,\sigma\cos i\sin^2 i}{\mu^2\cos^2 i - \sin^2 i - (\mu^2 - 1)^3\,\sigma^2\sin^4 i} \end{aligned}\right\}\dots(112).$$

The developed form of Q/P here given is found by putting

$$,\!\rho/\rho = \mu^2;\quad ub = \cos i;\quad u,\!b = \mu\cos,\!i;\quad \cos^2,\!i = 1 - \frac{\sin^2 i}{\mu^2}\ \dots(113),$$

and by dividing numerator and denominator by $(\mu^2 - 1)/\mu^2$.

---

* According to the notation of § 94' we have $Q/P = E/T$, and

$$T = \frac{P}{u^2\,(,\!\rho b + \rho,\!b)^2 + (N\sin^2 i\,q/u)^2}\ \text{ and }\ E = \frac{Q}{u^2\,(,\!\rho b + \rho,\!b)^2 + (N\sin^2 i\,q/u)^2}\,.$$

† This alternative form (111') comes from (111) by resolving $P^2 + Q^2$ into two factors according to the algebraic identity

$$(a^2 - b^2 - c^2)^2 + 4a^2c^2 = [(a+b)^2 + c^2]\,[(a-b)^2 + c^2].$$

But it is found directly by treating (108) as we treated a similar formula in (75), following Green, on a plan which is simpler in respect to the resultant magnitude, but less simple in respect to the phase, than the plan of (111).

§ 177. By (111′) and the notation of § 94′, we have

$$T^2 + E^2 - \left(\frac{{}_\prime\rho b - \rho{}_\prime b}{{}_\prime\rho b + \rho{}_\prime b}\right)^2 = T^2 + E^2 - \left[\frac{\tan\,(i - {}_\prime i)}{\tan\,(i + {}_\prime i)}\right]^2$$

$$\doteq \frac{4\rho{}_\prime\rho b{}_\prime b\,(N\sigma\sin^2 i)^2}{({}_\prime\rho b + \rho{}_\prime b)^2\,[u^2\,({}_\prime\rho b + \rho{}_\prime b)^2 + (N\sigma\sin^2 i)^2]}\quad\ldots(114),$$

or approximately

$$T^2 + E^2 - \left[\frac{\tan\,(i - {}_\prime i)}{\tan\,(i + {}_\prime i)}\right]^2 = \frac{4\rho{}_\prime\rho b{}_\prime b\,(N\sigma\sin^2 i)^2}{u^2({}_\prime\rho b + \rho{}_\prime b)^4}\quad\ldots(115).$$

In (114) and (115) $T^2 + E^2$ denotes the whole intensity of the reflected light, due to incident light of unit intensity vibrating in the plane of incidence, and the smallness of the right-hand members of these equations shows how little this exceeds that calculated according to Fresnel's formula

$$\left[\frac{\tan\,(i - {}_\prime i)}{\tan\,(i + {}_\prime i)}\right]^2.$$

Compare §§ 103—105 above.

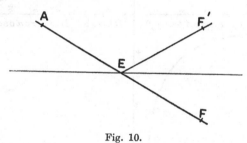

Fig. 10.

§ 178. In (110) and (111), s denotes a length AEF (fig. 10) in the line of the incident ray produced through E the point of incidence, or AEF' an equal length in the path of the incident and the reflected ray. Thus we see that the reflection causes a phasal set-back ϕ which increases from 0° through 90° to 180° when the angle of incidence is increased from 0° to 90°, as shown by (112); because (112) shows that $\tan\,\phi$ increases through positive values from 0 to $+\infty$ when i increases from 0 to a value slightly less than $\tan^{-1}\mu$: and, when i increases farther up to 90°, $\tan\,\phi$ increases through negative values from $-\infty$ to 0. These variations are illustrated by the accompanying diagrams (figs. 11, 12) drawn according to the following Table of values of $\tan\,\phi$ and ϕ calculated

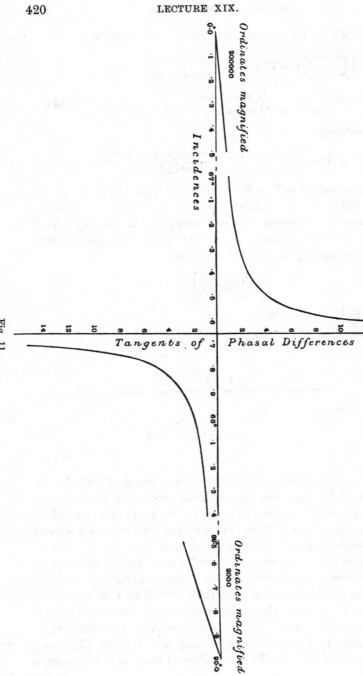

Fig. 11.

Diamond $\mu = 2\cdot434,\quad \sigma = \cdot00293$

from (112) with $\sigma = \cdot00293$ as for diamond according to § 181 Molar. below. Note how very slowly $\tan\phi$ and ϕ increase for increasing

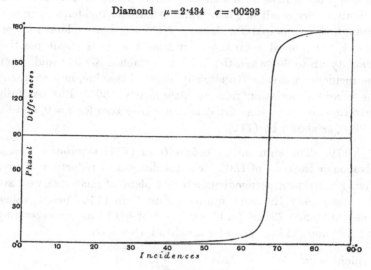

Diamond $\mu = 2\cdot434$ $\sigma = \cdot00293$

Fig. 12.

i	$\tan\phi$	ϕ	i	$\tan\phi$	ϕ
0°	·0	0	I	$\pm\infty$	90°
1°	·000008798	0° ·03′	67°·7	− 14·87770	93° 51′
10°	·0008890	0° 3′	67°·75	− 6·95857	98° 11′
20°	·003678	0° 13′	67°·8	− 4·53967	102° 25′
30°	·008835	0° 30′	67°·9	− 2·67607	110° 29′
40°	·01772	1° 1′	68°·0	− 1·89599	117° 49′
50°	·03469	1° 59″	68°·1	− 1·46724	124° 17′
60°	·08784	5° 1″	68°·2	− 1·19615	129° 54′
65°	·25164	14° 7′	68°·3	− 1·00921	134° 44′
67°	1·00386	45° 7′	68°·4	− ·872514	138° 54′
67°·1	1·18360	49° 48′	70°	− ·27083	164° 51′
67°·2	1·44216	55° 16′	75°	− ·07694	175° 36′
67°·3	1·85085	61° 37′	80°	− ·01640	179° 4′
67°·4	2·56306	68° 41′	89°	− ·00299	179° 50′
67°·5	4·20197	76° 37′	90°	·00000	180° 0′
67°·55	6·18014	80° 49′			
67°·6	11·70150	85° 7′			
$\tan^{-1}\mu - 0°\cdot00883 = I$		90° 0′			

Molar. incidences, except those very slightly less than, and very slightly greater than, the Principal Incidence, 67°·65. Note also how very suddenly tan ϕ rises from a small positive value to $+\infty$ and from $-\infty$ to a very small negative value when the incidence increases through the Principal Incidence. Note also how·suddenly therefore ϕ, the phasal retardation, increases from a small positive quantity up to 90°, when the incidence reaches 67°·65 ; and, when the incidence increases to slightly above this value, how suddenly the phasal retardation rises to very nearly 180°. But carefully bear this in mind that tan ϕ is essentially zero for $i = 0$, and for $i = 90°$, as shown by (112).

§ 179. The sign minus before G in (111) signifies a phasal advance or back-set of 180°. For incidence and reflection of rays having vibrations perpendicular to the plane of incidence, we have also essentially the sign minus before C in (116) below, taken from (20) and (23) of § 117 and (39) of § 123 as corresponding to (110) and (111). Thus for amplitudes we have

incident wave $\qquad C \cos \omega \left(t - \dfrac{s}{u} \right),$

reflected wave $\qquad -C' \dfrac{b-b}{,b+b} \cos \left[\omega \left(t - \dfrac{s}{u} \right) \right]$ $\Bigg\} \ldots\ldots\ldots\ldots(116).$

Here there is neither retardation nor acceleration corresponding to the ϕ of (111). We infer that if plane polarized light falls on the reflecting surface with its vibrational plane inclined at any angle α to the plane of incidence, and if we resólve it ideally into two components having their vibrational planes in and perpendicular to that plane, the component of the reflected light due to the former will be phasally behind that due to the latter by the retardation ϕ given by (112). Hence for every incidence the reflected light will be elliptically polarized; or circularly in case of phasal difference 90°, and equal amplitudes, of the two reflected components. The values of i and of α which make the reflected light circularly polarized, are called the "Principal Incidence" and "Principal Azimuth" (see § 97 above).

§ 180. The Principal Incidence, I, is the value of i which annuls the denominator of E/T as seen in (112). It is therefore given by the equation

$$\tan^2 I = \mu^2 - (\mu^2 - 1)^3 \sigma^2 \tan^2 I \sin^2 I \ldots\ldots\ldots(117).$$

Notwithstanding the greatness of $(\mu^2 - 1)^3$ for substance of high Molar. refractivity (Example 126·7 for Sulphuret of Arsenic) the second term of the second member of (117) is very small relatively to the first because of the smallness of $(q/u)^2$ for all transparent substances for which we have the observational data from Jamin. (See § 182 below.) Hence in that second term we may take μ^2 for $\tan^2 I$ and $\mu^2/(\mu^2 + 1)$ for $\sin^2 I$. Thus instead of (117) to determine the Principal Incidence I we have

$$\tan I \doteqdot \mu \left[1 - \frac{1}{2} \frac{\mu^2 (\mu^2 - 1)^3}{\mu^2 + 1} \sigma^2 \right] \dots\dots\dots\dots(118).$$

§ 181. To find the Principal Azimuth α; let D be the vibrational amplitude of plane polarized incident light; and $D \cos \alpha$, $D \sin \alpha$ the vibrational amplitudes of the components in and perpendicular to the plane of incidence. Thus with the notation of §§ 176, 179 we may take

$$G = D \cos \alpha, \quad C = D \sin \alpha \dots\dots\dots\dots\dots(119).$$

These must be such as to make the two components of the reflected light equal when the angle of incidence is that given by (117) (the Principal Incidence); that is to say, the value of i which makes $P = 0$; or approximately according to (109), $_{,}\rho b = \rho_{,}b$. This makes the amplitude of (111) approximately equal to $- GE$, or

$$- G \frac{Q}{4 (u_{,}\rho b)^2}, \quad \text{or} \quad - G \frac{(\mu^2 - 1)^2}{2 \sqrt{(\mu^2 + 1)}} \sigma \dots\dots\dots(120).$$

Equating this to the second line of (116) and using (119), we see that the condition for circular polarization is

$$\cos \alpha \frac{(\mu^2 - 1)^2}{2 \sqrt{(\mu^2 + 1)}} \sigma = \sin \alpha \frac{_{,}b - b}{_{,}b + b} \dots\dots\dots\dots(121).$$

Now the angle of incidence is approximately $\tan^{-1} \mu$; for which we have

$$\frac{_{,}b - b}{_{,}b + b} = \frac{\mu^2 - 1}{\mu^2 + 1} \dots\dots\dots\dots\dots(122),$$

and therefore for α, the principal azimuth, we have

$$\tan \alpha = \tfrac{1}{2} (\mu^2 - 1) \sqrt{(\mu^2 + 1)} \frac{q}{u} \dots\dots\dots\dots(123).$$

§ 182. The Table of § 105 above gives Jamin's values of tan α (his "k") for eight substances, and the result of a very rigorous examination by Rayleigh for water with specially purified surface to be substituted for Jamin's which was probably vitiated by natural impurity on the water surface. Omitting fluorine because of its exceptional negative value for tan α, which if genuine must be explained (see § 158 above) otherwise than by my assumption of $,v = \iota q$ and $v = \infty$, and omitting water because of the practical nullity of the Adamantine Property for it proved by Rayleigh, we have positive values of tan α for six substances, from which, for these substances, the following Table of values of q/u; and of $\tan^{-1} \mu - I$, the differences of the Principal Incidence from the Brewsterian angle; has been calculated according to (123) and (118). The smallness of these last mentioned differences is *very remarkable*.

Substance	$\sigma = q/u$	$\tan^{-1} \mu - I$	$\tan^{-1} \mu$
"Sulfure d'arsenic transparent" (Realgar)......	·01277	0°·08873	67°·83
"Blende transparente" (Zinc sulphide).........	·00706	0°·04289	67°·13
Diamond	·00293	0°·00883	67°·67
Flint Glass	·00936	0°·00593	59°·89
"Verre"	·00553	0°·000496	56°·08
Absolute Alcohol	·00284	0°·0000465	53°·13

§ 183. Another great fundamental province of Optics, luminous waves travelling through transparent crystals, was successfully explored by Fresnel more than eighty years ago. This he did with utterly imperfect dynamical leadings; but nevertheless he discovered what we now know to be, in every detail except his equivoluminal condition, the true laws of light-waves in a crystal.

§ 184. One notable detail of Fresnel's, which I described in my introductory lecture (Oct. 1st, 1884), was that he made the propagational velocity of light in a crystal depend on the direction of the vibration, and not on the axis of the shearing rotational

strain as the elastic solid theory has seemed to require. See Molar.
Lecture I., pages 17, 18, particularly the words "*If the effect
"depends upon the return force in an elastic solid.*" In the same
lecture (page 19) I told of an explanation of this difficulty
suggested first by Rankine and afterwards by Stokes and by
Rayleigh, to the effect that the different propagational velocities
of light in different directions through a crystal are due, not to
æolotropy of elastic action, but to æolotropy of effective inertia.
But I had also to say that Stokes working on this idea, which
had occurred to himself independently of Rankine's suggestion,
had been compelled to abandon it for the reasons stated in the
following reproduction of a short paper of twenty-one lines which
appeared in the *Philosophical Magazine* for October, 1872; and
which is all that he published on the subject:—

"It is now some years since I carried out, in the case of Iceland
" spar, the method of examination of the law of refraction which
" I described in my report on Double Refraction, published in the
" *Report* of the British Association for the year 1862, p. 272.
" A prism, approximately right-angled isosceles, was cut in such
" a direction as to admit of scrutiny, across the two acute angles,
" in directions of the wave-normal within the crystal comprising
" respectively inclinations of 90° and 45° to the axis. The directions
" of the cut faces were referred by reflection to the cleavage-
" planes, and thereby to the axis. The light observed was the
" bright D of a soda-flame.

" The result obtained was, that Huyghens' construction gives
" the true law of double refraction within the limits of errors of
" observation. The error, if any, could hardly exceed a unit in
" the fourth place of decimals of the index or reciprocal of the
" wave-velocity, the velocity in air being taken as unity. This
" result is sufficient absolutely to disprove the law resulting from
" the theory which makes double refraction depend on a difference
" of inertia in different directions.

" I intend to present to the Royal Society a detailed account
" of the observations; but in the meantime the publication of
" this preliminary notice of the result obtained may possibly be
" useful to those engaged in the theory of double refraction."

It is well that the essence of the result of this very important
experimental investigation was published: it is sad that we have

Molar. not the Author's intended communication to the Royal Society describing his work.

The corresponding experimental test for a Biaxal Crystal, resulting also in a minutely accurate verification of Fresnel's wave-surface, was carried out a few years later by Glazebrook*. Must we therefore give up all idea of explaining the different velocities of light in different directions through a crystal by æolotropic inertia? Yes, certainly, *if, as assumed by Green and Stokes, ether in a crystal is incompressible.*

§ 185. But Glazebrook has pointed out as a consequence of my suggestion of approximately zero velocity of condensational-rarefactional waves in a transparent solid or liquid, that with this assumption æolotropic inertia gives precisely Fresnel's shape of wave-surface and Fresnel's dependence of velocity on direction of vibration, irrespectively of the direction of the strain-axis. Thus after all we have a dynamical explanation of Fresnel's laws of light in a crystal which we may accept as in all probability absolutely true. To prove this let us first investigate the conditions for a plane wave in an isotropic elastic solid with any given values for its two moduluses of elasticity; k bulk-modulus and n rigidity-modulus; and with æolotropic inertia in respect to the motion of any small part of its substance. This æolotropy of effective inertia of ether through the substance of a transparent crystal follows naturally, we may almost say inevitably, from the Molecular Theory of §§ 162—168. In Lecture XX. details on which we need not at present enter will be carefully considered.

§ 186. Meantime we may simply assume that $B\rho_x$, $B\rho_y$, $B\rho_z$ are the virtual masses, or inertia-equivalents, relatively to motions parallel to x, y, z, of ether within a very small volume B containing a large number of the ponderable atoms concerned: so that

$$B\rho_x \frac{d\xi}{dt}, \quad B\rho_y \frac{d\eta}{dt}, \quad B\rho_z \frac{d\zeta}{dt} \quad \ldots\ldots\ldots\ldots\ldots(124)$$

are the forces which must act upon the ether to produce component accelerations $d\xi/dt$, etc. Hence the equations of molar

---

* "An experimental determination of the Values of the Velocities of Normal Propagation of Plane Waves in different directions in a Biaxal Crystal, and a Comparison of the Results with theory." By R. T. Glazebrook, Communicated by J. Clerk Maxwell. *Phil. Trans. Roy. Soc.*, 1879, Vol. 170.

motions of the ether will be (8) of § 113, modified by substituting Molar. for ρ in their first members ρ_x, ρ_y, ρ_z respectively.

§ 187.   To express a regular train of plane waves, take

$$\xi = \alpha f\left(t - \frac{\lambda x + \mu y + \nu z}{u}\right);$$

$$\eta = \beta f\left(t - \frac{\lambda x + \mu y + \nu z}{u}\right); \left.\begin{array}{c}\\\\\\\end{array}\right\} \dots\dots\dots(125),$$

$$\zeta = \gamma f\left(t - \frac{\lambda x + \mu y + \nu z}{u}\right)$$

where λ, μ, ν denote the direction cosines of a perpendicular to the wave-planes; and α, β, γ the direction cosines of the lines of vibration.   These, used in (7) of § 113, give

$$\delta = \frac{-\cos\vartheta}{u}\frac{d}{dt} f\left(t - \frac{\lambda x + \mu y + \nu z}{u}\right)\dots\dots\dots(126),$$

where
$$\cos\vartheta = \alpha\lambda + \beta\mu + \gamma\nu\dots\dots\dots\dots(127).$$

Thus ϑ is the inclination of the direction of the displacement, to the wave-normal.

Using (124), (125), (126) in the equations of motion, and removing from each side of each the factor

$$\frac{d^2}{dt^2} f\left(t - \frac{\lambda x + \mu y + \nu z}{u}\right)\dots\dots\dots(128),$$

we find

$$\rho_x\alpha = (k + \tfrac{1}{3}n)\frac{\lambda\cos\vartheta}{u^2} + n\frac{\alpha}{u^2};$$

$$\rho_y\beta = (k + \tfrac{1}{3}n)\frac{\mu\cos\vartheta}{u^2} + n\frac{\beta}{u^2}; \left.\begin{array}{c}\\\\\\\end{array}\right\}\dots\dots\dots(129).$$

$$\rho_z\gamma = (k + \tfrac{1}{3}n)\frac{\nu\cos\vartheta}{u^2} + n\frac{\gamma}{u^2}$$

§ 187'.   From these equations we determine the direction-cosines (α, β, γ) of the vibration; and the propagational velocity, u, of the plane wave (λ, μ, ν); thus: First solving for α, β, γ, and putting

$$\rho_x = n/a^2, \quad \rho_y = n/b^2, \quad \rho_z = n/c^2\dots\dots\dots(129'),$$

Molar. we find

$$\alpha = \frac{k + \frac{1}{3}n}{n} \frac{\lambda a^2 \cos \vartheta}{u^2 - a^2} \; ;$$

$$\beta = \frac{k + \frac{1}{3}n}{n} \frac{\mu b^2 \cos \vartheta}{u^2 - b^2} \; ; \quad \left.\rule{0pt}{60pt}\right\} \dots\dots\dots\dots(130),$$

$$\gamma = \frac{k + \frac{1}{3}n}{n} \frac{\nu c^2 \cos \vartheta}{u^2 - c^2}$$

whence by $\qquad \alpha^2 + \beta^2 + \gamma^2 = 1,$

$$\sec^2 \vartheta = \left(\frac{k + \frac{1}{3}n}{n}\right)^2 \left[\left(\frac{\lambda a^2}{u^2 - a^2}\right)^2 + \left(\frac{\mu b^2}{u^2 - b^2}\right)^2 + \left(\frac{\nu c^2}{u^2 - c^2}\right)^2\right] \dots(131).$$

Multiplying the first of (130) by λ, the second by μ, the third by ν and adding; and removing the common factor $\cos \vartheta$; we find

$$1 = \frac{k + \frac{1}{3}n}{n} \left(\frac{\lambda^2 a^2}{u^2 - a^2} + \frac{\mu^2 b^2}{u^2 - b^2} + \frac{\nu^2 c^2}{u^2 - c^2}\right) \dots\dots\dots\dots(132).$$

This is a cubic for determining u^2, the square of the propagational velocity. The three roots are all obviously real; one greater than the greatest of a^2, b^2, c^2, and the other two between the values of these quantities. For each value of u^2 the corresponding direction of vibration is given by (130).

§ 187″. Calling $u_1{}^2$, $u_2{}^2$ two of the three roots of (132) we find, by writing down the equation for each of these and subtracting one equation from the other, the following

$$0 = \frac{k + \frac{1}{3}n}{n} \left[\frac{\lambda^2 a^2}{(u_1{}^2 - a^2)(u_2{}^2 - a^2)} + \frac{\mu^2 b^2}{(u_1{}^2 - b^2)(u_2{}^2 - b^2)}\right.$$
$$\left. + \frac{\nu^2 c^2}{(u_1{}^2 - c^2)(u_2{}^2 - c^2)}\right](u_1{}^2 - u_2{}^2) \dots\dots(132').$$

And taking the corresponding notation in (130) we find

$$\alpha_1 \alpha_2 + \beta_1 \beta_2 + \gamma_1 \gamma_2 = \frac{k + \frac{1}{3}n}{n} \left[\frac{\lambda^2 a^4}{(u_1{}^2 - a^2)(u_2{}^2 - a^2)} + \frac{\mu^2 b^4}{(u_1{}^2 - b^2)(u_2{}^2 - b^2)}\right.$$
$$\left. + \frac{\nu^2 c^4}{(u_1{}^2 - c^2)(u_2{}^2 - c^2)}\right] \cos \vartheta_1 \cos \vartheta_2 \dots\dots(132'').$$

Comparing these two equations, we see that $\alpha_1 \alpha_2 + \beta_1 \beta_2 + \gamma_1 \gamma_2$ cannot generally be zero: that is to say the vibrational lines corresponding to parallel wave-planes travelling with different velocities cannot generally be perpendicular to one another. This result for the new theory, in which differences of velocity are due Molar. to inertial æolotropy, presents an interesting contrast to the theorem of Lecture XII., page 136, that the vibrational lines in any two of the three waves are mutually perpendicular; when there is no inertial æolotropy, and the differences of velocity are due to æolotropy of elasticity.

§ 188.   In each of the two extreme cases of $k = \infty$ and $k = -\frac{4}{3}n$ the cubic sinks to a quadratic, as we most readily* see by (in virtue of $\lambda^2 + \mu^2 + \nu^2 = 1$) writing (132) thus

$$\frac{\lambda^2 u^2}{u^2 - a^2} + \frac{\mu^2 u^2}{u^2 - b^2} + \frac{\nu^2 u^2}{u^2 - c^2} = \frac{k + \frac{4}{3}n}{n}\left(\frac{a^2\lambda^2}{u^2 - a^2} + \frac{b^2\mu^2}{u^2 - b^2} + \frac{c^2\nu^2}{u^2 - c^2}\right)$$
$$\dots\dots\dots(133).$$

§ 189.   From this we see that *if $k + \frac{4}{3}n$ is very great*, one root of the cubic for u^2 is very great, being given approximately by

$$u^2 \doteqdot \frac{k + \frac{4}{3}n}{n}(\lambda^2 a^2 + \mu^2 b^2 + \nu^2 c^2) \dots\dots\dots(134);$$

while the other two roots are approximately the roots of the quadratic,

$$\frac{a^2\lambda^2}{u^2 - a^2} + \frac{b^2\mu^2}{u^2 - b^2} + \frac{c^2\nu^2}{u^2 - c^2} = 0 \dots\dots\dots\dots(135).$$

This agrees with the propagational velocities for a given wave-plane (λ, μ, ν), found by Stokes and by Rayleigh from Rankine's hypothesis of æolotropic inertia, and Green's assumption of a virtually infinite resistance against compression.   It implies a wave-surface proved observationally by Stokes for Iceland spar (uniaxal), and by Glazebrook for arragonite (biaxal), to differ from the truth by far greater differences than could be accounted for by errors of observation.

§ 190.   But *if $k + \frac{4}{3}n$ is very small*, one of the three values of u^2 given by (133) is very small positive, being given approximately by

$$\left(\frac{\lambda^2}{a^2} + \frac{\mu^2}{b^2} + \frac{\nu^2}{c^2}\right)u^2 \doteqdot \frac{k + \frac{4}{3}n}{n} \dots\dots\dots\dots(136);$$

---

* Another way of managing this detail will be found in the investigation of chiral waves in § 206 of Lec. XX.

Molar. while the other two are approximately the roots of the quadratic

$$\frac{\lambda^2}{u^2 - a^2} + \frac{\mu^2}{u^2 - b^2} + \frac{\nu^2}{u^2 - c^2} = 0 \ \ldots\ldots\ldots(137).$$

This is Fresnel's equation [see (94) of Lecture XV.] for the two propagational velocities for a given direction of wave-plane (λ, μ, ν). It implies precisely Fresnel's celebrated wave-surface

$$\frac{a^2x^2}{r^2 - a^2} + \frac{b^2y^2}{r^2 - b^2} + \frac{c^2z^2}{r^2 - c^2} = 0 \ \ldots\ldots\ldots(138),$$

or $\ r^2 (a^2x^2 + b^2y^2 + c^2z^2) - a^2 (b^2 + c^2) x^2 - b^2 (c^2 + a^2) y^2 - c^2 (a^2 + b^2) z^2$

$$+ a^2b^2c^2 = 0 \ldots(138').$$

This equation is got, not now "after a very troublesome algebraic process," as said by Airy[*] in 1831, but by a very short and easy symmetrical method given by Archibald Smith in 1835[†]; the problem being to find the envelope of all the planes given by the equation

$$\lambda x + \mu y + \nu z = u$$

with $\qquad\qquad\qquad \lambda^2 + \mu^2 + \nu^2 = 1 \quad \left.\vphantom{\begin{matrix}1\\1\\1\end{matrix}}\right\} \ \ldots\ldots\ldots\ldots(138'').$

and $\qquad\qquad\qquad$ (137) for u

§ 191. The theorems of Glazebrook and Basset, stated in § 195 below, are readily proved by using (139) in connection with Archibald Smith's now well-known investigation of Fresnel's wave-surface. But the theory on which it is founded implies essentially condensation and rarefaction and therefore a direction of vibration *not in the wave-plane*; and not agreeing with Fresnel's which is *exactly in the wave-plane* (corresponding as it does to strictly equivoluminal waves). For our present case of $k + \frac{4}{3}n = 0$, the vibrational direction (α, β, γ) as given by (130) is

$$\alpha = -\frac{a^2\lambda \cos \vartheta}{u^2 - a^2}, \quad \beta = -\frac{b^2\mu \cos \vartheta}{u^2 - b^2}, \quad \gamma = -\frac{c^2\nu \cos \vartheta}{u^2 - c^2} \ \ldots\ldots(139);$$

and (131) for ϑ, the angle between the vibrational direction and the wave-normal, becomes

$$\sec^2 \vartheta = \left(\frac{a^2\lambda}{u^2 - a^2}\right)^2 + \left(\frac{b^2\mu}{u^2 - b^2}\right)^2 + \left(\frac{c^2\nu}{u^2 - c^2}\right)^2 \ \ldots\ldots\ldots(140).$$

---

[*] Airy's *Mathematical Tracts*, p. 353.

[†] *Trans. Cambridge Phil. Soc.* Vol. vi. p. 85; also *Phil. Mag.* Vol. xii. 1838 (1st half-year), p. 335.

§ 192.   For an example, take arragonite with, as in Lec. XV. Molar.
§ 37, (and $n = 1$ for simplicity)

$$\left. \begin{array}{lll} 1/a = 1\cdot5301 ; & a = \cdot65355 ; & a^2 = \cdot42713 \\ 1/b = 1\cdot6816 ; & b = \cdot59467 ; & b^2 = \cdot35363 \\ 1/c = 1\cdot6859 ; & c = \cdot59315 ; & c^2 = \cdot35183 \end{array} \right\} \dots (141);$$

and let us find the two propagational velocities (u_1, u_2) and the
inclinations $(\vartheta_1, \vartheta_2)$ of the vibrational lines to the wave-normal, for
the case in which the wave-normal is equally inclined to the three
principal axes $(\lambda = \mu = \nu = 1/\sqrt{3})$.   By the solution of the quadratic
(137) and by using the roots in (139) we find

$$\left. \begin{array}{lll} u_1^2 = \cdot40234 ; & u_1^{-1} = 1\cdot57653 ; & \vartheta_1 = 81° 37'\cdot68 \\ u_2^2 = \cdot35272 ; & u_2^{-1} = 1\cdot68377 ; & \vartheta_2 = 89° 49'\cdot24 \end{array} \right\} \dots (142).$$

Remarking that u_1^{-1} and u_2^{-1} are the indices of refraction for the
two waves of which the normals are each equally inclined to the
three principal axes (x, y, z) it is interesting to notice how nearly
the greater of them is equal to $\frac{1}{2}(1\cdot6816 + 1\cdot6859)$, the mean of the
two greater of the three principal refractive indices.   This, and the
nearness of ϑ_2 to 90°, are due to the smallness of the difference
between the two greater principal indices: that is to say, the
smallness of the difference between arragonite and a uniaxal
crystal.   In fact u_2^2, the second of our solutions of the quadratic,
corresponds to what would be the ordinary ray if the two greater
principal indices were equal.

§ 193.   To help in thoroughly understanding the condensations
and rarefactions which the theory gives us *in any plane wave
through a biaxal crystal*, take as wave-plane *any plane parallel to
one of three principal axes, OY* for instance.   It is clear without
algebra that the directions of vibration in the two waves for
every such direction of wave-plane are respectively parallel and
perpendicular to OY.   The former corresponds to the ordinary
ray: its vibrational direction is parallel to OY: it has propagational velocity $\sqrt{(n/\rho_y)}$, or b, according to (129').   It is a strictly
equivoluminal wave.   All this we verify readily in our algebra by
putting $\mu = 0$; which gives for one root of the quadratic (135)
$u^2 = b^2$, and gives by (139) $\sec^2 \vartheta = \infty$ and therefore $\vartheta = 90°$.

§ 194.   For the other root of the quadratic (135) we have

$$\frac{\lambda^2}{u^2 - a^2} + \frac{\nu^2}{u^2 - c^2} = 0 \; ;$$

whence
$$\left. \begin{aligned} u^2 &= c^2\lambda^2 + a^2\nu^2 \; ; \\ u^2 - a^2 &= (c^2 - a^2)\,\lambda^2 \; ; \\ u^2 - c^2 &= (a^2 - c^2)\,\nu^2 \end{aligned} \right\} \dots \dots\dots\dots\dots(143).$$

Using these in (139) we find

$$\frac{\gamma}{-\alpha} = \frac{c^2\lambda}{a^2\nu} \dots\dots\dots\dots\dots\dots(144).$$

Putting now $y = 0$ in (138′), the equation of Fresnel's wave-surface, we find for its intersection with the plane XOZ

$$(r^2 - b^2)\,(a^2x^2 + c^2z^2 - a^2c^2) = 0 \dots\dots\dots\dots(145).$$

This expresses, for the ordinary ray, a circle $r = b$; and for the extraordinary ray an ellipse,

$$\frac{x^2}{c^2} + \frac{z^2}{a^2} = 1 \dots\dots\dots\dots\dots\dots(146).$$

The wave-plane touches this ellipse at the point (x, z); hence

$$\frac{\lambda}{\nu} = \frac{a^2x}{c^2z} \dots\dots\dots\dots\dots\dots\dots(147).$$

Taking this for λ/ν in (144) we see that the vibrational line is perpendicular to the ray-direction, which is the radius vector of the ellipse through the point (x, z). This is only a particular case of the general theorem given by Glazebrook, *Phil. Mag.* Dec. 1888 page 528, that the direction of vibration is perpendicular to the ray in the new theory of double refraction founded on æolotropic inertia.

§ 195.   Moreover in this theory* Basset has given a most interesting theorem† to the effect that the direction of the vibration is a line drawn perpendicular to the radius vector from the foot of the perpendicular to any plane touching the wave-surface (the perpendicular to the radius vector being drawn from the centre of

---

* Basset's *Physical Optics* (Cambridge, 1892), § 265.

† Given also for a very different dynamical theory involving zero velocity of con-densational-rarefactional wave, by Sarrau in his Second Paper on the Propagation and the Polarization of Light in Crystals; *Liouville's Journal*, Vol. XIII. 1868, p. 86.

the wave-surface to the tangent-plane and to its point of contact). Molar.
This includes Glazebrook's theorem and appends to it a simpler
completion of the problem of drawing the vibrational line than
that given by Glazebrook in pages 529, 530 of the volume of the
Philosophical Magazine already referred to. The construction is
illustrated in figure 13, drawn exactly to scale for the principal
section through greatest and least principal diameters of wave-
surface for arragonite. P is the point of contact of the tangent-
plane KM. OP is the radius vector (optically the ray). F is the
foot of the perpendicular from the centre of the wave-surface.
FN, perpendicular to OP, not shown in the diagram, is the direction of the vibration. It would be interesting to construct on
a tenfold scale a portion of figure 13 around FP; but it is perhaps
more instructive to calculate the angle NFP (which is equal to
FOP) and the radius of curvature, at P, of the ellipse. Figure 13
illustrates the construction for any wave-plane whatever touching
the wave-surface in the point P; though it is drawn exactly to
scale only for the extraordinary ray in the principal section of
arragonite, through the greatest and least principal diameters.

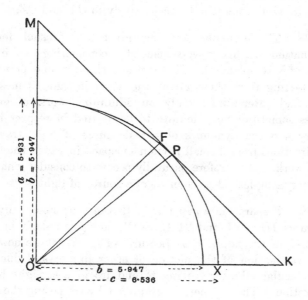

Fig. 13.

§ 196. Going back now to § 194 we see that the intersections of Fresnel's wave-surface with the three planes YOZ, ZOX, XOY, are expressed respectively by the following equations

$$\left.\begin{aligned}(r^2 - a^2)(b^2y^2 + c^2z^2 - b^2c^2) &= 0 \\ (r^2 - b^2)(c^2z^2 + a^2x^2 - c^2a^2) &= 0 \\ (r^2 - c^2)(a^2x^2 + b^2y^2 - b^2a^2) &= 0\end{aligned}\right\} \dots\dots\dots(148),$$

which prove that each intersection consists of an ellipse and a circle; the ellipse corresponding to the extraordinary ray, and the circle to the ordinary.

§ 197. Lastly consider wave-planes perpendicular to one or other of OX, OY, OZ. Take for example OX; we see that the two propagational velocities are b and c, with vibrational lines respectively parallel to OY and OZ. The physical explanation is much more easily understood than anything which we thought of in Lecture I. pages 17 to 20, when we were believing that differences of velocity were due to æolotropy of elasticity. We now see that the two waves travelling x-wards have different velocities because of greater or less effective inertias of the moving ether in its vibrations parallel respectively to OY and OZ.

§ 198. The fundamental view given by Fresnel for the determination of his wave-surface by considering an infinite number of wave-planes in all directions through one point, and waves starting from them all at the same instant, is most important and interesting; truly an admirable work of genius! It leaves something very definite to be desired in respect to the geometry and the dynamics of a real source of light travelling in all directions from a small portion of space in which the source does its work. It therefore naturally occurs to consider what may be the very simplest ideal element of a source of light.

§ 199. Preparation was made for this in the "molar" divisions of Lectures III....VI. and VIII....XIV. and particularly in pages 190 to 219 of the addition to Lecture XIV. What we now want is an investigation of the motion of ether in a crystal due to an ideal molecular vibrator moving to and fro in a straight line in any direction. The problem is simplified by supposing the direction to be one of the three lines OX, OY, OZ, of minimum, and of minimax, and of maximum effective inertia of ether in the crystal.

It seems to me that this should be found to be a practicable Molar. problem. Towards its solution we have Fresnel's wave-surface as the isophasal surface of the outward travelling disturbance or wave-motion : and we have the *direction* of the vibration in each part of the surface by the theorems of Glazebrook and Basset (§ 195 above). What remains to be found in our present problem is the amplitude of the vibration at any point of the wave-surface. One thing we see without calculation is, that at distances from the origin great in comparison with the greatest diameter of the source, the vibrational amplitude is zero at the four points in which the wave-surface is cut by the vibrational line produced in both directions from the source; when this line coincides with one of the three axes of symmetry OX, OY, OZ. The general solution of the problem is to be had by mere superposition of motions from the solutions for vibrations of the source in the three principal directions. For the present I must regretfully leave the problem hoping to be able to return to it later.

LECTURE XX.

FRIDAY, *October* 17, 5 P.M., 1884. *Written afresh*, 1903.

Molecular. §200. CONSIDERING how well Rankine's old idea of æolotropic inertia has served us for the theory of double refraction, it naturally occurs to try if we can found on it also a thorough dynamical explanation of the rotation of the plane of polarization of light in a transparent liquid, or crystal, possessing the chiral property. I prepared the way for working out this idea in a short paper communicated to the Royal Society of Edinburgh in Session 1870—71 under the title "On the Motion of Free Solids through a Liquid" which was re-published in the *Philosophical Magazine* for November 1871 as part of an article entitled "Hydrokinetic Solutions and Observations," and which constitutes the greater part of Appendix G of the present volume. The extreme difficulty of seeing how atoms or molecules embedded in (ether), an elastic solid could experience resistance to change of motion practically analogous to the quasi-inertia conferred on a solid moving through an incompressible liquid has, until a few weeks ago, prevented me from attempting to explain chiral polarization of light by æolotropic inertia. Now, the explanation is rendered easy and natural by the hypothesis explained in §§ 162—164 above and in §§ 204, 205 below and in Appendix A.

§ 201. To explain æolotropic inertia, whether chiral or not, of molecules in ether, from the rudimentary statements in Appendix A, take first the very simplest case; a diatomic molecule (A_1, A_2) consisting of two equal and similar atoms held together by powerful attraction; so as, with a single electrion in each, to constitute a rigid system when, as we shall at first suppose, the forces and motions with which we are concerned are so small that the

electrions have only negligible motion relatively to the atoms. Molecular.
This supposition will be definitely modified when we come, in
§§ 232—242, to explain chromatic dispersion on the new theory.

§ 202.   In fig. 14 the circles represent the bounding spherical
surfaces of the two atoms.   According to the details suggested for
the sake of definiteness in Appendix A and illustrated by its
diagram of stream-lines (fig. 5), the two atoms must overlap as
indicated in our present diagram fig. 14, if the stream-lines of
ether through each atom are disturbed by the presence of the
other.   Without attempting any definite solution of the extremely
difficult problem of determining stream-lines of ether through our
double atom we may be at present contented to know that the
quasi-inertias of the disturbed motion of ether within the molecule,
in the two cases in which the motion of the ether outside is parallel
to A_1A_2, and is perpendicular to A_1A_2, must be different.   It seems

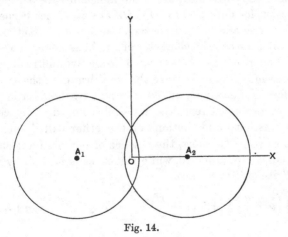

Fig. 14.

to me probable that the former must be less than the latter: I shall
only assume however that they are different, which is certainly
true: and I denote the former by α and the latter by β.   This
means that if the molecule is at rest, and the ether outside it is
moving uniformly according to velocity-components $(\dot{\xi}, \dot{\eta})$ respectively parallel to A_1A_2, and perpendicular to it in the plane of
the diagram, the kinetic energy of the whole etherial motion will
be greater by $\frac{1}{2}(\alpha\dot{\xi}^2 + \beta\dot{\eta}^2)$ than if the ether had the uniform
motion $(\dot{\xi}, \dot{\eta})$ everywhere.   Thus if B denote any volume of space

Molecular. completely surrounding one molecule but not including another, and if ρ be the undisturbed density of ether, the kinetic energy of etherial motion within B is

$$\tfrac{1}{2}\left[B\rho\,(\dot{\xi}^2 + \dot{\eta}^2) + \alpha\dot{\xi}^2 + \beta\dot{\eta}^2\right] \dots\dots\dots\dots(149).$$

Throughout §§ 202—205 we are supposing the different constituent molecules of the assemblage to occupy separate spaces. In § 225 we shall find ourselves obliged to assume overlapping complex molecules of silica in a quartz crystal.

§ 203. This gives a clear and definite explanation of the æolotropy of inertia (§ 184 above) suggested by Rankine for explaining double refraction. First for a uniaxal crystal, consider any homogeneous assemblage (Appendix H, §§ 3, 6, 15, 16, 17, 18, 19) of our diatomic molecules and let B be the volume of space allotted to each of them. The homogeneousness of the assemblage implies that the lines $A_1 A_2$ in all the molecules are parallel. Take this direction for OX; and for OY, OZ any two lines perpendicular to it and to one another. In respect to double refraction it is of no consequence what the character of the homogeneous assemblage may be: though it may be expected to be symmetrical relatively to the direction of $A_1 A_2$ because the equilibrium of the assemblage, and the forces of elasticity called into play by deformation, depend on mutual forces between the molecules not directly concerned with the elasticity and motions of the ether called into play in luminous waves. In short, the essence of our assumption is that the molecules are unmoved, while the ether is moved, by waves of light. Write (149) as follows

$$B\left[\tfrac{1}{2}\left(\rho + \frac{\alpha}{B}\right)\dot{\xi}^2 + \tfrac{1}{2}\left(\rho + \frac{\beta}{B}\right)\dot{\eta}^2\right] = B \cdot \tfrac{1}{2}\,(\rho_x\dot{\xi}^2 + \rho_y\dot{\eta}^2)\dots (150),$$

where

$$\rho_x = \rho + \frac{\alpha}{B}; \qquad \rho_y = \rho + \frac{\beta}{B}\dots\dots\dots\dots(151).$$

This shows clearly the meaning, and the physical explanation, of the ρ_x, ρ_y, ρ_z assumed in § 186 as virtual densities of ether relatively to motions in the three rectangular directions. With diatomic molecules having their axes parallel to OZ, we have essentially $\rho_y = \rho_x$. This gives the optical properties of a uniaxal crystal, which essentially present no difference between different directions perpendicular to the axis; though the homogeneous assemblage of molecules, constituting the crystal, presents es- sentially the differences in different directions corresponding to tactics of square order, or of equilateral-triangle order.

§ 204. When instead of the mere *diatomic* molecule of § 201 we have in each molecule a molecular structure which is isotropically symmetrical (square order or equilateral-triangle order) in planes perpendicular to one line OZ, we still have $\rho_y = \rho_x$, giving the axially isotropic optical properties of a uniaxal crystal : and this quite independently of any *symmetry* of the homogeneous assemblage of molecules constituting the crystal, if the virtual inertia contributed to the ether by each molecule is independent of its neighbours. If the structure of each molecule has no chirality (Appendix H, § 22, footnote) the homogeneous assemblage has no chirality. And if each molecule is geometrically and dynamically symmetrical with reference to three rectangular axes (OX, OY, OZ), but not isotropic in respect to these axes, we have generally three different values for ρ_x, ρ_y, ρ_z; and (§§ 187, 190 above) exactly Fresnel's wave-surface. This is quite independent of any *symmetry* of the homogeneous assemblage constituting the crystal : it is merely because the axes of symmetry of all the molecules are parallel in virtue of the assemblage being homogeneous. For brevity I now call a molecule which has chirality, a chiroid.

§ 205. When each molecule is a chiroid it may (Appendix G, Part 1) contribute a chiral property to the inertia of ether oscillating to and fro in the space occupied by the assemblage. To understand this chiral inertia, consider a volume B of ether, very small in all its diameters in comparison with a wave-length of light, and exactly equal in volume to the volume of space allotted (§ 201) to each molecule. Let (ξ, η, ζ) and $(\dot{\xi}, \dot{\eta}, \dot{\zeta})$ be the components of displacements and velocity of ether within B but not within the part or parts of B occupied by the atoms of the molecule. The components round x, y, z, of rotational velocity (commonly, but perhaps less conveniently, called angular velocity) of ether in space in the neighbourhood of B and not within any atom, are

$$\frac{1}{2}\left(\frac{d\dot{\zeta}}{dy} - \frac{d\dot{\eta}}{dz}\right); \quad \frac{1}{2}\left(\frac{d\dot{\xi}}{dz} - \frac{d\dot{\zeta}}{dx}\right); \quad \frac{1}{2}\left(\frac{d\dot{\eta}}{dx} - \frac{d\dot{\xi}}{dy}\right) \quad \dots\dots(152).$$

Let χ_x, χ_y, χ_z be coefficients which we may call the inertial chiralities of the molecule relative to x, y, z respectively. The

Molecular. chiral inertia with which we are concerned may be defined by
asserting that acceleration of any one of the three components
(152) of angular velocity of the ether, implies a mutual force
between the molecule and the ether, in the direction of the axis
of the rotational acceleration, which may be expressed by the
following equations* .

$$P = -\chi_x \frac{d^2}{dt^2}\left(\frac{d\zeta}{dy} - \frac{d\eta}{dz}\right); \quad Q = -\chi_y \frac{d^2}{dt^2}\left(\frac{d\xi}{dz} - \frac{d\zeta}{dx}\right); \quad R = -\chi_z \frac{d^2}{dt^2}\left(\frac{d\eta}{dx} - \frac{d\xi}{dy}\right)$$
$$\dots\dots(153);$$

where P, Q, R, denote components of force per unit of volume,
exerted by the molecules on the moving ether. Hence the
x, y, z-components of the elastic force on ether per unit of its
volume acting against inertial reaction are equal to

$$\rho_x \frac{d^2\xi}{dt^2} - P, \quad \rho_y \frac{d^2\eta}{dt^2} - Q, \quad \rho_z \frac{d^2\zeta}{dt^2} - R\dots\dots(154).$$

The equations of motion of ether occupying the same space as
our homogeneous assemblage of molecules will be found by substituting (154) for the first members of (8) in § 113. Thus with (153)
we find

$$\rho_x \frac{d^2\xi}{dt^2} - \chi_x \frac{d^2}{dt^2}\left(\frac{d\zeta}{dy} - \frac{d\eta}{dz}\right) = (k + \tfrac{1}{3}n)\frac{d}{dx}\left(\frac{d\xi}{dx} + \frac{d\eta}{dy} + \frac{d\zeta}{dz}\right) + n\nabla^2\xi$$
$$\dots\dots(155),$$

and the symmetricals in relation to y and z; three equations to
determine the three unknowns ξ, η, ζ.

§ 206. The method of treatment by an arbitrary function, as
in §§ 115, 120, 121, 127, 130, would be interesting; but because of
the triple differentiations in the chiral terms it is not convenient.
All that it can give is in reality, in virtue of Fourier's theorems,

---

* The negative sign is prefixed to χ in these equations in order to make χ
positive for a medium in which right-handed circularly polarized light travels faster
than left-handed (see § 215 below). Such a medium is by all writers on the subject
called a right-handed medium, because the vibrational line of plane polarized light
travelling through it turns clockwise as seen by a person testing it with a Nicol's
prism next his eye. The molecule which, according to our inertial theory, produces
this result is analogous to a left-handed screw-propeller in water and is therefore
properly to be called left-handed. Thus left-handed molecules produce an optically
right-handed medium.

included in the more convenient method of periodic functions; of Molar. which the most convenient form for our present problem is given by the use of imaginaries with the assumption ξ, η, ζ equal to constants multiplied into

$$\epsilon^{\iota\omega\left(t-\frac{\lambda x+\mu y+\nu z}{u}\right)}\ \dots\dots\dots\dots\dots\dots(156),$$

where λ, μ, ν denote the direction-cosines of a perpendicular to the wave-plane, and u the propagational velocity. This gives

$$\frac{d}{dx}=-\frac{\iota\omega\lambda}{u};\quad \frac{d}{dy}=-\frac{\iota\omega\mu}{u};\quad \frac{d}{dz}=-\frac{\iota\omega\nu}{u};\quad \nabla^2=-\frac{\omega^2}{u^2};\quad \frac{d^2}{dt^2}=-\omega^2$$

$$\dots\dots\dots(157),$$

which reduces (155) and its symmetricals to

$$\rho_x\omega^2\xi+\iota\chi_x\,\frac{\omega^3}{u}\,(\nu\eta-\mu\zeta)=(k+\tfrac{1}{3}n)\,\frac{\omega^2\lambda}{u^2}\,(\lambda\xi+\mu\eta+\nu\zeta)+n\,\frac{\omega^2}{u^2}\,\xi$$

$$\rho_y\omega^2\eta+\iota\chi_y\,\frac{\omega^3}{u}\,(\lambda\zeta-\nu\xi)=(k+\tfrac{1}{3}n)\,\frac{\omega^2\mu}{u^2}\,(\lambda\xi+\mu\eta+\nu\zeta)+n\,\frac{\omega^2}{u^2}\,\eta$$

$$\rho_z\omega^2\zeta+\iota\chi_z\,\frac{\omega^3}{u}\,(\mu\xi-\lambda\eta)=(k+\tfrac{1}{3}n)\,\frac{\omega^2\nu}{u^2}\,(\lambda\xi+\mu\eta+\nu\zeta)+n\,\frac{\omega^2}{u^2}\,\zeta$$

$$\dots\dots(158)^*.$$

Multiplying each member of these equations by $u^2\omega^{-2}$ and arranging in order of ξ, η, ζ, we find

$$(A-k'\lambda^2)\,\xi+(\iota\chi_x u\omega\nu-k'\lambda\mu)\,\eta+(-\iota\chi_x u\omega\mu-k'\nu\lambda)\,\zeta=0$$

$$(-\iota\chi_y u\omega\nu-k'\lambda\mu)\,\xi+(B-k'\mu^2)\,\eta+(+\iota\chi_y u\omega\lambda-k'\mu\nu)\,\zeta=0\ \}\dots(159),$$

$$(+\iota\chi_z u\omega\mu-k'\nu\lambda)\,\xi+(-\iota\chi_z u\omega\lambda-k'\mu\nu)\,\eta+(C-k'\nu^2)\,\zeta=0$$

where

$$A=\rho_x u^2-n;\quad B=\rho_y u^2-n;\quad C=\rho_z u^2-n;\quad k'=(k+\tfrac{1}{3}n)\dots(160).$$

Forming the determinant for elimination of the ratios ξ, η, ζ; and simplifying as much as possible, with reductions involving

$$\lambda^2+\mu^2+\nu^2=1\ \dots\dots\dots\dots\dots\dots(161);$$

and putting

$$E=(\chi_z-\chi_y)\,A+(\chi_x-\chi_z)\,B+(\chi_y-\chi_x)\,C$$
$$=u^2\,[(\chi_z-\chi_y)\,\rho_x+(\chi_x-\chi_z)\,\rho_y+(\chi_y-\chi_x)\,\rho_z]$$
$$=u^2\,[\chi_x\,(\rho_y-\rho_z)+\chi_y\,(\rho_z-\rho_x)+\chi_z\,(\rho_x-\rho_y)]\ \dots(162),$$

---

* These formulas, implying as they do that the chirality expressed by them is due, not to chirality of elasticity but to chirality of virtual inertia, are given by Boussinesq on p. 456 of Vol. ii. (1903) of his *Théorie Analytique de la Chaleur*.

Molar. we find

$$ABC\left[1-k'\left(\frac{\lambda^2}{A}+\frac{\mu^2}{B}+\frac{\nu^2}{C}\right)\right]$$

$$-u^2\omega^2\left[\chi_y\chi_z\lambda^2(A-k')+\chi_z\chi_x\mu^2(B-k')+\chi_x\chi_y\nu^2(C-k')\right]$$

$$+\iota uk'E\omega\lambda\mu\nu=0 \dots\dots\dots\dots\dots\dots\dots\dots\dots (163).$$

In (163) we have an equation of the sixth degree for the determination of u, the propagational velocity of waves whose wave-normal has λ, μ, ν for its direction-cosines.

This equation is greatly simplified by our assumption of practically zero propagational velocity of condensational-rarefactional waves; which makes $k'=-n$, and therefore by (160), and (129')

$$-k'\frac{\lambda^2}{A}=\frac{a^2\lambda^2}{u^2-a^2}=\frac{u^2\lambda^2}{u^2-a^2}-\lambda^2.$$

This, with corresponding formulas for μ^2/B and ν^2/C, gives

$$1-k'\left(\frac{\lambda^2}{A}+\frac{\mu^2}{B}+\frac{\nu^2}{C}\right)=u^2\left(\frac{\lambda^2}{u^2-a^2}+\frac{\mu^2}{u^2-b^2}+\frac{\nu^2}{u^2-c^2}\right).$$

By $k'=-n$, we also have

$$A-k'=\rho_xu^2;\quad B-k'=\rho_yu^2;\quad C-k'=\rho_zu^2.$$

Thus (163) is reduced to

$$u^2\left[\rho_x\rho_y\rho_z(u^2-a^2)(u^2-b^2)(u^2-c^2)\left(\frac{\lambda^2}{u^2-a^2}+\frac{\mu^2}{u^2-b^2}+\frac{\nu^2}{u^2-c^2}\right)\right.$$

$$\left.-\omega^2u^2(\chi_y\chi_z\rho_x\lambda^2+\chi_z\chi_x\rho_y\mu^2+\chi_x\chi_y\rho_z\nu^2)\right]-\iota unE\omega\lambda\mu\nu=0\dots(163').$$

§ 207. The imaginary term in (163), unless it vanishes, makes every one of the six values of u imaginary. The realisation of the corresponding result according to the principles of §§ 150, 151, above will be very interesting. It essentially demands an extension of the dynamical theory to include the conversion of the energy of wave-motion into thermal energy,—energy of irregular intermolecular motions; that is to say the dynamics of the absorption of luminous waves travelling through an assemblage of atoms or of groups of atoms.

§ 208. Remark first that when E does not vanish, the imaginary term in (163) vanishes if, and only if, one of the three direction-cosines λ, μ, ν, vanishes; that is to say if the wave-plane Molar. is parallel to one of the three principal axes. It is certainly a curious result that plane waves can be propagated without absorption if their plane is parallel to any one of the principal axes; while there is absorption for all waves not fulfilling this condition: in other words that the crystal should be perfectly transparent for all rays perpendicular to a principal axis, and somewhat absorptive for all rays not perpendicular to a principal axis. No such crystal is known in Nature or in chemical art. Time forbids us to go farther at present into this most interesting subject.

§ 209. Considering now cases in which E vanishes and therefore (163) has all its terms real, we see that in all such cases it becomes a cubic in u^2. A first and simplest case of this condition is indicated by the third member of (162), which shows that E vanishes when $\rho_x = \rho_y = \rho_z$. For this case (163) becomes

$$(\rho u^2 - n - k')[(\rho u^2 - n)^2 - u^2\omega^2(\chi_y\chi_z\lambda^2 + \chi_z\chi_x\mu^2 + \chi_x\chi_y\nu^2)] = 0 \left.\right\} (164).$$

where $\qquad \rho = \rho_x = \rho_y = \rho_z$

The three roots of this cubic are given, one of them by equating the first factor to zero, and the two others by the quadratic in u^2 obtained by equating the second factor to zero. No crystals are known to present the optical properties thus indicated for unequal values of χ_x, χ_y, χ_z. But it is conceivable that these properties may be found in some crystals of the cubic class, which might conceivably, while isotropic in respect to ordinary refraction, give different degrees of rotation of the plane of polarization of polarized rays travelling in different directions relatively to the three rectangular lines of geometrical symmetry.

§ 210. For the case of $\chi_x = \chi_y = \chi_z$ the quadratic of (164) becomes $(\rho u^2 - n)^2 = \omega^2\chi^2 u^3$ whence

$$\rho u^2 - n = \pm\ \omega\chi u \dots\dots\dots\dots\dots(165).$$

In all known cases, I think we may safely say in all conceivable cases, whether of chiral crystals, or of chiral liquids, $\omega\chi u$ is very small in comparison with n; in other words (§ 213 below) the difference between the propagational velocities of right-handed and left-handed circularly polarized light is exceedingly small in comparison with the mean of the two velocities. Hence we lose practically nothing of accuracy by taking $u = \sqrt{(n/\rho)}$ in the

Molar. second member of (165). Thus if we denote by u_1, u_2 the two positive values of u given by (165) and by u a mean of the two, we find by (165)

$$\frac{u_1{}^2 - u_2{}^2}{u^2} \fallingdotseq 2\,\frac{\omega\chi u}{n}; \quad \text{and} \quad \frac{u_1 - u_2}{u} \fallingdotseq \frac{\omega\chi u}{n} = \frac{\omega\chi}{\rho u}\ldots\ldots(166).$$

And, in accordance with (182), putting

$$\frac{\omega\chi}{\rho} = g\ldots\ldots\ldots\ldots\ldots\ldots\ldots\ldots(166'),$$

we have

$$u_1 - u_2 \fallingdotseq g\ldots\ldots\ldots\ldots\ldots\ldots\ldots(166'').$$

§ 211. To interpret this result go back to (158) simplified by making $\rho_x = \rho_y = \rho_z$ and $\chi_x = \chi_y = \chi_z$. The medium being isotropic in respect to all directions of the wave-plane, we lose nothing of generality by putting $\lambda = 1$, $\mu = 0$, $\nu = 0$. With these simplifications, the first of the three equations (158) gives $\xi = 0$ and the second and third multiplied by u^2/ω^2 become

$$(\rho u^2 - n)\,\eta = -\iota\chi\omega u\zeta; \quad (\rho u^2 - n)\,\zeta = \iota\chi\omega u\eta\ldots\ldots(167).$$

Equating the product of the first members to the product of the second members of these equations we find $(\rho u^2 - n)^2 = \chi^2\omega^2 u^2$, which verifies the determinantal quadratic as given at the beginning of § 210.

Using (165) to eliminate $(\rho u^2 - n)$ from (167) we find

$$\zeta = \pm\,\iota\eta\ldots\ldots\ldots\ldots\ldots\ldots\ldots(168).$$

Hence as an imaginary solution according to (156) of § 206 we have for ξ, η, ζ respectively

$$0; \quad C\epsilon^{\iota\omega\left(t-\frac{x}{u}\right)}; \quad \pm\,\iota C\epsilon^{\iota\omega\left(t-\frac{x}{u}\right)}\ldots\ldots\ldots(169).$$

Changing the sign of ι gives another imaginary solution; and taking the half-sum of the two imaginary solutions, we find as a real solution

$$\xi = 0; \quad \eta = C\cos\omega\left(t-\frac{x}{u}\right); \quad \zeta = \mp\,C\sin\omega\left(t-\frac{x}{u}\right)\ldots(170);$$

where

$$u \fallingdotseq \sqrt{\frac{n}{\rho} \pm \frac{\chi\omega}{2\rho}} \fallingdotseq \sqrt{\frac{n}{\rho}\left(1 \pm \frac{\chi\omega u}{2n}\right)}\ldots\ldots(171).$$

§ 212. The interpretation of this is, sinusoidal vibrations of equal amplitudes, parallel respectively to OY and OZ; the former being a quarter of a period behind or before the latter, according Molar. as we choose the upper or the lower of the two signs in each formula. The resultant (y, z) motion of the ether*, in the case

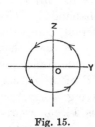

Fig. 15.

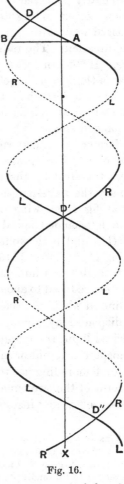

Fig. 16.

* This (y, z) component is irrotational. The (x, y) component of the etherial motion is rotational having OZ as axis; and the (x, z) component is rotational having OY as axis. We are not concerned with this view at present. But we meet it necessarily when we think out the geometry of § 205 (152); and it brings to us a very simple synthetic investigation of the velocity of circularly polarized light in a chiral liquid or in an isotropic chiral solid.

Molar. represented by the lower signs, is in circular orbits in the direction shown as anti-clockwise in the annexed diagram (fig. 15). This motion is the same in phase for all points of the ether in every plane perpendicular to OX, that is to say every wave-plane; and it varies from wave-plane to wave-plane so as to constitute a wave of circularly polarized light travelling x-wise with velocity u. The motion corresponding to the upper signs is another circularly polarized wave with opposite orbital motion, that is to say clockwise direction. The radius of the orbit is the same, C, in the two cases. If* when looking to fig. 15, we take the positive OX as towards the eye, a line of particles of the ether which is parallel to OX when undisturbed, becomes, in the wave of clockwise orbits, a right-handed spiral; and in the wave of anti-clockwise orbits a left-handed spiral. Thus the two waves are of opposite chiralities: the former are called right-handed, the latter left-handed. The steps of the two spirals (screws) are slightly different, being the spaces travelled by the two waves in their common period; that is to say the wave-lengths of the two waves. To understand this look at fig. 16 showing a right-handed spiral RRR of step 6 cm. and a left-handed spiral LLL of step 5 cm.; having a common axis OX, and both wound on a cylinder of radius $1\frac{1}{2}$ cm. representing C of (170). If the radius of this cylinder were reduced to a thousandth or a millionth of the step of either screw, and if the step were reduced to about 4.10^{-5} cm., either spiral might represent the line in which particles of ether lying in OX when undisturbed are displaced by homogeneous yellow or yellow-green circularly polarized light, travelling in the direction of OX positive, through a transparent liquid or solid of refractive index of about 1.5. With this understanding as to the scale of the diagram consider the resultant of the two equal coexisting displacements of wave-planes represented at any instant by the two spirals. At points of inter-

---

* This is a convention which I have uniformly followed for sixty years in respect to the positive and negative directions in three mutually perpendicular lines OX, OY, OZ. It makes the positive direction for angular velocities be from OX to OY, from OY to OZ, and from OZ to OX. This agrees with the ordinary conventions of English as well as foreign books on trigonometry, and geometry of two dimensions, which make anti-clockwise rotation be positive. It is convenient for inhabitants of the Northern Hemisphere as it makes positive the anti-clockwise orbital and rotational motions of the sun and planets as viewed from space above our North Pole, which we may call the north side of the mean plane of the planetary motions.

section of the two spirals, for example D, D', D'', the resultant Molar. displacement of each of the corresponding wave-planes is $2C$, the sum of the two equal components. The resultant displacement of the wave-plane through AB is twice the distance from OX of the middle point of AB, that is to say $2\sqrt{(C^2 - \frac{1}{4}AB^2)}$.

§ 213. Suppose now the two spirals to rotate in opposite directions, with equal angular velocity ω, round the axis OX; the right-handed spiral clockwise and the left-handed anti-clockwise when viewed from X towards O. This may be realised in an instructive model having one spiral wound on a brass or wooden cylinder, and the other on a glass tube fitting easily around it. The two moving spirals will represent the motions of the ether in two circularly polarized waves travelling x-wards with velocities

$$u_1 = \frac{\omega\lambda_1}{2\pi}, \quad u_2 = \frac{\omega\lambda_2}{2\pi} \dots\dots\dots (172),$$

λ_1, λ_2 denoting the steps of the two screws. The motion of the ether in any fixed plane perpendicular to OX, the plane through AB for instance in the diagram, will be found by the geometrical construction of § 212. Thus we see that a very short time after the time of the configuration shown in the diagram, the point D will come to the plane through AB, and the points A, B will come together at D: immediately after this they will separate, A leftwards in the diagram, B rightwards. Considering the whole movement we see that in any one fixed plane through OX, a wave-plane of the ether moves to and fro in a fixed straight line. The point D will travel x-wards with a velocity equal to

$$\frac{\omega/2\pi}{\frac{1}{2}\left(\frac{1}{\lambda_1}+\frac{1}{\lambda_2}\right)} = \frac{1}{\frac{1}{2}\left(\frac{1}{u_1}+\frac{1}{u_2}\right)} \dots\dots\dots (173),$$

being the harmonic mean of the velocities of the two compounded circular waves; and will revolve slowly clockwise round the cylinder of radius C at a rate, in radians per unit of distance travelled x-wise, equal to

$$\pi\left(\frac{1}{\lambda_2}-\frac{1}{\lambda_1}\right) = \frac{\omega}{2}\left(\frac{1}{u_2}-\frac{1}{u_1}\right) \dots\dots\dots (174).$$

Compare with (179) and (180) below.

§ 214.   The short algebraic expression and proof of all this is found by taking the two solutions represented by (170), (171) and writing down their sum with modification as follows :—

$$\frac{\eta}{C} = \cos \omega \left(t - \frac{x}{u_1} \right) + \cos \omega \left(t - \frac{x}{u_2} \right)$$

$$= 2 \cos \omega \left(t - \frac{x}{2u_2} - \frac{x}{2u_1} \right) \cos \frac{\omega}{2} \left(\frac{x}{u_2} - \frac{x}{u_1} \right) \quad \ldots \ldots (175);$$

$$\frac{\zeta}{C} = - \sin \omega \left(t - \frac{x}{u_1} \right) + \sin \omega \left(t - \frac{x}{u_2} \right)$$

$$= - 2 \cos \omega \left(t - \frac{x}{2u_2} - \frac{x}{2u_1} \right) \sin \frac{\omega}{2} \left(\frac{x}{u_2} - \frac{x}{u_1} \right) \ldots \ldots (176).$$

For magnitude and direction of the resultant of these we find

$$\sqrt{(\eta^2 + \zeta^2)} = 2C \cos \omega \left(t - \frac{x}{2u_2} - \frac{x}{2u_1} \right) \quad \ldots \ldots (177),$$

and

$$\frac{\zeta}{\eta} = - \tan \frac{\omega}{2} \left(\frac{x}{u_2} - \frac{x}{u_1} \right) \ldots \ldots \ldots \ldots \ldots (178).$$

These equations express rectilinear vibration through a total range $4C$ in period $\dfrac{2\pi}{\omega}$, in the line whose azimuth is

$$- \frac{\omega}{2} \left(\frac{x}{u_2} - \frac{x}{u_1} \right) \ldots \ldots \ldots \ldots \ldots \ldots (179).$$

This shows that the vibrational line in plane polarized light revolves clockwise in the wave-plane at a rate r, in turns per unit of space travelled, expressed by

$$r = \frac{\omega}{4\pi} \left(\frac{1}{u_2} - \frac{1}{u_1} \right) = \frac{\omega}{4\pi} \frac{u_1 - u_2}{u_1 u_2} = \tfrac{1}{2} \frac{u_1 - u_2}{\tau a^2} \ldots \ldots (180).$$

In the last member τ denotes the period of the vibrations, and a denotes $\sqrt{u_1 u_2}$, or the "geometric mean" of the propagational velocities, of the two waves.

§ 215.   Returning now to figure 16 we see that, as virtually said in §§ 212, 213 ;

I.   The orbital motions in right-handed circularly polarized light coming towards the eye are clockwise, and in left-handed anti-clockwise.

II. The vibrational line of plane polarized light traversing Molar. a chiral medium towards the eye, turns clockwise when the velocity of right-handed circularly polarized light is greater, and anti-clockwise when it is less, than that of left-handed circularly polarized light.

III. A chiral medium is called optically right-handed or left-handed according as the propagational velocity of right-handed or of left-handed circularly polarized light travelling through it is the greater.

All this (§§ 212—215) is Fresnel, pure and simple. It is his kinematics of the optical right-handed and left-handed chirality discovered by Arago and Biot in quartz, and by Biot in turpentine and in a vast number of other liquids.

§ 216. For the dynamical explanation take the third member of (180); and look back to (152) and (153) which show that, according to our notation, twice the acceleration of angular velocity of moving ether multiplied by χ is, when χ is negative, a force per unit bulk of the ether in the positive direction in the axis of the angular velocity, exerted on the ether by the fixed chiral molecules. If χ is positive, the chiral molecule is, as said in the footnote on § 205, analogous to a left-handed screw and its chirality is properly to be called left-handed : because (§ 205 above) if the ether concerned is viewed in the direction of the axis of the angular velocity, the force of left-handed molecule on ether is toward the eye (or positive) when the acceleration of the angular velocity of the ether is clockwise (or negative according to my convention regarding direction of rotation, stated in the footnote on § 212). With this statement the simple synthetic investigation of the velocity of circularly polarized light in a chiral medium, indicated in the first footnote of § 212, is almost completed; and by completing it we easily arrive, by a short cut, at the solution expressed in the third member of (180), for an isotropic medium; without the more comprehensive analytical investigation of §§ 205 to 214.

By (166) we saw that the left-handed molecule (χ positive) gives the greater propagational velocity (u_1) for right-handed circularly polarized light, and therefore (§ 215, III.) the medium is optically right-handed. Thus our conventions necessarily result in left-handed molecules making a right-handed medium and right-handed molecules a left-handed medium.

Molar.   § 217. Going back now to §§ 206, 207, 208, for light traversing
a chiral medium with three rectangular axes of symmetry corresponding to maximum, minimax and minimum wave velocities,
let us work out the solution for wave-plane perpendicular to one
of the principal axes, OZ for example. This makes $\lambda = 0$, $\mu = 0$,
$\nu = 1$; and, with (160), reduces the determinantal equation (162)
to

$$(C - k')\,[(\rho_x u^2 - n)\,(\rho_y u^2 - n) - u^2 \omega^2 \chi_x \chi_y] = 0 \ldots (181).$$

Hence, removing the first factor (which, when $k' = -n$, gives $u = 0$,
for the condensational-rarefactional wave) and putting

$$\frac{n}{\rho_x} = a^2; \quad \frac{n}{\rho_y} = b^2; \quad \frac{\omega^2 \chi_x \chi_y}{\rho_x \rho_y} = g^2 \ldots\ldots\ldots\ldots (182),$$

we find

$$(u^2 - a^2)\,(u^2 - b^2) = g^2 u^2 \ \ldots\ldots\ldots\ldots (183);$$

a quadratic of which the two roots are the squares of the velocities
of the z-ward waves. The third of equations (158) makes $z = 0$
for each of these waves; and therefore they are both exactly
equivoluminal. When $\chi_x \chi_y$ is positive the two roots of the quadratic are both positive; one of them $> a^2$, the other $< b^2$ if $a^2 > b^2$.

§ 218. In these formulas a and b denote the velocities of
light having its vibrational lines parallel to OX and OY respectively; and g is a comparatively very small velocity measuring
the chiral quality of the crystal. Judging by all that has been
hitherto discovered from observation of chiro-optic properties of
gases, liquids and solids, we may feel sure that g is in every case
exceedingly small in comparison with either a or b: it is about
one twenty-thousandth in quartz: and in cinnabar, which has the
greatest optic chirality hitherto recorded for any liquid or solid so
far as I know, g seems to be about fifteen times as great as in
quartz. Suppose now that g^2 is very small in comparison with
the difference between a^2 and b^2, we see, by the form of (183), that
the two values of u^2 given by the quadratic must be to a first
approximation equal to a^2 and b^2 respectively: hence to a second
approximation we have (taking forms convenient for the case of
$a^2 > b^2$),

$$u_1{}^2 - a^2 \fallingdotseq \frac{g^2 a^2}{a^2 - b^2}; \quad b^2 - u_2{}^2 \fallingdotseq \frac{g^2 b^2}{a^2 - b^2} \ldots\ldots\ldots (184).$$

§ 219. To find the character of the two waves corresponding
to the two roots of (183), put $\lambda = 0$, $\mu = 0$, $\nu = 1$ in the first and

second of (158), each multiplied by $u^2\omega^{-2}$; which, with (182), Molar. reduces them to

$$(u^2 - a^2)\, \xi = \frac{-\iota\chi_x\omega u}{\rho_x}\, \eta\,; \quad (u^2 - b^2)\, \eta = \frac{\iota\chi_y\omega u}{\rho_y}\, \xi \dots(185).$$

Equating the product of the first members to product of the second members of these equations we verify the determinantal quadratic (183). With either value of u^2 given by (183), either one or other of equations (185) may be used to complete the solution. The first of (185), however, is the more convenient for the root approximately equal to b^2; and the second for the root approximately equal to a^2. Taking them accordingly and realising in the usual manner, we find as follows for two completed independent solutions with arbitrary constants C_1, C_2:

$$u_1 > a\,; \quad \xi = C_1 \cos\omega \left(t - \frac{x}{u_1}\right); \quad \eta = -\, C_1 \frac{u_1\chi_y\omega/\rho_y}{u_1{}^2 - b^2} \sin\omega \left(t - \frac{x}{u_1}\right)$$

$$u_2 < b\,; \quad \xi = -\, C_2 \frac{u_2\chi_x\omega/\rho_x}{a^2 - u_2{}^2} \sin\omega \left(t - \frac{z}{u_2}\right); \quad \eta = C_2 \cos\omega \left(t - \frac{z}{u_2}\right)$$

$$\dots\dots\dots(186).$$

§ 220. The exceeding smallness of g/a^* has rendered fruitless all attempts hitherto made, so far as I know, to discover optic chirality in a " biaxal crystal," that is to say a crystal having three principal axes at right angles to one another of minimum, minimax and maximum wave-velocities. If it were discoverable at all, it certainly would be perceptible in light travelling along one of these three principal axes.

[*Dec.* 19, 1903. I have only to-day seen in *Phil. Mag.*, Oct. 1901 that Dr H. C. Pocklington has found rotations, per cm., 22° anti-clockwise, and 64° clockwise, of the vibrational line in polarized light travelling through crystallized sugar along two " axes," of which the first is nearly perpendicular to the cleavage plane : *a most interesting and important discovery.*]

§ 221. For light travelling along the axis, OZ, of a uniaxal crystal, take $a = b$ in (183)—(186), in this case, and we have by (183)

$$u^2 - a^2 = \pm\, gu \dots\dots\dots\dots(187).$$

---

* For sodium-light in quartz it is $\cdot 4605 . 10^{-4}$; and it may be expected to be correspondingly small in biaxal crystals.

Molar.   And, taking $\rho_y = \rho_x$, and $\chi_y = \chi_x$, we have, by (182),

$$g = \frac{\omega \chi_x}{\rho_x} \dots\dots\dots\dots\dots\dots(188).$$

With this notation and with (187), (186) becomes

$$u_1 > a \; ; \; \left. \begin{aligned} \xi &= \; C_1 \cos \omega \left(t - \frac{x}{u_1} \right) ; \; \eta = - \, C_1 \sin \omega \left(t - \frac{x}{u_1} \right) \\ u_2 < a \; ; \; \xi &= - C_2 \sin \omega \left(t - \frac{z}{u_2} \right) ; \; \eta = \; C_2 \cos \omega \left(t - \frac{z}{u_2} \right) \end{aligned} \right\} \dots (188').$$

Thus we see that for light travelling along the axis of a uniaxal crystal the velocities of right-handed and of left-handed circularly polarized light, and the rotation of the vibrational line of plane polarized light, are in every detail the same as we found in §§ 211, 212 for a medium wholly isotropic. But we shall see presently (§ 226 below) that two or three degrees of deviation from the axis produces a great change from the phenomena of a chiral isotropic medium; and that for rays inclined 30° or more up to 90°, the chirality is almost wholly swamped by the æolotropy, even when the æolotropy is as small as it is for quartz (§ 223 below). As remarked in § 220, there is no direction of light in a crystal, having three unequal values for a, b, c, in which the chiro-optic effect is not masked by the æolotropy; and therefore it is not so interesting to work out for a biaxal crystal the realised details of the solution (159), (162), (163). But it is exceedingly interesting to work them out for a uniaxal crystal, because of the great exaltation of the chiral phenomena when the ray is nearly in the direction of the axis, and because of the beautiful phenomena of Airy's spirals due to this exaltation; and because of the admirable experimental investigation of the wave-surface in quartz crystal by McConnel referred to in § 228 below.

§ 222. For waves transmitted in any direction through a uniaxal crystal; choose OZ as the axis, and therefore let $\rho_x = \rho_y$, and $\chi_x = \chi_y$, in (158)—(163) of § 206. This reduces (163) to $E = 0$ and therefore makes the waves in all directions real: that is to say transmissible with no change of motional configuration. Corresponding to (182) we may now put

$$\frac{n}{\rho_x} = \frac{n}{\rho_y} = a^2 ; \; \frac{n}{\rho_z} = c^2 ; \; \frac{\omega \chi_x}{\rho_x} = g ; \; \frac{\omega \chi_z}{\rho_z} = h \dots\dots(189).$$

Without loss of generality, we may simplify (158)—(162) by Molar. taking

$$\mu^2 = 0; \quad \lambda^2 = \sin^2 \theta; \quad \nu^2 = \cos^2 \theta \;\; \dots\dots\dots(190);$$

with these, (163') of § 206 gives

$$(u^2 - a^2)\,[u^2 - a^2 + (a^2 - c^2)\sin^2\theta] = g\,[g - (g-h)\sin^2\theta]\,u^2 \dots(191).$$

This is a quadratic equation for the determination of the squares of the velocities of the two plane waves whose wave-normals are inclined at the same angle θ to OZ. For $\theta = 90°$ we fall back on the case of §§ 217, 218, 219, but with c instead of b; and for $\theta = 0°$ we fall back directly on § 221. The exceedingly interesting transition from the subject of § 221 to our present subject is fully represented by the solution of the quadratic equation for u^2: and is best explained by tables of values of the two roots from $\theta = 0°$ to $\theta = 90°$ for some particular case or cases; or graphic representation by curves as in figs. 17, 18, 19. I have chosen the case of quartz crystal traversed by sodium-light; for which observation shows the rotation of the vibrational line to be 217° per centimetre of space travelled along the axis. This makes $g = 4·605 . 10^{-5} . a$; as we find by putting in the first member of (180), $r = 217/360$: and in the last member $u_1 - u_2 = g$; and $a = ·647593$, the reciprocal of 1·54418, the smallest refractive index of quartz at 18°, according to Rudberg; and $r = ·58932$, the period in decimal of a michron, of the mean of sodium-lights $D_1 D_2$; the michron being the unit of time which makes the velocity of light unity when the unit of space is the michron (or millionth of a metre). (See footnote on p. 150 above.)

As two sub-cases I have chosen $h = 0$ and $h = -g$; because for all that our theory tells us, (§ 223), h/g might be negative; or might be zero or might have any positive value less than, or equal to, or greater than, unity. McConnel's experimental investigation seems (§ 228 below) to make it certain that h/g for quartz is less than unity and probable that it is negative, and as small as -1, or perhaps smaller. As for g^2, (189) shows that its value is inversely proportional to the square of the period of the light, if χ_x is the same for light of all periods; g is positive or negative according as the crystal is optically right-handed or left-handed.

§ 223. c/a is the ratio of the smallest to the greatest refractive index of quartz; that is 1·54418/1·55328, according to the

Molar. figures used by McConnel; being (as I see in Landolt and Börnstein's Tables) Rudberg's results for temperature 18° and sodium-light. From this we have $(c/a)^2 = \cdot 98832$ and (191) becomes

$$\left.\begin{array}{c}\left(\dfrac{u^2}{a^2}-1\right)\left(\dfrac{u^2}{a^2}-w^2\right)=q\,\dfrac{u^2}{a^2}\\[2mm]\dfrac{u^4}{a^4}-(1+w^2+q)\,\dfrac{u^2}{a^2}=-w^2\end{array}\right\}\dots\dots\dots\dots(192),$$

or

where

$$1-w^2 = \cdot 01168\,\sin^2\theta = \frac{\cdot 01168}{2}\,(1-\cos 2\theta)\ \dots\ (193),$$

and

$$q = a^{-2}g\,[g-(g-h)\sin^2\theta]\dots\dots\dots\dots(194).$$

If u_1^2, u_2^2 denote the greater and less roots of (191), we have

$$a^{-2}(u_1^2+u_2^2) = 1 + w^2 + q\dots\dots\dots\dots(195);$$
$$a^{-2}(u_1^2-u_2^2) = \sqrt{[(1-w^2)+2(1+w^2)\,q+q^2]}\ \dots(196).$$

Remark that when $\theta = 90°$, $1-w^2 = \cdot 001168$, $q = a^{-2}gh$. Hence $-a^{-2}gh$ might be positive, and as great as $\frac{1}{4}.10^{-6}(1\cdot 168)^2$, or a little greater, without making the radical imaginary.

§ 224. In fig. 17, Curves 1 and 2 represent, from $\theta = 0°$ to $\theta = 30°$, $\frac{1}{2}[a^{-2}(u_1^2-u_2^2)-(1-w^2)]$ for the sub-cases $h = g$, and $h = -g$; and Curve 3 represents $\frac{1}{2}(1-w^2)$ on the same scale from $\theta = 0°$ to $\theta = 5°\cdot 3$. In fig. 18, Curves 1 and 2 represent, from $\theta = 30°$ to $\theta = 90°$, $\frac{1}{2}[a^{-2}(u_1^2-u_2^2)-(1-w^2)]$ for the sub-cases $h = g$ and $h = -g$, on a scale of ordinates ten times, and abscissas half, that of fig. 17.

Throughout the whole range of figs. 17 and 18, Curves 1 and 2 represent chiral differences from the squares of the æolotropic wave-velocities calculated according to the non-chiral constituent of the æolotropy of quartz crystal; Curve 3 of fig. 17, and equations (193)—(196), show that through the range from $\theta = 0°$ to $\theta = 5°\cdot 3$, the values of $a^{-1}u_1$, $a^{-1}u_2$, and q differ from unity by less than $5\cdot 1.10^{-5}$. Hence, through this range, we have, *very approximately*,

$$\tfrac{1}{2}\left[a^{-2}(u_1^2-u_2^2)-(1-w^2)\right] \doteqdot \frac{u_1-u_2}{a}-(1-w)\ \dots\ (197).$$

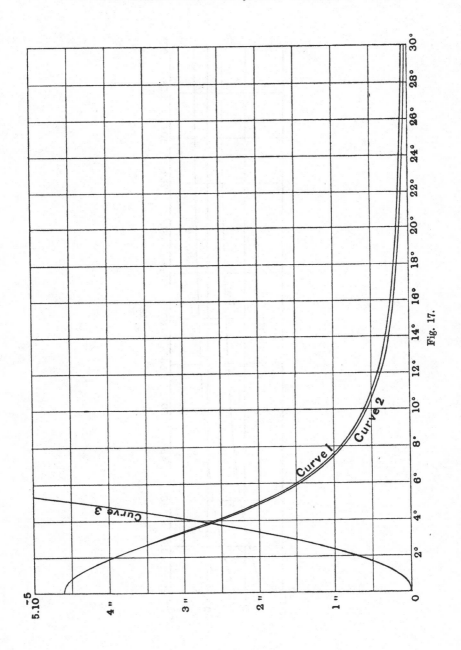

Fig. 17.

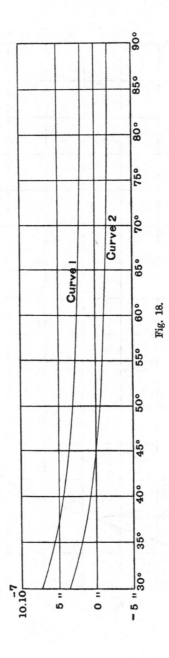

Fig. 18.

§ 225. Fig. 19 illustrates the critical features of the crystal- Molar. line influence, and of the combined crystalline and chiral influence, of quartz-crystal on light traversing it with wave-normal inclined to the axis at any angle up to 14°. The crystalline influence alone is represented in Curve 3 by downward ordinates equal to the excess of the velocity of the ordinary ray above the velocity

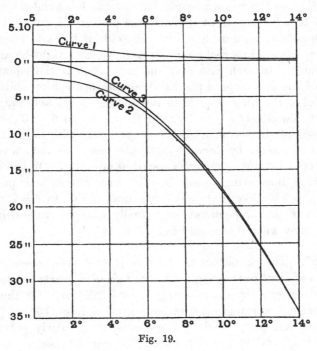

Fig. 19.

of the extraordinary ray, divided by the former; in an ideal crystal, corresponding to the mean of a right-handed and a left-handed quartz-crystal. Curve 1 represents the excess of the greater of the two wave-velocities in a real quartz-crystal above the velocity of the ordinary ray in the ideal mean crystal, divided by the latter. Curve 2 represents the excess (negative) of the less of the two wave-velocities in a real quartz-crystal above the velocity of the extraordinary ray in the ideal mean crystal, divided by the latter. According to our notation of § 223, the ordinates of Curves 3, 1, 2 are equal respectively to

$$w-1 ; \quad \frac{u_1}{a}-1 ; \quad \frac{u_2}{a}-w \quad \dots\dots\dots\dots\dots(198).$$

Molar.   § 226. From (198) we see that Curve 3 continued through the whole range from $\theta = 0°$ to $\theta = 90°$, represents the distance between a tangent plane on a prolate ellipsoid of revolution of unit axial semi-diameter, and the parallel tangent plane on the circumscribed spherical surface of unit radius (the ellipsoid and the sphere constituting the wave-surface for the ideal uniaxal crystal corresponding to the mean of right-handed and left-handed quartz). Curves 1 and 2 represent similarly the projections, outward from the sphere and inwards from the ellipsoid, of the two sheets of the wave-surface in a real quartz-crystal whether right-handed or left-handed. In each case θ is the inclination to the equatorial plane, of the two tangent planes whose distance is represented by ordinates in the diagram. It is interesting to see and judge by Curve 1 how closely at $\theta = 14°$, and thence up to $\theta = 90°$, one of the sheets of the wave-surface in quartz agrees with the spherical surface: and to see by Curves 2 and 3 how nearly, at and above 14°, the other sheet agrees with the inscribed ellipsoid. On the other hand, it is interesting to see by the three curves how preponderating is chirality over æolotropy, from $\theta = 0°$ to $\theta = 2°$; and to see how the preponderance gradually changes from chirality to æolotropy when θ increases from 2° to 14°.

§ 227. The characters of the two plane waves, whose wave-normals are inclined at angle θ to the axis of a quartz crystal, are to be discovered by commencing as in § 222; and, for the case there defined, working out a realised solution from (159) of § 206. We thus find that, for $\theta = 0$ each wave is circularly polarized; and, for all values of θ between 0° and 90°, each wave is elliptically polarized; the axes of the elliptic orbit being, one of them perpendicular to, and the other in, the plane through the wave-normal and the axis of the crystal. The former is the greater of the two axes for the wave which has the greater velocity (u_1); the latter is the greater for the wave having the less velocity (u_2). For all values of θ greater than six or seven degrees the less axis of the elliptic orbit is very small in comparison with the greater: that is to say each wave consists of very nearly rectilinear vibrations, or is very nearly "plane polarized"; and one of the two waves approximates closely to the "ordinary ray," the other to the "extraordinary ray" in the ideal non-chiral crystal corresponding to the mean of right-handed and left-handed quartz.

§ 228.   Looking at the two sub-cases represented by Curve 1 Molar.
and Curve 2 of figures 17 and 18, we see that the difference
between them is very small, probably quite imperceptible to the
most delicate observation practicable, when $\theta < 6°$.   At $\theta = 10°$
the difference of the ordinates for the two curves is about 1/17
of the ordinates for either; and might be perceptible to observation, though it represents an exceedingly small proportion, about
1/500, of the whole difference of velocities between the two rays.
This difference, as shown in figure 19, is itself very small, being
only about 18.10^{-5} of a, the mean velocity of the right-handed and
left-handed axial rays.   How exceedingly searching McConnel's
experimental investigation was may be judged by the fact shown
in his figure 3, page 321*, that from $\theta = 14°$ to $\theta = 30°$ he found
definite systematic differences between "MacCullagh's Theory"
and "Sarrau's Theory."   The results of MacCullagh's Theory are
expressed by our Curve 1; and our Curve 2 shows results differing
from MacCullagh's in the same direction as Sarrau's but somewhat more.   Thus McConnel's investigation was more than amply
sensitive to distinguish between our two sub-cases represented by
Curves 1 and 2 of figures 17 and 18.   Looking at McConnel's
figure 3, and remarking that ·003794 is the value of $a - b$ (our
$a - c$), which he took as correct according to his statement of
Rudberg's results, we see that what he takes as Sarrau's Theory
under-corrects the error of MacCullagh's: and our sub-case 2,
which I chose on this account, must make the correction very
nearly perfect.   But I see on looking to Sarrau's paper† that
his theory was not, as supposed by McConnel, confined to rela-
tively small deviations from $q = \dfrac{g^2}{a^2} \cos^4 \theta$ (McConnel's notation of
his page 314 translated into ours of § 226); and that proper
values of his constants (without the restriction of f_1 and g_1
relatively small, stated by McConnel), may be found to give a
perfect agreement with McConnel's observations.   The same may
be said of Voigt's theory as pointed out by McConnel (p. 314).
Thus we may take it that Sarrau's and Voigt's theories lead
to results perfectly consistent with McConnel's observations.
Theories of MacCullagh, Clebsch, Lang, and Boussinesq, are referred to by McConnel as giving a constant for what we have

---

* *Phil. Trans. Roy. Soc.*, Part I., 1886.

† Sarrau, *Liouville*, sér. 2, tome XIII. (1868), p. 101.

Molar. denoted by q, and as giving a fairly good approximation to the results of his observations; but decidedly less good than Sarrau and Voigt, who allowed q to vary with θ. All these theories agree in taking the mutual forces between different parts of the vibrating medium as the origin of the chiral property, and differ essentially from my theory which (§§ 162—166, 200—205) finds it in virtual inertia of the ether as disturbed by chiral groups of atoms.

§ 229. The following Table shows, in degrees per centimetre, the rotation of the vibrational line of polarized light travelling through various substances; crystalline solid, and liquid; taken from Landolt and Börnstein's Tables, Edition 1894.

Substance	Quality of Light	Direction of Rotation *	Rotation of vibrational line in degrees per centimetre	Observer
Solid crystals:				
Cinnabar....................	Red		2700 to 3000	Descloizeaux
Quartz.......................	$\begin{cases} D_1 \\ D \text{ (mean)} \\ D_2 \end{cases}$		$\left. \begin{array}{c} 217{\cdot}27 \\ 217{\cdot}05 \\ 216{\cdot}84 \end{array} \right\}$	Soret & Sarasin
Lead Hyposulphate + 4 aq	D		55·31	Pape
Potash Hyposulphate ...	D		83·85	Pape
Potassium Sulphate— Lithium Chromate $K_2SO_4 + Li_2CrO_4$	D		19·3	H. Traube
Sodium Bromate	D		21·7	H. Traube
Sodium Chlorate	D		31·04	Guye
,, ,, ...	D		31·6	Sohncke
Sodium Periodate + 3 aq	D		233·0	Groth
Liquids:				
Turpentine $C_{10}H_{16}$.........	D	R.	$1{\cdot}4147 \times d$†	Landolt
Nicotine $C_{10}H_{14}N_2$.........	D	L.	$16{\cdot}155 \times d$†	Landolt

* Clockwise as seen by the observer is called right-handed (R.), anti-clockwise, left-handed (L.).

† Where d denotes the specific gravity of the fluid. See p. 450 of Landolt and Börnstein. I do not see any good reason for the necessity of introducing d in the manner indicated in my table in the text, and rendered necessary by the notation adopted in the tables of Landolt and Börnstein. Some other embarrassing peculiarities in their notation have prevented me from venturing to quote any one of the numerous examples of the rotatory effect of "active" substances dissolved in "non-active" liquids given in pp. 450—458 of these tables.

§ 230. Important and interesting information regarding chiro- Molar. optic properties of liquids and solids is to be found in Mascart's *Traité d'Optique*, Volume II., Edition 1891, pages 247 to 369 ; and regarding Faraday's magneto-optic rotation of the vibrational line in pages 370 to 392. The first section of Appendix I in the present volume contains an important statement, given to me by the late Sir George Stokes, regarding chirality in crystals, and in crystalline molecules. Appendix H (Molecular Tactics of a Crystal, § 22 footnote, §§ 47—52) contains statements of fundamental principles in the pure geometry of chirality. Appendix G contains a complete mathematical theory of the quasi-inertia of a solid of any shape moving through a perfect liquid, with special remarks on chirality of this quasi-inertia.

§ 231. Lecture XX. as originally given, and fully reported in the papyrograph edition, and an appendix to it entitled "Improved Gyrostatic Molecule," contained unsatisfactory dynamical efforts to illustrate or explain Faraday's magneto-optic rotation. These are not reproduced in the present volume: but instead, an old paper of date 1856 entitled "Dynamical Illustration of the Magnetic and the Helicoidal Rotatory Effects of Transparent Bodies on Polarized Light" is reproduced as Appendix F. This paper contains a statement of dynamical principles concerned in the two kinds of rotation of the vibrational line of plane polarized light travelling through transparent solids or fluids; which I believe may even now be accepted as fundamentally correct. When we have a true physical theory of the disturbance produced by a magnet in pure ether, and in ether in the space occupied by ponderable matter, fluid or solid, there will probably be no difficulty in giving as thoroughly satisfactory explanation of the magneto-optic rotation as we now have of the chiro-optic.

§ 232. In conclusion, let us consider what modification of the original Maxwell-Sellmeier dynamics of ordinary and anomalous dispersion must be made when we adopt the atomic hypothesis of Appendices A and E and of Lec. XIX., §§ 162—168.

In App. A it is temporarily assumed, for the sake of a definite illustration, that the enormous variation of the etherial density within an atom is due to a purely Boscovichian force acting on the ether, in lines through the centre of the atom and varying

Molecular. as a function of the distance.   This makes no provision for vibrator or vibrators within an atom ; and, for the explanation of molecular vibrators, it only grants such molecular groups of atoms, as we have had for fifty years in the kinetic theory of gases, according to Clausius' impregnable doctrine of specific heats with regard to the partition of energy between translational and other than translational movements of the molecules.   Now, in App. E, and in applications of it suggested in §§ 162—168 of Lec. XIX., we have foundation for something towards a complete electro-etherial theory, of the Stokes-Kirchhoff vibrators * in the dynamics of spectrum-analysis, and of the Maxwell-Sellmeier explanation of dispersion.

§ 233.   In our new theory, every single electrion within a mono-electrionic atom, and every group of two, three, or more, electrions, within a poly-electrionic atom, is a vibrator which, in a source of light, takes energy from its collision with other atoms, and radiates out energy in waves travelling through the surrounding ether.   But at present we are not concerned with the source ; and in bringing this last of our twenty lectures to an end, I must limit myself to finding the effect of the presence of electrionic vibrators in ether, on the velocity of light traversing it.

§ 234.   The "fundamental modes" of which, in Lec. X., p. 120, we have denoted the periods by κ, $\kappa_,$, $\kappa_{,,}$, ... are now modes of vibration of the electrions within a fixed atom, when the ether around it and within it has no other motion than what is produced by vibrations of the electrions.   It is to be remarked however that a steady motion of the atom through space occupied by the ether, will not affect the vibrations of the electrions within it, relatively to the atom.

§ 235.   To illustrate, consider first the simple case of a mono-electrionic atom having a single electrion within it.   There is just one mode of vibration, and its period is

$$\kappa = 2\pi \sqrt{\frac{m}{c}} = \frac{2\pi}{e} \sqrt{ma^3}\dots\dots\dots(199),$$

where a denotes the radius of the atom, e the quantity of resinous electricity in an electrion, and m its virtual mass ; and c denotes

* See Lec. IX. pp. 101, 102, 103.

ELECTRO-ETHERIAL EXPLANATION OF UNIFORM RIGIDITY.

$e^2\alpha^{-3}$. This we see because the atom, being mono-electrionic, has Molecular. the same quantity of vitreous electricity as an electrion has of resinous; and therefore (App. E, § 4) the force towards the centre, experienced by an electrion held at a distance x from the centre, is $e^2\alpha^{-3}x$; which is denoted in § 240 by cx.

§ 236. Consider next a group of i electrions in equilibrium, or disturbed from equilibrium, within an i-electrionic atom. The force exerted by the atom on any one of the electrions is $ie^2\alpha^{-3}D$, towards the centre, if D is its distance from the centre. Let now the group be held in equilibrium with its constituents displaced through equal parallel distances, x, from their positions of equilibrium. Parallel forces each equal to $ie^2\alpha^{-3}x$, applied to the electrions, will hold them in equilibrium *; and if let go, they will vibrate to and fro in parallel lines, all in the same period

$$\frac{1}{\sqrt{i}}\frac{2\pi}{e}\sqrt{m\alpha^3}\ \dots\dots\dots\dots\dots\dots(200).$$

This therefore is one of the fundamental modes of vibration of the group; and it is clearly the mode of longest period. Thus we see that the periods of the gravest vibrational modes of different electrionic vibrators are directly as the square roots of the cubes of the radii of the atoms and inversely as the square roots of the numbers of the electrions; provided that in each case the atom is electrically neutralised by an integral number of electrions. Compare App. E, § 6.

§ 237. I now propose an assumption which, while greatly simplifying the theory of the quasi inertia-loading of ether when it moves through space occupied by ponderable matter as set forth in App. A, perfectly explains the practical equality of the rigidity of ether through all space, whether occupied also by, or void of, ponderable matter. My proposal is that *the radius of an electrion is so extremely small that the quantity of ether within its sphere of condensation* (Lec. XIX., § 166) *is exceedingly small in comparison with the quantity of undisturbed ether in a volume equal to the volume of the smallest atom.*

This assumption, in connection with §§ 164, 166 of Lec. XIX., makes the density of the ether exceedingly nearly constant through

---

* Compare App. E, § 23.

Molecular. all space outside the spheres of condensation of electrions. This is true of space whether void of atoms, or occupied by closely packed, or even overlapping, atoms; and the spheres of condensation occupy but a very small proportion of the whole space even where most densely crowded with poly-electrionic atoms. The highly condensed ether within the sphere of condensation close around each electrion might have either greater or less rigidity than ether of normal density, without perceptibly marring the agreement between the normal rigidity of undisturbed ether, and the working rigidity of the ether within the atom. This seems to me in all probability the true explanation of what everyone must have felt to be one of the greatest difficulties in the dynamical theory of light;—the equality of the rigidity of ether inside and outside a transparent body.

§ 238.   The smallness of the rarefaction of the ether within an atom and outside the sphere or spheres of condensation around its electrions, implies exceedingly small contribution to virtual inertia of vibrating ether, by that rarefaction; so small that I propose to neglect it altogether. Thus if an atom is temporarily deprived of its electrion or electrions (rendering it vitreously electrified to the highest degree possible), ether vibrating to and fro through it will experience the same inertial resistance as if undisturbed by the atom. Its presence will not be felt in any way by the ether existing in the same place. Thus the actual inertia-loading of ether to which the refraction of light is due, is produced practically by the electrions, and but little if at all perceptibly by the atoms, of the transparent body.

§ 239.   For the present I assume an electrion to be massless, that is to say devoid of intrinsic inertia, and to possess virtual inertia only on account of the kinetic energy which accompanies its steady motion through still ether. This is in reality an energy of relative motion; and does not exist when electrion and ether are moving at the same speed.   See App. A *passim*, and equation (202) § 240 below.

§ 240.   Come now to the wave-velocity problem and begin with the simplest possible case,—only one electrion in each atom. Consider waves of x-vibration travelling y-wards according to the formula (203) below. Take a sample atom in the wave-plane at distance y from XOZ. The atom is unmoved by the ether-waves;

while the electrion is set vibrating to and fro through its Molecular. centre.

At time t, let x be the displacement of the electrion, from the centre of the atom (or its absolute displacement because at present we assume the atom to be absolutely fixed):

ξ the displacement of the ether around the atom:

ρ the mean density of the ether within and around the atom, being, according to our assumptions, *exactly* the same as the normal density of undisturbed ether:

n the rigidity of the ether within and around the atoms, being, according to our assumptions, *very approximately* the same at every point as the rigidity of undisturbed ether:

N the number of atoms per unit of volume:

cx the electric attraction towards the centre of its atom, experienced by the electrion in virtue of its displacement, x:

m the virtual mass of an electrion:

E a cube of ether equal to $1/N$ of the unit of volume, having the centre of one, and only one, atom within it.

The equation of motion of E, multiplied by N, is

$$\rho \frac{d^2\xi}{dt^2} = n \frac{d^2\xi}{dy^2} - Ncx \dots\dots\dots\dots\dots(201);$$

and the equation of motion of the electrion within it, is

$$m \frac{d^2(x-\xi)}{dt^2} = -cx \quad \dots\dots\dots\dots\dots(202).$$

§ 241. The solution of these two equations for the regular regime of wave-motion is of the form

$$\xi = C \sin \omega \left(t - \frac{y}{v}\right); \quad x = C' \sin \omega \left(t - \frac{y}{v}\right) \dots\dots(203),$$

where ω is given. Our present object is to find the two unknowns C/C' (or ξ/x), and v. By (203) we see that

$$\frac{d^2}{dt^2} = -\omega^2; \quad \frac{d^2}{dy^2} = -\frac{\omega^2}{v^2} \dots\dots\dots\dots\dots(204).$$

This reduces (201) and (202) to

$$\left(\rho - \frac{n}{v^2}\right)\xi = \frac{Nc}{\omega^2}x = Nm(x-\xi)\dots\dots\dots(205),$$

T. L.   30

from which we find

$$\frac{x}{\xi} = \frac{m\omega^2}{m\omega^2 - c} = \frac{-\kappa^2}{\tau^2 - \kappa^2} \dots\dots\dots\dots(206),$$

and $$\frac{1}{v^2} = \frac{\rho}{n} + \frac{Nm}{n}\left(1 - \frac{x}{\xi}\right) = \frac{\rho}{n} + \frac{Nm}{n}\frac{\tau^2}{\tau^2 - \kappa^2} \dots\dots(207).$$

The last member is introduced with the notation

$$\tau = \frac{2\pi}{\omega}; \quad \kappa = 2\pi\sqrt{\frac{m}{c}} \dots\dots\dots\dots(208);$$

where τ denotes the period of the waves, and κ the period of an electrion displaced from the centre of its atom, and left vibrating inside, while the surrounding ether is all at rest except for the outward travelling waves, by which its energy is carried away at some very small proportionate rate per period; perhaps not more than 10^{-6}. It is clear that the greater the wave-length of the outgoing waves, in comparison with the radius of the sphere of condensation of the vibrating electrion, the smaller is the proportionate loss of energy per period. (Compare with the more complex problem, in which there are outgoing waves of two different velocities, worked out in the Addition to Lec. XIV., pp. 190—219. See particularly the examples in pp. 217, 218, 219.)

§ 242. Look back now to the diagram of Lec. XII., p. 145, representing our complex molecular vibrator of Lec. I., pp. 12, 13, reduced to a single free mass, m; connected by springs with the rigid sheath, the lining of an ideal spherical cavity in ether. In respect to that old diagram, let x now denote what was denoted on p. 145 by $\xi - x$; that is to say the displacement of the ether, relatively to m. Thus in the old illustrative ideal mechanism, cx denotes a resultant force of springs acting on m : in the new suggestion of an electro-etherial reality cx denotes simply the electric attraction of the atom on its electrion m, when displaced to a distance x from its centre. In the old mechanism it is the pulls on ether by the springs, equal and opposite to their forces on m, by which m acts on the ether (always admittedly an unreal kind of agency, invoked only by way of dynamical illustration). In the new electric design, m acts directly on the ether, in simple proportion to acceleration of relative motion. It does so because, in virtue of the ether's inertia when m is being relatively accelerated the ether is less dense before than behind m, and therefore the Molecular. resultant of m's attraction on it is backwards.

It is interesting to see that every one of the formulas of §§ 240, 241 (with the new notation of x, in the old dynamical problem), are applicable to both the old and the new subjects: and to know that the solution of the problem in terms of periods is the same in the two cases, notwithstanding the vast difference between the artificial and unreal details of the mechanism thought of and illustrated by models in 1884, and the probably real details of ether, electricity and ponderable matter, suggested in 1900— 1903.

§ 243. The interesting question of energy referred to in Lec. X., ll. 18—21 of p. 111 becomes more and more interesting now when we seem to understand its real quadruple character in

(I)   kinetic energy of pure ether,

(II)   potential energy of elasticity of ether,

(III)   electric potential energy of mutual repulsions of electrions and of attractions between electrions and atoms,

(IV)   potential energy of attraction of electrions on ether.

It is slightly and imperfectly treated in App. C. It must, when fully worked out, include a dynamical theory of phosphorescence. For the present I must leave it with much regret, to allow this Volume to be prepared for publication.

APPENDIX A.

ON THE MOTION PRODUCED IN AN INFINITE ELASTIC SOLID BY THE MOTION THROUGH THE SPACE OCCUPIED BY IT OF A BODY ACTING ON IT ONLY BY ATTRACTION OR REPULSION*.

§ 1. THE title of the present communication describes a pure problem of abstract mathematical dynamics, without indication of any idea of a physical application. For a merely mathematical journal it might be suitable, because the dynamical subject is certainly interesting both in itself and in its relation to waves and vibrations. My reason for occupying myself with it, and for offering it to the Royal Society of Edinburgh, is that it suggests a conceivable explanation of the greatest difficulty hitherto presented by the undulatory theory of light;—the motion of ponderable bodies through infinite space occupied by an elastic solid†.

§ 2. In consideration of the confessed object, and for brevity, I shall use the word atom to denote an ideal substance occupying a given portion of solid space, and acting on the ether within it and around it, according to the old-fashioned eighteenth century idea of attraction and repulsion. That is to say, every infinitesimal volume A of the atom acts on every infinitesimal volume B of the ether with a force in the line PQ joining the centres of these two volumes, equal to

$$Af(P, PQ)\rho B \quad \ldots\ldots\ldots\ldots\ldots\ldots(1),$$

* Communicated to the *Phil. Mag.* by the author, having been read before the Royal Society of Edinburgh, July 16th, 1900; and before the "Congrès" of the Paris Exhibition in August 1900.

† The so-called "electro-magnetic theory of light" does not cut away this foundation from the old undulatory theory of light. It adds to that primary theory an enormous province of transcendent interest and importance; it demands of us not merely an explanation of all the phenomena of light and radiant heat by transverse vibrations of an elastic solid called ether, but also the inclusion of electric currents, of the permanent magnetism of steel and lodestone, of magnetic force, and of electrostatic force, in a comprehensive etherial dynamics.

where ρ denotes the density of the ether at Q, and $f(P, PQ)$ denotes a quantity depending on the position of P and on the distance PQ. The whole force exerted by the atom on the portion ρB of the ether at Q, is the resultant of all the forces calculated according to (1), for all the infinitesimal portions A into which we imagine the whole volume of the atom to be divided.

§ 3. According to the doctrine of the potential in the well-known mathematical theory of attraction, we find rectangular components of this resultant as follows:—

$$X = \rho B \frac{d}{dx} \phi(x, y, z); \quad Y = \rho B \frac{d}{dy} \phi(x, y, z);$$
$$Z = \rho B \frac{d}{dz} \phi(x, y, z) \right\} \quad \dots(2),$$

where x, y, z denote coordinates of Q referred to lines fixed with reference to the atom, and ϕ denotes a function (which we call the potential at Q due to the atom) found by summation as follows:—

$$\phi = \iiint A \int_{PQ}^{\infty} dr\, f(P, r) \dots\dots\dots(3),$$

where $\iiint A$ denotes integration throughout the volume of the atom.

§ 4. The notation of (1) has been introduced to signify that no limitation as to admissible law of force is essential; but no generality, that seems to me at present practically desirable, is lost if we assume, henceforth, that it is the Newtonian law of the inverse square of the distance. This makes

$$f(P, PQ) = \frac{\alpha}{PQ^2} \dots\dots\dots(4),$$

and therefore $\int_{PQ}^{\infty} dr\, f(P, r) = \frac{\alpha}{PQ}$ $\dots\dots\dots(5),$

where α is a coefficient specifying for the point, P, of the atom, the intensity of its attractive quality for ether. Using (5) in (3) we find

$$\phi = \iiint A \frac{\alpha}{PQ} \dots\dots\dots(6),$$

and the components of the resultant force are still expressed by (2). We may suppose α to be either positive or negative

(positive for attraction and negative for repulsion); and in fact
in our first and simplest illustration of the problem we suppose
it to be positive in some parts and negative in other parts of the
atom, in such quantities as to fulfil the condition

$$\iiint A\alpha = 0 \quad\ldots\ldots\ldots\ldots\ldots\ldots\ldots(7).$$

§ 5.   As a first and very simple illustration, suppose the atom
to be spherical, of radius unity, with concentric interior spherical
surfaces of equal density.   This gives, for the direction of the
resultant force on any particle of the ether, whether inside or
outside the spherical boundary of the atom, a line through the
centre of the atom.   We may now take $A = 4\pi r^2 dr$.   The further
assumption of (7) may thus be expressed by

$$\int_0^1 dr\, r^2\alpha = 0 \quad\ldots\ldots\ldots\ldots\ldots\ldots(8);$$

and this, as we are now supposing the forces between every
particle of the atom and every particle of the ether to be subject
to the Newtonian law, implies, that the resultant of its attractions
and repulsions is zero for every particle of ether outside the
boundary of the atom.   To simplify the case to the utmost, we
shall further suppose the distribution of positive and negative
density of the atom, and the law of compressibility of the ether,
to be such, that the average density of the ether within the atom
is equal to the undisturbed density of the ether outside.   Thus
the attractions and repulsions of the atom in lines through its
centre produce, at different distances from its centre, condensations and rarefactions of the ether, with no change of the total
quantity of it within the boundary of the atom; and therefore
produce no disturbance of the ether outside.   To fix the ideas,
and to illustrate the application of the suggested hypothesis to
explain the refractivity of ordinary isotropic transparent bodies
such as water or glass, I have chosen a definite particular case in
which the distribution of the ether when at rest within the atom
is expressed by the following formula, and partially shown in
the accompanying diagram (fig. 1), and tables of calculated
numbers :—

$$r^3 = \frac{r'^3}{1 + K(1 - r')^2} \quad\ldots\ldots\ldots\ldots\ldots(9).$$

Here, r' denotes the undisturbed distance from the centre of the
atom, of a particle of the ether which is at distance r when at

rest under the influence of the attractive and repulsive forces. According to this notation $\frac{4\pi}{3}\,\delta\,(r^3)$ is the disturbed volume of a spherical shell of ether whose undisturbed radius is r' and thickness $\delta r'$ and volume $\frac{4\pi}{3}\,\delta\,(r'^3)$. Hence, if we denote the disturbed and undisturbed densities of the ether by ρ and unity respectively, we have

$$\rho\delta\,(r^3) = \delta\,(r'^3)\dots\dots\dots\dots\dots\dots(10);$$

whence, by (9),

$$\rho = \frac{3\,[1 + K\,(1 - r')^2]^2}{3 + K\,(3 - r')\,(1 - r')}\dots\dots\dots\dots(11).$$

This gives $1 + K$ for the density of the ether at the centre of the atom. In order that the disturbance may suffice for refractivities such as those of air, or other gases, or water, or glass, or other transparent liquids or isotropic solids, according to the dynamical theory explained in § 16 below, I find that K may for some cases be about equal to 100, and for others must be considerably greater. I have therefore taken $K = 100$, and calculated and drawn the accompanying tables and diagram accordingly.

TABLE I.

Col. 1	Col. 2	Col. 3	Col. 3'	Col. 4	Col. 5
r'	$\dfrac{r'^3}{r^3} = 1 + K\,(1 - r')^2$	r	$r' - r$	ρ	$(\rho - 1)\,r^2$
0·00	101·0	0·000	0·000	101·0	0·000
·05	91·25	·011	·039	88·1	·011
·10	82·0	·023	·077	75·3	·039
·20	65·0	·049	·151	55·8	·132
·30	50·0	·082	·218	39·1	·256
·40	37·0	·120	·280	25·8	·357
·50	26·0	·169	·331	15·8	·423
·60	17·0	·233	·367	8·76	·423
·70	10·0	·325	·375	4·17	·338
·80	5·0	·468	·332	1·60	·131
·85	3·25	·578	·272	0·90	− 0·033
·90	2·00	·715	·185	0·50	− ·256
·95	1·25	·865	·085	·35	− ·486
·96	1·16	·897	·063	·36	− ·515
·97	1·09	·928	·042	·39	− ·525
·98	1·04	·957	·023	·46	− ·495
·99	1·01	·982	·008	·61	− ·376
1·00	1·00	1·000	·000	1·00	− ·000

Table II.

Col. 1	Col. 2	Col. 3	Col. 4	Col. 5
r	r'	$r - r'$	ρ	$(\rho-1)r^2$
0·00	0·000	0·000	101·00	0·000
·02	·091	·071	78·5	·030
·04	·169	·129	64·4	·191
·06	·235	·175	49·6	·175
·08	·297	·217	39·5	·246
·10	·351	·251	31·8	·308
·20	·551	·351	11·8	·432
·30	·677	·377	5·00	·360
·40	·758	·358	2·46	·234
·50	·816	·316	1·34	·085
·60	·858	·258	0·82	− 0·065
·70	·895	·195	0·53	− ·231
·80	·929	·129	0·38	− ·397
·90	·961	·061	0·36	− ·518
1·00	1·000	·000	1·00	·000

§ 6. The diagram (fig. 1) helps us to understand the displacement of ether and the resulting distribution of density, within the atom. The circular arc marked 1·00 indicates a spherical portion of the boundary of the atom; the shorter of the circular arcs marked ·95, ·90, ... ·20, ·10 indicate spherical surfaces of undisturbed ether of radii equal to these numbers. The positions of the spherical surfaces of the same portions of ether under the influence of the atom, are indicated by the arc marked 1·00, and the longer of the arcs marked ·95, ·90, ... ·50, and the complete circles marked ·40, ·30, ·20, ·10. It may be remarked that the average density of the ether within any one of the disturbed spherical surfaces, is equal to the cube of the ratio of the undisturbed radius to the disturbed radius, and is shown numerically in column 2 of Table I. Thus, for example, looking at the table and diagram, we see that the cube of the radius of the short arc marked ·50 is 26 times the cube of the radius of the long arc marked ·50, and therefore the average density of the ether within the spherical surface corresponding to the latter is 26 times the density (unity) of the undisturbed ether within the spherical surface corresponding to the former. The densities shown in column 4 of each table are the densities of the ether at (not the average density of the ether within) the concentric spherical surfaces of radius r in the atom. Column 5 in each table shows $1/4\pi e$ of the excess (positive or negative) of the quantity of ether in a shell of radius r and infinitely small thickness e as disturbed by the atom above the quantity in a

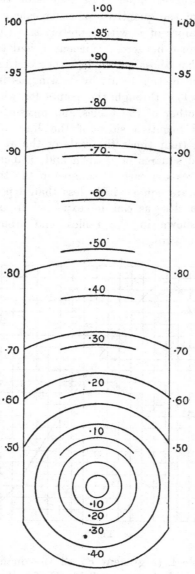

Fig. 1.

shell of the same dimensions of undisturbed ether. The formula of col. 2 makes $r = 1$ when $r' = 1$, that is to say the total quantity of the disturbed ether within the radius of the atom is the same as that of undisturbed ether in a sphere of the same radius. Hence the sum of the quantities of ether calculated from col. 5 for consecutive values of r, with infinitely small differences from $r = 0$ to $r = 1$, must be zero. Without calculating for smaller differences of r than those shown in either of the tables, we find a close verification of this result by drawing, as in fig. 2, a curve to represent $(\rho - 1) r^2$ through the points for which its value is given in one or other of the tables, and measuring the areas on the positive and negative sides of the line of abscissas. By drawing on paper (four times the scale of the annexed diagram), showing engraved squares of ·5 inch and ·1 inch, and counting the smallest squares and parts of squares in the two areas, I have verified that they are equal within less than 1 per cent. of either sum, which is as close as can be expected from the numerical approximations shown in the tables, and from the accuracy attained in the drawing.

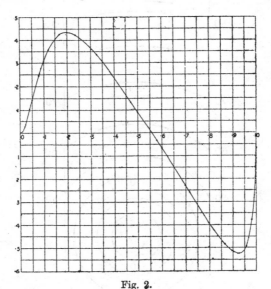

Fig. 2.

§ 7. In Table I. (argument r') all the quantities are shown for chosen values of r', and in Table II. for chosen values of r.

The calculations for Table I. are purely algebraic, involving merely cube roots beyond elementary arithmetic. To calculate in terms of given values of r the results shown in Table II. involves the solution of a cubic equation. They have been actually found by aid of a curve drawn from the numbers of col. 3 Table I., showing r in terms of r'. The numbers in col. 2 of Table II. showing, for chosen values of r, the corresponding values of r', have been taken from the curve; and we may verify that they are approximately equal to the roots of the equation shown at the head of col. 2 of Table I., regarded as a cubic for r' with any given values of r and K.

Thus, for example, taking $r' = \cdot929$ we calculate $r = \cdot811$,

	$r' = \cdot816$		$r = \cdot498$,
	$r' = \cdot677$		$r = \cdot301$,
	$r' = \cdot091$		$r = \cdot0208$,

where we should have $r = \cdot8$, $\cdot5$, $\cdot3$, and $\cdot02$ respectively. These approximations are good enough for our present purpose.

§ 8. The diagram of fig. 2 is interesting, as showing how, with densities of ether varying through the wide range of from $\cdot35$ to 101, the whole mass within the atom is distributed among the concentric spherical surfaces of equal density. We see by it, interpreted in conjunction with col. 4 of the tables, that from the centre to $\cdot56$ of the radius the density falls from 101 to 1. For radii from $\cdot56$ to 1, the values of $(\rho - 1)\, r^2$ decrease to a negative minimum of $\cdot525$ at $r = \cdot93$, and rise to zero at $r = 1$. The place of minimum density is of course inside the radius at which $(\rho - 1)\, r^2$ is a minimum; by cols. 4 and 3 of Table I., and cols. 4 and 1 of Table II., we see that the minimum density is about $\cdot35$, and at distance approximately $\cdot87$ from the centre.

§ 9. Let us suppose now our atom to be set in motion through space occupied by ether, and kept in motion with a uniform velocity v, which we shall first suppose to be infinitely small in comparison with the propagational velocity of equivoluminal* waves through pure ether undisturbed by any other substance than that of the atom. The velocity of the earth in its orbit

* That is to say, waves of transverse vibration, being the only kind of wave in an isotropic solid in which every part of the solid keeps its volume unchanged during the motion. See *Phil. Mag.*, May, August, and October, 1899.

round the sun being about 1/10,000 of the velocity of light, is small enough to give results, kinematic and dynamic, in respect to the relative motion of ether and the atoms constituting the earth closely in agreement with this supposition. According to it, the position of every particle of the ether at any instant is the same as if the atom were at rest; and to find the motion produced in the ether by the motion of the atom, we have a purely kinematic problem of which an easy graphic solution is found by marking on a diagram the successive positions thus determined for any particle of the ether, according to the positions of the atom at successive times with short enough intervals between them, to show clearly the path and the varying velocity of the particle.

§ 10. Look, for example, at fig. 3, in which a semi-circumference of the atom at the middle instant of the time we are going to consider, is indicated by a semicircle $C_{20} A C_0$, with diameter $C_0 C_{20}$ equal to two units of length. Suppose the centre of the atom to move from right to left in the straight line $C_0 C_{20}$ with velocity ·1, taking for unit of time the time of travelling 1/10 of the radius. Thus, reckoning from the time when the centre is at C_0, the times when it is at C_2, C_5, C_{10}, C_{18}, C_{20} are 2, 5, 10, 18, 20. Let Q' be the undisturbed position of a particle of ether before time 2 when the atom reaches it, and after time 18 when the atom leaves it. This implies that $Q'C_2 = Q'C_{18} = 1$, and $C_2 C_{10} = C_{10} C_{18} = ·8$, and therefore $C_{10} Q' = ·6$. The position of the particle of ether, which when undisturbed is at Q', is found for any instant t of the disturbance as follows :—

Take $C_0 C = t/10$; draw $Q'C$, and calling this r' find $r' - r$ by formula (9), or Table I. or II.: in $Q'C$ take $Q'Q = r' - r$. Q is the position at time t of the particle whose undisturbed position is Q'. The drawing shows the construction for $t = 2$, and $t = 5$, and $t = 18$. The positions at times 2, 3, 4, 5, ... 15, 16, 17, 18 are indicated by the dots marked 2, 3, 4, 5, 6, 7, 8, 9, 0, 1, 2, 3, 4, 5, 6, 7, 8 on the closed curve with a corner at Q', which has been found by tracing a smooth curve through them. This curve, which, for brevity, we shall call the orbit of the particle, is clearly tangential to the lines $Q'C_2$ and $Q'C_{18}$. By looking to the formula (9), we see that the velocity of the particle is zero at the instants of leaving Q' and returning to it. Fig. 4 shows the

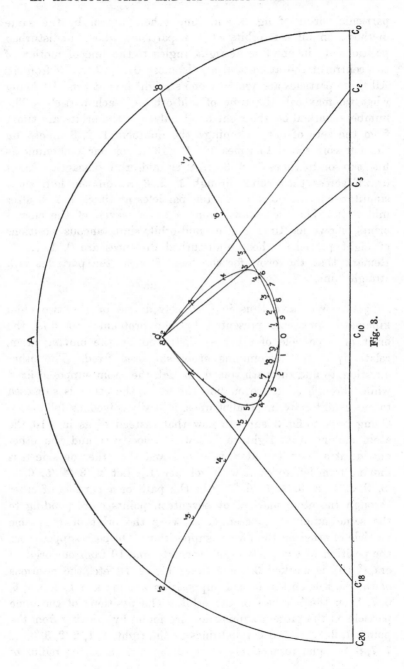

Fig. 3.

particular orbit of fig. 3, and nine others drawn by the same method; in all ten orbits of ten particles whose undisturbed positions are in one line at right angles to the line of motion of the centre of the atom, and at distances 0, ·1, ·2, ... ·9 from it. All these particles are again in one straight line at time 10, being what we may call the time of mid-orbit of each particle. The numbers marked on the right-hand halves of the orbits are times from the zero of our reckoning; the numbers 1, 2, 3 ... etc. on the left correspond to times 11, 12, 13 ... of our reckoning as hitherto, or to times 1, 2, 3 ... after mid-orbit passages. Lines drawn across the orbits through 1, 2, 3 ... on the left, show simultaneous positions of the ten particles at times 1, 2, 3 after mid-orbit. The line drawn from 4 across seven of the curved orbits, shows for time 4 after mid-orbit, simultaneous positions of eight particles, whose undisturbed distances are 0, ·1, ... ·7. Remark that the orbit for the first of these ten particles is a straight line.

§ 11. We have thus in § 10 solved one of the two chief kinematic questions presented by our problem:—to find the orbit of a particle of ether as disturbed by the moving atom, relatively to the surrounding ether supposed fixed. The other question, to find the path traced through the atom supposed fixed while, through all space outside the atom, the ether is supposed to move uniformly in parallel lines, is easily solved, as follows:— Going back to fig. 3, suppose now that instead of, as in § 10, the atom moving from right to left with velocity ·1 and the ether outside it at rest, the atom is at rest and the ether outside it is moving from left to right with velocity ·1. Let ′2, ′3, ′4, ′5, ′6, ′7, ′8, ′9, 0, ′1, ′2, ′3, ′4, ′5, ′6, ′7, ′8 be the path of a particle of ether through the atom marked by seventeen points corresponding to the same numbers unaccented showing the orbit of the same particle of ether on the former supposition. On both suppositions, the position of the particle of ether at time 10 from our original era, (§ 10), is marked 0. For times 11, 12, 13, etc., the positions of the particle on the former supposition are marked 1, 2, 3, 4, 5, 6, 7, 8 on the left half of the orbit. The positions of the same particle on the present supposition are found by drawing from the points 1, 2, 3, ... 7, 8 parallel lines to the right, 1 ′1, 2 ′2, 3 ′3, ... 7 ′7, 8 ′8, equal respectively to ·1, ·2, ·3, ... ·7, ·8 of the radius of

the atom, being our unit of length. Thus we have the latter half
of the passage of the particle through the atom; the first half is
equal and similar on the left-hand side of the atom. Applying

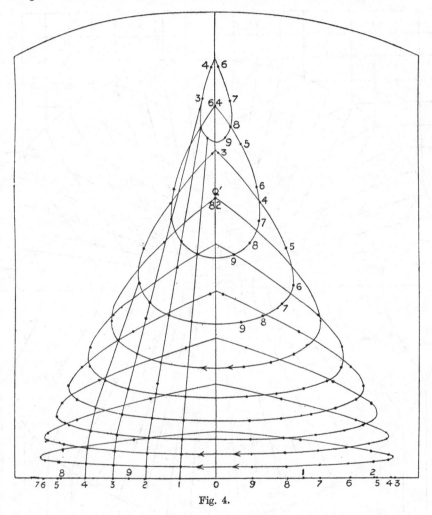

Fig. 4.

the same process to every one of the ten orbits shown in
fig. 4, and to the nine orbits of particles whose undisturbed
distances from the central line on the other side are ·1, ·2, ... ·9,
we find the set of stream-lines shown in fig. 5 (p. 480). The
dots on these lines show the positions of the particles at times

0, 1, 2, ... 19, 20 of our original reckoning (§ 10). The numbers
on the stream-line of the particle whose undisturbed distance
from the central line is ·6 are marked for comparison with fig. 3.

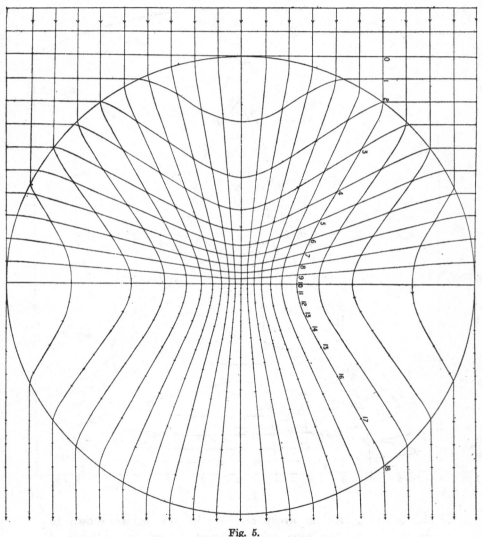

Fig. 5.

The lines drawn across the stream-lines on the left-hand side of
fig. 5 show simultaneous positions of rows of particles of ether
which, when undisturbed, are in straight lines perpendicular to

the direction of motion. The quadrilaterals thus formed within the left-hand semicircle show the figures to which the squares of ether, seen entering from the left-hand end of the diagram, become altered in passing through the atom. Thus we have completed the solution of our second chief kinematic question.

§ 12. The first dynamic question that occurs to us, returning to the supposition of moving atom and of ether outside it at rest, is :—What is the total kinetic energy (κ) of the portion of the ether which at any instant is within the atom? To answer it, think of an infinite circular cylinder of the ether in the space traversed by the atom. The time-integral from any era $t = 0$ of the total kinetic energy of the ether in this cylinder is $t\kappa$; because the ether outside the cylinder is undisturbed by the motion of the atom according to our present assumptions. Consider any circular disk of this cylinder of infinitely small thickness e. After the atom has passed it, it has contributed to $t\kappa$, an amount equal to the time-integral of the kinetic energies of all the orbits of small parts into which we may suppose it divided, and it contributes no more in subsequent time. Imagine the disk divided into concentric rings of rectangular cross-section $e\,dr'$. The mass of one of these rings is $2\pi r'\,dr'\,e$ because its density is unity; and all its parts move in equal and similar orbits. Thus we find that the total contribution of the disk amounts to

$$2\pi e \int_0^1 dr'\,r' \int ds^2/dt \dots\dots\dots\dots\dots(12),$$

where $\int ds^2/dt$ denotes integration over one-half the orbit of a particle of ether whose undisturbed distance from the central line is r'; (because $\frac{1}{2}ds^2/dt^2$ is the kinetic energy of an ideal particle of unit mass moving in the orbit considered). Now the time-integral κt is wholly made up by contributions of successive disks of the cylinder. Hence (12) shows the contribution per time e/q, q being the velocity of the atom; and (κ being the contribution per unit of time) we therefore have

$$\kappa = 2\pi q \int_0^1 dr'\,r' \int ds^2/dt \dots\dots\dots\dots(13).$$

§ 13. The double integral shown in (13) has been evaluated with amply sufficient accuracy for our present purpose by seemingly rough summations; firstly, the summations $\int ds^2/dt$ for the ten orbits shown in fig. 4, and secondly, summation of these sums each multiplied by $dr'r'$. In the summations for each half-orbit, ds has been taken as the lengths of the curve between the consecutive points from which the curve has been traced. This implies taking $dt = 1$ throughout the three orbits corresponding to undisturbed distances from the central line equal respectively to 0, ·6, ·8; and throughout the other semi-orbits, except for the portions next the corner, which correspond essentially to intervals each < 1. The plan followed is sufficiently illustrated by the accompanying Table III., which shows the whole process of calculating and summing the parts for the orbit corresponding to undisturbed distance ·7.

Table IV. shows the sums for the ten orbits and the products of each sum multiplied by the proper value of r', to prepare for the final integration, which has been performed by finding the area of a representative curve drawn on conveniently squared paper as described in § 6 above. The result thus found is ·02115. It is very satisfactory to see that, within ·1 per cent., this agrees with the simple sum of the widely different numbers shown in col. 3 of Table IV.

TABLE III.

Orbit $r' = \cdot 7$.

ds	ds^2	dt	ds^2/dt
·006	·000036	0·14	·000257
·137	·018769	1·00	·018769
·112	·012544	1·00	·012544
·077	·005929	1·00	·005929
·050	·002500	1·00	·002500
·048	·002304	1·00	·002304
·050	·002500	1·00	·002500
·052	·002704	1·00	·002704
		Sum......	·047507

TABLE IV.

r'	$\int ds^2/dt$	$\cdot 1 . r' . \int ds^2/dt$
·0	·0818	·00000
·1	·0804	·00080
·2	·0781	·00156
·3	·0769	·00231
·4	·0722	·00289
·5	·0670	·00335
·6	·0567	·00340
·7	·0475	·00332
·8	·0310	·00248
·9	·0114	·00102
	Sum......	·02113

§ 14. Using in (13) the conclusion of § 13, and taking $q = 1$, we find

$$\kappa = 2\pi \cdot \cdot 002115 \quad\dots\dots\dots\dots\dots\dots(14).$$

A convenient way of explaining this result is to remark that it is

·634 of the kinetic energy $\left(\dfrac{4\pi}{3}\dfrac{1}{2}(\cdot1)^2\right)$ of an ideal globe of rigid matter of the same bulk as our atom, moving with the same velocity. Looking now at the definition of κ in the beginning of § 12, we may put our conclusion in words, thus:—The distribution of etherial density within our ideal spherical atom represented by (11) with $K = 100$, gives rise to kinetic energy of the ether within it at any instant, when the atom is moving slowly through space filled with ether, equal to ·634 of the kinetic energy of motion with the same velocity through ideal void space, of an ideal rigid globe of the same bulk as the atom, and the same density as the undisturbed density of the ether. Thus if the atom, which we are supposing to be a constituent of real ponderable matter, has an inertia of its own equal to I per unit of its volume, the effective inertia of its motion through space occupied by ether will be $\dfrac{\pi}{6}s^3(I + \cdot634)$; the diameter of the atom being now denoted by s (instead of 2 as hitherto), and the inertia of unit bulk of the ether being still (as hitherto) taken as unit of inertia. In all that follows we shall suppose I to be very great, much greater than 10^6; perhaps greater than 10^{12}.

§ 15. Consider now, as in § 11 above, our atom at rest; and the ether moving uniformly in the space around the atom, and through the space occupied by the atom, according to the curved stream-lines and the varying velocities shown in fig. 5. The effective inertia of any portion of the ether containing the atom will be greater than the simple inertia of an equal volume of the ether by the amount $\dfrac{\pi}{6}s^3 \times \cdot634$. This follows from the well-known dynamical theorem that the total kinetic energy of any moving body or system of bodies is equal to the kinetic energy due to the motion of its centre of inertia, plus the sum of the kinetic energies of the motions of all its parts relative to the centre of inertia.

§ 16. Suppose now a transparent body—solid, liquid, or gaseous—to consist of an assemblage of atoms all of the same magnitude and quality as our ideal atom defined in § 2, and with I enormously great as described in § 14. The atoms may be all motionless as in an absolutely cold solid, or they may have the thermal motions of the molecules of a solid, liquid, or gas at

31—2

any temperature not so high but that the thermal velocities are everywhere small in comparison with the velocity of light. The effective inertia of the ether per unit volume of the assemblage will be exceedingly nearly the same as if the atoms were all absolutely fixed, and will therefore, by § 15, be equal to

$$1 + N\frac{\pi}{6}s^3 \cdot 634 \dots\dots\dots\dots\dots\dots(15),$$

where N denotes the number of atoms per cubic centimetre of the assemblage, one centimetre being now our unit of length. Hence, if we denote by V the velocity of light in undisturbed ether, its velocity through the space occupied by the supposed assemblage of atoms will be

$$V\left/\left(1 + N\frac{\pi}{6}s^3 \cdot 634\right)^{\frac{1}{2}}\right. \dots\dots\dots\dots(16).$$

§ 17. For example, let us take $N = 4 \times 10^{20}$*; and, as I find suits the cases of oxygen and argon, $s = 1\cdot42 \times 10^{-8}$, which gives $N\frac{\pi}{6}s^3 = \cdot60 \times 10^{-3}$. The assemblage thus defined would, if condensed one-thousand-fold, have ·6 of its whole volume occupied by the atoms and ·4 by undisturbed ether; which is somewhat denser than the cubic arrangement of globes

$$\left(\text{space unoccupied} = 1 - \frac{\pi}{6} = \cdot4764\right),$$

and less dense than the densest possible arrangement

$$\left(\text{space unoccupied} = 1 - \frac{\pi}{3\sqrt{2}} = \cdot2595\right).$$

Taking now $N\frac{\pi}{6}s^3 = \cdot60 \times 10^{-3}$ in (16), we find for the refractive index of our assemblage 1·00019, which is somewhat smaller than the refractive index of oxygen (1·000273). By taking for K a larger value than 100 in (11), we could readily fit the formula to give, in

---

* I am forced to take this very large number instead of Maxwell's 19×10^{18}, as I have found it otherwise impossible to reconcile the known viscosities and the known condensations of hydrogen, oxygen, and nitrogen with Maxwell's theoretical formulas. [In § 50 of Lect. XVII. of the present volume we saw that the smaller value 10^{20} is admissible and probably may be not far from the truth.] It must be remembered that Avogadro's law makes N the same for all gases.

an assemblage in which ·6 × 10⁻³ of the whole space is occupied by the atom, exactly the refractive index of oxygen, nitrogen, or argon, or any other gas. It is remarkable that according to the particular assumptions specified in § 5, a density of ether in the centre of the atom considerably greater than 100 times the density of undisturbed ether is required to make the refractivity as great as that of oxygen. There is, however, no difficulty in admitting so great a condensation of ether by the atom, if we are to regard our present problem as the basis of a physical hypothesis worthy of consideration,

§ 18. There is, however, one serious, perhaps insuperable, difficulty to which I must refer in conclusion: the reconciliation of our hypothesis with the result that ether in the earth's atmosphere is motionless relatively to the earth, seemingly proved by an admirable experiment designed by Michelson, and carried out with most searching care to secure a trustworthy result, by himself and Morley*. I cannot see any flaw either in the idea or in the execution of this experiment. But a possibility of escaping from the conclusion which it seemed to prove may be found in a brilliant suggestion made independently by Fitzgerald†, and by Lorentz‡ of Leyden, to the effect that the motion of ether through matter may slightly alter its linear dimensions; according to which if the stone slab constituting the sole plate of Michelson and Morley's apparatus has, in virtue of its motion through space occupied by ether, its lineal dimensions shortened one one-hundred-millionth§ in the direction of motion, the result of the experiment would not disprove the free motion of ether through space occupied by the earth.

* *Phil. Mag.*, December, 1887.

† Public Lectures in Trinity College, Dublin.

‡ *Versuch einer Theorie der electrischen und optischen Erscheinungen in bewegten Körpern.* Leiden, 1895.

§ This being the square of the ratio of the earth's velocity round the sun (30 kilometres per sec.) to the velocity of light (300,000 kilometres per sec.).

APPENDIX B.

NINETEENTH CENTURY CLOUDS OVER THE DYNAMICAL THEORY OF HEAT AND LIGHT*.

(Friday evening Lecture, Royal Institution, April 27, 1900.)

[IN the present article the substance of the lecture is reproduced—with large additions, in which work commenced at the beginning of last year and continued after the lecture, during thirteen months up to the present time, is described—with results confirming the conclusions and largely extending the illustrations which were given in the lecture. I desire to take this opportunity of expressing my obligations to Mr William Anderson, my secretary and assistant, for the mathematical tact and skill, the accuracy of geometrical drawing, and the unfailingly faithful perseverance in the long-continued and varied series of drawings and algebraic and arithmetical calculations, explained in the following pages. The whole of this work, involving the determination of results due to more than five thousand individual impacts, has been performed by Mr Anderson.—K., Feb. 2, 1901.]

§ 1.   The beauty and clearness of the dynamical theory, which asserts heat and light to be modes of motion, is at present obscured by two clouds.   I.   The first came into existence with the undulatory theory of light, and was dealt with by Fresnel and Dr Thomas Young; it involved the question, How could the earth move through an elastic solid, such as essentially is the luminiferous ether?   II.   The second is the Maxwell-Boltzmann doctrine regarding the partition of energy.

§ 2.   CLOUD I.—RELATIVE MOTION OF ETHER AND PONDERABLE BODIES; such as movable bodies at the earth's surface, stones, metals, liquids, gases; the atmosphere surrounding the earth; the earth itself as a whole; meteorites, the moon, the sun,

---

* *Journal of the Royal Institution.*   Also *Phil. Mag.* July, 1901.

and other celestial bodies. We might imagine the question satisfactorily answered, by supposing ether to have practically perfect
elasticity for the exceedingly rapid vibrations, with exceedingly
small extent of distortion, which constitute light; while it behaves
almost like a fluid of very small viscosity, and yields with exceedingly small resistance, practically no resistance, to bodies
moving through it as slowly as even the most rapid of the
heavenly bodies. There are, however, many very serious objections
to this supposition; among them one which has been most noticed,
though perhaps not really the most serious, that it seems in-
compatible with the known phenomena of the aberration of light.
Referring to it, Fresnel, in his celebrated letter* to Arago, wrote
as follows:

"Mais il paraît impossible d'expliquer l'aberration des étoiles
"dans cette hypothèse; je n'ai pu jusqu'à présent du moins con-
"cevoir nettement ce phénomène qu'en supposant que l'éther
"passe librement au travers du globe, et que la vitesse communi-
"quée à ce fluide subtil n'est qu'une petite partie de celle de la
"terre; n'en excède pas le centième, par exemple.

"Quelque extraordinaire que paraisse cette hypothèse au premier
"abord, elle n'est point en contradiction, ce me semble, avec l'idée
"que les plus grands physiciens se sont faite de l'extrème porosité
"des corps."

The same hypothesis was given by Thomas Young, in *his*
celebrated statement that ether passes through among the molecules or atoms of material bodies like wind blowing through a
grove of trees. It is clear that neither Fresnel nor Young had
the idea that the ether of their undulatory theory of light, with
its transverse vibrations, is essentially an elastic solid, that is to
say, matter which resists change of shape with permanent or
sub-permanent force. If they had grasped this idea they must
have noticed the enormous difficulty presented by the laceration
which the ether must experience if it moves through pores or
interstices among the atoms of matter.

§ 3. It has occurred to me that, without contravening anything we know from observation of nature, we may simply deny
the scholastic axiom that two portions of matter cannot jointly

---

* *Annales de Chimie*, 1818; quoted in full by Larmor in his recent book, *Æther
and Matter*, pp. 320—322.

occupy the same space, and may assert, as an admissible hypothesis, that ether does occupy the same space as ponderable
matter, and that ether is not displaced by ponderable bodies
moving through space occupied by ether. But how then could
matter act on ether, and ether act on matter, to produce the
known phenomena of light (or radiant heat), generated by the
action of ponderable bodies on ether, and acting on ponderable
bodies to produce its visual, chemical, phosphorescent, thermal,
and photographic effects? There is no difficulty in answering
this question if, as it probably is, ether is a compressible and
dilatable* solid. We have only to suppose that the atom exerts
force on the ether, by which condensation or rarefaction is produced within the space occupied by the atom. At present† I
confine myself, for the sake of simplicity, to the suggestion of a
spherical atom producing condensation and rarefaction, with concentric spherical surfaces of equal density, but the same total
quantity of ether within its boundary as the quantity in an
equal volume of free undisturbed ether.

§ 4. Consider now such an atom given at rest anywhere in
space occupied by ether. Let force be applied to it to cause it to
move in any direction, first with gradually increasing speed, and
after that with uniform speed. If this speed is anything less
than the velocity of light, the force may be mathematically proved
to become zero at some short time after the instant when the
velocity of the atom becomes uniform, and to remain zero for
ever thereafter. What takes place is this:

§ 5. During all the time in which the velocity of the atom is
being augmented from zero, two sets of non-periodic waves, one
of them equi-voluminal, the other irrotational (which is therefore
condensational-rarefactional), are being sent out in all directions
through the surrounding ether. The rears of the last of these
waves leave the atom, at some time after its acceleration ceases.
This time, if the motion of the ether outside the atom, close

---

* To deny this property is to attribute to ether indefinitely great resistance
against forces tending to condense it or to dilate it—which seems, in truth, an
infinitely difficult assumption.

† Further developments of the suggested idea have been contributed to the
Royal Society of Edinburgh, and to the Congrès International de Physique held in
Paris in August. (*Proc. R.S.E.* July 1900; Vol. of reports, in French, of the *Cong.
Inter.*; and *Phil. Mag.*, Aug., Sept., 1900.)

beside it, is infinitesimal, is equal to the time taken by the slower wave (which is the equi-voluminal) to travel the diameter of the atom, and is the short time referred to in § 4. When the rears of both waves have got clear of the atom, the ether within it and in the space around it, left clear by both rears, has come to a steady state of motion relatively to the atom. This steady motion approximates more and more nearly to uniform motion in parallel lines, at greater and greater distances from the atom. At a distance of twenty diameters it differs exceedingly little from uniformity.

§ 6. But it is only when the velocity of the atom is very small in comparison with the velocity of light, that the disturbance of the ether in the space close round the atom is infinitesimal. The propositions asserted in § 4 and the first sentence of § 5 are true, however little the final velocity of the atom falls short of the velocity of light. If this uniform final velocity of the atom exceeds the velocity of light, by ever so little, a non-periodic conical wave of equi-voluminal motion is produced, according to the same principle as that illustrated for sound by Mach's beautiful photographs of illumination by electric spark, showing, by changed refractivity, the condensational-rarefactional disturbance produced in air by the motion through it of a rifle bullet. The semi-vertical angle of the cone, whether in air or ether, is equal to the angle whose sine is the ratio of the wave velocity to the velocity of the moving body*.

* On the same principle we see that a body moving steadily (and, with little error, we may say also that a fish or water-fowl propelling itself by fins or web-feet) through calm water, either floating on the surface or wholly submerged at some moderate distance below the surface, produces no wave disturbance if its velocity is less than the minimum wave velocity due to gravity and surface tension (being about 23 cms. per second, or ·44 of a nautical mile per hour, whether for sea water or fresh water); and if its velocity exceeds the minimum wave velocity, it produces a wave disturbance bounded by two lines inclined on each side of its wake at angles each equal to the angle whose sine is the minimum wave velocity divided by the velocity of the moving body. It is easy for anyone to observe this by dipping vertically a pencil or a walking-stick into still water in a pond (or even in a good-sized hand-basin), and moving it horizontally, first with exceeding small speed, and afterwards faster and faster. I first noticed it nineteen years ago, and described observations for an experimental determination of the minimum velocity of waves, in a letter to William Froude, published in *Nature* for Oct., in *Phil. Mag.* for Nov. 1871, and in App. G below, from which the following is extracted. "[Recently, in the "schooner yacht *Lalla Rookh*], being becalmed in the Sound of Mull, I had an "excellent opportunity, with the assistance of Professor Helmholtz, and my brother

§ 7. If, for a moment, we imagine the steady motion of the atom to be at a higher speed than the wave velocity of the condensational-rarefactional wave, two conical waves, of angles corresponding to the two wave velocities, will be steadily produced; but we need not occupy ourselves at present with this case because the velocity of the condensational-rarefactional wave in ether is, we are compelled to believe, enormously great in comparison with the velocity of light.

§ 8. Let now a periodic force be applied to the atom so as to cause it to move to and fro continually, with simple harmonic motion. By the first sentence of § 5 we see that two sets of periodic waves, one equi-voluminal, the other irrotational, are continually produced. Without mathematical investigation we see that if, as in ether, the condensational-rarefactional wave velocity is very great in comparison with the equi-voluminal wave velocity, the energy taken by the condensational-rarefactional wave is exceedingly small in comparison with that taken by the equi-voluminal wave; how small we can find easily enough by regular mathematical investigation. Thus we see how it is that the hypothesis of § 3 suffices for the answer suggested in that section to the question, How could matter act on ether so as to produce light?

§ 9. But this, though of primary importance, is only a small part of the very general question pointed out in § 3 as needing answer. Another part, fundamental in the undulatory theory of

"from Belfast [the late Professor James Thomson], of determining by observation "the minimum wave-velocity with some approach to accuracy. The fishing-line "was hung at a distance of two or three feet from the vessel's side, so as to cut the "water at a point not sensibly disturbed by the motion of the vessel. The speed "was determined by throwing into the sea pieces of paper previously wetted, and "observing their times of transit across parallel planes, at a distance of 912 centi-"metres asunder, fixed relatively to the vessel by marks on the deck and gunwale. "By watching carefully the pattern of ripples and waves which connected the "ripples in front with the waves in rear, I had seen that it included a set of "parallel waves slanting off obliquely on each side and presenting appearances "which proved them to be waves of the critical length and corresponding minimum "speed of propagation." When the speed of the yacht fell to but little above the critical velocity, the front of the ripples was very nearly perpendicular to the line of motion, and when it just fell below the critical velocity the ripples disappeared altogether, and there was no perceptible disturbance on the surface of the water. The sea was "glassy"; though there was wind enough to propel the schooner at speed varying between $\frac{1}{4}$ mile and 1 mile per hour.

optics, is, How is it that the velocity of light is smaller in transparent ponderable matter than in pure ether? Attention was called to this particular question in my address, to the Royal Institution, of last April; and a slight explanation of my proposal for answering it was given, and illustrated by a diagram. The validity of this proposal is confirmed [in App. A] by a somewhat elaborate discussion and mathematical investigation of the subject worked out since that time and communicated under the title, " On the Motion produced in an Infinite Elastic Solid by the Motion through the Space occupied by it of a Body acting on it only by Attraction or Repulsion," to the Royal Society of Edinburgh on July 17, and to the Congrès International de Physique for its meeting at Paris in the beginning of August.

§ 10. The other phenomena referred to in § 3 come naturally under the general dynamics of the undulatory theory of light, and the full explanation of them all is brought much nearer if we have a satisfactory fundamental relation between ether and matter, instead of the old intractable idea that atoms of matter displace ether from the space before them, when they are in motion relatively to the ether around them. May we then suppose that the hypothesis which I have suggested clears away the first of our two clouds? It certainly would explain the " aberration of light " connected with the earth's motion through ether in a thoroughly satisfactory manner. It would allow the earth to move with perfect freedom through space occupied by ether without displacing it. In passing through the earth the ether, an elastic solid, would not be lacerated as it would be according to Fresnel's idea of porosity and ether moving through the pores as if it were a fluid. Ether would move relatively to ponderables with the perfect freedom wanted for what we know of aberration, instead of the imperfect freedom of air moving through a grove of trees suggested by Thomas Young. According to it, and for simplicity neglecting the comparatively very small component due to the earth's rotation (only ·46 of a kilometre per second at the equator where it is a maximum), and neglecting the imperfectly known motion of the solar system through space towards the constellation Hercules, discovered by Herschel*, there

* The splendid spectroscopic method originated by Huggins thirty-three years ago, for measuring the component in the line of vision of the relative motion of the

would be at all points of the earth's surface a flow of ether at the
rate of 30 kilometres per second in lines all parallel to the tangent
to the earth's orbit round the sun.   There is nothing inconsistent
with this in all we know of the ordinary phenomena of terrestrial
optics; but, alas! there is inconsistency with a conclusion that
ether in the earth's atmosphere is motionless relatively to the
earth, seemingly proved by an admirable experiment designed by
Michelson, and carried out, with most searching care to secure a
trustworthy result, by himself. and Morley*.   I cannot see any
flaw either in the idea or in the execution of this experiment.
But a possibility of escaping from the conclusion which it seemed
to prove may be found in a brilliant suggestion made independently by Fitzgerald† and by Lorentz‡ of Leyden, to the effect
that the motion of ether through matter may slightly alter its
linear dimensions, according to which if the stone slab constituting
the sole plate of Michelson and Morley's apparatus has, in virtue
of its motion through space occupied by ether, its lineal dimensions
shortened one one-hundred-millionth§ in the direction of motion,
the result of the experiment would not disprove the free motion
of ether through space occupied by the earth.

§ 11.   I am afraid we must still regard Cloud No. I. as very
dense.

earth, and any visible star, has been carried on since that time with admirable
perseverance and skill by other observers, who have from their results made
estimates of the velocity and direction of the motion through space of the centre
of inertia of the solar system.   My Glasgow colleague, Professor Becker, has kindly
given me the following information on the subject of these researches :
   " The early (1888) Potsdam photographs of the spectra of 51 stars brighter than
" $2\frac{1}{2}$ magnitude have been employed for the determination of the apex and velocity
" of the solar system.   Kempf (*Astronomische Nachrichten*, Vol. 132) finds for the
" apex : right ascension, 206° ± 12°; declination, 46° ± 90°; velocity, 19 kilometres
" per second ; and Risteen (*Astronomical Journal*, 1893) finds practically the same
" quantities.   The proper motions of the fixed stars assign to the apex a position
" which may be anywhere in a narrow zone parallel to the Milky-way, and extending
" 20° on both sides of a point of Right Ascension 275° and Declination +30°.   The
" authentic mean of 13 values determined by the methods of Argélander or Airy
" gives 274° and +35° (André, *Traité d'Astronomie Stellaire*)."
   * *Phil. Mag.*, December 1887.
   † Public Lectures in Trinity College, Dublin.
   ‡ *Versuch einer Theorie der electrischen und optischen Erscheinungen in bewegten
Körpern.*   Leiden, 1895.
   § This being the square of the ratio of the earth's velocity round the sun
(30 kilometres per sec.) to the velocity of light (300,000 kilometres per sec.).

§ 12. CLOUD II. Waterston (in a communication to the Royal
Society, now famous, which, after lying forty-five years buried
and almost forgotten in the archives, was rescued from oblivion
by Lord Rayleigh and published, with an introductory notice of
great interest and importance, in the *Transactions* of the Royal
Society for 1892), enunciated the following proposition: " In mixed
"media the mean square molecular velocity is inversely propor-
"tional to the specific weight of the molecule. This is the law
"of the equilibrium of vis viva." Of this proposition Lord
Rayleigh in a footnote* says, "This is the first statement of a
"very important theorem (see also *Brit. Assoc. Rep.*, 1851). The
"demonstration, however, of § 10 can hardly be defended. It
"bears some resemblance to an argument indicated and exposed
"by Professor Tait (*Edinburgh Trans.*, Vol. 33, p. 79, 1886). There
"is reason to think that this law is intimately connected with
"the Maxwellian distribution of velocities of which Waterston
"had no knowledge."

§ 13. In Waterston's statement the "specific weight of a
molecule" means what we now call simply the mass of a molecule; and "molecular velocity" means the translational velocity
of a molecule. Writing on the theory of sound in the *Phil. Mag.*
for 1858, and referring to the theory developed in his buried
paper†, Waterston said, "The theory...assumes...that if the
"impacts produce rotatory motion the vis viva thus invested bears
"a constant ratio to the rectilineal vis viva." This agrees with
the very important principle or truism given independently about
the same time by Clausius to the effect that the mean energy,
kinetic and potential, due to the relative motion of all the parts
of any molecule of a gas, bears a constant ratio to the mean
energy of the motion of its centre of inertia when the density and
pressure are constant.

§ 14. Without any knowledge of what was to be found in
Waterston's buried paper, Maxwell, at the meeting of the British
Association at Aberdeen, in 1859‡, gave the following proposition

---

* *Phil. Trans.* A, 1892, p. 16.

† " On the Physics of Media that are Composed of Free and Perfectly Elastic
Molecules in a State of Motion." *Phil. Trans.* A, 1892, p. 13.

‡ "Illustrations of the Dynamical Theory of Gases," *Phil. Mag.*, January and
July 1860, and *Collected Papers*, Vol. I. p. 378.

regarding the motion and collisions of perfectly elastic spheres:
"Two systems of particles move in the same vessel; to prove
"that the mean vis viva of each particle will become the same
"in the two systems." This is precisely Waterston's proposition
regarding the law of partition of energy, quoted in § 12 above;
but Maxwell's 1860 proof was certainly not more successful than
Waterston's. Maxwell's 1860 proof has always seemed to me
quite inconclusive, and many times I urged my colleague, Professor Tait, to enter on the subject. This he did, and in 1886
he communicated to the Royal Society of Edinburgh a paper*
on the foundations of the kinetic theory of gases, which contained
a critical examination of Maxwell's 1860 paper, highly appreciative
of the great originality and splendid value, for the kinetic theory
of gases, of the ideas and principles set forth in it; but showing
that the demonstration of the theorem of the partition of energy
in a mixed assemblage of particles of different masses was
inconclusive, and successfully substituting for it a conclusive
demonstration.

§ 15. Waterston, Maxwell, and Tait, all assume that the
particles of the two systems are thoroughly mixed (Tait, § 18),
and their theorem is of fundamental importance in respect to
the specific heats of mixed gases. But they do not, in any of
the papers already referred to, give any indication of a proof of
the corresponding theorem, regarding the partition of energy
between two sets of equal particles separated by a membrane
impermeable to the molecules, while permitting forces to act
across it between the molecules on its two sides†, which is the
simplest illustration of the molecular dynamics of Avogadro's
law. It seems to me, however, that Tait's demonstration of the
Waterston-Maxwell law may possibly be shown to virtually include,
not only this vitally important subject, but also the very interesting, though comparatively unimportant, case of an assemblage of
particles of equal masses with a single particle of different mass
moving about among them.

* *Phil. Trans. R.S.E.*, "On the Foundations of the Kinetic Theory of Gases,"
May 14 and December 6, 1886, and January 7, 1887.

† A very interesting statement is given by Maxwell regarding this subject in his
latest paper regarding the Boltzmann-Maxwell doctrine. "On Boltzmann's Theorem
on the Average Distribution of Energy in a System of Material Points," *Camb. Phil.
Trans.*, May 6, 1878; *Collected Papers*, Vol. II. pp. 713—741.

§ 16. In §§ 12, 14, 15, " particle " has been taken to mean what is commonly, not correctly, called an elastic sphere, but what is in reality a Boscovich atom acting on other atoms in lines exactly through its centre of inertia (so that no rotation is in any case produced by collisions), with, as law of action between two atoms, *no force at distance greater than the sum of their radii, infinite force at exactly this distance.* None of the demonstrations, unsuccessful or successful, to which I have referred would be essentially altered if, instead of this last condition, we substitute a repulsion increasing with diminishing distance, according to any law for distances less than the sum of the radii, subject only to the condition that it would bē infinite before the distance became zero. In fact the impact, oblique or direct, between two Boscovich atoms thus defined, has the same result after the collision is completed (that is to say, when their spheres of action get outside one another) as collision between two conventional elastic spheres, imagined to have radii dependent on the lines and velocities of approach before collision (the greater the relative velocity the smaller the effective radii); and the only assumption essentially involved in those demonstrations is, that the radius of each sphere is very small in comparison with the average length of free path.

§ 17. But if the particles are Boscovich atoms, having centre of inertia not coinciding with centre of force; or quasi-Boscovich atoms, of non-spherical figure; or (a more acceptable supposition) if each particle is a cluster of two or more Boscovich atoms; rotations and changes of rotation would result from collisions. Waterston's and Clausius' leading principle, quoted in § 13 above, must now be taken into account, and Tait's demonstration is no longer applicable. Waterston and Clausius, in respect to rotation, both wisely abstained from saying more than that the average kinetic energy of rotation bears a constant ratio to the average kinetic energy of translation. With magnificent boldness Boltzmann and Maxwell declared that the ratio is equality; Boltzmann having found what seemed to him a demonstration of this remarkable proposition, and Maxwell having accepted the supposed demonstration as valid.

§ 18. Boltzmann went further*, and extended the theorem

* " Studien über das Gleichgewicht der lebendigen Kraft zwischen bewegten materiellen Punkten." *Sitzb. K. Akad. Wien,* October 8, 1868.

of equality of mean kinetic energies to any system of a finite number of material points (Boscovich atoms) acting on one another, according to any law of force, and moving freely among one another; and finally, Maxwell* gave a demonstration extending it to the generalised Lagrangian coordinates of any system whatever, with a finite or infinitely great number of degrees of freedom. The words in which he enunciated his supposed theorem are as follows :

"The only assumption which is necessary for the direct proof "is that the system, if left to itself in its actual state of motion, " will, sooner or later, pass [infinitely nearly †] through every phase "which is consistent with the equation of energy" (p. 714), and again (p. 716) :

"It appears from the theorem, that in the ultimate state of "the system the average‡ kinetic energy of two portions of the "system must be in the ratio of the number of degrees of freedom "of those portions.

"This, therefore, must be the condition of the equality of "temperature of the two portions of the system."

I have never seen validity in the demonstration§ on which Maxwell founds this statement, and it has always seemed to me exceedingly improbable that it can be true. If true, it would be very wonderful, and most interesting in pure mathematical dynamics. Having been published by Boltzmann and Maxwell it would be worthy of most serious attention, even without con-

---

* "On Boltzmann's Theorem on the Average Distribution of Energy in a System of Material Points," Maxwell's *Collected Papers*, Vol. II. pp. 713—741, and *Camb. Phil. Trans.*, May 6, 1878.

† I have inserted these two words as certainly belonging to Maxwell's meaning. —K.

‡ The average here meant is a time-average through a sufficiently long time.

§ The mode of proof followed by Maxwell, and its connection with antecedent considerations of his own and of Boltzmann, imply, as included in the general theorem, that the average kinetic energy of any one of three rectangular components of the motion of the centre of inertia of an isolated system, acted upon only by mutual forces between its parts is equal to the average kinetic energy of each generalised component of motion relatively to the centre of inertia. Consider, for example, as "parts of the system" two particles of masses m and m' free to move only in a fixed straight line, and connected to one another by a massless spring. The Boltzmann-Maxwell doctrine asserts that the average kinetic energy of the motion of the inertial centre is equal to the average kinetic energy of the motion relative to the inertial centre. This is included in the wording of Maxwell's statement in the text if, but not unless, $m = m'$. See footnote in § 7 of my paper, "On some Test-Cases for the Boltzmann-Maxwell Doctrine regarding Distribution of Energy," *Proc. Roy. Soc.*, June 11, 1891.

sideration of its bearing on thermo-dynamics. But, when we consider its bearing on thermo-dynamics, and in its first and most obvious application we find it destructive of the kinetic theory of gases, of which Maxwell was one of the chief founders, we cannot see it otherwise than as a cloud on the dynamical theory of heat and light.

§ 19. For the kinetic theory of gases, let each molecule be a cluster of Boscovich atoms. This includes every possibility ("dynamical," or "electrical," or "physical," or "chemical") regarding the nature and qualities of a molecule and of all its parts. The mutual forces between the constituent atoms must be such that the cluster is in stable equilibrium if given at rest; which means, that if started from equilibrium with its constituents in any state of relative motion, no atom will fly away from it, provided the total kinetic energy of the given initial motion does not exceed some definite limit. A gas is a vast assemblage of molecules thus defined, each moving freely through space, except when in collision with another cluster, and each retaining all its own constituents unaltered, or only altered by interchange of similar atoms between two clusters in collision.

§ 20. For simplicity we may suppose that each atom, A, has a definite radius of activity, α, and that atoms of different kinds, A, A', have different radii of activity, α, α'; such that A exercises no force on any other atom, A', A'', when the distance between their centres is greater than $\alpha + \alpha'$ or $\alpha + \alpha''$. We need not perplex our minds with the inconceivable idea of "virtue," whether for force or for inertia, residing in a mathematical point* the centre of the atom; and without mental strain we can distinctly believe that the substance (the "substratum" of qualities) resides not in a point, nor vaguely through all space, but definitely in the spherical volume of space bounded by the spherical surface whose radius is the radius of activity of the atom, and whose centre is the centre of the atom. In our intermolecular forces thus defined we have no violation of the old scholastic law, "Matter cannot act where it is not," but we explicitly violate the other scholastic law, "Two portions of matter cannot simul-"taneously occupy the same space." We leave to gravitation,

---

* See *Math. and Phys. Papers*, Vol. III. Art. xcvii. "Molecular Constitution of Matter," § 14.

and possibly to electricity (probably not to magnetism), the at present very unpopular idea of action at a distance.

§ 21. . We need not now (as in § 16, when we wished to keep as near as we could to the old idea of colliding elastic globes) suppose the mutual force to become infinite repulsion before the centres of two atoms, approaching one another, meet. Following Boscovich, we may assume the force to vary according to any law of alternate attraction and repulsion, but without supposing any infinitely great force, whether of repulsion or attraction, at any particular distance; but we must assume the force to be zero when the centres are coincident. We may even admit the idea of the centres being absolutely coincident, in at all events some cases of a chemical combination of two or more atoms; although we might consider it more probable that in most cases the chemical combination is a cluster, in which the volumes of the constituent atoms overlap without any two centres absolutely coinciding.

§ 22. The word "collision" used without definition in § 19 may now, in virtue of §§ 20, 21, be unambiguously defined thus: Two atoms are said to be in collision during all the time their volumes overlap after coming into contact. They necessarily in virtue of inertia separate again, unless some third body intervenes with action which causes them to remain overlapping; that is to say, causes combination to result from collision. Two clusters of atoms are said to be in collision when, after being separate, some atom or atoms of one cluster come to overlap some atom or atoms of the other. In virtue of inertia the collision must be followed either by the two clusters separating, as described in the last sentence of § 19, or by some atom or atoms of one or both systems being sent flying away. This last supposition is a matter-of-fact statement belonging to the magnificent theory of dissociation, discovered and worked out by Sainte-Clair Deville without any guidance from the kinetic theory of gases. In gases approximately fulfilling the gaseous laws (Boyle's and Charles'), two clusters must in general fly asunder after collision. Two clusters could not possibly remain permanently in combination without at least one atom being sent flying away after collision between two clusters with no third body intervening*.

* See Kelvin's *Math. and Phys. Papers*, Vol. III. Art. xcvII. § 33. In this reference, for " scarcely" substitute " not."

§ 23. Now for the application of the Boltzmann-Maxwell doctrine to the kinetic theory of gases: consider first a homogeneous single gas, that is, a vast assemblage of similar clusters of atoms moving and colliding as described in the last sentence of § 19; the assemblage being so sparse that the time during which each cluster is in collision is very short in comparison with the time during which it is unacted on by other clusters, and its centre of inertia, therefore, moves uniformly in a straight line. If there are i atoms in each cluster, it has $3i$ freedoms to move, that is to say, freedoms in three rectangular directions for each atom. The Boltzmann-Maxwell doctrine asserts that the mean kinetic energies of these $3i$ motions are all equal, whatever be the mutual forces between the atoms. From this, when the durations of the collisions are not included in the time-averages, it is easy to prove algebraically (with exceptions noted below) that the time-average of the kinetic energy of the component translational velocity of the inertial centre*, in any direction, is equal to any one of the $3i$ mean kinetic energies asserted to be equal to one another in the preceding statement. There are exceptions to the algebraic proof corresponding to the particular exception referred to in the last footnote to § 18 above; but, nevertheless, the general Boltzmann-Maxwell doctrine includes the proposition, even in those cases in which it is not deducible algebraically from the equality of the $3i$ energies. Thus, without exception, the average kinetic energy of any component of the motion of the inertial centre is, according to the Boltzmann-Maxwell doctrine, equal to $\frac{1}{3i}$ of the whole average kinetic energy of the system. This makes the total average energy, potential and kinetic, of the whole motion of the system, translational and relative, to be $3i(1+P)$ times the mean kinetic energy of one component of the motion of the inertial centre, where P denotes the ratio of the mean potential energy of the relative displacements of the parts to the mean kinetic energy of the whole system. Now, according to Clausius' splendid and easily proved theorem regarding the partition of energy in the kinetic theory of gases, the ratio of the difference of the two thermal capacities to the constant-volume thermal capacity is equal to the ratio of twice a single

---

* This expression I use for brevity to signify the kinetic energy of the whole mass ideally collected at the centre of inertia.

component of the translational energy to the total energy. Hence, if according to our usual notation we denote the ratio of the thermal capacity pressure-constant to the thermal capacity volume-constant by k, we have,

$$k - 1 = \frac{2}{3i\,(1 + P)}.$$

§ 24. *Example* 1. For first and simplest example, consider a monatomic gas. We have $i = 1$, and according to our supposition (the supposition generally, perhaps universally, made) regarding atoms, we have $P = 0$. Hence, $k - 1 = \frac{2}{3}$.

This is merely a fundamental theorem in the kinetic theory of gases for the case of no rotational or vibrational energy of the molecule; in which there is no scope either for Clausius' theorem or for the Boltzmann-Maxwell doctrine. It is beautifully illustrated by mercury vapour, a monatomic gas according to chemists, for which many years ago Kundt, in an admirably designed experiment, found $k - 1$ to be very approximately $\frac{2}{3}$: and by the newly discovered gases argon, helium, and krypton, for which also $k - 1$ has been found to have approximately the same value, by Rayleigh and Ramsay. But each of these four gases has a large number of spectrum lines, and therefore a large number of vibrational freedoms, and therefore, if the Boltzmann-Maxwell doctrine were true, $k - 1$ would have some exceedingly small value, such as that shown in the ideal example of § 26 below. On the other hand, Clausius' *theorem* presents no difficulty; it merely asserts that $k - 1$ is necessarily less than $\frac{2}{3}$ in each of these four cases, as in every case in which there is any rotational or vibrational energy whatever; and proves, from the values found experimentally for $k - 1$ in the four gases, that in each case the total of rotational and vibrational energy is exceedingly small in comparison with the translational energy. It justifies admirably the chemical doctrine that mercury vapour is *practically a monatomic gas*, and it proves that argon, helium, and krypton, are also *practically monatomic*, though none of these gases has hitherto shown any chemical affinity or action of any kind from which chemists could draw any such conclusion.

But Clausius' theorem, taken in connection with Stokes' and Kirchhoff's dynamics of spectrum analysis, throws a new light on what we are now calling a " practically monatomic gas." It shows that, unless we admit that the atoms can be set into rotation or vibration by mutual collisions (a most unacceptable hypothesis), each atom must have satellites connected with it (or ether condensed into it or around it) and kept, by the collisions, in motion relatively to it with total energy exceedingly small in comparison with the translational energy of the whole system of atom and satellites. The satellites must in all probability be of exceedingly small mass in comparison with that of the chief atom. Can they be the "ions" by which J. J. Thomson explains the electric conductivity induced in air and other gases by ultra-violet light, Röntgen rays and Becquerel rays?

Finally, it is interesting to remark that all the values of $k-1$ found by Rayleigh and Ramsay are somewhat less than $\frac{2}{3}$; argon ·64, ·61; helium ·652; krypton ·666. If the deviation from ·667 were accidental they would probably have been some in defect and some in excess.

Example 2. As a next simplest example let $i = 2$, and as a very simplest case let the two atoms be in stable equilibrium when concentric, and be infinitely nearly concentric when the clusters move about, constituting a homogeneous gas. This supposition makes $P = \frac{1}{2}$, because the average potential energy is equal to the average kinetic energy in simple harmonic vibrations; and in our present case half the whole kinetic energy, according to the Boltzmann-Maxwell doctrine, is vibrational, the other half being translational. We find $k-1 = \frac{2}{9} = ·2222$.

Example 3. Let $i = 2$; let there be stable equilibrium, with the centres C, C' of the two atoms at a finite distance a asunder, and let the atoms be always very nearly at this distance asunder when the clusters are not in collision. The relative motions of the two atoms will be according to three freedoms, one vibrational, consisting of very small shortenings and lengthenings of the distance CC', and two rotational, consisting of rotations round one or other of two lines perpendicular to each other and perpendicular to CC' through the inertial centre. With these conditions and limitations, and with the supposition that half the average kinetic energy of the rotation is comparable with the average kinetic energy of the vibrations, or exactly equal to it as according to the Boltzmann-Maxwell doctrine, it is easily proved that in rotation the excess of CC' above the equilibrium distance a, due to centrifugal force, must be exceedingly small in comparison with the maximum value of $CC' - a$ due to the vibration. Hence the average potential energy of the rotation is negligible in comparison with the potential energy of the vibration. Hence, of the three freedoms for relative motion there is only one contributory to P, and therefore we have $P = \frac{1}{6}$. Thus we find

$$k - 1 = \frac{2}{7} = \cdot 2857.$$

The best way of experimentally determining the ratio of the two thermal capacities for any gas is by comparison between the observed and the Newtonian velocities of sound. It has thus been ascertained that, at ordinary temperatures and pressures, $k - 1$ differs but little from $\cdot 406$ for common air, which is a mixture of the two gases nitrogen and oxygen, each diatomic according to modern chemical theory; and the greatest value that the Boltzmann-Maxwell doctrine can give for a diatomic gas is the $\cdot 2857$ of Ex. 3. This notable discrepance from observation suffices to absolutely disprove the Boltzmann-Maxwell doctrine. What is really established in respect to partition of energy is what Clausius' theorem tells us (§ 23 above). We find, as a result of observation and true theory, that the average kinetic energy of translation of the molecules of common air is $\cdot 609$ of the total energy, potential and kinetic, of the relative motion of the constituents of the molecules.

§ 25. The method of treatment of Ex. 3 above, carried out for a cluster of any number of atoms greater than two not in one line, $j + 2$ atoms, let us say, shows us that there are three translational freedoms; three rotational freedoms, relatively to axes through the inertial centre; and $3j$ vibrational freedoms. Hence we have $P = \dfrac{j}{j + 2}$, and we find $k - 1 = \dfrac{1}{3(1 + j)}$. The values of $k - 1$ thus calculated for a triatomic and tetratomic gas, and calculated as above in Ex. 3 for a diatomic gas, are shown in the following table, and compared with the results of observation for several such gases.

It is interesting to see how the dynamics of Clausius' theorem is verified by the results of observation shown in the table. The values of $k - 1$ for all the gases are less than $\frac{2}{3}$, as they must be when there is any appreciable energy of rotation or vibration in the molecule. They are different for different diatomic gases;

ranging from ·42 for oxygen to ·32 for chlorine, which is quite as might be expected, when we consider that the laws of force between the two atoms may differ largely for the different kinds of atoms. The values of $k-1$ are, on the whole, smaller for the

Gas	Values of $k-1$	
	According to the B.-M. doctrine.	By Observation.
Air	$\frac{2}{7} = ·2857$	·406
H_2	,, ,,	·40
O_2	,, ,,	·41
Cl_2	,, ,,	·32
CO	,, ,,	·39
NO	,, ,,	·39
CO_2	$\frac{1}{6} = ·1667$	·30
N_2O	,, ,,	·331
NH_3	$\frac{1}{9} = ·1111$	·311

tetratomic and triatomic than for the diatomic gases, as might be expected from consideration of Clausius' principle. It is probable that the differences of $k-1$ for the different diatomic gases are real, although there is considerable uncertainty with regard to the observational results for all or some of the gases other than air. It is certain that the discrepancies from the values, calculated according to the Boltzmann-Maxwell doctrine, are real and great; and that in each case, diatomic, triatomic, and tetratomic, the doctrine gives a value for $k-1$ much smaller than the truth.

§ 26. But, in reality, the Boltzmann-Maxwell doctrine errs enormously more than is shown in the preceding table. Spectrum analysis showing vast numbers of lines for each gas makes it certain that the number of freedoms of the constituents of each molecule is enormously greater than those which we have been counting, and therefore that unless we attribute vibratile quality to each individual atom, the molecule of every one of the ordinary gases must have a vastly greater number of atoms in its constitution than those hitherto reckoned in regular chemical doctrine. Suppose, for example, there are forty-one atoms in the molecule of any particular gas; if the doctrine were true we should have $j = 39$. Hence there are 117 vibrational freedoms, so that there might be 117 visible lines in the spectrum of the gas; and we have $k - 1 = \dfrac{1}{120} = \cdot0083$. There is, in fact, no possibility of reconciling the Boltzmann-Maxwell doctrine with the truth regarding the specific heats of gases.

§ 27. It is, however, not quite possible to rest contented with the mathematical verdict not proven, and the experimental verdict not true, in respect to the Boltzmann-Maxwell doctrine. I have always felt that it should be mathematically tested by the consideration of some particular case. Even if the theorem were true, stated as it was somewhat vaguely, and in such general terms that great difficulty has been felt as to what it is really meant to express, it would be very desirable to see even one other simple case, besides that original one of Waterston's, clearly stated and tested by pure mathematics. Ten years ago* I suggested a number of test cases, some of which have been courteously considered by Boltzmann; but no demonstration either of the truth or untruth of the doctrine as applied to any one of them has hitherto been given. A year later, I suggested what seemed to me a decisive test case disproving the doctrine; but my statement was quickly and justly criticised by Boltzmann and Poincaré; and more recently Lord Rayleigh† has shown very clearly that my simple test case was quite indecisive. This last article of Rayleigh's has led me to resume the consideration of several different classes of dynamical problems, which had occupied me more or less at various times during the last twenty years, each presenting exceedingly interesting features in connection with the double question : Is this a case which admits of the application of the Boltzmann-Maxwell doctrine; and, if so, is the doctrine true for it ?

§ 28. Premising that the mean kinetic energies with which the Boltzmann-Maxwell doctrine is concerned are time-integrals of energies divided by totals of the times, we may conveniently divide the whole class of problems, with reference to which the doctrine comes into question, into two classes.

Class I. Those in which the velocities considered are either

---

* "On some Test-Cases for the Maxwell-Boltzmann Doctrine regarding Distribution of Energy," *Proc. Roy. Soc.*, June 11, 1891.

† *Phil. Mag.*, Vol. xxxiii. 1892. p. 356, " Remarks on Maxwell's Investigation respecting Boltzmann's Theorem."

constant or only vary suddenly—that is to say, in infinitely small times—or in times so short that they may be omitted from the time-integration. To this class belong:

(*a*) The original Waterston-Maxwell case and the collisions of ideal rigid bodies of any shape, according to the assumed law that the translatory and rotatory motions lose no energy in the collisions.

(*b*) The frictionless motion of one or more particles constrained to remain on a surface of any shape, this surface being either closed (commonly called finite though really endless), or being a finite area of plane or curved surface, bounded like a billiard-table, by a wall or walls, from which impinging particles are reflected at angles equal to the angles of incidence.

(*c*) A closed surface, with non-vibratory particles moving within it freely, except during impacts of particles against one another or against the bounding surface.

(*d*) Cases such as (*a*), (*b*), or (*c*), with impacts against boundaries and mutual impacts between particles, softened by the supposition of finite forces during the impacts, with only the condition that the durations of the impacts are so short as to be practically negligible, in comparison with the durations of free paths.

Class II.   Cases in which the velocities of some of the particles concerned sometimes vary gradually; so gradually that the times during which they vary must be included in the time-integration. To this class belong examples such as (*d*) of Class I. with durations of impacts not negligible in the time-integration.

§ 29.   Consider first Class I. (*b*) with a finite closed surface as the field of motion and a single particle moving on it. If a particle is given, moving in any direction through any point I of the field, it will go on for ever along one determinate geodetic line. The question that first occurs is, Does the motion fulfil Maxwell's condition (see § 18 above); that is to say, for this case, if we go along the geodetic line long enough, shall we pass infinitely nearly to any point Q whatever, including I, of the surface an infinitely great number of times in all directions? This question cannot be answered in the affirmative without reservation. For example, if the surface be exactly an ellipsoid it must be answered in the negative, as is proved in the following §§ 30, 31, 32.

§ 30.   Let AA', BB', CC' be the ends of the greatest, mean, and least diameters of an ellipsoid.   Let U_1, U_2, U_3, U_4 be the umbilics in the arcs AC, CA', $A'C'$, $C'A$.   A known theorem in the geometry of the ellipsoid tells us that every geodetic through U_1 passes through U_3, and every geodetic through U_2 passes through U_4.   This statement regarding geodetic lines on an ellipsoid of three unequal axes is illustrated by fig. 1, a diagram showing for the extreme case in which the shortest axis is zero, the exact construction of a geodetic through U_1 which is a focus of the ellipse shown in the diagram. U_3, C', U_4 being infinitely near to U_1, C, U_2 respectively are indicated by double letters at the same

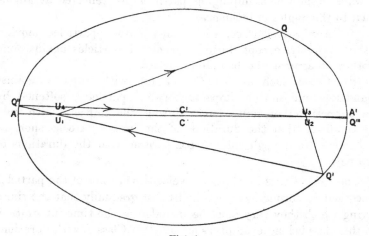

Fig. 1.

points.   Starting from U_1 draw the geodetic U_1QU_3; the two parts of which U_1Q and QU_3 are straight lines.   It is interesting to remark that in whatever direction we start from U_1 if we continue the geodetic through U_3, and on through U_1 again and so on endlessly, as indicated in the diagram by the straight lines U_1Q, U_3Q', U_1Q'', U_3Q''', and so on, we come very quickly to lines approaching successively more and more nearly to coincidence with the major axis.   At every point where the path strikes the ellipse it is reflected at equal angles to the tangent.   The construction is most easily made by making the angle between the reflected path and a line to one focus, equal to the angle between the incident path and a line to the other focus.

§ 31. Returning now to the ellipsoid:—From any point I, between U_1 and U_2, draw the geodetic IQ, and produce it through Q on the ellipsoidal surface. It must cut the arc $A'C'A$ at some point between U_3 and U_4, and, if continued on and on, it must cut the ellipse $ACA'C'A$ successively between U_1 and U_2, or between U_3 and U_4; never between U_2 and U_3, or U_4 and U_1. This, for the extreme case of the smallest axis zero, is illustrated by the path $IQQ'Q''Q'''Q^{iv}Q^v$ in fig. 2.

§ 32. If now, on the other hand, we commence a geodetic through any point J between U_1 and U_4, or between U_2 and U_3, it will never cut the principal section containing the umbilics, either between U_1 and U_2 or between U_3 and U_4. This for the extreme case of $CC' = 0$ is illustrated in fig. 3.

§ 33. It seems not improbable that if the figure deviates by ever so little from being exactly ellipsoidal, Maxwell's condition might be fulfilled. It seems indeed quite probable that Maxwell's condition (see §§ 13, 29, above) is fulfilled by a geodetic on a closed surface of any shape in general, and that exceptional cases,

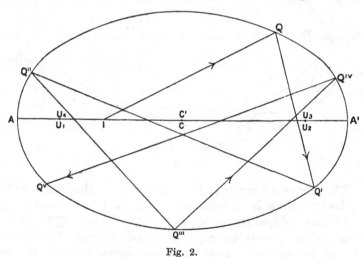

Fig. 2.

in which the question of § 29 is to be answered in the negative, are merely particular surfaces of definite shapes, infinitesimal deviations from which will allow the question to be answered in the affirmative.

§ 34. Now with an affirmative answer to the question—is Maxwell's condition fulfilled?—what does the Boltzmann-Maxwell doctrine assert in respect to a geodetic on a closed surface? The mere wording of Maxwell's statement, quoted in § 13 above, is not applicable to this case, but the meaning of the doctrine as interpreted from previous writings both of Boltzmann and Maxwell, and subsequent writings of Boltzmann, and of Rayleigh*, the most recent supporter of the doctrine, is that a single geodetic drawn long enough will not only fulfil Maxwell's condition of passing infinitely near to every point of the surface in all directions, but will pass with equal frequencies in all directions; and as many times within a certain infinitesimal distance $\pm \delta$ of any one point P as of any other point P' anywhere over the whole surface. This, if true, would be an exceedingly interesting theorem.

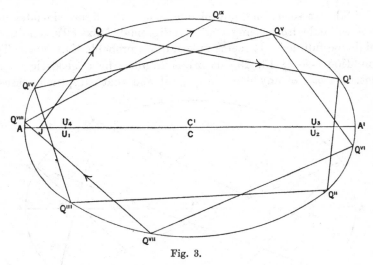

Fig. 3.

§ 35. I have made many efforts to test it for the case in which the closed surface is reduced to a plane with other boundaries than an exact ellipse (for which as we have seen in §§ 30, 31, 32, the investigation fails through the non-fulfilment of Maxwell's preliminary condition). Every such case gives, as we have seen, straight lines drawn across the enclosed area turned on meeting the boundary, according to the law of equal angles of incidence and reflection, which corresponds also to the case of an ideal

* *Phil. Mag.*, January 1900.

perfectly smooth non-rotating billiard-ball moving in straight lines except when it strikes the boundary of the table ; the boundary being of any shape whatever, instead of the ordinary rectangular boundary of an ordinary billiard-table, and being perfectly elastic. An interesting illustration, easily seen through a large lecture-hall, is had by taking a thin wooden board, cut to any chosen shape, with the corner edges of the boundary smoothly rounded, and winding a stout black cord round and round it many times, beginning with one end fixed to any point, I, of the board. If the pressure of the cord on the edges were perfectly frictionless the cord would, at every turn round the border, place itself so as to fulfil the law of equal angles of incidence and reflection, modified in virtue of the thickness of the board. For stability it would be necessary to fix points of the cord to the board by staples pushed in over it at sufficiently frequent intervals, care being taken that at no point is the cord disturbed from its proper straight line by the staple. [Boards of a considerable variety of shape with cords thus wound on them were shown as illustrations of the lecture.]

§ 36. A very easy way of drawing accurately the path of a particle moving in a plane and reflected from a bounding wall of any shape, provided only that it is not concave externally in any part, is furnished by a somewhat interesting kinematical method illustrated by the accompanying diagram (fig. 4). It is easily

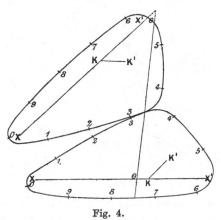

Fig. 4.

realised by using two equal and similar pieces of board, cut to any desired figure one of them being turned upside down relatively to the other, so that when the two are placed together with corresponding points in contact, each is the image of the other relative to the plane of contact regarded as a mirror. Sufficiently close corresponding points should be accurately marked on the boundaries of the two figures, and this allows great accuracy to be obtained in the drawing of the free path after each reflection. The diagram shows consecutive free paths 74·6—32·9 given, and 32·9—54·7, found by producing 74·6—32·9 through the point of contact. The process involves the exact measurement of the length (l)—say to three significant figures—and its inclination (θ) to a chosen line of reference XX'. The summations $\Sigma l \cos 2\theta$ and $\Sigma l \sin 2\theta$ give, as explained below, the difference of time-integrals of kinetic energies of component motions parallel and perpendicular respectively to XX', and parallel and perpendicular respectively to KK', inclined at 45° to XX'. From these differences we find (by a procedure equivalent to that of finding the principal axes of an ellipse) two lines at right angles to one another, such that the time-integrals of the components of velocity parallel to them are respectively greater than and less than those of the components parallel to any other line. [This process was illustrated by models in the lecture.]

§ 37. Virtually the same process as this, applied in the case of a scalene triangle ABC (in which $BC = 20$ centimetres and the angles $A = 97°$, $B = 29\cdot5°$, $C = 53\cdot5°$), was worked out in the Royal

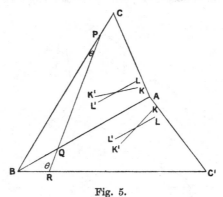

Fig. 5.

Institution during the fortnight after the lecture, by Mr Anderson, with very interesting results. The length of each free path (l), and its inclination to BC (θ), reckoned acute or obtuse according to the indications in the diagram (fig. 5), were measured to the nearest millimetre and the nearest integral degree. The first free path was drawn at random, and the continuations, after 599 reflections (in all 600 paths), were drawn in a manner illustrated by fig. 5, which shows, for example, a path PQ on one triangle continued to QR on the other. The two when folded together round the line AB shows a path PQ, continued on QR after reflection. For each path $l \cos 2\theta$ and $l \sin 2\theta$ were calculated and entered in tables with the proper algebraic signs. Thus, for the whole 600 paths, the following summations were found:

$$\Sigma l = 3298; \quad \Sigma l \cos 2\theta = +128 \cdot 8; \quad \Sigma l \sin 2\theta = -201 \cdot 9.$$

Remark, now, if the mass of the moving particle is 2, and the velocity one centimetre per second, $\Sigma l \cos 2\theta$ is the excess of the time-integral of kinetic energy of component motion parallel to BC above that of component motion perpendicular to BC, and $\Sigma l \sin 2\theta$ is the excess of the time-integral of kinetic energy of component motion perpendicular to KK' above that of component motion parallel to KK'; KK' being inclined at 45° to BC in the direction shown in the diagram. Hence the positive value of $\Sigma l \cos 2\theta$ indicates a preponderance of kinetic energy due to component motion parallel to BC above that of component motion perpendicular to BC; and the negative sign of $\Sigma l \sin 2\theta$ shows preponderance of kinetic energy of component motion parallel to KK', above that of component motion perpendicular to KK'. Deducing a determination of two axes at right angles to each other, corresponding respectively to maximum and minimum kinetic energies, we find LL', being inclined to KK' in the direction shown, at an angle $= \frac{1}{2} \tan^{-1} \dfrac{128 \cdot 8}{201 \cdot 9}$, is what we may call the axis of maximum energy, and a line perpendicular to LL' the axis of minimum energy; and the excess of the time-integral of the energy of component velocity parallel to LL' exceeds that of the component perpendicular to LL' by $239 \cdot 4$, being

$$\sqrt{128 \cdot 8^2 + 201 \cdot 9^2}.$$

This is $7 \cdot 25$ per cent. of the total of Σl which is the time-integral of the total energy. Thus, in our result, we find a very notable deviation from the Boltzmann-Maxwell doctrine, which asserts for the present case that the time-integrals of the component kinetic energies are the same for all directions of the component. The percentage which we have found is not very large; and, most probably, summations for several successive 600 flights would present considerable differences, both of the amount of the deviation from equality and the direction of the axes of maximum and minimum energy. Still, I think there is a strong probability that the disproof of the Boltzmann-Maxwell doctrine is genuine, and the discrepance is somewhat approximately of the amount and direction indicated. I am supported in this view by scrutinising the thirty sums for successive sets of twenty flights: thus I find $\Sigma l \cos 2\theta$ to be positive for eighteen out of thirty, and $\Sigma l \sin 2\theta$ to be negative for nineteen out of the thirty.

§ 38. A very interesting test-case is represented in the accompanying diagram, fig. 6—a circular boundary of semicircular corrugations. In this case it is obvious from the symmetry that the time-integral of kinetic energy of component motion parallel

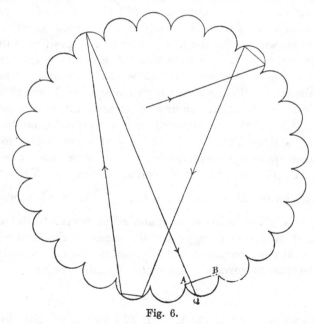

Fig. 6.

to any straight line must, in the long run, be equal to that parallel to any other. But the Boltzmann-Maxwell doctrine asserts, that the time-integrals of the kinetic energies of the two components, radial and transversal, according to polar coordinates, would be equal. To test this I have taken the case of an infinite number of the semicircular corrugations, so that in the time-integral it is not necessary to include the times between successive impacts of the particle on any one of the semicircles. In this case the geometrical construction would, of course, fail to show the precise point Q at which the free path would cut the diameter AB of the semicircular hollow to which it is approaching; and I have evaded the difficulty in a manner thoroughly suitable for thermodynamic application, such as the kinetic theory of gases. I arranged to draw lots for one out of the 199 points dividing AB into 200 equal parts. This was done by taking 100 cards*, 0, 1...98, 99, to represent distances from the middle point, and, by the toss of a coin, determining on which side of the middle point it was to be (plus or minus for head or tail, frequently changed to avoid possibility of error by bias). The draw for one of the hundred numbers (0...99) was taken after very thorough shuffling of the cards in each case. The point of entry having been found, a large scale geometrical construction was used to determine the successive points of impact and the inclination θ of the emergent path to the diameter AB. The inclination of the entering path to the diameter of the semicircular hollow struck at the end of the flight has the same value θ. If we call the diameter of the large circle unity the length of each flight is $\sin \theta$. Hence, if the velocity is unity and the mass of the particle 2, the time-integral of the whole kinetic energy is $\sin \theta$; and it is easy to prove that the time-integrals of the squares of the components of the velocity, perpendicular to and along the line from each point of the path to the centre of the large circle, are respectively $\theta \cos \theta$ and $\sin \theta - \theta \cos \theta$. By summation for 143 flights we have found

$$\Sigma \sin \theta = 121\cdot3; \quad \Sigma \theta \cos \theta = 54\cdot15;$$

whence $\quad \Sigma (\sin \theta - \theta \cos \theta) = \Sigma \theta \cos \theta + 13\cdot0.$

This is a notable deviation from the Boltzmann-Maxwell doctrine, which makes $\Sigma (\sin \theta - \theta \cos \theta)$ equal to $\Sigma \theta \cos \theta$. We have

---

* I had tried numbered billets (small squares of paper) drawn from a bowl, but found this very unsatisfactory. The best mixing we could make in the bowl seemed to be quite insufficient to secure equal chances for all the billets. Full-sized cards like ordinary playing-cards, well shuffled, seemed to give a very fairly equal chance to every card. Even with the full-sized cards, electric attraction sometimes intervenes and causes two of them to stick together. In using one's fingers to mix dry billets of card, or of paper, in a bowl, very considerable disturbance may be expected from electrification.

found the former to exceed the latter by a difference which amounts to 10·7 of the whole $\Sigma \sin \theta$.

Out of fourteen sets of ten flights I find that the time-integral of the transverse component is less than half the whole in twelve sets, and greater in only two. This seems to prove beyond doubt that the deviation from the Boltzmann-Maxwell doctrine is genuine; and that the ultimate time-integral of the transverse component is certainly smaller than the time-integral of the radial component.

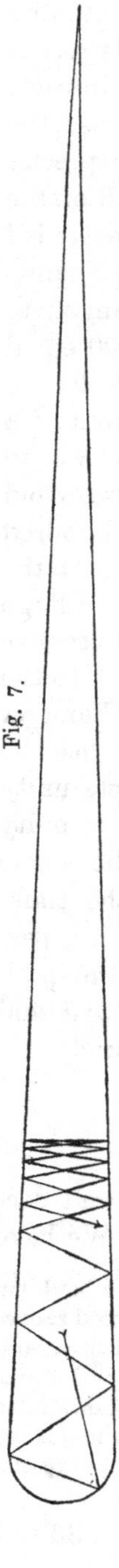

§ 39.   It is interesting to remark that, on Brewster's kaleidoscopic principle of successive images, our present result is applicable (see § 38 above). to the motion of a particle, flying about in an enclosed space, of the same shape as the surface of a marlin-spike (fig. 7), with its angle any exact submultiple of 360°, $(360° \div i)$ *. Symmetry shows that the axes of maximum or minimum kinetic energy must be in the direction of the middle line of the length of the figure and perpendicular to it. Our conclusion is that the time-integral of kinetic energy is maximum for the longitudinal component and minimum for the transverse.   In the series of flights, corresponding to the 143 of fig. 6 which we have investigated, the number of flights is of course many times 143 in fig. 7, because of the reflections at the straight sides of the marlin-spike.   It will be understood, that we are considering merely motion in one plane through the axis of the marlin-spike.

§ 40.   The most difficult and seriously troublesome statistical investigation in respect to the partition of energy which I have hitherto attempted, has been to find the proportions of translational and rotational energies in various cases, in each of which a rotator experiences multitudinous reflections at two fixed parallel planes between which it moves, or at one plane to which it is brought back by a constant force through its centre of inertia, or by a force varying directly as the distance from the plane.   Two

Fig. 7.

* See my *Electrostatics and Magnetism*, § 208, for same principle as to electric images.

different rotators were considered, one of them consisting of two equal balls, fixed at the ends of a rigid massless rod, and each ball reflected on striking either of the planes; the other consisting of two balls, 1 and 100, fixed at the ends of a rigid massless rod, the smaller ball passing freely across the plane without experiencing any force, while the greater is reflected every time it strikes. The second rotator may be described, in some respects more simply, as a hard massless ball having a mass = 1 fixed anywhere eccentrically within it, and another mass = 100 fixed at its centre. It may be called, for brevity, a biassed ball.

§ 41. In every case of a rotator whose rotation is changed by an impact, a transcendental problem of pure kinematics essentially occurs to find the time and configuration of the first impact; and another such problem to find if there is a second impact, and, if so, to determine it. Chattering collisions of one, two, three, four, five or more impacts, are essentially liable to occur, even to the extreme case of an infinite number of impacts and a collision consisting virtually of a gradually varying finite pressure. Three is the greatest number of impacts we have found in any of our calculations. The first of these transcendental problems, occurring essentially in every case, consists in finding the smallest value of θ which satisfies the equation

$$\theta - i = \frac{\omega a}{v}(1 - \sin \theta);$$

where ω is the angular velocity of the rotator before collision; a is the length of a certain rotating arm; i its inclination to the reflecting plane at the instant when its centre of inertia crosses a plane F, parallel to the reflecting plane and distant a from it; and v is the velocity of the centre of inertia of the rotator. This equation is, in general, very easily solved by calculation (trial and error), but more quickly by an obvious kinematic method, the simplest form of which is a rolling circle carrying an arm of adjustable length. In our earliest work we performed the solution arithmetically, after that kinematically. If the distance between the two parallel planes is moderate in comparison with $2a$ (the effective diameter of the rotator), i for the beginning of the collision with one plane has to be calculated from the end of the preceding collision against the other plane by a transcendental

33—2

equation, on the same principle as that which we have just been
considering. But I have supposed the distance between the two
planes to be very great, practically infinite, in comparison with $2a$,
and we have therefore found i by lottery for each collision, using
180 cards corresponding to 180° of angle. In the case of the
biassed globe, different equally probable values of i through a
range of 360° was required, and we found them by drawing from
the pack of 180 cards and tossing a coin for plus or minus.

§ 42.   Summation for 110 flights of the rotator, consisting of
two equal masses, gave as the time-integral of the whole energy
200·03, and an excess of rotatory above translatory 42·05. This
is just 21 per cent. of the whole; a large deviation from the
Boltzmann-Maxwell doctrine, which makes the time-integrals of
translatory and rotatory energies equal.

§ 43.   In the solution for the biassed ball (masses 1 and 100)
we found great irregularities due to "runs of luck" in the toss for
plus or minus, especially when there was a succession of five or six
pluses or five or six minuses. We therefore, after calculating a
sequence of 200 flights with angles each determined by lottery,
calculated a second sequence of 200 flights with the equally
probable set of angles given by the same numbers with altered
signs.   The summation for the whole 400 gave 555·55 as the
time-integral of the whole energy, and an excess, 82·5, of the
time-integral of the translatory, over the time-integral of the
rotatory energy.   This is nearly 15 per cent. We cannot, however, feel great confidence in this result, because the first set of
200 made the translatory energy less than the rotatory energy by
a small percentage (2·3) of the whole, while the second 200 gave
an excess of translatory over rotatory amounting to 35·9 per cent.
of the whole.

§ 44.   All our examples considered in detail or worked out,
hitherto, belong to Class I. of § 28. As a first example of Class II.
consider a case merging into the geodetic line on a closed surface S.
Instead of the point being constrained to remain on the surface,
let it be under the influence of a field of force, such that it is
attracted towards the surface with a finite force, if it is placed
anywhere very near the surface on either side of it, so that if the
particle be placed on S and projected perpendicularly to it, either

inwards or outwards, it will be brought back before it goes farther from the surface than a distance h, small in comparison with the shortest radius of curvature of any part of the surface. The Boltzmann-Maxwell doctrine asserts that the time-integral of kinetic energy of component motion normal to the surface would be equal to one-third of the kinetic energy of component motion at right angles to the normal; by normal being meant, a straight line drawn from the actual position of the point at any time perpendicular to the nearest part of the surface S. This, if true, would be a very remarkable proposition. If h is infinitely small we have simply the mathematical condition of constraint to remain on the surface, and the path of the particle is exactly a geodetic line. If the force towards S is zero, when the distance on either side of S is $\pm h$, we have the case of a particle placed between two guiding surfaces with a very small distance $2h$ between them. If S and therefore each of the guiding surfaces, is in every normal section convex outwards, and if the particle is placed on the outer guide-surface and projected in any direction in it with any velocity, great or small, it will remain on that guide-surface for ever, and travel along a geodetic line. If now it be deflected very slightly from motion in that surface, so that it will strike against the inner guide-surface, we may be quite ready to learn that the energy of knocking about between the two surfaces will grow up from something very small in the beginning till, in the long run, its time-integral is comparable with the time-integral of twice the energy of component motion parallel to the tangent plane of either surface. But will its ultimate value be exactly one-third that of the tangential energy, as the doctrine tells us it would be? We are, however, now back to Class I.; we should have kept to Class II. by making the normal force on the particle always finite, however great.

§ 45. Very interesting cases of Class II. § 28 occur to us readily in connection with the cases of Class I. worked out in §§ 38, 41, 42, 43.

§ 46. Let the radius of the large circle in § 38 become infinitely great: we have now a plane F (floor) with *semicircular cylindric hollows*, or semicircular hollows as we shall say for brevity; the motion being confined to one plane perpendicular to F, and to the edges of the hollows. For definiteness we shall take for F the plane of the edges of the hollows. Instead now of a particle after collision flying along the chord of the circle of § 38 it would go on for ever in a straight line. To bring it back to the plane F, let it be acted on either (α) by a force towards the plane in simple proportion to the distance, or (β) by a constant force. This latter supposition (β) presents to us the very interesting case of an elastic ball bouncing from a corrugated floor, and describing gravitational parabolas in its successive flights, the durations of

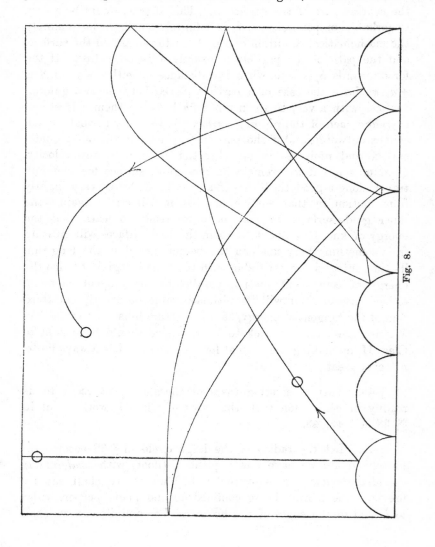

Fig. 8.

the different flights being in simple proportion to the component of velocity perpendicular to the plane F. The supposition (α) is purely ideal; but it is interesting because it gives a half curve of sines for each flight, and makes the times of flight from F after a collision and back again to F the same for all the flights, whatever be the inclination on leaving the floor and returning to it. The supposition (β) is illustrated in fig. 8, with only the variation that the corrugations are convex instead of concave, and that two vertical planes are fixed to reflect back the particle instead of allowing it to travel indefinitely, either to right or to left.

§ 47.   Let the rotator of §§ 41 to 43, instead of bouncing to and fro between two parallel planes, impinge only on one plane F, and let it be brought back by a force through its centre of inertia, either (α) varying in simple proportion to the distance of the centre of inertia from F, or (β) constant. Here, as in § 46, the times of flight in case (α) are all the same, and in (β) they are in simple proportion to the velocity of its centre of inertia when it leaves F or returns to it.

§ 48.   In the cases of §§ 46, 47 we have to consider the time-integral for each flight of the kinetic energy of the component velocity of the particle perpendicular to F, and of the whole velocity of the centre of inertia of the rotator, which is itself perpendicular to F. If q denotes the velocity perpendicular to F of the particle, or of the centre of inertia of the rotator, at the instants of crossing F at the beginning and end of the flight, and if 2 denotes the mass of the particle or of the rotator so that the kinetic energy is the same as the square of the velocity, the time-integral is in case (α) $\frac{1}{2}q^2T$, and in case (β) $\frac{1}{3}q^2T$, the time of the flight being denoted in each case by T. In both (α) and (β), § 46, if we call 1 the velocity of the particle, which is always the same, we have $q^2 = \sin^2\theta$, and the other component of the energy is $\cos^2\theta$. In § 47 it is convenient to call the total energy 1; and thus $1 - q^2$ is the total rotational energy which is constant throughout the flight. Hence, remembering that the times of flight are all the same in case (α) and are proportional to the value of q in case (β); in case (α), whether of § 46 or § 47, the time-integrals of the kinetic energies to be compared are as $\frac{1}{2}\Sigma q^2$ to $\Sigma(1 - q^2)$, and in case (β) they are as $\frac{1}{3}\Sigma q^3$ to $\Sigma q(1 - q^2)$.

§ 49.   Hence with the following notation :—

In § 46 $\begin{cases} \text{time-integral of kinetic energy perpendicular to } F, = V \\ \quad\text{,,} \qquad\qquad\qquad\text{,,} \qquad\qquad \text{parallel to} \qquad F, = U \end{cases}$

In § 47 $\begin{cases} \quad\text{,,} \qquad\qquad \text{translatory energy} = V \\ \quad\text{,,} \qquad\qquad \text{rotatory} \qquad\text{,,} \quad = R \end{cases}$

we have

$$\frac{V-U}{V+U}\begin{cases} = \dfrac{\Sigma\,(\frac{3}{2}q^2 - 1)}{\Sigma\,(1 - \frac{1}{2}q^2)} \text{ in case } (\alpha) \\[3mm] = \dfrac{\Sigma\,(\frac{4}{3}q^3 - q)}{\Sigma\,(q - \frac{2}{3}q^3)} \quad\text{,,} \quad (\beta) \end{cases}$$

$$\frac{V-R}{V+R}\begin{cases} = \dfrac{\Sigma\,(\frac{3}{2}q^2 - 1)}{\Sigma\,(1 - \frac{1}{2}q^2)} \quad\text{,,} \quad (\alpha) \\[3mm] = \dfrac{\Sigma\,(\frac{4}{3}q^3 - q)}{\Sigma\,(q - \frac{2}{3}q^3)} \quad\text{,,} \quad (\beta) \end{cases}$$

§ 49′.   By the processes described above q was calculated for the single particle and corrugated floor (§ 46), and for the rotator of two equal masses each impinging on a fixed plane (§§ 41, 42), and for the biassed ball (central and eccentric masses 100 and 1 respectively, §§ 41, 43).   Taking these values of q, summing q, q^2 and q^3 for all the flights, and using the results in § 48, we find the following six results :

Single particle bounding from corrugated floor (semicircular hollows), 143 flights :—

$\dfrac{V-U}{V+U}\begin{cases} = + \cdot197 \text{ for isochronous sinusoidal flights.} \\ = + \cdot136 \text{ for gravitational parabolic} \quad\text{,,} \end{cases}$

Rotator of two equal masses, bounding from plane floor, 110 flights :—

$\dfrac{V-R}{V+R}\begin{cases} = - \cdot179 \text{ for isochronous sinusoidal flights.} \\ = - \cdot150 \text{ for gravitational parabolic} \quad\text{,,} \end{cases}$

Biassed ball, bounding from plane floor, 400 flights :—

$\dfrac{V-R}{V+R}\begin{cases} = + \cdot025 \text{ for isochronous sinusoidal flights.} \\ = - \cdot014 \text{ for gravitational parabolic} \quad\text{,,} \end{cases}$

The smallness of the deviation of the last two results, from what the Boltzmann-Maxwell doctrine makes them, is very remarkable when we compare it with the 15 per cent. which we have found (§ 43 above) for the biassed ball bounding free from force, to and fro between two parallel planes.

§ 50. The last case of partition of energy which we have
worked out statistically relates to an impactual problem, belonging partly to Class I. § 28, and partly to Class II. It was designed
as a nearer approach to practical application in thermodynamics
than any of those hitherto described. It is, in fact, a one-dimensional illustration of the kinetic theory of gases. Suppose a row
of a vast number of atoms, of equal masses, to be allowed freedom
to move only in a straight line between fixed bounding planes L
and K. Let P the atom next K be caged between it and a parallel
plane C, at a distance from it very small in comparison with the
average of the free paths of the other particles; and let Q, the
atom next to P, be perfectly free to cross the cage-front C, without
experiencing force from it. Thus, while Q gets freely into the cage
to strike P, P cannot follow it out beyond the cage-front. The
atoms being all equal, every simple impact would produce merely
an interchange of velocities between the colliding atoms, and no
new velocity could be introduced, if the atoms were perfectly hard
(§ 16 above), because this implies that no three can be in collision
at the same time. I do not, however, limit the present investigation
to perfectly hard atoms. But, to simplify our calculations, we shall
suppose P and Q to be infinitely hard. All the other atoms we
shall suppose to have the property defined in § 21 above. They
may pass through one another in a simple collision, and go asunder
each with its previous velocity unaltered, if the differential velocity
be sufficiently great; they must recoil from one another with interchanged velocities if the initial differential velocity was not great
enough to cause them to go through one another. Fresh velocities
will generally be introduced by three atoms being in collision at
the same time, so that even if the velocities were all equal, to
begin with, inequalities would supervene in virtue of three or
more atoms being in collision at the same time; whether the
initial differential velocities be small enough to result in two
recoils, or whether one or both the mutual approaches lead to a
passage or passages through one another. Whether the distribution of velocities, which must ultimately supervene, is or is not
according to the Maxwellian law, we need not decide in our minds;
but, as a first example, I have supposed the whole-multitude to be
given with velocities distributed among them according to that
law (which, if they were infinitely hard, they would keep for ever
after); and we shall further suppose equal average spacing in

different parts of the row, so that we need not be troubled with the consideration of waves, as it were of sound, running to and fro along the row because of inequalities of density.

§ 51. For our present problem we require two lotteries to find the influential conditions at each instant, when Q enters P's cage —lottery I. for the velocity (v) of Q at impact; lottery II. for the phase of P's motion. For lottery I. (after trying 837 small squares of paper with velocities written on them and mixed in a bowl, and finding the plan unsatisfactory) we took nine stiff cards, numbered 1, 2...9, of the size of ordinary playing-cards with rounded corners, with one hundred numbers written on each in ten lines of ten numbers. The velocities on each card are shown on the following table.

TABLE SHOWING THE NUMBER OF THE DIFFERENT VELOCITIES ON THE DIFFERENT CARDS.

Velocity	·1	·2	·3	·4	·5	·6	·7	·8	·9	1·0	1·1	1·2	1·3	1·4	1·5	1·6	1·7	1·8	1·9	2·0	2·1	2·2	
Card 1	100																						
„ 2	7	93																					
„ 3		10	90																				
„ 4			9	91																			
„ 5				1	84	15																	
„ 6						60	40																
„ 7							26	57	17														
„ 8									31	40	29												
„ 9											3	26	19	15	11	9	6	4	3	2	1	1	Σv
Sums of velocities	107	103	99	92	84	75	66	57	48	40	32	26	19	15	11	9	6	4	3	2	1	1	900

The number of times each velocity occurs was chosen to fulfil as nearly as may be the Maxwellian law, which is $Cdve^{-\frac{v^2}{k}} = $ the number of velocities between $v + \frac{1}{2}dv$ and $v - \frac{1}{2}dv$. We took $k = 1$, which, if dv were infinitely small, would make the mean of the squares of the velocities equal exactly to ·5; we took $dv = ·1$ and $Cdv = 108$ to give, as nearly as circumstances would allow, the Maxwellian law, and to make the total number of different velocities 900. The sum of the squares of all these 900 velocities is 468·4, which divided by 900 is ·52. In the practice of this lottery the numbered cards were well shuffled and then one was drawn; the particular one of the hundred velocities on this card to be chosen was found by drawing one card from a pack of one hundred numbered 1, 2...99, 100. In lottery II. a pack of one hundred cards is used to draw one of one hundred decimal numbers from ·01 to 1·00. The decimal drawn, called α, shows the proportion of the whole period of P from the cage-front C, to K, and back to C, still unperformed at the instant when Q crosses C. Now remark, that if Q overtakes P in the first half of its period it gives its velocity v to P, and follows it inwards; and therefore there must be a second impact when P meets it after reflection from K, and gives it back the velocity v which it had on entering. If Q meets P in the second half of its period, Q will, by the first impact, get P's original velocity, and may with this velocity escape from the cage. But it may be overtaken by P before it gets out of the cage, in which case it will go away from the cage with its own original velocity v unchanged. This occurs always if, and never unless, u is less than $v\alpha$; P's velocity being denoted by u, and Q's by v. This case of Q overtaken by P can only occur if the entering velocity of Q is greater than the speed of P before collision. Except in this case P's speed is unchanged by the collision. Hence we see that it is only when P's speed is greater than Q's before collision that there can be interchange, and this interchange leaves P with less speed than Q. If every collision involved interchange, the average velocity of P would be equalised by the collisions to the average velocity of Q, and the average distribution of different velocities would be identical for Q and P. Non-fulfilment of this equalising interchange can, as we have seen, only occur when Q's speed is less than P's, and therefore the average speed and the average kinetic energy of P must be less than the average kinetic energy of Q.

§ 52. We might be satisfied with this, as directly negativing the Boltzmann-Maxwell doctrine for this case. It is however, interesting to know, not only that the average kinetic energy of Q is greater than that of the caged atom, but, further, to know how much greater it is. We have therefore worked out sum-mations for 300 collisions between P and Q, beginning with

$u^2 = \cdot5$ ($u = \cdot71$), being approximately the mean of v^2 as given by the lottery. It would have made no appreciable difference in the result if we had begun with any value of u, large or small, other than zero. Thus, for example, if we had taken 100 as the first value of u, this speed would have been taken by Q at the first impact, and sent away along the practically infinite row, never to be heard of again; and the next value of u would have been the first value drawn by lottery for v. Immediately before each of the subsequent impacts, the velocity of P is that which it had from Q by the preceding impact. In our work, the speeds which P actually had at the first sixteen times of Q's entering the cage were $\cdot71$, $\cdot5$, $\cdot3$, $\cdot2$, $\cdot2$, $\cdot1$, $\cdot1$, $\cdot2$, $\cdot2$, $\cdot5$, $\cdot7$, $\cdot2$, $\cdot3$, $\cdot6$, $1\cdot5$, $\cdot5$—from which we see how little effect the choice of $\cdot71$ for the first speed of P had on those that follow. The summations were taken in successive groups of ten; in every one of these Σv^2 exceeded Σu^2. For the 300 we found $\Sigma v^2 = 148\cdot53$ and $\Sigma u^2 = 61\cdot62$, of which the former is $2\cdot41$ times the latter. The two ought to be equal according to the Boltzmann-Maxwell doctrine. Dividing Σv^2 by 300 we find $\cdot495$; which chances to be more nearly equal to the intended $\cdot5$ than the $\cdot52$ which is on the cards (§ 51 above). A still greater deviation from the B.-M. equality ($2\cdot71$ instead of $2\cdot41$) was found by taking Σv^3 and $\Sigma u'^2 v$ to allow for greater probability of impact with greater than with smaller values of v; u' being the velocity of P after collision with Q.

§ 53. We have seen in § 52 that Σu^2 must be less than Σv^2, but it seemed interesting to find how much less it would be with some other than the Maxwellian law of distribution of velocities. We therefore arranged cards for a lottery, with an arbitrarily chosen distribution, quite different from the Maxwellian. Eleven cards, each with one of the eleven numbers 1, 3...19, 21, to correspond to the different velocities $\cdot1$, $\cdot3$...$1\cdot9$, $2\cdot1$, were prepared and used instead of the nine cards in the process described in § 51 above. In all except one of the eleven tens, Σv^2 was greater than Σu^2, and for the whole 110 impacts we found $\Sigma v^2 = 179\cdot90$, and $\Sigma u^2 = 97\cdot66$; the former of these is $1\cdot84$ times the latter. In this case we found the ratio of Σv^3 to $\Sigma u'^2 v$ to be $1\cdot87$.

§ 54. In conclusion, I wish to refer, in connection with Class II. § 28, to a very interesting and important application of the doctrine, made by Maxwell himself, to the equilibrium of a tall column of gas under the influence of gravity.   Take, first, our one-dimensional gas of § 50, consisting of a straight row of a vast number of equal and similar atoms.   Let now the line of the row be vertical, and let the atoms be under the influence of terrestrial gravity, and suppose, first, the atoms to resist mutual approach, sufficiently to prevent any one from passing through another with the greatest relative velocity of approach that the total energy given to the assemblage can allow.   The Boltzmann-Maxwell doctrine (§ 18 above) asserting as it does that the time-integral of the kinetic energy is the same for all the atoms, makes the time-average of the kinetic energy the same for the highest as for the lowest in the row.   This, if true, would be an exceedingly interesting theorem.   But now, suppose two approaching atoms not to repel one another with infinite force at any distance between their centres, and suppose energy to be given to the multitude sufficient to cause frequent instances of two atoms passing through one another.   Still the doctrine can assert nothing but that the time-integral of the kinetic energy of any one atom is equal to that of any other atom, which is now a self-evident proposition, because the atoms are of equal masses, and each one of them in turn will be in every position of the column, high or low.   (If in the row there are atoms of different masses, the Waterston-Maxwell doctrine of equal average energies would, of course, be important and interesting.)

§ 55.   But now, instead of our ideal one-dimensional gas, consider a real homogeneous gas, in an infinitely hard vertical tube, with an infinitely hard floor and roof, so that the gas is under no influence from without, except gravity.   First, let there be only two or three atoms, each given with sufficient velocity to fly against gravity from floor to roof.   They will strike one another occasionally, and they will strike the sides and floor and roof of the tube much more frequently than one another.   The time-averages of their kinetic energies will be equal.   So will they be if there are twenty atoms, or a thousand atoms, or a million, million, million, million, million atoms.   *Now* each atom will strike another atom much more frequently than the sides or floor or roof of the tube.   In the long run each atom will be in every part of the tube as often as is every other atom.   The time-integral of the kinetic energy of any one atom will be equal to the time-integral of the kinetic energy of any other atom. This truism is simply and solely all that the Boltzmann-Maxwell doctrine asserts for a vertical column of a homogeneous monatomic gas. It is, I believe, a general impression that the Boltzmann-Maxwell doctrine, asserting a law of partition of the kinetic part of the whole energy, includes obviously a theorem that the average kinetic energies of the atoms in the upper parts of a vertical column of gas, are equal to those of the atoms in the lower parts of the column. Indeed, with the wording of Maxwell's statement, § 18, before us, we might suppose it to assert that two parts of our vertical column of gas, if they contain the same number of atoms, must have the same kinetic energy, though they be situated, one of them near the bottom of the column, and the other near the top. Maxwell himself, in his 1866 paper ("The Dynamical Theory of Gases")*, gave an independent synthetical demonstration of this proposition, and did not subsequently, so far as I know, regard it as immediately deducible from the partitional doctrine generalised by Boltzmann and himself several years after the date of that paper.

§ 56. Both Boltzmann and Maxwell recognised the experimental contradiction of their doctrine presented by the kinetic theory of gases; and felt that an explanation of this incompatibility was imperatively called for. For instance, Maxwell, in a lecture on the dynamical evidence of the molecular constitution of bodies, given to the Chemical Society, Feb. 18, 1875, said: "I have put "before you what I consider to be the greatest difficulty yet en- "countered by the molecular theory. Boltzmann has suggested "that we are to look for the explanation in the mutual action "between the molecules and the ethereal medium which surrounds "them. I am afraid, however, that if we call in the help of this "medium we shall only increase the calculated specific heat, "which is already too great." Rayleigh, who has for the last twenty years been an unwavering supporter of the Boltzmann-Maxwell doctrine, concludes a paper "On the Law of Partition of Energy," published a year ago in the *Phil. Mag.*, Jan. 1900, with the following words: "The difficulties connected with the appli- "cation of the law of equal partition of energy to actual gases "have long been felt. In the case of argon and helium and

* Addition, of date December 17, 1866. *Collected Papers*, Vol. II. p. 76.

ABANDON THE DOCTRINE CONSTITUTING CLOUD II.   527

"mercury vapour, the ratio of specific heats (1·67) limits the
"degrees of freedom of each molecule to the three required for
"translatory motion.   The value (1·4) applicable to the principal
"diatomic gases, gives room for the three kinds of translation and
"for two kinds of rotation.   Nothing is left for rotation round
"the line joining the atoms, nor for relative motion of the atoms
"in this line.   Even if we regard the atoms as mere points, whose
"rotation means nothing, there must still exist energy of the
"last-mentioned kind, and its amount (according to law) should
"not be inferior.

   " We are here brought face to face with a fundamental difficulty,
"relating not to the theory of gases merely, but rather to general
"dynamics.   In most questions of dynamics, a condition whose
"violation involves a large amount of potential energy may be
"treated as a constraint.   It is on this principle that solids are
"regarded as rigid, strings as inextensible, and so on.   And it is
"upon the recognition of such constraints that Lagrange's method
"is founded.   But the law of equal partition disregards potential
"energy.   However great may be the energy required to alter
"the distance of the two atoms in a diatomic molecule, practical
"rigidity is never secured, and the kinetic energy of the relative
"motion in the line of junction is the same as if the tie were of
"the feeblest.   The two atoms, however related, remain two atoms,
"and the degrees of freedom remain six in number.

   "What would appear to be wanted is some escape from the
"destructive simplicity of the general conclusion."

   The simplest way of arriving at this desired result is to deny
the conclusion; and so, in the beginning of the twentieth century,
to lose sight of a cloud which has obscured the brilliance of the
molecular theory of heat and light during the last quarter of the
nineteenth century.

APPENDIX C.

ON THE DISTURBANCE PRODUCED BY TWO PARTICULAR
FORMS OF INITIAL DISPLACEMENT IN AN INFINITELY
LONG MATERIAL SYSTEM FOR WHICH THE VELOCITY
OF PERIODIC WAVES DEPENDS ON THE WAVE-LENGTH.

§ 1. ONE of the chosen forms of initial displacement is

$$ce^{-x^2/a^2} \quad\dots\dots\dots\dots\dots\dots\dots\dots(1),$$

where x is distance from O, the centre of the origin of the
disturbance; a is a length-parameter of the initial disturbance;
and c is a length very small (regarded in the mathematics as
infinitely small) in comparison with a.

§ 2. Let the displacement at (x, t) in an infinite train of
periodic waves be expressed by

$$D \cos q\,(x - \alpha - vt) \quad\dots\dots\dots\dots\dots\dots(2).$$

Here $2\pi/q$ is the wave-length; and v is the propagational
velocity which we may regard as a function of q; and α is the
distance from O of a point where the displacement is a positive
maximum at time, $t = 0$. Any number of solutions such as (2)
with different values of D, q, and α, or with $-v$ for v, superimposed, give a complex solution, found by simple summation
because the displacements and their sums are infinitely small.
Hence a solution, *the* solution for standing vibrations of period
$2\pi/qv$, is found by taking the half-sum of (2) for $\pm v$; being
this:—

$$D \cos q\,(x - \alpha) \cos vt \quad\dots\dots\dots\dots\dots(3).$$

And a summation of (3) for our present purpose gives a solution,

$$\eta = \frac{c}{\pi} \int_0^\infty dq \int_{-\infty}^\infty d\alpha e^{-\alpha^2/a^2} \cos q\,(x - \alpha) \cos vt \quad\dots\dots(4).^{*}$$

When $t = 0$ this gives $d/dt\ \eta = 0$; and, by " Fourier's theorem,"
the second member becomes ce^{-x^2/a^2} which agrees with (1). Hence
(4) expresses *the solution of our problem*.

* See Lec. X. Part I., top of p. 116.

§ 3. If in (4) we expand $\cos q (x - \alpha)$, the factor of $\sin qx$ disappears in the summation $\int_{-\infty}^{\infty} d\alpha$; and the factor of $\cos qx$, being $\int_{-\infty}^{\infty} d\alpha \epsilon^{-\alpha^2/a^2} \cos q\alpha$, is equal to $\sqrt{\pi} a \epsilon^{-q^2 a^2/4}$ according to an evaluation given by Laplace in 1810*. Thus we reduce (4) to the following simpler form

$$\eta = \frac{ca}{\sqrt{\pi}} \int_0^{\infty} dq \epsilon^{-q^2 a^2/4} \cos qx \cos qvt \ \ldots\ldots\ldots(5).$$

The verification that this becomes $c\epsilon^{-x^2/a^2}$ when $t = 0$, is *most interesting*: it is done by a second application of the same evaluation of Laplace's with a curiously inverted notation.

§ 4. Whatever function v may be of the wave-length, $2\pi/q$, we have in (5) a thoroughly convenient solution of our problem; calculable by highly convergent quadratures whether the definite integral is reducible to finite terms or not. For one very interesting case, an infinitely long elastic rod, taken as an example by Fourier†, the definite integral *is reducible to finite terms*.

In this case the velocity of periodic waves is inversely as the wave-length, and, by choosing our units conveniently, we may put $v = q$, in (5). Thus, and by taking $a = 1$, and by substituting $\frac{1}{2}\int_{-\infty}^{\infty} dq$ for $\int_0^{\infty} dq$, we find

$$\eta = \frac{c}{2\sqrt{\pi}} \int_{-\infty}^{\infty} dq \epsilon^{-q^2/4} \cos q^2 t \cos qx \ \ldots\ldots\ldots(6).$$

Denoting by R the half-sum for $\pm \iota$, we may write this thus;

$$\eta = \frac{c}{2\sqrt{\pi}} R \int_{-\infty}^{\infty} dq \epsilon^{-[(\frac{1}{4} - \iota t) q^2 - \iota qx]} \ \ldots\ldots\ldots(7).$$

Putting $\frac{1}{4} - \iota t = h^2$, we reduce the index of the exponential to $-(hq - \iota x/2h)^2 - (x/2h)^2$; and putting $hq - \iota x/2h = z$, we find for the definite integral,

$$h^{-1} \int_{-\infty}^{\infty} dz \epsilon^{-z^2} \cdot \epsilon^{-(x/2h)^2} = \frac{\sqrt{\pi}}{h} \cdot \epsilon^{-(x/2h)^2} \ \ldots\ldots\ldots(8).$$

---

* *Mémoires de l'Institut*, 1810. See Gregory's *Examples*, p. 480.
† *Chaleur*, Chap. xix. Article 406.

Hence (7) becomes

$$\eta = \tfrac{1}{2}cRh^{-1}\,\epsilon^{-(x/2h)^2} = \tfrac{1}{2}cRh^{-1}\epsilon^{\frac{-x^2(1+4ti)}{1+16t^2}} \quad \ldots\ldots\ldots\ldots(9).$$

Now $$1/h = (\tfrac{1}{16} + t^2)^{-\frac{1}{4}}\,(\cos \tfrac{1}{2}\phi + \iota \sin \tfrac{1}{2}\phi),$$

if $$\phi = \tan^{-1} 4t \quad \ldots\ldots\ldots\ldots\ldots\ldots(10).$$

Hence, if we put $$\frac{x^2\,.\,4t}{1 + 16t^2} = \theta \quad \ldots\ldots\ldots\ldots\ldots(11),$$

we find by (9), $\quad \eta = c\,(1 + 16t^2)^{-\frac{1}{4}}\,\epsilon^{\frac{-x^2}{1+16t^2}} \cos\,(\tfrac{1}{2}\phi - \theta) \quad \ldots\ldots\ldots(12).$

§ 5. This is a very interesting solution in "finite terms." To illustrate it let first $x = 0$; we have, $\theta = 0$, and

$$\eta = c\,(1 + 16t^2)^{-\frac{1}{4}} \cos \tfrac{1}{2}\phi = \sqrt{\frac{1 + \tau}{2\tau^2}}, \text{ where } \tau = \surd(1 + 16t^2)\ldots(13).$$

Hence the middle point of the originally displaced part of the rod subsides to its undisturbed position *not vibrationally*, but continuously, and ultimately in proportion to $t^{-\frac{1}{2}}$ when t is very great. And when t is very great we may neglect ϕ in (12), and we have $\theta \fallingdotseq x/\tau$; so that we find

$$\eta \fallingdotseq c\tau^{-\frac{1}{2}}\,\epsilon^{-(x/\tau)^2} \cos \frac{x^2}{\tau} \quad \ldots\ldots\ldots\ldots(14).$$

This shows that, in going from the origin we first find a zero of displacement at $x = \surd(\pi\tau/2)$; and beyond this there is an infinite number of zeros, and vibratory subsidence to repose. Interesting graphic illustrations will require but little labour.

§ 6. As another even more interesting case, take the two-dimensional case of the old deep-water wave problem of Poisson and Cauchy. In it the velocity of periodic waves is proportional to $\surd\lambda$ and we may therefore conveniently take $v = q^{-\frac{1}{2}}$. This in (5), with $a = 1$, gives

$$\eta = \frac{c}{\surd\pi} \int_0^\infty dq\,\epsilon^{-q^2/4} \cos qx \cos \surd qt \ldots\ldots\ldots\ldots(15),$$

an irreducible definite integral which expresses the displacement of the water at time t and distance x from the origin of the disturbance; by a convergent formula thoroughly convenient for calculation by quadratures for all values large or small of the two variables. Analytical expedients may no doubt be found to diminish the labour: but it is satisfactory to know that the method of quadratures, with moderate labour, gives very interesting illustrations; though great labour would be needed for full graphical illustrations by time-curves and space-curves.

§ 7. My assistant Mr Witherington has evaluated η by (15) with $c/\sqrt{\pi} = 2$; for four cases as follows:

Case 1. $x = 0$, $t = 0$; $\eta = 1·763$; $\sqrt{\pi} = 1·77245$,
  ,,   2. $x = 0$, $t = 1$; $\eta = ·911$;
  ,,   3. $x = 0$, $t = 2$; $\eta = -·451$;
  ,,   4. $x = 2$, $t = 2$; $\eta = ·559$.

Case 1 was worked as a test for accuracy in a first trial of quadrature. The result is about 2/3 per cent. too small; the correct result being $\sqrt{\pi}$. The accuracy of the quadratures is sufficient for merely illustrative purposes.

§ 8. Interesting as is the solution by quadratures for water-waves resulting from the initial disturbance (1), it is less important than the following solution in finite terms *for another form of initial disturbance*, which was given in my short paper "On the Front and Rear of a Procession of free Waves in Deep Water"*:—

$$\xi = \frac{d\phi}{dx}, \quad \eta = \frac{d\phi}{dy}, \quad \phi = \frac{2c}{r} \left\{ (r+y)^{\frac{1}{2}} \cos \frac{gt^2 x}{4r^2} + (r-y)^{\frac{1}{2}} \sin \frac{gt^2 x}{4r^2} \right\} \epsilon^{-\frac{gt^2 y}{4r^2}}$$

$$\dotsc\dotsc(16).$$

Here (ξ, η) denote the displacement of the particle of water whose equilibrium-position is (x, y); r denotes $(x^2 + y^2)^{\frac{1}{2}}$; and g, gravity. According to a beautiful discovery of Cauchy and Poisson, for waves in infinitely deep water, the pressure is constant at all the particles $(x + \xi_0, y + \eta_0)$ for any constant value, y_0, of y; and therefore the water above every such surface may be removed without disturbing the motion of the water below it. Hence η_0 is the vertical displacement of any point in a free surface of particles whose undisturbed position is $y = y_0$. The value of η_0 for $t = 0$, from (16), is

$$c (r_0^2 - r_0 y_0 - 2y_0^2)/r_0^3 (r_0 + y_0)^{\frac{1}{2}} \dotsc\dotsc\dotsc(17).$$

This, with any value we please to assign to y_0, is the initiating free-surface for (16); r_0 denotes $(x^2 + y_0^2)^{\frac{1}{2}}$.

I hope to reproduce, with extensions, for early publication, curves of the wave-motion expressed by (16), shown in Section A of the British Association, Sep. 1886, and in the Royal Society of Edinburgh, Dec. 20, 1886.

* *Proc. R.S.E.* Jan. 7, 1887; *Phil. Mag.* Feb. 1887.

APPENDIX D.

ON THE CLUSTERING OF GRAVITATIONAL MATTER IN ANY PART OF THE UNIVERSE*.

GRAVITATIONAL matter, according to our ideas of universal gravitation, would be all matter. Now is there any matter which is not subject to the law of gravitation? I think I may say with absolute decision that there is. We are all convinced, with our President, that ether is matter, but we are forced to say that we must not expect to find in ether the ordinary properties of molar matter which are generally known to us by action resulting from force between atoms and matter, ether and ether, and atoms of matter and ether. Here I am illogical when I say between matter and ether, as if ether were not matter. It is to avoid an illogical phraseology that I use the title "gravitational matter." Many years ago I gave strong reason to feel certain that ether was outside the law of gravitation. We need not absolutely exclude, as an idea, the possibility of there being a portion of space occupied by ether beyond which there is absolute vacuum—no ether and no matter. We admit that that is something that one could think of; but I do not believe any living scientific man considers it in the slightest degree probable that there is a boundary around our universe beyond which there is no ether and no matter. Well, if ether extends through all space, then it is certain that ether cannot be subject to the law of mutual gravitation between its parts, because if it were subject to mutual attraction between its parts its equilibrium would be unstable, unless it were infinitely incompressible. But here, again, I am reminded of the critical character of the ground on which we stand in speaking of properties of matter beyond what we see or feel by experiment. I am afraid I must here express a view different from that which

---

* Communicated by the Author to the *Phil. Mag.*, having been read before the British Association at the Glasgow meeting.

Professor Rücker announced in his Address, when he said that continuity of matter implied absolute resistance to condensation. We have no right to bar condensation as a property of ether. While admitting ether not to have any atomic structure, it is postulated as a material which performs functions of which we know something, and which may have properties allowing it to perform other functions of which we are not yet cognisant. If we consider ether to be matter, we postulate that it has rigidity enough for the vibrations of light, but we have no right to say that it is absolutely incompressible. We must admit that sufficiently great pressure all round could condense the ether in a given space, allowing the ether in surrounding space to come in towards the ideal shrinking surface. When I say that ether must be outside the law of gravitation, I assume that it is not infinitely incompressible. I admit that if it were infinitely incompressible, it might be subject to the law of mutual gravitation between its parts; but to my mind it seems infinitely improbable that ether is infinitely incompressible, and it appears more consistent with the analogies of the known properties of molar matter, which should be our guides, to suppose that ether has not the quality of exerting an infinitely great force against compressing action of gravitation. Hence, if we assume that it extends through all space, ether must be outside the law of gravitation—that is to say, truly imponderable. I remember the contempt and self-complacent compassion with which sixty years ago I myself—I am afraid—and most of the teachers of that time looked upon the ideas of the elderly people who went before us, and who spoke of " the imponderables." I fear that in this, as in a great many other things in science, we have to hark back to the dark ages of fifty, sixty, or a hundred years ago, and that we must admit there is something which we cannot refuse to call matter, but which is not subject to the Newtonian law of gravitation. That the sun, stars, planets, and meteoric stones are all of them ponderable matter is true, but the title of my paper implies that there is something else. Ether is not any part of the subject of this paper; what we are concerned with is gravitational matter, ponderable matter. Ether we relegate, not to a limbo of imponderables, but to distinct species of matter which have inertia, rigidity, elasticity, compressibility, but not heaviness. In a paper I have already published I gave strong reasons for limiting to a

definite amount the quantity of matter in space known to astro-
nomers. I can scarcely avoid using the word "universe," but I
mean our universe, which may be a very small affair after all,
occupying a very small portion of all the space in which there is
ponderable matter.

Supposing a sphere of radius $3·09.10^{16}$ kilometres (being the
distance at which a star must be to have parallax $0''·001$) to have
within it, uniformly distributed through it, a quantity of matter
equal to one thousand million times the sun's mass; the velocity
acquired by a body placed originally at rest at the surface would,
in five million years, be about 20 kilometres per second, and in
twenty-five million years would be 108 kilometres per second (if
the acceleration remained sensibly constant for so long a time).
Hence, if the thousand million suns had been given at rest
twenty-five million years ago, uniformly distributed throughout
the supposed sphere, many of them would now have velocities of
20 or 30 kilometres per second, while some would have less and
some probably greater velocities than 108 kilometres per second;
or, if they had been given thousands of million years ago at rest
so distributed that now they were equally spaced throughout the
supposed sphere, their mean velocity would now be about 50 kilometres per second*. This is not unlike the measured velocities
of stars, and hence it seems probable that there might be as
much matter as one thousand million suns within the distance
$3·09.10^{16}$ kilometres. The same reasoning shows that ten thousand
million suns in the same sphere would produce, in twenty-
five million years, velocities far greater than the known star
velocities: and hence there is probably much less than ten
thousand million times the sun's mass in the sphere considered.
A general theorem discovered by Green seventy-three years ago
regarding force at a surface of any shape, due to matter (gravitational, or ideal electric, or ideal magnetic) acting according to
the Newtonian law of the inverse square of the distance, shows
that a non-uniform distribution of the same total quantity of
matter would give greater velocities than would the uniform distribution. Hence we cannot, by any non-uniform distribution of
matter within the supposed sphere of $3·09.10^{16}$ kilometres radius,
escape from the conclusion limiting the total amount of the

* *Phil. Mag.* August 1901, pp. 169, 170.

matter within it to something like one thousand million times the sun's mass.

If we compare the sunlight with the light from the thousand million stars, each being supposed to be of the same size and brightness as our sun, we find that the ratio of the apparent brightness of the star-lit sky to the brightness of our sun's disc would be $3 \cdot 87.10^{-13}$. This ratio* varies directly with the radius of the containing sphere, the number of equal globes per equal volume being supposed constant; and hence to make the sum of the apparent areas of discs $3 \cdot 87$ per cent. of the whole sky, the radius must be $3 \cdot 09.10^{27}$ kilometres. With this radius light would take $3\frac{1}{4}.10^{14}$ years to travel from the outlying stars to the centre. Irrefragable dynamics proves that the life of our sun as a luminary may probably be between twenty-five and one hundred million years; but to be liberal, suppose each of our stars to have a life of one hundred million years as a luminary: and it is found that the time taken by light to travel from the outlying stars to the centre of the sphere is three and a quarter million times the life of a star. Hence it follows that to make the whole sky aglow with the light of all the stars at the same time the commencements of the stars must be timed earlier and earlier for the more and more distant ones, so that the time of the arrival of the light of every one of them at the earth may fall within the durations of the lights of all the others at the earth. My supposition as to uniform density is quite arbitrary; but nevertheless I think it highly improbable that there can be enough of stars (bright or dark) to make a total of star-disc area more than 10^{-12} or 10^{-11} of the whole sky.

To help to understand the density of the supposed distribution of one thousand million suns in a sphere of $3 \cdot 09.10^{16}$ kilometres radius, imagine them arranged exactly in cubic order, and the volume per sun is found to be $123 \cdot 5.10^{39}$ cubic kilometres, and the distance from one star to any one of its six nearest neighbours would be $4 \cdot 98.10^{13}$ kilometres. The sun seen at this distance would probably be seen as a star of between the first and second magnitudes; but supposing our thousand million suns to be all of such brightness as to be stars of the first magnitude at distance corresponding to parallax $1'' \cdot 0$, the brightness at distance $3 \cdot 09.10^{16}$ kilometres would be one one-millionth of this; and the most

* *Phil. Mag.* August 1901, p. 175.

distant of our stars would be seen through powerful telescopes as stars of the sixteenth magnitude. Newcomb estimated from thirty to fifty million as the number of stars visible in modern telescopes. Young estimated at one hundred million the number visible through the Lick telescope. This larger estimate is only one-tenth of our assumed one thousand million masses equal to the sun, of which, however, nine hundred million might be either non-luminous, or, though luminous, too distant to be seen by us at their actual distances from the earth. Remark, also, that it is only for facility of counting that we have reckoned our universe as one thousand million suns; and that the meaning of our reckoning is that the total amount of matter within a sphere of $3.09.10^{16}$ kilometres radius is one thousand million times the sun's mass. The sun's mass is $1.99.10^{27}$ metric tons, or $1.99.10^{33}$ grammes. Hence our reckoning of our supposed spherical universe is that the ponderable part of it amounts to $1.99.10^{42}$ grammes, or that its average density is $1.61.10^{-23}$ of the density of water.

Let us now return to the question of sum of apparent areas. The ratio of this sum to 4π, the total apparent area of the sky viewed in all directions, is given by the formula*:

$$\alpha = \frac{3N}{4}\left(\frac{a}{r}\right)^2,$$

provided its amount is so small a fraction of unity that its diminution by eclipses, total or partial, may be neglected. In this formula, N is a number of globes of radius a uniformly distributed within a spherical surface of radius r. For the same quantity of matter in N' globes of the same density, uniformly distributed through the same sphere of radius r, we have

$$\frac{N'}{N} = \left(\frac{a}{a'}\right)^3$$

and therefore

$$\frac{a'}{a} = \frac{\alpha}{\alpha'}.$$

With $N = 10^9$, $r = 3.09.10^{16}$ kilometres; and a (the sun's radius) $= 7.10^5$ kilometres; we had $\alpha = 3.87.10^{-13}$. Hence $a' = 7$ kilometres gives $\alpha' = 3.87.10^{-8}$; and $a'' = 1$ centimetre gives $\alpha'' = 1/36.9$. Hence if the whole mass of our supposed universe were reduced

* *Phil. Mag.* August 1901, p. 175.

to globules of density 1·4 (being the sun's mean density), and of 2 centimetres diameter, distributed uniformly through a sphere of 3·09.10^{16} kilometres radius, an eye at the centre of this sphere would lose only 1/36·9 of the light of a luminary outside it! The smallness of this loss is easily understood when we consider that there is only one globule of 2 centimetres diameter per 364,000,000 cubic kilometres of space, in our supposed universe reduced to globules of 2 centimetres diameter. Contrast with the total eclipse of the sun by a natural cloud of water spherules, or by the cloud of smoke from the funnel of a steamer.

Let now all the matter in our supposed universe be reduced to atoms (literally brought back to its probable earliest condition). Through a sphere of radius r let atoms be distributed uniformly in respect to gravitational quality. It is to be understood that the condition "uniformly" is fulfilled if equivoluminal globular or cubic portions, small in comparison with the whole sphere, but large enough to contain large numbers of the atoms, contain equal total masses, reckoned gravitationally, whether the atoms themselves are of equal or unequal masses, or of similar or dissimilar chemical qualities. As long as this condition is fulfilled, each atom experiences very approximately the same force as if the whole matter were infinitely fine-grained, that is to say, utterly homogeneous.

Let us therefore begin with a uniform sphere of matter of density ρ, gravitational reckoning; with no mutual forces except gravitation between its parts, given with every part at rest at the initial instant: and let it be required to find the subsequent motion. Imagining the whole divided into infinitely thin concentric spherical shells, we see that every one of them falls inwards, as if attracted by the whole mass within it collected at the centre. Hence our problem is reduced to the well known students' exercise of finding the rectilinear motion of a particle attracted according to the inverse square of the distance from a fixed point. Let x_0 be the initial distance, $\dfrac{4\pi\rho}{3}x_0{}^3$ the attracting mass, v and x velocity and distance from the centre at time t. The solution of the problem for the time during which the particle is falling towards the centre is

$$\tfrac{1}{2}v^2 = \frac{4\pi\rho}{3}x_0{}^3\left(\frac{1}{x} - \frac{1}{x_0}\right)$$

and $t = \sqrt{\dfrac{3}{8\pi\rho}}\left(\dfrac{\pi}{2} - \theta + \tfrac{1}{2}\sin 2\theta\right) = \dfrac{\pi}{2}\sqrt{\dfrac{3}{8\pi\rho}}\left[1 - \dfrac{2\theta}{\pi}\left(1 - \dfrac{\sin 2\theta}{2\theta}\right)\right],$

where θ denotes the acute angle whose sine is $\sqrt{\dfrac{x}{x_0}}$. This shows
that the time of falling through any proportion of the initial
distance is the same whatever be the initial distance; and that
the time (which we shall denote by T) of falling to the centre is
$\tfrac{1}{2}\pi\sqrt{\dfrac{3}{8\pi\rho}}$. Hence in our problem of homogeneous gravitational
matter given at rest within a spherical surface and left to fall
inwards, the augmenting density remains homogeneous, and the
time of shrinkage to any stated proportion of the initial radius
is inversely as the square root of the density.

To apply this result to the supposed spherical universe of
radius $3{\cdot}09.10^{16}$ kilometres, and mass equal to a thousand million
times the mass of our sun, we find the gravitational attraction on
a body at its surface gives acceleration of $1{\cdot}37.10^{-13}$ kilometres
per second per second. This therefore is the value of $\dfrac{4\pi\rho}{3}x_0$,
with one second as the unit of time and one kilometre as the
unit of distance; and we find $T = 52{\cdot}8.10^{13}$ seconds $= 16{\cdot}8$ million
years. Thus our formulas become

$$\tfrac{1}{2}v^2 = 1{\cdot}37.10^{-13}\,x_0\left(\dfrac{x_0}{x} - 1\right)$$

giving

$$v = 5{\cdot}23.19^{-7}\sqrt{x_0\left(\dfrac{x_0}{x} - 1\right)}$$

and

$$t = 52{\cdot}8.10^{13}\left[1 - \dfrac{2\theta}{\pi}\left(1 - \dfrac{\sin 2\theta}{2\theta}\right)\right];$$

whence, when $\sin\theta$ is very small,

$$t = 52{\cdot}8.10^{13}\left(1 - \dfrac{4\theta^3}{3\pi}\right).$$

Let now, for example, $x_0 = 3{\cdot}09.10^{16}$ kilometres, and $\dfrac{x_0}{x} = 10^7$;
and, therefore, $\sin\theta = \theta = 3{\cdot}16.10^{-4}$; whence, $v = 291{,}000$ kilometres per second, and $t = T - 7080$ seconds $= T - 2$ hours approximately.

By these results it is most interesting to know that our supposed sphere of perfectly compressible fluid, beginning at rest
with density $1{\cdot}61.10^{-23}$ of that of water, and of any magnitude

large or small, and left unclogged by ether to shrink under the influence of mutual gravitation of its parts, would take nearly seventeen million years to reach ·0161 of the density of water, and about two hours longer to shrink to infinite density at its centre. It is interesting also to know that if the initial radius is $3·09.10^{16}$ kilometres, the inward velocity of the surface is 291,000 kilometres per second at the instant when its radius is $3·09.10^9$ and its density ·0161 of that of water. If now, instead of an ideal compressible fluid, we go back to atoms of ordinary matter of all kinds as the primitive occupants of our sphere of $3·09.10^{16}$ kilometres radius, all these conclusions (provided all the velocities are less than the velocity of light) would still hold, notwithstanding the ether occupying the space through which the atoms move. This would, I believe*, exercise no resistance whatever to uniform motion of an atom through it; but it would certainly add quasi-inertia to the intrinsic Newtonian inertia of the atom itself moving through ideal space void of ether; which, according to the Newtonian law, would be exactly in proportion to the amount of its gravitational quality. The additional quasi-inertia must be exceedingly small in comparison with the Newtonian inertia, as is demonstrated by the Newtonian proofs, including that founded on Kepler's laws for the groups of atoms constituting the planets, and movable bodies experimented on at the earth's surface.

In one thousand seconds of time after the density ·0161 of the density of water is reached, the inward surface velocity would be 305,000 kilometres per second, or greater than the velocity of light; and the whole surface of our condensing globe of gas or vapour or crowd of atoms would begin to glow, shedding light inwards and outwards. All this is absolutely realistic, except the assumption of uniform distribution through a sphere of the enormous radius of $3·09.10^{16}$ kilometres, which we adopted temporarily for illustrational purpose. The enormously great velocity (291,000 kilometres per second) and rate of acceleration (13·7 kilometres per second per second) of the boundary inwards, which we found at the instant of density ·0161 of that of water, are due to greatness of the primitive radius, and the uniformity of density in the primitive distribution.

---

* See App. A, "On the Motion produced in an Infinite Elastic Solid by the Motion through the Space occupied by it of a Body acting on it only by Attraction and Repulsion."

To come to reality, according to the most probable judgment present knowledge allows us to form, suppose at many millions, or thousands of millions, or millions of millions of years ago, all the matter in the universe to have been atoms very nearly at rest* or quite at rest; more densely distributed in some places than in others; of infinitely small average density through the whole of infinite space. In regions where the density was then greater than in neighbouring regions, the density would become greater still; in places of less density, the density would become less; and large regions would quickly become void or nearly void of atoms. These large void regions would extend so as to completely surround regions of greater density. In some part or parts of each cluster of atoms thus isolated, condensation would go on by motions in all directions not generally convergent to points, and with no perceptible mutual influence between the atoms until the density becomes something like 10^{-6} of our ordinary atmospheric density, when mutual influence by collisions would begin to become practically effective. Each collision would give rise to a train of waves in ether. These waves would carry away energy, spreading it out through the void ether of infinite space. The loss of energy, thus taken away from the atoms, would reduce large condensing clusters to the condition of gas in equilibrium† under the influence of its own gravity only, or rotating like our sun or moving at moderate speeds as in spiral nebulas, &c. Gravitational condensation would at first produce rise of temperature, followed later by cooling and ultimately freezing, giving solid bodies; collisions between which would produce meteoric stones such as we see them. We cannot regard as probable that these lumps of broken-looking solid matter (something like the broken stones used on our macadamised roads) are primitive forms in which matter was created. Hence we are forced, in this twentieth century, to views regarding the atomic origin of all things closely resembling those presented by Democritus, Epicurus, and their majestic Roman poetic expositor, Lucretius.

* "On Mechanical Antecedents of Motion, Heat, and Light," *Brit. Assoc. Rep.*, Part 2, 1854; *Edin. New Phil. Journ.* Vol. I. 1855; *Comptes Rendus*, Vol. XL. 1855; Kelvin's *Collected Math. and Phys. Papers*, Vol. II. Art. lxix.

† Homer Lane, *American Journal of Science*, 1870, p. 57; Sir W. Thomson, *Phil. Mag.* March 1887, p. 287.

APPENDIX E.

AEPINUS ATOMIZED*.

§ 1. ACCORDING to the well-known doctrine of Aepinus, commonly referred to as the one-fluid theory of electricity, positive and negative electrifications consist in excess above, and deficiency below, a natural quantum of a fluid, called the electric fluid, permeating among the atoms of ponderable matter. Portions of matter void of the electric fluid repel one another; portions of the electric fluid repel one another; portions of the electric fluid and of void matter attract one another.

§ 2. My suggestion is that the Aepinus' fluid consists of exceedingly minute equal and similar atoms, which I call electrions†, much smaller than the atoms of ponderable matter; and

---

* From the Jubilee Volume presented to Prof. Boscha in November, 1901.

† I ventured to suggest this name in a short article published in *Nature*, May 27, 1897, in which, after a slight reference to an old idea of a "one-fluid theory of electricity" with *resinous electricity as the electric fluid*, the following expression of my views at that time occurs : " I prefer to consider an atomic theory " of electricity foreseen as worthy of thought by Faraday and Clerk Maxwell, very " definitely proposed by Helmholtz in his last lecture to the Royal Institution, and " largely accepted by present-day workers and teachers. Indeed Faraday's law of " electro-chemical equivalence seems to necessitate something atomic in electricity, " and to justify the very modern name *electron* [given I believe originally by Johnstone Stoney, and now largely used to denote an atom of either vitreous or resinous electricity]. The older, and at present even more popular, name *ion* " given sixty years ago by Faraday, suggests a convenient modification of it; *elec-* " *trion*, to denote an atom of resinous electricity. And now, adopting the essentials " of Aepinus' theory, and dealing with it according to the doctrine of Father Bos- " covich, each atom of ponderable matter is an electron of vitreous electricity; " which, with a neutralizing electron of resinous electricity close to it, produces a " resulting force on every distant electron and electrion which varies inversely as the " cube of the distance, and is in the direction determined by the well-known re- " quisite application of the parallelogram of forces." It will be seen that I had not then thought of the hypothesis suggested in the present communication, that while electrions permeate freely through all space, whether occupied only by ether or occupied also by the volumes of finite spheres constituting the atoms of ponderable matter, each electrion in the interior of an atom of ponderable matter experiences electric force towards the centre of the atom, just as if the atom contained within it, fixed relatively to itself, a uniform distribution of ideal electric matter.

that they permeate freely through the spaces occupied by these
greater atoms and also freely through space not occupied by them.
As in Aepinus' theory we must have repulsions between the
electrions; and repulsions between the atoms independently of the
electrions; and attractions between electrions and atoms without
electrions.   For brevity, in future by atom I shall mean an atom of
ponderable matter, whether it has any electrions within it or not.

§ 3.   In virtue of the discovery and experimental proof by
Cavendish and Coulomb of the law of inverse square of distance
for both electric attractions and repulsions, we may now suppose
that the atoms, which I assume to be all of them spherical, repel
other atoms outside them with forces inversely as the squares of
distances between centres; and that the same is true of electrions,
which no doubt occupy finite spaces, although at present we are
dealing with them as if they were mere mathematical points,
endowed with the property of electric attraction and repulsion.
We must now also assume that every atom attracts every electrion
outside it with a force inversely as the square of the distance
between centres.

§ 4.   My assumption that the electrions freely permeate the
space occupied by the atoms requires a knowledge of the law of
the force experienced by an electrion within an atom.   As a tentative hypothesis, I assume for simplicity that the attraction experienced by an electrion approaching an atom varies exactly
according to the inverse square of the distance from the centre,
as long as the electrion is outside; has no abrupt change when
the electrion enters the atom; but decreases to zero simply as the
distance from the centre when the electrion, approaching the
centre, is within the spherical boundary of the atom.   This is
just as it would be if the electric virtue of the atom were due
to uniform distribution through the atom of an ideal electric
substance of which each infinitely small part repels infinitely
small portions of the ideal substance in other atoms, and attracts
electrions, according to the inverse square of the distance.   But
we cannot make the corresponding supposition for the mutual
force between two *overlapping* atoms; because we must keep ourselves free to add a repulsion or attraction according to any law of
force, that we may find convenient for the explanation of electric,
elastic, and chemical properties of matter.

§ 5. The neutralizing quantum of electrions for any atom or group of atoms has exactly the same quantity of electricity of one kind as the atom or group of atoms has of electricity of the opposite kind. The quantum for any single atom may be one or two or three or any integral number, and need not be the same for all atoms. The designations mono-electrionic, dielectrionic, trielectrionic, tetraelectrionic, polyelectrionic, etc., will accordingly be convenient. It is possible that the differences of quality of the atoms of different substances may be partially due to the quantum-numbers of their electrions being different; but it is possible that the differences of quality are to be wholly explained in merely Boscovichian fashion by differences in the laws of force between the atoms, and may not imply any differences in the numbers of electrions constituting their quantums.

§ 6. Another possibility to be kept in view is that the neutralizing quantum for an atom may not be any integral number of electrions. Thus for example the molecule of a diatomic gas, oxygen, or nitrogen, or hydrogen, or chlorine, might conceivably have three electrions or some odd number of electrions for its quantum so that the single atoms, O, N, H, Cl, if they could exist separately, must be either vitreously or resinously electrified and cannot be neutral.

§ 7. The present usage of the designations, positive and negative, for the two modes of electrification originated no doubt with the use of glass globes or cylinders in ordinary electric machines giving vitreous electricity to the insulated prime conductor, and resinous electricity to the not always insulated rubber. Thus Aepinus and his followers regarded the prime conductors of their machines as giving the true electric fluid, and leaving a deficiency of it in the rubbers to be supplied from the earth. It is curious, in Beccaria's account of his observations made about 1760 at Garzegna in Piedmont on atmospheric electricity, to read of " The mild excessive electricity of the air in fair weather." This in more modern usage would be called mild positive electricity. The meaning of either expression, stated in non-hypothetical language, is, the mild vitreous electricity of the air in fair weather.

§ 8. In the mathematical theory of electricity in equilibrium, it is a matter of perfect indifference which of the opposite electric

manifestations we call positive and which negative. But the
great differences in the disruptive and luminous effects, when the
forces are too strong for electric equilibrium, presented by the two
modes of electrification, which have been known from the earliest
times of electric science, show physical properties not touched by
the mathematical theory. And Varley's comparatively recent
discovery * of the molecular torrent of resinously electrified
particles from the "kathode" or resinous electrode in apparatus
for the transmission of electricity through vacuum or highly
rarefied air, gives strong reason for believing that the mobile
electricity of Aepinus' theory is resinous, and not vitreous as he
accidentally made it. I shall therefore assume that our electrions
act as extremely minute particles of *resinously* electrified matter;
that a void atom acts simply as a little globe of atomic substance,
possessing as an essential quality vitreous electricity uniformly
distributed through it or through a smaller concentric globe; and
that ordinary ponderable matter, not electrified, consists of a vast
assemblage of atoms, not void, but having within the portions of
space which they occupy, just enough of electrions to annul electric
force for all places of which the distance from the nearest atom is
large in comparison with the diameter of an atom, or molecular
cluster of atoms.

§ 9. This condition respecting distance would, because of the

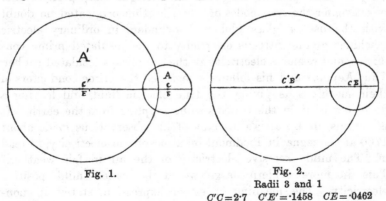

Fig. 1.   Fig. 2.
                           Radii 3 and 1
              $C'C = 2\cdot7$ $C'E' = \cdot1458$ $CE = \cdot0462$

inverse square of the distance law for the forces, be unnecessary
and the electric force would be rigorously null throughout all

* *Proc. Roy. Soc.* Vol. xix. 1871, pp. 239, 240.

space outside the atoms, if every atom had only a single electrion at its centre, provided that the electric quantities of the opposite electricities (reckoned according to the old definition of mathematical electrostatics) are equal in the atom and in the electrion. But even if every neutralized separate atom contains just one electrion in stable equilibrium at its centre, it is obvious that, when two atoms overlap so far that the centre of one of them is within the spherical boundary of the other, the previous equilibrium of the two electrions is upset, and they must find positions of equilibrium elsewhere than at the centres. Thus in fig. 1 each electrion is at the centre of its atom, and is attracted and repelled with equal forces by the neighbouring atom and electrion at *its* centre. In fig. 2, if E and E' were at the centres C, C', of the two atoms, E would be repelled by E' more than it would be attracted by the atom A'. Hence both electrions being supposed free, E will move to the right; and because of its diminished repulsion on E', E' will follow it in the same direction. The equations of equilibrium of the two are easily written down, not so easily solved without some slight arithmetical artifice. The solution is correctly shown in fig. 2, for the case in which one radius is three times the other, and the distance between the centres is 2·7 times the smaller radius*. The investigation in

---

* Calling e the quantity of electricity, vitreous or resinous, in each atom or electrion: ζ the distance between the centres of the atoms; a, a' the radii of the two atoms; x, x' the displacements of the electrions from the centres; X, X' the forces experienced by the electrions; we have

$$X = e^2 \left[-\frac{x}{a^3} + \frac{1}{(\zeta + x - x')^2} - \frac{\zeta + x}{a'^3} \right];$$

$$X' = e^2 \left[-\frac{x'}{a'^3} + \frac{1}{(\zeta - x')^2} - \frac{1}{(\zeta + x - x')^2} \right].$$

Each of these being equated to zero for equilibrium gives us two equations which are not easily dealt with by frontal attack for the determination of two unknown quantities x, x'; but which may be solved by a method of successive approximations, as follows:—Let x_0, x_1, ... x_i, x_0', x_1', ... x_i', be successive approximations to the values of x and x', and take

$$x_i + 1 = \frac{1}{\frac{1}{a^3} + \frac{1}{a'^3}} \left(\frac{1}{D_i^2} - \frac{\zeta}{a'^3} \right); \quad x_i' + 1 = a'^3 \left\{ \frac{1}{(\zeta - x_i')^2} - \frac{1}{(\zeta + x_i + 1 - x_i')^2} \right\};$$

where $D_i^2 = (\zeta + x_i - x_i')^2$. As an example, take $a = 1$, $a' = 3$. To find solutions for gradual approach between centres, take successively $\zeta = 2\cdot9$, 2·8, 2·7, 2·6. Begin with $x_0 = 0$, $x_0' = 0$, we find $x_4 = \cdot01243$, $x_4' = \cdot0297$, and the same values for x_5, and x_5'. Take next $\zeta = 2\cdot8$, $x_0 = \cdot01243$, $x_0' = \cdot0297$; we find $x_4 = x_5 = \cdot0269$, $x_4' = x_5' = \cdot0702$. Thus we have the solution for the second distance between

the footnote shows that if the atoms are brought a little nearer, the equilibrium becomes unstable; and we may infer that both electrions jump to the right, E' to settle at a point within the atom A on the left-hand side of its centre; and E outside A', to

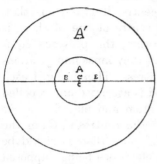

settle at a point still within A. If, lastly, we bring the centres closer and closer together till they coincide, E comes again within A', and the two electrions settle, as shown in fig. 3 at distances on the two sides of the common centre, each equal to

$$\frac{1}{2} \sqrt[3]{\frac{2}{\frac{1}{\alpha^3} + \frac{1}{\alpha'^3}}},$$

Fig. 3.
$EC' = CE = \cdot622$

which for the case $\alpha' = 3\alpha$ is

$$\frac{1}{2} \alpha \sqrt[3]{\frac{2 \cdot 27}{28}} = \cdot622\alpha.$$

§ 10. Mutual action of this kind might probably be presented in such binary combinations as O_2, N_2, H_2, Cl_2, CO, SO, $NaCl$ (dry common salt) if each single atom, O, N, H, Cl, C*, S, Na†, had just one electrion for its neutralizing quantum. If the combination is so close that the centres coincide, the two electrions will rest stably at equal distances on the two sides of the common centre as at the end of § 9. I see at present no reason for considering it excessively improbable that this may be the case for SO, or for any other binary combinations of *two atoms of different quality* for neither of which there is reason to believe that its neutralizing quantum is not exactly one electrion. But for the binary combinations of two atoms of identical quality which the

centres. Next take $\zeta = 2\cdot7$, $x_0 = \cdot0269$, $x_0' = \cdot0702$; we find $x_6 = x_7 = \cdot0462$, $x_6' = x_7' = \cdot1458$. Working similarly for $\zeta = 2\cdot6$, we do not find convergence, and we infer that a position of unstable equilibrium is reached by the electrions for some value of ζ between $2\cdot7$ and $2\cdot6$.

* The complexity of the hydrocarbons and the Van 't Hoff and Le Bel doctrine of the asymmetric results (chirality) produced by the quadrivalence of carbon makes it probable that the carbon atom takes at least four electrions to neutralize it electrically.

† The fact that sodium, solid or liquid, is a metallic conductor of electricity makes it probable that the sodium atom, as all other metallic elements, takes a large number of electrions to neutralize it. (See below § 30.)

chemists have discovered in diatomic gases (O_2, N_2, etc.) there must, over and above the electric repulsion of the two similar electric globes, be a strong atomic repulsion preventing stable equilibrium with coincident centres, however strongly the atoms may be drawn together by the attractions of a pair of mutually repellent electrions within them; because without such a repulsion the two similar atoms would become one, which no possible action in nature could split into two.

§ 11. Returning to § 9, let us pull the two atoms gradually asunder from the concentric position to which we had brought them. It is easily seen that they will both remain within the smaller atom A, slightly disturbed from equality of distance on the two sides of its centre by attractions towards the centre of A'; and that when A' is infinitely distant they will settle at distances each equal to $\frac{1}{2}\alpha \sqrt[3]{2} = \cdot 62996\alpha$ on the two sides of C, the centre of A. If, instead of two mono-electrionic atoms, we deal as in § 9 with two polyelectrionic atoms, we find after separation the number of electrions in the smaller atom increased and in the larger decreased; and this with much smaller difference of magnitude than the three to one of diameters which we had for our mono-electrionic atoms of § 9. This is a very remarkable conclusion, pointing to what is probably the true explanation of the first known of the electric properties of matter; attractions and repulsions produced by rubbed amber. Two ideal solids consisting of assemblages of mono-electrionic atoms of largely different sizes would certainly, when pressed and rubbed together and separated, show the properties of oppositely electrified bodies; and the preponderance of the electrionic quality would be in the assemblage of which the atoms are the smaller. Assuming as we do that the electricity of the electrions is of the resinous kind, we say that after pressing and rubbing together and separating the two assemblages, the assemblage of the smaller atoms is resinously electrified and the assemblage of the larger atoms is vitreously electrified. This is probably the true explanation of the old-known fact that ground glass is resinous relatively to polished glass. The process of polishing might be expected to smooth down the smaller atoms, and to leave the larger atoms more effective in the surface.

§ 12. It probably contains also the principle of the explana-

35—2

tion of Erskine Murray's* experimental discovery that surfaces of metals, well cleaned by rubbing with glass-paper or emery-paper, become more positive or less negative in the Volta contact electricity scale by being burnished with a smooth round hard steel burnisher. Thus a zinc plate brightened by rubbing on glass-paper rose by ·23 volt by repeated burnishing with a hard steel burnisher, and fell again by the same difference when rubbed again with glass-paper. Copper plates showed differences of about the same amount and in the same direction when similarly treated. Between highly burnished zinc and emery-cleaned copper, Murray found a Volta-difference of 1·13 volts, which is, I believe, considerably greater than the greatest previously found Volta-difference between pure metallic surfaces of zinc and copper.

§ 13. To further illustrate the tendency (§ 9) of the smaller atom to take electrions from the larger, consider two atoms; A', of radius α', the greater, having an electrion in it to begin with; and A, radius α, the smaller, void.

By ideal forces applied to the atoms while the electrion is free let them approach gradually from a very great distance apart. The attraction of A draws the electrion from the centre of A'; at first very slightly, but farther and farther as the distance between the atoms is diminished. What 'will be the position of the

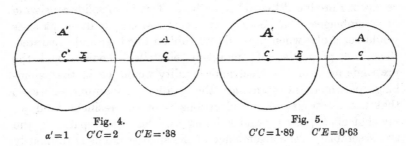

Fig. 4.

$\alpha' = 1$ $C'C = 2$ $C'E = ·38$

Fig. 5.

$C'C = 1·89$ $C'E = 0·63$

electrion when the distance between the centres is, as in fig. 4, $2\alpha'$? Without calculation we see that the electrion would be in equilibrium if placed at the point in which the surface of A' is cut by the line of centres; but the equilibrium would be obviously

* "On Contact Electricity of Metals," *Proc. Roy. Soc.* Vol. LXIII. 1898, p. 113. See also Lord Kelvin, "Contact Electricity of Metals," *Phil. Mag.* Vol. XLVI. 1898, pp. 96—98.

unstable, and a simple calculation* shows that the stable position actually taken by the electrion is ·38a' from C', when the distance between the centres is 2a' (fig. 4). If the distance between the centres is now diminished from 2a' to 1·89a' (a being now supposed to be anything less than ·89a') the electrion comes gradually to distance ·63a' from C' (fig. 5); its equilibrium there becomes unstable; and it jumps out of A' towards A (like a cork jumping out of a bottle). It will shoot through A (A' and A being held fixed); and after several oscillations to and fro, perhaps† ten or twenty [or perhaps a million (K., Aug. 9, 1903)], if it has only quasi-inertia due to condensation or rarefaction‡ produced by it in ether; or perhaps many times more if it has intrinsic inertia of its own; it will settle, with decreasing range of excursions, sensibly to rest within A, attracted somewhat from the centre by A'. If, lastly, A' and A be drawn asunder to their original great distance, the electrion will not regain its original position in A', but will come to the centre of A and rest there. Here then we have another illustration of the tendency found in § 9, of the smaller atom to take electrions from the larger.

§ 14. In preventing the two atoms from rushing together by holding them against the attractive force of the electrion, we shall have gained more work during the approach than we afterwards spent on the separation; and we have now left the system

---

* Denoting by ζ the distance between the centres, and by X the force on E when its distance from C' is x', we have

$$X = e^2 \left[\frac{1}{(\zeta - x')^2} - \frac{x'}{a'^2} \right].$$

Hence for equilibrium $\dfrac{1}{(\zeta - x')^2} = \dfrac{x'}{a'^3}$. This is a cubic for x' of which the proper root (the smallest root) for the case $\zeta = 2a'$ is ·38a'. The formula for X has a minimum value when $\zeta - x' = a' \sqrt[3]{2}$, which makes

$$X = \frac{e^2}{a'^2} \left[\frac{3}{2} \sqrt[3]{2} - \frac{\zeta}{a'} \right].$$

Hence the value of x' for equilibrium coincides with the value of X, a minimum, and the equilibrium becomes unstable, when ζ is diminished to $\dfrac{3 \sqrt[3]{2}}{2} a' = 1\cdot890a'$. For this, the value of x' is $\dfrac{\sqrt[3]{2}}{2} a' = \cdot63a'$.

† "On the Production of Wave Motion in an Elastic Solid," *Phil. Mag.* Oct. 1899, § 44.

‡ "On the Motion of Ponderable Matter through Space occupied by Ether," *Phil. Mag.* Aug. 1900, §§ 15, 17. [Reproduced in the present Volume as App. A.]

deprived of the farther amount of energy carried away by etherial waves into space.

§ 15. The system in its final state with the electrion at the centre of the smaller atom has less potential energy in it than it had at the beginning (when the electrion was at the centre of A'), by a difference equal to the excess of the work which we gained during the approach above that which we spent on the final separation of A' and A, plus the amount carried away by the etherial waves. All these items except the last are easily calculated from the algebra of the footnote on § 13; and thus we find how much is our loss of energy by the etherial waves.

§ 16. Very interesting statical problems are presented to us by consideration of the equilibrium of two or more electrions within one atom, whether a polyelectrionic atom with its saturating number, or an atom of any electric strength with any number of electrions up to the greatest number that it can hold. To help to clear our ideas, first remark that if the number of electrions is infinite, that is to say if we go back to Aepinus' electric fluid, but assume it to permeate freely through an atom of any shape whatever and having any arbitrarily given distribution of electricity of the opposite kind fixed within it, the greatest quantity of fluid which it can take is exactly equal to its own, and lodges with density equal to its own in every part. Hence if the atom is spherical, and of equal electric density throughout as we have supposed it, and if its neutralizing quantum of electrions is a very large number, their configuration of equilibrium will be an assemblage of more and more nearly uniform density from surface to centre, the greater the number. Any Bravais homogeneous assemblage whatever would be very nearly in equilibrium if all the electrions in a surface-layer of thickness a hundred times the shortest distance from electrion to electrion were held fixed; but the equilibrium would be unstable except in certain cases. It may seem probable that it is stable if the homogeneous assemblage is of the species which I have called* equilateral, being that in which each electrion with any two of its twelve next neighbours forms an equilateral triangle. If now all the electrions in the

---

* *Molecular Tactics of a Crystal*, § 4, being the second Robert Boyle Lecture, delivered before the Oxford University Junior Scientific Club, May 16, 1893 (Clarendon Press, Oxford). [Reproduced in the present Volume as App. H.]

surface layer are left perfectly free, a slight rearrangement among themselves and still slighter among the neighbouring electrions in the interior will bring the whole multitude (of thousands or millions) to equilibrium. The subject is of extreme interest, geometrical, dynamical and physical, but cannot be pursued further at present.

§ 17.   To guide our ideas respecting the stable equilibrium of moderate numbers of electrions within an atom, remark first that for any number of electrions there may be equilibrium with all the electrions on one spherical surface concentric with the atom. To prove this, discard for a moment the atom and imagine the electrions, whatever their number, to be attached to ends of equal inextensible strings of which the other ends are fixed to one point *C*. Every string will be stretched in virtue of the mutual repulsions of the electrions; and there will be a configuration or configurations of equilibrium with the electrions on a spherical surface.   Whatever their number there is essentially at least one configuration of stable equilibrium.   Remark also that there is always a configuration of equilibrium in which all the strings are in one plane, and the electrions are equally spaced round one great circle of the sphere.   This is the sole configuration for two electrions or for three electrions: but for any number exceeding three it is easily proved to be unstable, and is therefore not the sole configuration of equilibrium.   For four electrions it is easily seen that, besides the unstable equilibrium in one plane, there is only one stable configuration, and in this the four electrions are at the four corners of an equilateral tetrahedron.

§ 18.   For five electrions we have clearly stable equilibrium with three of them in one plane through *C*, and the other two at equal distances in the line through *C* perpendicular to this plane. There is also at least one other configuration of equilibrium: this we see by imagining four of the electrions constrained to remain in a freely movable plane, which gives stable equilibrium with this plane at some distance from the centre and the fifth electrion in the diameter perpendicular to it.   And similarly for any greater number of electrions; we find a configuration of equilibrium by imagining all but one of them to be constrained to remain in a freely movable plane.   But it is not easy, without calculation, to see, at all events for the case of only five electrions, whether that equilibrium would be stable if the constraint of all of them but one, to one plane is annulled. For numbers greater than five it seems certain that that equilibrium is unstable.

§ 19. For six we have a configuration of stable equilibrium with the electrions at the six corners of a regular octahedron; for eight at the corners of a cube. For ten, as for any even number, we should have two configurations of equilibrium (both certainly unstable for large numbers) with two halves of the number in two planes at equal distances on the two sides of the centre. For twelve we have a configuration of stable equilibrium with the electrions at positions of the twelve nearest neighbours to C in an equilateral homogeneous assemblage of points*; for twenty at the twenty corners of a pentagonal dodecahedron. All these configurations of § 19 except those for ten electrions are stable if, as we are now supposing, the electrions are constrained to a spherical surface on which they are free to move.

§ 20. Except the cases of § 18, the forces with which the strings are stretched are the same for all the electrions of each case. Hence if we now discard the strings and place the electrions in an atom on a spherical surface concentric with it, its attraction on the electrions towards the centre takes the place of the tension of the string, provided it is of the proper amount. But it does not secure, as did the strings, against instability relatively to radial displacements, different for the different electrions. To secure the proper amount of the radial force the condition is $\dfrac{ie^2r}{a^3} = T$; where i denotes the number of electrions; e the electric quantity on each (and therefore, § 8, ie the electric quantity of vitreous electricity in the atom); r denotes the radius of the spherical surface on which the electrions lie: a the radius of the atom; and T the tension of the string in the arrangement of § 17. We have generally $T = q\dfrac{e^2}{r^2}$ where q is a numeric depending on the number and configuration of the electrions found in each case by geometry. Hence we have $\dfrac{r}{a} = \sqrt[3]{\dfrac{q}{i}}$ for the ratio of the radius of the smaller sphere on which the electrions lie to the radius of the atom. For example, take the case of eight electrions

* App. H, " Molecular Tactics of a Crystal," § 4.

at the eight corners of a cube. T is the resultant of seven repulsions, and we easily find $q = \frac{3}{4}(\sqrt{3} + \sqrt{\frac{3}{2}} + \frac{1}{3})$ and finally $\frac{r}{a} = \cdot 6756$. Dealing similarly with the cases of two, three, four, and six electrons, we have the following table of values of $\left(\frac{r}{a}\right)^3$ and $\frac{r}{a}$; to which is added a last column showing values of $i^2 \frac{3a^2 - r^2}{2a^2} - \Sigma \frac{a}{D}$, being $\frac{a}{e^2}$ of the work required to remove the electrons to infinite distance. D is the distance between any two of the electrons.

Number of Electrons	Configuration	$\left(\frac{r}{a}\right)^3$	$\frac{r}{a}$	$\frac{a}{e^2} \times$ work required to remove the electrons to infinite distance $= w$
1	At the centre	0	0	1·500
2	At quarter points from the ends of a diameter	$\frac{1}{8}$	·5000	4·500
3	At the corners of an equilateral triangle	$\frac{1}{3\sqrt{3}}$	·5774	9·000
4	At the corners of a square	$\frac{\sqrt{2}}{8} + \frac{1}{16}$	·6208	14·750
4	At the corners of an equilateral tetrahedron	$\frac{3}{16}\sqrt{\frac{3}{2}}$	·6124	15·000
6	At the corners of an equilateral octahedron	$\frac{1 + 4\sqrt{2}}{24}$	·6522	33·335
8	At the corners of a cube	$\frac{3}{32}\left(\sqrt{3} + \sqrt{\frac{3}{2}} + \frac{1}{3}\right)$	·6756	52·180

§ 21. In the configurations thus expressed the equilibrium is certainly stable for the cases of two electrons, three electrons, and four electrons at the corners of a tetrahedron. It seems to me, without calculation, also probably stable for the case of six, and possibly even for the case of eight. For the case of twenty at the corners of a pentagonal dodecahedron the equilibrium is probably not stable; and even for the cases of twelve electrons and ten electrons, the equilibrium in the configurations described in §§ 18, 19 may probably be unstable, when, as now, we have the attraction of the atom towards the centre instead of the inextensible strings.

§ 22. In fact when the number of electrons exceeds four,

we must think of the tendency to be crowded out of one spherical surface, which with very large numbers gives a tendency to uniform distribution throughout the volume of the atom as described in § 16 above. Thus, in the case of five electrions, § 18 shows a configuration of equilibrium in which the two electrions lying in one diameter are, by the mutual repulsions, pushed very slightly further from the centre than are the three in the equatorial plane. In this case the equilibrium is clearly stable. Another obvious configuration, also stable, of five electrions within an atom is one at the centre, and four on a concentric spherical surface at the corners, of a tetrahedron. From any case of any number of electrions all on one spherical surface, we may pass to another configuration with one more electrion placed at the centre and the proper proportionate increase in the electric strength of the atom. Thus from the cases described in § 19, we may pass to configurations of equilibrium for seven, nine, eleven, thirteen, and twenty-one electrions. All these cases, with questions of stability or instability and of the different amounts of work required to pluck all the electrions out of the atom and remove them to infinite distances, present most interesting subjects for not difficult mathematical work ; and I regret not being able to pursue them at present.

§ 23. Consider now the electric properties of a real body, gaseous, liquid, or solid, constituted by an assemblage of atoms with their electrions. It follows immediately from our hypothesis, that in a monatomic gas or in any sufficiently sparse assemblage of single atoms, fixed or moving, Faraday's "*conducting power for lines of electric force,*" or *what is now commonly called the specific electro-inductive capacity, or the electro-inductive permeability, exceeds unity by three times the ratio of the sum of the volumes of the atoms to the whole volume of space occupied by the assemblage,* whether the atoms be mono-electrionic or polyelectrionic, and however much the electrion, or group of electrions, within each atom is set to vibrate or rotate with each collision, according to the kinetic theory of gases. To prove this, consider, in a uniform field of electrostatic force of intensity F, a single atom of radius a; and, at rest within it, a group of i electrions in stable equilibrium. The action of F produces simply displacements of the electrions relatively to the atom, equal and in parallel lines, with therefore no change of shape and no rotation ; and, x denoting the amount of this displacement, the equation for the equilibrium of each electrion is $\dfrac{iex}{\alpha^3} = F$. This gives $iex = \alpha^3 F$ for the electric moment of the electrostatic polarization induced in the atom by F. In passing, remark that $\alpha^3 F$ is also equal to the electric moment of the polarization produced in an insulated unelectrified metal globe of radius α, when brought into an electrostatic field of intensity F: and conclude that the electric inductive capacity of a uniformly dense assemblage of fixed metallic globules, so sparse that their mutual influence is negligible, is the same as that of an equal and similar assemblage of our hypothetical atoms, whatever be the number of electrions in each, not necessarily the same in all. Hence our hypothetical atom realizes perfectly for sparse assemblages Faraday's suggestion of "small globular conductors, as shot" to explain the electro-polarization which he discovered in solid and liquid insulators. (*Experimental Researches*, § 1679.)

§ 24. Denoting now by N the number of atoms per unit volume we find $N V \alpha^3 F$ as the electric moment of any sparse enough assemblage of uniform density occupying volume V in a uniform electric field of intensity F. Hence $N \alpha^3$ is what (following the analogy of electro-magnetic nomenclature) we may call the electro-inductive susceptibility[*] of the assemblage; being the electric moment per unit bulk induced by an electric field of unit intensity. Denoting this by μ, and the electro-inductive permeability by ω we have [*Electrostatics and Magnetism*, § 629, (14)],

$$\omega = 1 + 4\pi\mu = 1 + 3\left(N\frac{4\pi\,\alpha^3}{3}\right),$$

which proves the proposition stated at the commencement of § 23.

§ 25. To include vibrating and rotating groups of electrions in the demonstration, it is only necessary to remark that the time-average of any component of the displacement of the centre of inertia of the group relatively to the centre of the atom will, under the influence of F, be the same as if the assemblage were at rest in stable equilibrium.

§ 26. The consideration of liquids consisting of closely packed mobile assemblages of atoms or groups of atoms with their elec-

---

[*] Suggested in my *Electrostatics and Magnetism*, §§ 628, 629.

trions, forming compound molecules, as in liquid argon or helium (monatomic), nitrogen, oxygen, etc. (diatomic), or pure water, or water with salts or other chemical substances dissolved in it, or liquids of various complex chemical constitutions, cannot be entered on in the present communication, further than to remark that the suppositions we have made regarding forces, electric and other, between electrions and atoms, seem to open the way to a very definite detailed dynamics of electrolysis, of chemical affinity, and of heat of chemical combination. Estimates of the actual magnitudes concerned (the number of molecules per cubic centimetre of a gas, the mass in grammes of an atom of any substance, the diameters of the atoms, the absolute value of the electric quantity in an electrion, the effective mass or inertia of an electrion) seem to show that the intermolecular electric forces are more than amply great enough to account for heat of chemical combination, and every mechanical action manifested in chemical interactions of all kinds. We might be tempted to assume that all chemical action is electric, and that all varieties of chemical substance are to be explained by the numbers of the electrions required to neutralize an atom or a set of atoms (§ 6 above); but we can feel no satisfaction in this idea when we consider the great and wild variety of quality and affinities manifested by the different substances or the different "chemical elements"; and as we are assuming the electrions to be all alike, we must fall back on Father Boscovich, and require him to explain the difference of quality of different chemical substances, by different laws of force between the different atoms.

§ 27.   Consider lastly a solid; that is to say, an assemblage in which the atoms have no relative motions, except through ranges small in comparison with the shortest distances between their centres*. The first thing that we remark is that every solid would, at zero of absolute temperature (that is to say all its atoms and electrions at rest), be a perfect insulator of electricity under the influence of electric forces, moderate enough not to pluck electrions out of the atoms in which they rest stably when there is no disturbing force. The limiting value of F here indicated

---

* I need scarcely say that it is only for simplicity in the text that we conveniently ignore Roberts-Austen's admirable discovery of the interdiffusion of solid gold and solid lead, found after a piece of one metal is allowed to rest on a piece of the other for several weeks, months, or years.

for perfect insulation, I shall for brevity call the disruptional force or disruptional intensity. It is clear that this disruptional force is smaller the greater the number of electrions within an atom.

§ 28. The electro-inductive permeability of a solid at zero temperature is calculable by the static dynamics of § 24, modified by taking into account forces on the electrions of one atom due to the attractions of neighbouring atoms and the repulsions of their electrions. Without much calculation it is easy to see that generally the excess of the electro-inductive permeability above unity will be much greater than three times the sum of the volumes of the electric atoms per unit volume of space, which we found in § 24 for the electro-inductive permeability of an assemblage of single atoms, sparse enough to produce no disturbance by mutual actions. Also without much calculation, it is easy to see that now the induced electric moment will not be in simple proportion to F, the intensity of the electric field, as it was rigorously for a single atom through the whole range up to the disruptional value of F; but will tend to increase more than in simple proportion to the value of F; though for small practical values of F the law of simple proportion is still very nearly fulfilled.

§ 29. Raise the temperature now to anything under that at which the solid would melt. This sets the electrions to performing wildly irregular vibrations, so that some of them will occasionally be shot out of their atoms. Each electrion thus shot out will quickly either fall back into the atom from which it has been ejected, or will find its way into another atom. If the body be in an electric field F, a considerable proportion of the electrions which are shot out will find their way into other atoms in the direction in which they are pulled by F; that is to say, the body which was an infinitely perfect insulator at zero absolute temperature has now some degree of electric conductivity, which is greater the higher the temperature. There can be no doubt that this is a matter-of-fact explanation of the electric conductivity, which so nearly perfect an insulator as the flint glass of my quadrant electrometer at atmospheric temperatures shows, when heated to far below its melting point (according to Professor T. Gray[*], $\cdot 98 . 10^{-24}$ at 60° Cent.; $4\cdot9 . 10^{-24}$ at 100°; $8300 . 10^{-24}$ at

[*] *Proc. R. S.* Vol. xxxiv. Jan. 12, 1882. The figures of the text are in c. g. s.

200° Cent.). It explains also the enormous increase of electric conductivity of rare earths at rising temperatures above 800° C., so admirably taken advantage of by Professor Nernst in his now celebrated electric lamp.

§ 30. If the hypotheses suggested in the present communication are true, the electric conductivity of metals must be explained in the same way as that of glass, gutta-percha, vulcanite, Nernst filament, etc., with only this difference, that the metallic atom must be so crowded with electrions that some of them are always being spilt out of each atom by the intermolecular and electrionic thermal motions, not only at ordinary atmospheric temperatures, and higher, but even at temperatures of less than 16° Centigrade above the absolute zero of temperature. I say 16° because in Dewar's Bakerian Lecture to the Royal Society of London, June 13, 1901, *The Nadir of Temperature*, we find that platinum, gold, silver, copper and iron have exceedingly high electric conductivity at the temperature of liquid hydrogen boiling under 30 mms. of mercury, which must be something between 20°·5, the boiling point of hydrogen at 760 mms. pressure, and 16°, the temperature of melting solid hydrogen, both determined by Dewar with his helium thermometer. There is no difficulty in believing that the electrions in each of the metallic atoms are so numerous that though they rest in stable equilibrium within the atoms, closely packed to constitute the solid metal at 0° absolute, and may move about within the atom with their wildly irregular thermal motions at 1° of absolute temperature, they may between 1° and 2° begin to spill from atom to atom. Thus, like glass or a Nernst filament below 300° absolute, a metal may be an almost perfect insulator of electricity below 1° absolute: may, like glass at 333° absolute, show very notable conductivity at 2° absolute: and, like glass at 473° absolute as compared with glass at 333° absolute, may show 8000 times as much electric conductivity at 2°·8 as at 2°. And, like the Nernst filament at 1800° or 2000° absolute, our hypothetical metal may at 6° absolute show a high conductivity, comparable with that of lead or copper at ordinary temperatures. The electric conductivity in the Nernst filament goes on increasing as the temperature rises till the filament melts or evaporates. Nevertheless it is quite conceivable that in our hypothetical metal with rising temperature from 2° to 16° absolute the electric conductivity may come to a maximum and decrease with further rise

of temperature up to and beyond ordinary atmospheric temperatures. In fact, while some extent of thermal motions is necessary
for electric conductivity (because there can be no such thing as
"lability" in electrostatic equilibrium), too much of these motions
must mar the freedom with which an electrion can thread its
way through the crowd of atoms to perform the function of electric
conduction. It seems certain that this is the matter-of-fact
explanation of the diminution of electric conductivity in metals
with rise of temperature.

§ 31. Regretting much not to be able (for want of time) to
include estimates of absolute magnitudes in the present communication, I end it with applications of our hypothesis to the
pyro-electricity and piezo-electricity of crystals. A crystal is a
homogeneous assemblage of bodies. Conversely, a homogeneous
assemblage of bodies is not a crystal if the distance between
centres of nearest neighbours is a centimetre or more; it is a
crystal if the distance between nearest neighbours is 10^{-8} of a cm.
or less. Pyro-electricity and piezo-electricity are developments
of vitreous and resinous electric forces such as would result from
vitreous and resinous electrification on different parts of the
surface of a crystal, produced respectively by change of temperature and by stress due to balancing forces applied to the
surfaces.

§ 32. To see how such properties can or must exist in crystals
composed of our hypothetical atoms with electrions, consider first

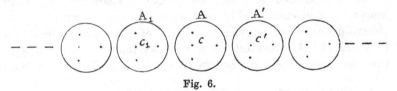

Fig. 6.

merely a row of equal tetra-electrionic atoms in a straight line,
each having its quantum of four electrions within it. Fig. 6
shows a configuration of stable equilibrium of the electrions not,
however, truly to scale. The sets of three dots indicate trios of
electrions at the corners of equilateral triangles, the middle dot
in each row being alternately on the far side and the near side of
the plane of the paper, which contains the centres of the atoms
and the remaining electrion of each four. Let C_1, C, C', be the

centres of the atoms A_1, A, A'. An easy calculation shows that the quartet of electrions within A, regarded for the moment as a group of four material points rigidly connected, is attracted to the left with a less force by A_1 than to the right by A' (in making the calculation remember that A_1 attracts all the electrions within A as if it were a quantity e of vitreous electricity collected at C_1, and similarly in respect to A'). There are corresponding smaller differences between the opposite attractions of the more and more remote atoms on the two sides of A. Let δ denote the excess of the sum of the rightwards of these attractions above the leftwards. The geometrical centre of the electrions within A is displaced rightwards from C to a distance, l, equal to $\dfrac{a^3\delta}{16e^2}$.

§ 33. Imagine now a crystal or a solid of any shape built up of parallel rows of atoms such as those of § 32. The amount of the displacing force on each quartet of electrions will be some-what altered by mutual action between the rows, but the general character of the result will be the same; and we see that throughout the solid, except in a thin superficial layer of perhaps five or tens atoms deep, the whole interior is in a state of homo-geneous electric polarization, of which the electric moment per unit of volume is $4eNl$; where N is the number of atoms per unit volume, and l is the displacement of the geometrical centre of each quartet from the centre of its atom. This is the interior molecular condition of a di-polar pyro-electric crystal, which I described in 1860* as probably accounting for their known pyro-electric quality, and as in accordance with the free electro-polarities of fractured surfaces of tourmaline discovered by Canton†. If a crystal, which we may imagine as given with the electrions wholly undisturbed from their positions according

---

* *Collected Mathematical and Physical Papers*, Vol. I. p. 315.

† Wiedemann (*Die Lehre von der Elektricität*, Second Edition, 1894, Vol. II. § 378) mentions an experiment without fully describing it by which a null result, seemingly at variance with Canton's experimental discovery and condemnatory of my suggested theory, was found. Interesting experiments might be made by pressing together and reseparating fractured surfaces of tourmaline, or by pressing and rubbing polished surfaces together and separating them. It would be very difficult to get trustworthy results by breakages, because it would be almost impossible to avoid irregular electrifications by the appliances used for making the breakage. The mode of electric measurement followed in the experiment referred to by Wiedemann is not described.

to § 32, is dipped in water and then allowed to dry, electrions would by this process be removed from one part of its surface and distributed over the remainder so as to wholly annul its external manifestation of electric quality. If now either by change of temperature or by mechanical stress the distances between the atoms are altered, the interior electro-polarization becomes necessarily altered; and the masking superficial electrification previously got by the dipping in water and drying, will now not exactly annul the electrostatic force in the air around the solid. If at the altered temperature or under the supposed stress the solid is again dipped in water and dried, the external electric force will be again annulled. Thus is explained the pyro-electricity of tourmaline discovered by Aepinus.

§ 34. But a merely di-polar electric crystal with its single axis presents to us only a small, and the very simplest, part of the whole subject of electro-crystallography. In boracite, a crystal of the cubic class, Haüy found in the four diagonals of the cube, or the perpendiculars to the pairs of faces of the regular octahedron, four di-polar axes: the crystal on being irregularly heated or cooled showed as it were opposite electricities on the surfaces in the neighbourhood of opposite pairs of corners of the cube, or around the centres of the opposite pairs of triangular faces of the octahedron. His discoveries allow us to conclude that in general the electric æolotropy of crystals is octo-polar with four axes, not merely di-polar as in the old-known electricity of the tourmaline. The intensities of the electric virtue are generally different for the four axes, and the directions of the axes are in general unsymmetrically oriented for crystals of the unsymmetrical classes. For crystals of the optically uniaxal class, one of the electro-polar axes must generally coincide with the optic axis, and the other three may be perpendicular to it. The intensities of the electro-polar virtue are essentially equal for these three axes: it may be null for each of them: it may be null or of any value for the so-called optic axis. Haüy found geometrical differences in respect to crystalline facets at the two ends of a tourmaline; and between the opposite corners of cubes, as leucite, which possess electro-polarity. There are no such differences between the two ends of a quartz crystal (hexagonal prism with hexagonal pyramids at the two ends) but there are structural differences (visible or invisible) between the opposite edges of the hexagonal prism. The electro-polar virtue is null for the axis of the prism, and is proved to exist between the opposite edges by the beautiful piezo-electric discovery of the brothers Curie, according to which a thin flat bar, cut with its faces and its length perpendicular to two parallel faces of the hexagonal prism and its breadth parallel to the edges of the prism, shows opposite electricities on its two faces, when stretched by forces pulling its ends. This proves the three electro-polar axes to bisect the 120° angles between the consecutive plane faces of the prism.

§ 35. For the present let us think only of the octo-polar electric æolotropy discovered by Haüy in the cubic class of crystals. The quartet of electrions at the four corners of a tetrahedron presents itself readily as possessing intrinsically the symmetrical octo-polar quality which is realized in the natural crystal. If we imagine an assemblage of atoms in simple cubic order each containing an equilateral quartet of electrions, all similarly oriented with their four faces perpendicular to the four diagonals of each structural cube, we have exactly the required æolotropy; but the equilibrium of the electrions all similarly oriented would probably be unstable; and we must look to a less simple assem-blage in order to have stability with similar orientation of all the electrionic quartets.

§ 36. This, I believe, we have in the doubled equilateral homogeneous assemblage of points described in § 69 of my paper on *Molecular Constitution of Matter* republished from the *Trans-actions of the Royal Society of Edinburgh* for 1889 in Volume III. of my *Collected Mathematical and Physical Papers* (p. 426); which may be described as follows for an assemblage of equal and similar globes. Beginning with an equilateral homogeneous assemblage of points, A, make another similar assemblage of points, B, by placing a B in the centre of each of the similarly oriented quartets of the assemblage of A's. It will be found that every A is at the centre of an *oppositely* oriented quartet of the B's. To understand this, let A_1, A_2, A_3, A_4, be an equilateral quartet of the A's; and imagine A_2, A_3, A_4, placed on a horizontal glass plate* with A_1 above it. Let B_1 be at the centre of

* Parallel glass plates carrying little white or black or coloured paper-circles are useful auxiliaries for graphic construction and illustrative models in the molecular theory of crystals.

A_1, A_2, A_3, A_4, and let B_1, B_2, B_3, B_4, be a quartet of the B's similarly oriented to A_1, A_2, A_3, A_4. We see that B_2, B_3, B_4, lie below the glass plate, and that the quartet B_1, B_2, B_3, B_4, has none of the A's at its centre. But the vertically opposite quartet B_1, B_2', B_3', B_4', contains A_1 within it; and it is oppositely oriented to the quartet A_1, A_2, A_3, A_4. Thus we see that, while the half of all the quartets of A's which are oriented oppositely to A_1, A_2, A_3, A_4, are void of B's, the half of the quartets of B's oppositely oriented to A_1, A_2, A_3, A_4, have each an A within it, while the other half of the quartets of the B's are all void of A's.

§ 37. Now let all the points A and all the points B, of § 36, be centres of equal and similar spherical atoms, each containing a quartet of electrions. The electrions will be in stable equilibrium under the influence of their own mutual repulsions, and the attractions of the atoms, if they are placed as equilateral quartets of proper magnitude, concentric with the atoms, and oriented all as any one quartet of the A's or B's. To see that this is true, confine attention first to the five atoms A_1, A_2, A_3, A_4, B_1. If the electrions within A_1, A_2, A_3, A_4 are all held similarly oriented to the quartet of the centres of these atoms, the quartet of electrions within B_1 must obviously be similarly oriented to the other quartets of electrions. If again, these be held oriented oppositely to the quartet of the atoms, the stable configuration of the electrions within B_1, will still be similar to the orientation of the quartets within A_1, A_2, A_3, A_4, though opposite to the orientation of the centres of these atoms. If, when the quartets of electrions are all thus similarly oriented either way, the quartet within B_1 is turned to reverse orientation, this will cause all the others to turn and settle in stable equilibrium according to this reversed orientation. Applying the same consideration to every atom of the assemblage and its four nearest neighbours, we have proof of the proposition asserted at the commencement of the present section. It is most interesting to remark that if, in a vast homogeneous assemblage of the kind with which we are dealing, the orientation of any one of the quartets of electrions be reversed and held reversed, all the others will follow and settle in stable equilibrium in the reversed orientation.

§ 38. This double homogeneous assemblage of tetra-electrionic atoms seems to be absolutely the simplest* molecular structure in which Haüy's octo-polar electric quality can exist. To see that it has octo-polar electric quality, consider an octahedron built up according to it. The faces of this octahedron, taken in proper order, will have, next to them, alternately points and triangular faces of the electrionic quartets within the atoms. This itself is the kind of electric æolotropy which constitutes octo-polar quality. Time prevents entering fully at present on any dynamical investigation of static or kinetic results.

§ 39. [*Added Oct.* 23, 1901.] Since what precedes was written, I have seen the explanation of a difficulty which had prevented me from finding what was wanted for octo-polar electric æolotropy in a homogeneous assemblage of single atoms. I now find (§ 40 below) that quartets of electrions will rest stably in equilibrium,

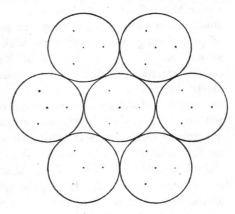

Fig. 7.

under the influence of the mutual repulsion between electrion and electrion and attraction between atom and electrion, in an equilateral homogeneous assemblage in the configuration indicated in fig. 7. The quartets of electrions are supposed to have their edges parallel to the six lines of symmetry of the assemblage. The plane of the paper is supposed to be that of the centres of the seven atoms. The central point in each circle represents a simple electrion which is at distance r, according to the notation of § 20 above, from the plane of the paper on the near side; and therefore the other three at the corners of an equilateral triangle at distance $\frac{1}{3}r$ on the far side to make the electric centre of gravity of the quartet coincide with the centre of its atom. The radius of the circle on which these three lie is $\frac{2\sqrt{2}}{3} r$ or ·94. The diagram is drawn correctly to scale according to the value ·612α given

---

* Not the simplest. See § 40 below.

for r in the table of § 20, on the supposition that the circles shown in the diagram represent the electric spheres of the atoms in contact.

§ 40. Imagine now the electrions of each quartet to be rigidly connected with one another and given freedom only to rotate about an axis perpendicular to the plane of the paper. To all of them apply torques; turning the central quartet of the diagram slowly and keeping all the others at rest. It is clear that the first 60° of turning brings the central quartet to a position of unstable equilibrium, and 60° more to a position of stable equilibrium corresponding to the first position, which we now see is stable when the others are all held fixed. We are now judging simply from the mutual actions between our central quartet and the six shown around it in the diagram; but it may be easily proved that our judgment is not vitiated by the mutual action between the central quartet and all around it, including the six in the diagram. Similarly we see that any one quartet of the assemblage, free to turn round an axis perpendicular to the plane of the paper while all the others are fixed, is in stable equilibrium when oriented as are those shown in the diagram. And similarly again we see the same conclusion in respect to three other diagrams in the three other planes parallel to the faces of the tetrahedrons or corresponding octahedrons of the assemblage. Hence we conclude that if the axial constraints are all removed, and the quartets left perfectly free, every one of them rests in stable equilibrium when oriented either as one set or as the other set of equilateral tetrahedronal quartets of the assemblage. It is interesting to remark that if, after we turned the central quartet through 60°, we had held it in that position and left all the others free to rotate, rotational vibrations would have spread out among them from the centre; and, after losing in waves spreading through ether outside the assemblage the energy which we gave them by our torque acting on the central quartet, they would come to stable equilibrium with every one of them turned 60° in one direction or the other from its primitive position, and oriented as the central quartet in the position in which we held it.

§ 41. We have thus found that an equilateral homogeneous assemblage of atoms each having four electrions within it, arranges these electrions in equilateral quartets all oriented in one or other of two ways. The assemblage of atoms and electrions thus produced is essentially octo-polar. Of the two elementary structural tetrahedrons, of the two orientations, one will have every one of its electrionic quartets pointing towards, the other from, its faces. The elementary structural octahedron has four of its faces next to corners, and four next to triangles, of its electrionic quartets. This is essentially a dynamically octo-polar* assemblage; and it supplies us with a perfect explanation of the piezo-electric quality to be inferred from the brothers Curie's experimental discovery, and Voigt's mathematical theory.

§ 42. Look at the diagram in § 39; and remember that it indicates a vast homogeneous assemblage consisting of a vast number of parallel plane layers of atoms on each side of the plane of the paper, in which seven atoms are shown. The quartets of electrions were described as all similarly oriented, and each of them equilateral, and having its geometrical centre at the centre of its atom; conditions all necessary for equilibrium.

§ 43. Let now the assemblage of atoms be homogeneously stretched from the plane on both sides to any extent, small or great, without any component motions of the centres of the atoms parallel to the planes of the layers. First let the stretch be very great; great enough to leave undisturbed by the other layers the layer for which the centres of atoms are, and the geometrical centres of the quartets were, in the plane of the paper. The geometrical centres of the quartets are not now in the plane of the paper. The single electrions on the near side seen in the diagram over the centres of the circles are drawn towards the plane of the paper; the equilateral triangles on the far side are also drawn nearer to the paper; and the equilateral triangles are enlarged in each atom by the attractions of the surrounding atoms. The contrary inward movements of the single atoms on one side of the plane, and of the triplets on the other side, cannot

---

* The octo-polar *pyro-electricity*, which is supposed to have been proved by Haüy's experiment, must have been due to something æolotropic in the heating. Uniform heating throughout a regular cube or octahedron could not give opposite electric manifestations in the four pairs of alternate corners of the cube, or alternate faces of the octahedron. Nevertheless the irregular finding of electric octo-polarity by Haüy is a splendid discovery; of which we only now know the true and full significance, through the experimental and mathematical labours of the brothers Curie, of Friedel, and of Voigt.

in general be in the proportion of three to one. Hence the geometrical centres of gravity of the quartets are now displaced perpendicularly to the plane of the paper to far side or near side; I cannot tell which without calculation. The calculation is easy but essentially requires much labour; involving as it does the determination of three unknowns, the length of each side of the equilateral triangle seen in the diagram, the distance of each of its corners from the electrion on the near side of the paper, and the displacement of the geometrical centre of gravity of the four to one side or other of the plane. Each one of the three equations involves summations of infinite convergent series, expressing force components due to all the atoms surrounding any chosen one in the plane. A method of approximation on the same general plan as that of the footnote to § 9 above would give a practicable method of calculation.

§ 44. Return to § 42; and consider the diagram as representing a crystal in its natural unstressed condition, consisting of a vast train of assemblages of atoms with centres in the plane of the paper, and in parallel planes on each side of it. We now see that the forces experienced by the electrions of one quartet from all the surrounding atoms in the plane of the paper would, if uncompensated, displace the geometrical centre of gravity of the quartet to one side or other of the plane of the paper, and we infer that the forces experienced from all the atoms on the two sides of this plane give this compensation to keep the centre of gravity of the quartet in the plane. Stretch now the assemblage to any degree equally in all directions. The quartets remain equilateral with their centres of gravity in the plane of the paper and parallel planes. Lastly stretch it farther equally in all directions parallel to the plane of the paper, with no component motion perpendicular to this plane. This last stretching diminishes the influence of all the atoms whose centres are in the plane of the paper tending to displace the centres of gravity of their electrions in one direction from this plane; and therefore leaves all the atoms out of this plane to predominate, and to cause a definite calculable displacement of the centres of gravity of all the quartets in the contrary direction to the former.

§ 45. To realize the operations of § 44, cut a thin hexagonal plate from the middle between two opposite corners of a cubic crystal, or parallel faces of an octahedron. Fix clamps to the six edges of this plate, and apply forces pulling their pairs equally in contrary directions. The whole material of the plate becomes electro-polar with electric moment per unit bulk equal to $4Nex$; of which the measurable result is uniform electrostatical potentials* in vacuous ether close to the two sides of the plate, differing by $4\pi \,.\, 4Next$; where t denotes the thickness of the plate, x the calculated displacement of the centre of gravity of each quartet from the centres of the atoms parallel to the two faces of the plate, e the electric mass of an electrion, and N the number of atoms per cubic centimetre of the substance. This crystal of the cubic class is, in Voigt's mathematical theory, the analogue to the electric effect discovered in quartz by the brothers Curie, and measured by aid of thin metal foils attached to the two faces of the plate and metrically connected to the two principal electrodes of an electrometer.

* See my *Electrostatics and Magnetism*, § 512, Cor. 3.

APPENDIX F.

(Having been in type more than twelve years as Articles XCIII.—XCVII.
for projected Vol. IV. of *Mathematical and Physical Papers.*)

ART. XCIII. DYNAMICAL ILLUSTRATIONS OF THE MAGNETIC
AND THE HELIÇOIDAL ROTATORY EFFECTS OF TRANSPARENT
BODIES ON POLARIZED LIGHT.

[From the *Proc. Roy. Soc.*, Vol. VIII., June 1856 ; *Phil. Mag.*, March 1857.]

THE elastic reaction of a homogeneously strained solid has a
character essentially devoid of all heliçoidal and of all dipolar
asymmetry. Hence the rotation of the plane of polarization of
light passing through bodies which either intrinsically possess the
heliçoidal property (syrup, oil of turpentine, quartz crystals, &c.),
or have the magnetic property induced in them, must be due to
elastic reactions dependent on the heterogeneousness of the strain
through the space of a wave, or to some heterogeneousness of the
luminous motions* dependent on a heterogeneousness of parts of
the matter of lineal dimensions not infinitely small in comparison
with the wave length. An infinitely homogeneous solid could
not possess either of those properties if the stress at any point of
it was influenced only by parts of the body touching it; but if
the stress at one point is directly influenced by the strain in
parts at distances from it finite in comparison with the wave
length, the heliçoidal property might exist, and the rotation of
the plane of polarization, such as is observed in many liquids and
in quartz crystals, could be explained as a direct dynamical
consequence of the statical elastic reaction called into play by
such a strain as exists in a wave of polarized light. It may,
however, be considered more probable that the matter of transparent bodies is really heterogeneous from one part to another of
lineal dimensions not infinitely small in comparison with a wave

---

* As would be were there different sets of vibrating particles, or were Rankine's
important hypothesis true, that the vibrations of luminiferous particles are directly
affected by pressure of a surrounding medium in virtue of its inertia. [In Lectures
XIX. and XX. we have seen reason to believe that this is true.]

length, than that it is infinitely homogeneous and has the property
of exerting finite direct "molecular" force at distances comparable
with the wave length: and it is certain that any spiral heterogeneousness of a vibrating medium must, if either right-handed
or left-handed spirals predominate, cause a finite rotation of the
plane of polarization of all waves of which lengths are not
infinitely great multiples of the steps of the structural spirals.
Thus a liquid filled homogeneously with spiral fibres, or a solid
with spiral passages through it of steps not less than the forty-
millionth of an inch, or a crystal with a right-handed or a left-
handed geometrical arrangement of parts of some such lineal
dimensions as the forty-millionth of an inch, might be certainly
expected to cause either a right-handed or a left-handed rotation
of ordinary light (the wave length being $\frac{1}{40,000}$th of an inch for
homogeneous yellow).

But the magnetic influence on light discovered by Faraday
depends on the direction of motion of moving particles. For
instance, in a medium possessing it, particles in a straight line
parallel to the lines of magnetic force, displaced to a helix round
this line as axis, and then projected tangentially with such velocities as to describe circles, will have different velocities according
as their motions are round in one direction (the same as the
nominal direction of the galvanic current in the magnetizing
coil), or in the contrary direction. But the elastic reaction of the
medium must be the same for the same displacements, whatever
be the velocities and directions of the particles; that is to say,
the forces which are balanced by centrifugal force of the circular
motions are equal, while the luminiferous motions are unequal.
The absolute circular motions being therefore either equal or such
as to transmit equal centrifugal forces to the particles initially
considered, it follows that the luminiferous motions are only components of the whole motion; and that a less luminiferous component in one direction, compounded with a motion existing in
the medium when transmitting no light, gives an equal resultant
to that of a greater luminiferous motion in the contrary direction
compounded with the same non-luminous motion. I think it is
not only impossible to conceive any other than this dynamical
explanation of the fact that circularly polarized light transmitted
through magnetized glass parallel to the lines of magnetizing force,
with the same quality, right-handed always, or left-handed always,

is propagated at different rates according as its course is in the direction or is contrary to the direction in which a north magnetic pole is drawn; but I believe it can be demonstrated that no other explanation of that fact is possible. Hence it appears that Faraday's optical discovery affords a demonstration of the reality of Ampère's explanation of the ultimate nature of magnetism; and gives a definition of magnetization in the dynamical theory of heat. The introduction of the principle of moments of momenta ("the conservation of areas") into the mechanical treatment of Mr Rankine's hypothesis of "molecular vortices," appears to indicate a line perpendicular to the plane of resultant rotatory momentum ("the invariable plane") of the thermal motions as the magnetic axis of a magnetized body, and suggests the resultant moment of momenta of these motions as the definite measure of the "magnetic moment." The explanation of all phenomena of electro-magnetic attraction or repulsion, and of electro-magnetic induction, is to be looked for simply in the inertia and pressure of the matter of which the motions constitute heat. Whether this matter is or is not electricity, whether it is a continuous fluid interpermeating the spaces between molecular nuclei, or is itself molecularly grouped; or whether all matter is continuous, and molecular heterogeneousness consists in finite vortical or other relative motions of contiguous parts of a body; it is impossible to decide, and perhaps in vain to speculate, in the present state of science.

I append the solution of a dynamical problem for the sake of the illustrations it suggests for the two kinds of effect on the plane of polarization referred to above.

Let the two ends of a cord of any length be attached to two points at the ends of a horizontal arm made to rotate round a vertical axis through its middle point at a constant angular velocity, ω, and let a second cord bearing a weight be attached to the middle of the first cord. The two cords being each perfectly light and flexible, and the weight a material point, it is required to determine its motion when infinitely little disturbed from its position of equilibrium*.

Let l be the length of the second cord, and m the distance from

---

* By means of this arrangement, but without the rotation of the bearing arm, a very beautiful experiment, due to Professor Blackburn, may be made by attaching to the weight a bag of sand discharging its contents through a fine aperture.

the weight to the middle point of the arm bearing the first. Let x and y be, at any time t, the rectangular coordinates of the position of the weight, referred to the position of equilibrium O, and two rectangular lines OX, OY, revolving uniformly in a horizontal plane in the same direction, and with the same angular velocity as the bearing arm; then, if we choose OX parallel to this arm, and if the rotation be in the direction with OY preceding OX, we have, for the equations of motion,

$$\frac{d^2x}{dt^2} - \omega^2 x - 2\omega \frac{dy}{dt} = -\frac{g}{l} x,$$

$$\frac{d^2y}{dt^2} - \omega^2 y + 2\omega \frac{dx}{dt} = -\frac{g}{m} y.$$

If for brevity we assume

$$\frac{1}{2}\left(\frac{g}{l} + \frac{g}{m}\right) = n^2, \text{ and } \frac{1}{2}\left(\frac{g}{l} - \frac{g}{m}\right) = \lambda^2,$$

we find, by the usual methods, the following solution:—

$$x = A \cos\left\{[\omega^2 + n^2 + (\lambda^4 + 4n^2\omega^2)^{\frac{1}{2}}]^{\frac{1}{2}} t + \alpha\right\}$$

$$+ B \cos\left\{[\omega^2 + n^2 - (\lambda^4 + 4n^2\omega^2)^{\frac{1}{2}}]^{\frac{1}{2}} t + \beta\right\},$$

$$y = -\frac{2\omega^2 - \lambda^2 + (\lambda^4 + 4n^2\omega^2)^{\frac{1}{2}}}{2\omega\,[\omega^2 + n^2 + (\lambda^4 + 4n^2\omega^2)^{\frac{1}{2}}]^{\frac{1}{2}}} A \sin\phi$$

$$-\frac{2\omega^2 - \lambda^2 - (\lambda^4 + 4n^2\omega^2)^{\frac{1}{2}}}{2\omega\,[\omega^2 + n^2 - (\lambda^4 + 4n^2\omega^2)^{\frac{1}{2}}]^{\frac{1}{2}}} B \sin\psi,$$

where A, α, B, β are arbitrary constants, and ϕ and ψ are used for brevity to denote the arguments of the cosines appearing in the expression for x.

The interpretation of this solution, when ω is taken equal to the component of the earth's angular velocity round a vertical at the locality, affords a full explanation of curious phenomena which have been observed by many in failing to repeat Foucault's admirable pendulum experiment. When the mode of suspension is perfect, we have $\lambda = 0$; but in many attempts to obtain Foucault's result, there has been an asymmetry in the mode of attachment of the head of the cord or wire used, or there has been a slight lateral unsteadiness in the bearings of the point of suspension, which has made the observed motion be the same as that expressed by the preceding solution, where λ has some small value either greater than or less than ω, and n has the value $\sqrt{\dfrac{g}{l}}$. The only case, however, that need be considered as illustrative of the subject of the present communication is that in which ω is very great in comparison with n. To obtain a form of solution readily interpreted in this case, let

$$[\omega^2 + n^2 + (\lambda^4 + 4n^2\omega^2)^{\frac{1}{2}}]^{\frac{1}{2}} = \omega + \rho, \quad [\omega^2 + n^2 - (\lambda^4 + 4n^2\omega^2)^{\frac{1}{2}}]^{\frac{1}{2}} = \omega - \sigma,$$

$$\frac{2\omega^2 - \lambda^2 + (\lambda^4 + 4n^2\omega^2)^{\frac{1}{2}}}{2\omega\,[\omega^2 + n^2 + (\lambda^4 + 4n^2\omega^2)^{\frac{1}{2}}]^{\frac{1}{2}}} = 1 + e,$$

$$\frac{2\omega^2 - \lambda^2 - (\lambda^4 + 4n^2\omega^2)^{\frac{1}{2}}}{2\omega\,[\omega^2 + n^2 - (\lambda^4 + 4n^2\omega^2)^{\frac{1}{2}}]^{\frac{1}{2}}} = 1 - f.$$

The preceding solution becomes

$$x = A \cos\{(\omega + \rho)\,t + \alpha\} + B \cos\{(\omega - \sigma)\,t + \beta\}$$

$$y = -A \sin\{(\omega + \rho)\,t + \alpha\} - B \sin\{(\omega - \sigma)\,t + \beta\}$$
$$\quad - eA \sin\{(\omega + \rho)\,t + \alpha\} + fB \sin\{(\omega - \sigma)\,t + \beta\}.$$

To express the result in terms of coordinates ξ, η, with reference to fixed axes, instead of the revolving axes OX, OY, we may assume

$$\xi = x \cos \omega t - y \sin \omega t,$$

$$\eta = x \sin \omega t + y \cos \omega t.$$

Then we have

$$\xi = A \cos(\rho t + \alpha) + B \cos(\sigma t - \beta)$$
$$\quad + (eA \sin\{(\omega + \rho)\,t + \alpha\} - fB \sin\{(\omega - \sigma)\,t + \beta\}) \sin \omega t$$

$$\eta = -A \sin(\rho t + \alpha) + B \sin(\sigma t - \beta)$$
$$\quad + (-eA \sin\{(\omega + \rho)\,t + \alpha\} + fB \sin\{(\omega - \sigma)\,t + \beta\}) \cos \omega t.$$

When ω is very large, e and f are both very small, and the last two terms of each of these equations become very small periodic terms, of very rapidly recurring periods, indicating a slight tremor in the resultant motion. Neglecting this, and taking $\alpha = 0$ and $\beta = 0$, as we may do without loss of generality, by properly choosing the axes of reference, and the era of reckoning for the time, we have finally, for an approximate solution of a suitable kind,

$$\xi = A \cos \rho t + B \cos \sigma t,$$

$$\eta = -A \sin \rho t + B \sin \sigma t.$$

The terms B, in this expression, represent a circular motion of period $\dfrac{2\pi}{\sigma}$, in the positive direction (that is, from the positive axis of ξ to the positive axis of η), or in the same direction as that of the rotation ω; and the terms A represent a circular motion, of period $\dfrac{2\pi}{\rho}$, in the contrary direction. Now, ω being very great, ρ and σ are very nearly equal to one another; but ρ is rather less than σ, as the following approximate expressions derived from their exact values expressed above, show :—

$$\rho = n + \frac{1}{8}\frac{\lambda^4}{\omega^2 n} - \frac{1}{8}\frac{\lambda^4}{\omega^3}, \qquad \sigma = n + \frac{1}{8}\frac{\lambda^4}{\omega^2 n} + \frac{1}{8}\frac{\lambda^4}{\omega^3}.$$

Hence the form of solution simply expresses that circular vibrations of the pendulum in the contrary directions have slightly different periods, the shorter, $\dfrac{2\pi}{\sigma}$, when the motion of the pendulum follows that of the arm supporting it, and the longer, $\dfrac{2\pi}{\rho}$, when it is in the contrary direction. The equivalent statement, *that if the pendulum be simply drawn aside from its position of equilibrium, and let go without initial velocity, the vertical plane of its motion will rotate slowly at the angular rate* $\frac{1}{2}(\sigma - \rho)$, is expressed most shortly by taking $A = B$, and reducing the preceding solution to the form

$$\xi = 2A \cos \varpi t \cos n't,$$

$$\eta = 2A \sin \varpi t \cos n't,$$

where $\qquad n' = \dfrac{1}{2}(\sigma + \rho)$, or, approximately, $n' = n + \dfrac{1}{8}\dfrac{\lambda^4}{\omega^2 n}$,

and $\qquad \varpi = \dfrac{1}{2}(\sigma - \rho)$, or, approximately, $\varpi = \dfrac{1}{8}\dfrac{\lambda^4}{\omega^3}$.

It is a curious part of the conclusion thus expressed, that the faster the bearing arm is carried round, the slower does the plane of a simple vibration of the pendulum follow it. When the bearing arm is carried round infinitely fast, the plane of a vibration of the pendulum will remain steady, and the period will be n; in other words, the motion of the pendulum will be the same as that of a simple pendulum whose length is $\dfrac{2}{\dfrac{1}{e}+\dfrac{1}{m}}$, or a harmonic mean between the effective lengths in the two principal planes of the actual pendulum.

It is easy to prove from this, that if a long straight rod, or a stretched cord possessing some rigidity, unequally elastic or of unequal dimensions, in different transverse directions, be made to rotate very rapidly round its axis, and if vibrations be maintained in a line at right angles to it through any point, there will result, running along the rod or cord, waves of sensibly rectilineal transverse vibrations, in a plane which in the forward progress of the wave, turns at a uniform rate in the same direction as the rotation of the substance; and that if $\dfrac{2\pi}{\omega}$ be the period of rotation of the substance, and l and m the lengths of simple pendulums respectively isochronous with the vibrations of two plane waves of the same length, a, in the planes of maximum and of minimum elasticity of the substance, when destitute of rotation, the period of vibration in a wave of the same length in the substance when made to rotate will be

$$\frac{2\pi}{n\left(1 + \dfrac{1}{8}\dfrac{\lambda^4}{\omega^2 n^2}\right)};$$

and the angle through which the plane of vibration turns, in the propagation through a wave length, will be

$$\frac{\pi}{4}\frac{\lambda^4}{n\omega^3};$$

or the number of wave lengths through which the wave is propagated before its plane turns once round, will be

$$\frac{8n\omega^3}{\lambda^4};$$

where, as before,

$$n = \sqrt{g\cdot\frac{1}{2}\left(\frac{1}{l}+\frac{1}{m}\right)}, \qquad \lambda = \sqrt{g\cdot\frac{1}{2}\left(\frac{1}{l}-\frac{1}{m}\right)},$$

and ω denotes the angular velocity with which the substance is made to rotate.

If next we suppose the rod or cord to be slightly twisted about its axis, so that its directions of maximum and minimum elasticity shall lie on two rectangular heliçoidal surfaces (*heliçoïdes gauches*), and if, while regular rectilineal vibrations are maintained at one point of it with a period to which the wave length corresponding is a very large multiple of the step of the screw, the substance be made to rotate so rapidly as to make the velocity of a point carried along one of the screw surfaces in a line parallel to the axis be equal to the velocity of propagation of a wave, it is clear that a series of sensibly plane waves will run along the rod or cord with no rotation of the plane of vibration. The period of vibration of a particle will be, approximately, the same as before, that is, approximately, equal to $\frac{2\pi}{n}$. Its velocity of propagation will therefore be $\frac{na}{2\pi}$, and, if s be the step of the screw, the period of rotation of the substance, to fulfil the stated condition, must be $\frac{2\pi s}{na}$, or its angular velocity $\frac{na}{s}$. Now it is easily seen that the effects of the rapid rotation, and the effects of the slight twist, may be considered as independently superimposed; and therefore the effect of the twist, with no rotation of the substance, must be to give a rotation to the plane of vibration equal and contrary to that which the rotation of the substance would give if there were no twist. But the effect on the plane of vibration, due to an angular velocity ω, of rotation of the substance, is, as we have seen, one turn in $\frac{8n\omega^3}{\lambda^4}$ wave lengths; and therefore it is one turn in $8\frac{n^4}{\lambda^4}\frac{a^3}{s^3}$ wave lengths when the angular velocity is $\frac{na}{s}$. Hence the effect of a twist amounting to one turn in a length, s, a small fraction of the wave length, is to cause the plane of vibration of a wave to turn round with the forward propagation of the wave, at the rate of one turn in $8\frac{n^4}{\lambda^4}\frac{a^3}{s^3}$ wave lengths, in the same direction as that of a point kept on one of the screw surfaces.

From these illustrations it is easy to see in an infinite variety of ways how to make structures, homogeneous when considered on a large enough scale, which (1) with certain rotatory motions of component parts having, in portions large enough to be sensibly homogeneous, resultant axes of momenta arranged like lines of magnetic force, *shall have the dynamical property by which the optical phenomena of transparent bodies in the magnetic field are explained*; (2) with spiral arrangements of component parts, having axes all ranged parallel to a fixed line, *shall have the axial rotatory property corresponding to that of quartz crystal*; and (3) with spiral arrangements of component groups, having axes totally un-arranged, *shall have the isotropic rotatory property possessed by solutions of sugar and tartaric acid, by oil of turpentine, and many other liquids.*

Art. XCIV. Sui fenomeni magnetocristalline.

[*Nuovo Cimento*, IV., 1856.]

[Electrostatics and Magnetism, Art. XXX.]

Art. XCV. On the Alteration of Temperature accompany-ing Changes of Pressure in Fluids.

[*Proc. Roy. Soc.*, June, 1857; *Phil. Mag.*, June Suppl., 1858.]

The subject of this paper is given in *Mathematical and Physical Papers*, Art. XLVIII. (Vol. I.).

Art. XCVI. Remarks on the interior Melting of Ice.

[Part of a letter to Prof. Stokes; *Proc. Roy. Soc.* IX., Feb. 1858.]

In the Number of the *Proceedings* just published, which I received yesterday, I see some very interesting experiments de-scribed in a communication by Dr Tyndall, "On some Physical Properties of Ice." I write to you to point out that they afford direct ocular evidence of my brother's theory of the plasticity of ice, published in the *Proceedings* of the 7th of May last; and to add, on my own part, a physical explanation of the blue veins in glaciers, and of the lamellar structure which Dr Tyndall has shown to be induced in ice by pressure, as described in the sixth section of his paper.

Thus, my brother, in his paper of last May, says, "If we " commence with the consideration of a mass of ice perfectly free " from porosity, and free from liquid particles diffused through its " substance, and if we suppose it to be kept in an atmosphere at " or above 0° Cent., then, as soon as pressure is applied to it, pores " occupied by liquid water must instantly be formed in the com-" pressed parts, in accordance with the fundamental principle of " the explanation I have propounded—the lowering, namely, of the

" freezing-point or melting-point, by pressure, and the fact that ice
" cannot exist at 0° Cent. under a pressure exceeding that of the
" atmosphere." Dr Tyndall finds that when a cylinder of ice is
placed between two slabs of box-wood, and subjected to gradually
increasing pressure, a dim cloudy appearance is observed, which
he finds is due to the melting of small portions of the ice in the
interior of the mass. The permeation into portions of the ice,
for a time clear, " by the water squeezed against it from such
parts as may be directly subjected to the pressure," theoretically
demonstrated by my brother, is beautifully illustrated by Dr Tyndall's statement, that " the hazy surfaces produced by the com-
" pression of the mass were observed to be in a state of intense
" commotion, which followed closely upon the edge of the surface
" as it advanced through the solid. It is finally shown that these
" surfaces are due to the liquefaction of the ice in planes perpen-
" dicular to the pressure."

There can be no doubt but that the " oscillations" in the
melting-point of ice, and the distinction between strong and weak
pieces in this respect, described by Dr Tyndall in the second
section of his paper, are consequences of the varying pressures
which different portions of a mass of ice must experience when
portions within it become liquefied.

The elevation of the melting temperature which my brother's
theory shows must be produced by diminishing the pressure of ice
below the atmospheric pressure, and to which I alluded as a
subject for experimental illustration, in the article describing my
experimental demonstration of the lowering effect of pressure
(*Proceedings, Roy. Soc. Edin.* Feb. 1850), demonstrates that a
vesicle of water cannot form in the interior of a solid of ice
except at a temperature higher than 0° Cent. This is a conclusion
which Dr Tyndall expresses as a result of mechanical considerations : thus, " Regarding heat as a mode of motion," " liberty of
" liquidity is attained by the molecules at the surface of a mass of
" ice before the molecules at the centre of the mass can attain this
" liberty."

The physical theory shows that a removal of the atmospheric
pressure would raise the melting-point of ice by $\frac{3}{400}$ths of a degree
Centigrade. Hence it is certain that the interior of a solid of
ice, heated by the condensation of solar rays by a lens, will rise
to at least that excess of temperature above the superficial parts.

It appears very nearly certain that cohesion will prevent the evolution of a bubble of vapour of water in a vesicle of water forming by this process in the interior of a mass of ice, until a high "negative pressure" has been reached, that is to say, until cohesion has been called largely into operation, especially if the water and ice contain little or no air by absorption (just as water freed from air may be raised considerably above its boiling-point under any non-evanescent hydrostatic pressure). Hence it appears nearly certain that the interior of a block of ice originally clear, and made to possess vesicles of water by the concentration of radiant heat, as in the beautiful experiments described by Dr Tyndall in the commencement of his paper, will rise very considerably in temperature, while the vesicles enlarge under the continued influence of the heat received by radiation through the cooler enveloping ice and through the fluid medium (air and a watery film, or water) touching it all round, which is necessarily at 0° Cent. where it touches the solid.

I find I have not time to execute my intention of sending you to-day a physical explanation of the blue veins of glaciers which occurred to me last May, but I hope to be able to send it in a short time.

ART. XCVII. ON THE STRATIFICATION OF VESICULAR ICE BY PRESSURE.

[Part of a letter to Prof. Stokes; *Proc. Roy. Soc.* IX., April, 1858.]

In my last letter to you I pointed out that my brother's theory of the effect of pressure in lowering the freezing-point of water, affords a perfect explanation of various remarkable phenomena involving the internal melting of ice, described by Professor Tyndall in the Number of the *Proceedings* which has just been published. I wish now to show that the stratification of vesicular ice by pressure observed on a large scale in glaciers, and the lamination of clear ice described by Dr Tyndall as produced in hand specimens by a Bramah's press, are also demonstrable as conclusions from the same theory.

Conceive a continuous mass of ice, with vesicles containing either air or water distributed through it; and let this mass be pressed together by opposing forces on two opposite sides of it.

The vesicles will gradually become arranged in strata perpendicular to the lines of pressure, *because of the melting of ice in the localities of greatest pressure and the regelation of the water in the localities of least pressure, in the neighbourhood of groups of these cavities.* For, any two vesicles nearly in the direction of the condensation will afford to the ice between them a relief from pressure, and will occasion an aggravated pressure in the ice round each of them in the places farthest out from the line joining their centres; while the pressure in the ice on the far sides of the two vesicles will be somewhat diminished from what it would be were their cavities filled up with the solid, although not nearly as much diminished as it is in the ice between the two. Hence, as demonstrated by my brother's theory and my own experiment, the melting temperature of the ice round each vesicle will be highest on its side nearest to the other vesicle, and lowest in the localities on the whole farthest from the line joining the centres. Therefore, ice will melt from these last-mentioned localities, and, if each vesicle have water in it, the partition between the two will thicken by freezing on each side of it. Any two vesicles, on the other hand, which are nearly in a line perpendicular to the direction of pressure will agree in leaving an aggravated pressure to be borne by the solid between them, and will each direct away some of the pressure from the portions of the solid next itself on the two sides farthest from the plane through the centres, perpendicular to the line of pressure. This will give rise to an increase of pressure on the whole in the solid all round the two cavities, and nearly in the plane perpendicular to the pressure, although nowhere else so much as in the part between them. Hence these two vesicles will gradually extend towards one another by the melting of the intervening ice, and each will become flattened in towards the plane through the centres perpendicular to the direction of pressure, by the freezing of water on the parts of the bounding surface farthest from this plane. It may be similarly shown that two vesicles in a line oblique to that of condensation will give rise to such variations of pressure in the solid in their neighbourhood, as to make them, by melting and freezing, to extend, each obliquely *towards* the other and *from* the parts of its boundary most remote from a plane midway between them, perpendicular to the direction of pressure.

The general tendency clearly is for the vesicles to become flattened and arranged in layers, in planes perpendicular to the direction of the pressure from without.

It is clear that the same general tendency must be experienced even when there are bubbles of air in the vesicles, although no doubt the resultant effect would be to some extent influenced by the running down of water to the lowest part of each cavity.

I believe it will be found that these principles afford a satisfactory physical explanation of the origin of that beautiful veined structure which Professor Forbes has shown to be an essential organic property of glaciers. Thus the first effect of pressure not equal in all directions, on a mass of snow, ought to be, according to the theory, to convert it into a stratified mass of layers of alternately clear and vesicular ice, perpendicular to the direction of maximum pressure. In his remarks " On the Conversion of the Névé into Ice*," Professor Forbes says, " *that the conversion into ice is simultaneous*" (and in a particular case referred to "*identical*") "*with the formation of the blue bands;...* and that these bands are " formed where the pressure is most intense, and where the dif- " ferential motion of the parts is a maximum, that is, near the walls " of a glacier." He farther states, that, after long doubt, he feels satisfied that the conversion of snow into ice is due to the effects of pressure on the loose and porous structure of the former; and he formally abandons the notion that the blue veins are due to the freezing of infiltrated water, or to any other cause than the kneading action of pressure. All the observations he describes seem to be in most complete accordance with the theory indicated above. Thus, in the thirteenth letter, he says, " the blue veins " are formed where the pressure is most intense and the differential " motion of the parts a maximum."

Now the theory not only requires pressure, but requires difference of pressure in different directions to explain the stratification of the vesicles. Difference of pressure in different directions produces the "differential motion" referred to by Professor Forbes. Further, the difference of pressure in different directions must be continued until a very considerable amount of this differential motion, or distortion, has taken place, to produce any sensible degree of stratification in the vesicles. The absolute amount of distortion experienced by any portion of the viscous mass is

* Thirteenth Letter on Glaciers, section (2), dated Dec. 1846.

therefore an index of the persistence of the differential pressure, by the continued action of which the blue veins are induced. Hence also we see why blue veins are not formed in any mass, ever so deep, of snow *resting* in a hollow or corner.

As to the direction in which the blue veins appear to lie, they must, according to the theory, be something intermediate between the surfaces perpendicular to the greatest pressure, and the surfaces of sliding; since they will commence being formed exactly perpendicular to the direction of greatest pressure, and will, by the differential motion accompanying their formation, become gradually laid out more and more nearly parallel to the sides of the channel through which the glacier is forced. This circumstance, along with the comparatively weak mechanical condition of the white strata (vesicular layers between the blue strata), must, I think, make these white strata become ultimately, in reality, the surfaces of "sliding" or of "tearing," or of chief differential motion, as according to Professor Forbes's observations they seem to be. His first statement on the subject, made as early as 1842, that "the blue veins seem to be perpendicular to the lines of maximum pressure," is, however, more in accordance with their mechanical origin, according to the theory I now suggest, than the supposition that they are *caused* by the tearing action which is found to take place along them when formed. It appears to me, therefore, that Dr Tyndall's conclusion, that the vesicular stratification is produced by pressure in surfaces perpendicular to the directions of maximum pressure, is correct as regards the mechanical origin of the veined structure; while there seems every reason, both from observation and from mechanical theory, to accept the view given by Professor Forbes of their function in glacial motion.

The mechanical theory I have indicated as the explanation of the veined structure of glacial ice is especially applicable to account for the stratification of the vesicles observed in ice originally clear, and subjected to differential pressure, by Dr Tyndall; the formation of the vesicles themselves being, as remarked in my last letter*, anticipated by my brother's theory, published in the *Proceedings* for May, 1857.

I believe the theory I have given above contains the true explanation of one remarkable fact observed by Dr Tyndall in

* See *Proceedings* for February 25, 1858.

connexion with the beautiful set of phenomena which he discovered to be produced by radiant heat, concentrated on an internal portion of a mass of clear ice by a lens; the fact, namely, that the planes in which the vesicles extend are generally parallel to the sides when the mass of ice operated on is a flat slab; for the solid will yield to the "negative" internal pressure due to the contractility of the melting ice, most easily in the direction perpendicular to the sides. The so-called negative pressure is therefore least, or which is the same thing, the positive pressure is greatest in this direction. Hence the vesicles of melted ice, or of vapour caused by the contraction of melted ice, must, as I have shown, tend to place themselves parallel to the sides of the slab.

The division of the vesicular layers into leaves like six-petaled flowers is a phenomenon which does not seem to me as yet so easily explained; but I cannot see that any of the phenomena described by Dr Tyndall can be considered as having been proved to be due to ice having mechanical properties of a uniaxal crystal.

[It now seems to me most probable Tyndall was right in attributing the six-rayed structure to the molecular mechanics of a uniaxal crystal.   K., *Dec.* 13, 1903.]

APPENDIX G.

HYDROKINETIC SOLUTIONS AND OBSERVATIONS.

PART I. *On the Motion of Free Solids through a Liquid.*

THIS paper commences with the following extract from the author's private journal, of date January 6, 1858:—

"Let $\mathfrak{X}$, $\mathfrak{Y}$, $\mathfrak{Z}$, $\mathfrak{L}$, $\mathfrak{M}$, $\mathfrak{N}$ be rectangular components of an impulsive force and an impulsive couple applied to a solid of invariable shape, with or without inertia of its own, in a perfect liquid, and let u, v, w, ϖ, ρ, σ be the components of linear and angular velocity generated. Then, if the *vis viva** (twice the mechanical value) of the whole motion be, as it cannot but be, given by the expression

$$Q = [u,\, u]\, u^2 + [v,\, v]\, v^2 + \ldots + 2\,[v,\, u]\, vu + 2\,[w,\, u]\, wu + \ldots$$
$$+ 2\,[\varpi,\, u]\, \varpi u + \ldots,$$

where $= [u,\, u]$, $[v,\, v]$, &c. denote 21 constant coefficients determinable by transcendental analysis from the form of the surface of the solid, probably involving only elliptic transcendentals when the surface is ellipsoidal: involving, of course, the moments of inertia of the solid itself: we must have

$$[u,\, u]\, u + [v,\, u]\, v + [w,\, u]\, w + [\varpi,\, u]\, \varpi + [\rho,\, u]\, \rho + [\sigma,\, u]\, \sigma = \mathfrak{X},\ \&c.,$$
$$[u,\, \varpi]\, u + [v,\, \varpi]\, v + [w,\, \varpi]\, w + [\varpi,\, \varpi]\, \varpi + [\rho,\, \varpi]\, \rho + [\sigma,\, \varpi]\, \sigma = \mathfrak{L},\ \&c.$$

If now a continuous force X, Y, Z, and a continuous couple L, M, N, referred to axes fixed in the body, are applied, and if $\mathfrak{X}, \ldots,$ &c. denote the impulsive force and couple capable of

---

* Parts I. and II. from the *Proceedings of the Royal Society of Edinburgh*, 1870–71. Parts III. and IV. from letters to Professor Tait of August 1871. Part V. from *Phil. Mag.* November, 1871.

In Part II., T, instead of $\frac{1}{2}Q$, is used to denote the "mechanical value," or, as it is now called, the "kinetic energy" of the motion.

generating from rest the motion u, v, w, ϖ, ρ, σ, which exists in reality at any time t; or, merely mathematically, if $\mathfrak{X}$ &c. denote for brevity the preceding linear functions of the components of motion, the equations of motion are as follow:—

$$\frac{d\mathfrak{X}}{dt} - \mathfrak{Y}\sigma + \mathfrak{Z}\rho = X, \qquad \frac{d\mathfrak{Y}}{dt} = \&c.,$$
$$\frac{d\mathfrak{L}}{dt} - \mathfrak{Y}w + \mathfrak{Z}v - \mathfrak{M}\sigma + \mathfrak{N}\rho = L,$$
$$\frac{d\mathfrak{M}}{dt} - \mathfrak{Z}u + \mathfrak{X}w - \mathfrak{N}\varpi + \mathfrak{L}\sigma = M,$$
$$\frac{d\mathfrak{N}}{dt} - \mathfrak{X}v + \mathfrak{Y}u - \mathfrak{L}\rho + \mathfrak{M}\varpi = N$$

$$\dotfill(1).$$

Three first integrals, when

$$X = 0, \quad Y = 0, \quad Z = 0, \quad L = 0, \quad M = 0, \quad N = 0 \dots(2),$$

must of course be, and obviously are,

$$\mathfrak{X}^2 + \mathfrak{Y}^2 + \mathfrak{Z}^2 = \text{const.} \dotfill(3),$$

resultant momentum constant;

$$\mathfrak{L}\mathfrak{X} + \mathfrak{M}\mathfrak{Y} + \mathfrak{N}\mathfrak{Z} = \text{const.} \dotfill(4),$$

resultant of moment of momentum constant; and

$$u\mathfrak{X} + v\mathfrak{Y} + w\mathfrak{Z} + \varpi\mathfrak{L} + \rho\mathfrak{M} + \sigma\mathfrak{N} = Q = \text{const.} \dots(5)."$$

These equations were communicated in a letter to Professor Stokes, of date (probably January) 1858, and they were referred to by Professor Rankine, in his first paper on Stream-lines, communicated to the Royal Society of London[*], July 1863.

They are now communicated to the Royal Society of Edinburgh, and the following proof is added:—

Let P be any point fixed relatively to the body; and at time t, let its coordinates relatively to axes OX, OY, OZ, fixed in space, be x, y, z. Let PA, PB, PC be three rectangular axes fixed relatively to the body, and (A, X), (A, Y), ... the cosines of the nine inclinations of these axes to the fixed axes OX, OY, OZ.

---

[*] These equations will be very conveniently called the Eulerian equations of the motion. They correspond precisely to Euler's equations for the rotation of a rigid body, and include them as a particular case. As Euler seems to have been the first to give equations of motion in terms of coordinate components of velocity and force referred to lines fixed relatively to the moving body, it will be not only convenient, but just, to designate as "Eulerian equations" any equations of motion in which the lines of reference, whether for position, or velocity, or moment of momentum, or force, or couple, move with the body, or the bodies, whose motion is the subject.

Let the components of the "impulse*" or generalized momentum parallel to the fixed axes be ξ, η, ζ, and its moments round the same axes λ, μ, ν; so that if X, Y, Z be components of force acting on the solid, in line through P, and L, M, N components of couple, we have

$$\left.\begin{array}{l} \dfrac{d\xi}{dt} = X, \quad \dfrac{d\eta}{dt} = Y, \quad \dfrac{d\zeta}{dt} = Z, \\[2mm] \dfrac{d\lambda}{dt} = L + Zy - Yz, \quad \dfrac{d\mu}{dt} = M + Xz - Zx, \quad \dfrac{d\nu}{dt} = N + Yx - Xy \end{array}\right\} \quad (6).$$

Let $\mathfrak{X}$, $\mathfrak{Y}$, $\mathfrak{Z}$ and $\mathfrak{L}$, $\mathfrak{M}$, $\mathfrak{N}$ be the components and moments of the impulse relatively to the axes PA, PB, PC moving with the body. We have

$$\left.\begin{array}{l} \xi = \mathfrak{X}\,(A, X) + \mathfrak{Y}\,(B, X) + \mathfrak{Z}\,(C, X), \\ \dotfill \\ \dotfill \\ \lambda = \mathfrak{L}\,(A, X) + \mathfrak{M}\,(B, X) + \mathfrak{N}\,(C, X) + \mathfrak{Z}y - \mathfrak{Y}z, \\ \dotfill \\ \dotfill \end{array}\right\} \dots (7).$$

Now let the fixed axes OX, OY, OZ be chosen coincident with the position at time t of the moving axes PA, PB, PC: we shall consequently have

$$\left.\begin{array}{l} x = 0, \quad y = 0, \quad z = 0, \\[2mm] \dfrac{dx}{dt} = u, \quad \dfrac{dy}{dt} = v, \quad \dfrac{dz}{dt} = w \end{array}\right\} \dots\dots\dots\dots\dots (8),$$

$$\left.\begin{array}{l} (A, X) = (B, Y) = (C, Z) = 1, \\ (A, Y) = (A, Z) = (B, X) = (B, Z) = (C, X) = (C, Y) = 0, \\[2mm] \dfrac{d(A, Y)}{dt} = \sigma, \quad \dfrac{d(B, X)}{dt} = -\sigma, \quad \dfrac{d(C, X)}{dt} = \rho, \\[2mm] \dfrac{d(A, Z)}{dt} = -\rho, \quad \dfrac{d(B, Z)}{dt} = \varpi, \quad \dfrac{d(C, Y)}{dt} = -\varpi \end{array}\right\} \dots (9).$$

Using (7), (8), and (9) in (6), we find (1).

One chief object of this investigation was to illustrate dynamical effects of heliçoidal property (that is, right or left-handed asymmetry). The case of complete isotropy, with heliçoidal quality, is that in which the coefficients in the quadratic expression Q fulfil the following conditions:—

---

* See "Vortex Motion," § 6, *Trans. Roy. Soc. Edin.* (1868).

$[u,\ u] = [v,\ v] = [w, w]$ (let m be their common value),

$[\varpi, \varpi] = [\rho,\ \rho] = [\sigma,\ \sigma]$ „ n „   „   „

$[u,\ \varpi] = [v,\ \rho] = [w, \sigma]$ „ h „   „   „

$[v,\ w] = [w, u] = [u,\ v] = 0;\ [\rho,\ \sigma] = [\sigma,\ \varpi] = [\varpi, \rho] = 0,$

and

$[u,\ \rho] = [u,\ \sigma] = [v,\ \sigma] = [v,\ \varpi] = [w,\ \varpi] = [w, \rho] = 0$

$\left.\begin{array}{c}\\\\\\\\\\\\\\\end{array}\right\}$ ...(10);

so that the formula for Q is

$$Q = m\,(u^2 + v^2 + w^2) + n\,(\varpi^2 + \rho^2 + \sigma^2) + 2h\,(u\varpi + v\rho + w\sigma)\ldots(11).$$

For this case, therefore, the Eulerian equations (1) become

$$\frac{d\,(mu + h\varpi)}{dt} - m\,(v\sigma - w\rho) = X,\ \&\mathrm{c.},$$

and $\dfrac{d\,(n\varpi + hu)}{dt} = L,\ \&\mathrm{c.}$ ...(11').

[Memorandum :—Lines of reference fixed relatively to the body.]

But inasmuch as (11') remains unchanged when the lines of reference are altered to any other three lines at right angles to one another through P, it is easily shown directly from (6), (7), and (9) that if, altering the notation, we take u, v, w to denote the components of the velocity of P parallel to three fixed rectangular lines, and ϖ, ρ, σ the components of the body's angular velocity round these lines, we have

$$\frac{d\,(mu + h\varpi)}{dt} = X,\ \&\mathrm{c.},\qquad \frac{d\,(mv + h\rho)}{dt} = Y,$$

and $\dfrac{d\,(n\varpi + hu)}{dt} + h\,(\sigma v - \rho w) = L,\ \&\mathrm{c.}$ ...(12),

[Memorandum :—Lines of reference fixed in space],

which are more convenient than the Eulerian equations.

The integration of these equations, when neither force nor couple acts on the body ($X = 0$ &c., $L = 0$ &c.), presents no difficulty; but its result is readily seen from § 21 (" Vortex Motion ") to be that, when the impulse is both translatory and rotational, the point P, round which the body is isotropic, moves uniformly in a circle or spiral so as to keep at a constant distance from the " axis of the impulse," and that the components of angular velocity round the three fixed rectangular axes are constant.

An isotropic heliçoid [chiroid as I now call it; Lec. XX., § 204] may be made by attaching projecting vanes to the surface of a globe in proper positions; for instance, cutting at 45° each, at the middles of the twelve quadrants of any three great circles dividing the globe into eight quadrantal triangles. By making the globe and the vanes of light paper, a body might probably be obtained rigid enough and light enough to illustrate by its motions through air the motions of an isotropic heliçoid through an incompressible liquid. But curious phenomena, not deducible from the present investigation, will, no doubt, on account of viscosity, be observed.

PART II.

Still considering only one movable rigid body, infinitely remote from disturbance of other rigid bodies, fixed or movable, let there be an aperture or apertures through it, and let there be irrotational circulation or circulations (§ 60, " Vortex Motion ") through them. Let ξ, η, ζ be the components of the " impulse " at time t, parallel to three fixed axes, and λ, μ, ν its moments round these axes, as above, with all notation the same, we still have (§ 26, " Vortex Motion ")

$$\left.\begin{aligned} \frac{d\xi}{dt} &= X, \text{ &c.,} \\ \frac{d\lambda}{dt} &= L + Zy - Yz, \text{ &c.} \end{aligned}\right\} \dots\dots(6) \text{ (repeated).}$$

But, instead of for $\frac{1}{2}Q$ a quadratic function of the components of velocity as before, we now have

$$T = E + \tfrac{1}{2}\left\{[u, u]\, u^2 + \dots + 2\,[u, v]\, uv + \dots\right\} \dots\dots(13),$$

where E is the kinetic energy of the fluid motion when the solid is at rest, and $\frac{1}{2}\{[u, u]\, u^2 + \dots\}$ is the same quadratic as before. The coefficients $[u, u]$, $[u, v]$, &c. are determinable by a transcendental analysis, of which the character is not at all influenced by the circumstance of there being apertures in the solid. And instead of $\xi = \dfrac{dT}{du}$, &c., as above, we now have

$$\xi = \frac{dT}{du} + Il, \quad \eta = \frac{dT}{dv} + Im, \quad \zeta = \frac{dT}{dw} + In, \left.\begin{array}{c} \\ \\ \end{array}\right\}...(14),$$

$$\lambda = \frac{dT}{d\varpi} + I\,(ny - mz) + Gl, \quad \mu = \&c., \quad \nu = \&c.$$

where I denotes the resultant "impulse" of the cyclic motion when the solid is at rest, l, m, n its direction-cosines, G its "rotational moment" (§ 6, "Vortex Motion"), and x, y, z the coordinates of any point in its "resultant axis." These, (14) with (13), used in (6) give the equations of the solid's motion referred to fixed rectangular axes. They have the inconvenience of the coefficients $[u, u]$, $[u, v]$, &c. being functions of the angular coordinates of the solid. The Eulerian equations (free from this inconvenience) are readily found on precisely the same plan as that adopted above for the old case of no cyclic motion in the fluid.

The formulæ for the case in which the ring is circular, has no rotation round its axis, and is not acted on by applied forces, though, of course, easily deduced from the general equations (14), (13), (6), are more readily got by direct application of first principles. Let P be such a point in the axis of the ring, and $\mathfrak{C}$, A, B such constants that $\frac{1}{2}(\mathfrak{C}\omega^2 + Au^2 + Bv^2)$ is the kinetic energy due to rotational velocity ω round D, any diameter through P, and translational velocities u along the axis and v perpendicular to it. The impulse of this motion, together with the supposed cyclic motion, is therefore compounded of

momentum in lines through P $\begin{cases} Au + I \text{ along the axis,} \\ Bv \text{ perpendicular to axis,} \end{cases}$

and moment of momentum $\mathfrak{C}\omega$ round the diameter D.

Hence if OX be the axis of resultant momentum, (x, y) the coordinates of P relatively to fixed axes OX, OY; θ the inclination of the axis of the ring to O; and ξ the constant value of the resultant momentum, we have

$$\xi \cos\theta = Au + I; \quad -\xi \sin\theta = Bv; \quad \xi y = \mathfrak{C}\omega; \left.\begin{array}{c} \\ \end{array}\right\}...(15).$$

and $\quad \dot{x} = u\cos\theta - v\sin\theta; \quad \dot{y} = u\sin\theta + v\cos\theta; \quad \dot{\theta} = \omega$

Hence for θ we have the differential equation

$$A\mathfrak{C}\frac{d^2\theta}{dt^2} + \xi\left[I\sin\theta + \frac{A-B}{2B}\xi\sin 2\theta\right] = 0......(16),$$

which shows that the ring oscillates rotationally according to the law of a horizontal magnetic needle carrying a bar of soft iron rigidly attached to it parallel to its magnetic axis.

When θ is and remains infinitely small, $\dot{\theta}$, y, and $\dot{y}$ are each infinitely small, $\dot{x}$ remains infinitely nearly constant, and the ring experiences an oscillatory motion in period

$$2\pi \sqrt{\frac{B\mathfrak{C}}{[I + (A - B)\,\dot{x}]\,(I + A\dot{x})}},$$

compounded of translation along OY and rotation round the diameter D. This result is curiously comparable with the well-known gyroscopic vibrations.

PART III.   *The Influence of frictionless Wind on Waves in water supposed frictionless. (Letter to Professor* TAIT, *of date August* 16, 1871.)

Taking OX vertically downwards and OY horizontal, let

$$x = h \sin n\,(y - at)\dots\dots\dots\dots\dots\dots(1)$$

be the equation of the section of the water by a plane perpendicular to the wave-ridges; and let h (the half wave-height) be infinitely small in comparison with $\dfrac{2\pi}{n}$ (the wave-length). The x-component of the velocity of the water at the surface is then

$$- nah \cos n\,(y - at) \dots\dots\dots\dots\dots(2).$$

This (because h is infinitesimal) must be the value of $\dfrac{d\phi}{dx}$ for the point $(0, y)$, if ϕ denote the velocity-potential at any point (x, y) of the water. Now because

$$\frac{d^2\phi}{dx^2} + \frac{d^2\phi}{dy^2} = 0,$$

and ϕ is a periodic function of y, and a function of x which (as we suppose the water infinitely deep) becomes zero when $x = \infty$, it must be of the form

$$P \cos (ny - e)\, \epsilon^{-nx},$$

where P and e are independent of x and y. Hence, taking $\dfrac{d\phi}{dx}$, putting $x = 0$ in it, and equating it to (2), we have

$$- Pn \cos (ny - e) = - nah \cos (ny - nat);$$

and therefore $P = \alpha h$, and $e = n\alpha t$; so that we have

$$\phi = \alpha h \epsilon^{-nx} \cos n\,(y - \alpha t) \quad\ldots\ldots\ldots\ldots\ldots(3).$$

This, it is to be remarked, results simply from the assumptions that the water is frictionless, that it has been at rest, and that its surface is moving in the manner specified by (1).

If the air were a frictionless liquid moving irrotationally, with a constant velocity V at heights above the water (that is to say, values of $-x$) considerably exceeding the wave-length, its velocity-potential ψ, found on the same principle, would be

$$\psi = (V - \alpha)\,h\epsilon^{nx} \cos n\,(y - \alpha t) + Vy\ldots\ldots\ldots\ldots(4).$$

Let now q denote the resultant velocity at any point (x, y) of the air. Neglecting infinitesimals of the order $(nh)^2$, we have

$$\tfrac{1}{2}q^2 = \tfrac{1}{2}V^2 - V(V - \alpha)\,nh\epsilon^{nx} \sin n\,(y - \alpha t)\ldots\ldots\ldots(5).$$

Now, if p denote the pressure at any point (x, y) in the air, and σ the density of the air, we have by the general equation for pressure in an irrotationally moving fluid,

$$C - p = \sigma \left(\frac{d\psi}{dt} + \tfrac{1}{2}q^2 - gx\right)\ldots\ldots\ldots\ldots(6).$$

Using (4) and (5) in this and putting $C = \tfrac{1}{2}\sigma V^2$, we find

$$p = \sigma\left\{nh\,(V - \alpha)^2 \epsilon^{nx} \sin n\,(y - \alpha t) + gx\right\}\ldots\ldots\ldots(7).$$

Similarly if p' denote the pressure at any point (x, y) of the water, since in this case q^2 is infinitesimal, we have

$$-p' = \frac{d\phi}{dt} - gx = nh\alpha^2 \epsilon^{-nx} \sin n\,(y - \alpha t) - gx \quad\ldots\ldots(8),$$

the density of the water being taken as unity.

Now let T be the cohesive tension of the separating surface of air and water. The curvature of this surface being $\dfrac{d^2x}{dy^2}$ derived from (1), is equal to

$$- n^2 h \sin n\,(x - \alpha t)\ldots\ldots\ldots\ldots\ldots\ldots(9).$$

Hence at any point in the water-surface,

$$p - p' = T n^2 h \sin n\,(x - \alpha t)\ldots\ldots\ldots\ldots(10);$$

and by (7) and (8), with for x its value by (1) (which, as h is infinitesimal, may be taken as zero except in their last terms), we have

$$p - p' = h\left\{n\left[\sigma\,(V - \alpha)^2 + \alpha^2\right] - g\,(1 - \sigma)\right\} \sin n\,(x - \alpha t)\ldots(11).$$

This, compared with (10), gives

$$n\left[\sigma\,(V - \alpha)^2 + \alpha^2\right] - g\,(1 - \sigma) = T n^2 \ldots\ldots\ldots\ldots(12).$$

Let
$$w = \sqrt{\frac{g(1-\sigma)+Tn^2}{(1+\sigma)n}} \quad \dots\dots\dots\dots(13),$$

which (being the value of α for $V=0$) is the velocity of propagation of waves with no wind, when the wave-length is $\frac{2\pi}{n}$. Then (12) becomes

$$\frac{\alpha^2 + \sigma(V-\alpha)^2}{1+\sigma} = w^2; \text{ whence } \alpha^2 = (1+\sigma)w^2 - \sigma(V-\alpha)^2\dots(14).$$

This determines α, the velocity of the waves when there is wind, of velocity V, in the direction of their advance. It shows that, for given wave-length, $2\pi/n$, the greatest wave-velocity is

$$w\sqrt{(1+\sigma)},$$

which is reached when this is the velocity of the wind. It is interesting to see that with wind of any other speed than that of the waves, and in the direction of the waves, their speed is less. For instance, the wave-speed with no wind, which is w, is less by approximately $\frac{1}{2}\sigma$ of w, (or about 1/1650,) than the speed when the wind is with the waves and of their speed. The explanation clearly is that when the air is motionless relatively to the wave crests and hollows its inertia is not called into play. Solving (14) for α, as a quadratic, we have

$$\alpha = \frac{\sigma V}{1+\sigma} \pm \left\{ w^2 - \frac{\sigma V^2}{(1+\sigma)^2} \right\}^{\frac{1}{2}} \quad \dots\dots\dots\dots(15).$$

This result leads to the following conclusions :—

(1) When $V/w = \sqrt{\dfrac{1+\sigma}{\sigma}} \fallingdotseq 28.7 \times \left(1 + \dfrac{1}{1650}\right),$

one of the values of α is zero, that is to say, static corrugations of wave-length $2\pi/n$, would be equilibrated by wind of velocity

$$w\sqrt{[(1+\sigma)/\sigma]}.$$

But the equilibrium would be unstable.

(2) When $V/w = \dfrac{1+\sigma}{\sqrt{\sigma}} \fallingdotseq 28.7 \times \left(1 + \dfrac{1}{825}\right),$

the two values of α are equal.

(3) When $V/w > \dfrac{1+\sigma}{\sqrt{\sigma}},$

both values of α are imaginary, and therefore the wind would blow into spin-drift, waves of length $2\pi/n$ or shorter.

Looking back to (13), we see that it gives a minimum value for w equal to

$$\sqrt{\frac{2\sqrt{gT}(1-\sigma)}{1+\sigma}} \quad\ldots\ldots\ldots\ldots\ldots(16).$$

Hence the water with a plane level surface would be unstable, even if air were frictionless, when the velocity of the wind exceeds

$$\sqrt{\frac{2\sqrt{gT}(1-\sigma^2)}{\sigma}}\ldots\ldots\ldots\ldots\ldots(16').$$

W. T.

Part IV. (*Letter to* Professor Tait, *of date August* 23, 1871.)

Defining a ripple as any wave on water whose length $< 2\pi\sqrt{\dfrac{T'}{g'}}$*, where

$$\left.\begin{array}{l} g' = g\,\dfrac{1-\sigma}{1+\sigma}, \\[2ex] T' = \dfrac{T}{1+\sigma} \end{array}\right\}\ldots\ldots\ldots\ldots\ldots(17);$$

and

$(\sigma = \cdot00121)$, you always see an exquisite pattern of ripples in front of any solid cutting the surface of water and moving horizontally at any speed, fast or slow [if not less than about 23 cm. per sec.]. The ripple-length is the smaller root of the equation

$$\frac{2\pi}{\lambda}\,T' + \frac{\lambda}{2\pi}\,g' = w^2 \ldots\ldots\ldots\ldots\ldots(18),$$

where w is the velocity of the solid. The latter may be a sailing-vessel or a row-boat, a pole held vertically and moved horizontally, an ivory pencil-case, a penknife-blade either edge or flat side foremost, or (best) a fishing-line kept approximately vertical by a lead weight hanging down below water, while carried along at about half a mile per hour by a becalmed vessel. The fishing-line shows both roots admirably; ripples in front, and waves of same velocity (λ the greater root of same equation) in rear. If so fortunate as to be becalmed again, I shall try to get a drawing of the whole pattern, showing the transition at the sides from ripples

* Which for pure water = 1·7 centim. (see Part V.).

to waves. When the speed with which the fishing-line is dragged is diminished towards the critical velocity

$$\sqrt{2\sqrt{g'T'}},$$

which is the minimum velocity of a wave, being [see Part V. below] for pure water 23 centims. per second (or $\dfrac{1}{2\cdot29}$ of a nautical mile per hour), the ripples in front elongate and become less curved, and the waves in rear become shorter, till at the critical velocity waves and ripples seem nearly equal, and with ridges nearly in straight lines perpendicular to the line of motion. (This is observation.) It seems that the critical velocity may be determined with some accuracy by experiment thus [see Part V. below]:—

Remark that the shorter the ripple-length the greater is the velocity of propagation: and that the moving force of the ripple-motion is partly gravity, but chiefly cohesion; and with very short ripple-length it is almost altogether cohesion, $i.e.$ the same force as that which makes a dew-drop tremble. The least velocity of frictionless air that can raise a ripple on rigorously quiescent frictionless water is [(16′) above]

$$660 \text{ centimetres per second}$$

$$\left(\text{being } \frac{1+\sigma}{\sqrt{\sigma}} \times \text{minimum wave-velocity}\right)$$

$$= 12\cdot8 \text{ nautical miles per hour.}$$

Observation shows the sea to be ruffled by wind of a much smaller velocity than this. Such ruffling, therefore, is due to viscosity of the air.

W. T.

Postscript to PART IV. (*October* 17, 1871).

The influence of viscosity gives rise to a greater pressure on the anterior than on the posterior side of a solid moving uniformly relatively to a fluid. A symmetrical solid, as for example a globe, moving uniformly through a frictionless fluid, experiences augmentation of pressure in front and behind equally; and diminished pressure over an intervening zone. Observation (as for instance in Mr J. R. Napier's experiments on his "pressure log," for measuring the speed of vessels, and experiments by Joule and myself *, on the pressure at different points of a solid globe exposed to wind) shows that, instead of being increased, the pressure is sometimes actually diminished on the posterior side of a solid moving through a real fluid such as air or water. Wind blowing across ridges and hollows of a fixed solid (such as the furrows of a field) must, because of the viscosity of the air, press with greater force on the slopes facing it than on the sheltered slopes. Hence if a regular series of waves at sea consisted of a solid body moving with the actual velocity of the waves, the wind would do work upon it, or it would do work upon the air, according as the velocity of the wind were greater or less than the velocity of the waves. This case does not afford an exact parallel to the influence of wind on waves, because the surface particles of water do not move forward with the velocity of the waves as those of the furrowed solid do. Still it may be expected that when the velocity of the wind exceeds the velocity of propagation of the waves, there will be a greater pressure on the posterior slopes than on the anterior slopes of the waves; and *vice versâ*, that when the velocity of the waves exceeds the velocity of the wind, or is in the direction opposite to that of the wind, there will be a greater pressure on the anterior than on the posterior slopes of the waves. In the first case the tendency will be to augment the wave, in the second case to diminish it. The question whether a series of waves of a certain height gradually augment with a certain force of wind or gradually subside through the wind not being strong enough to sustain them, cannot be decided offhand. Towards answering it Stokes's investigation of the work against viscosity of water required to maintain a wave †, gives a most important and suggestive instalment. But no theoretical solution, and very little of experimental investigation, can be referred to with respect to the eddyings of the air blowing across the tops of the waves, to which, by its giving rise to greater pressure on the posterior than on the

---

* "Thermal Effects of Fluids in Motion," *Trans. Roy. Soc.* 1860; *Phil. Mag.* 1860, Vol. xx. p. 552.

† *Trans. Camb. Phil. Soc.* 1851 ("Effect of Internal Friction of Fluids on the Motion of Pendulums," § V.).

anterior slopes, the influence of the wind in sustaining and maintaining waves is chiefly if not altogether due.

My attention having been called three days ago, by Mr Froude, to Scott Russell's Report on Waves (British Association, York, 1844), I find in it a remarkable illustration or indication of the leading idea of the theory of the influence of wind on waves, that the velocity of the wind must exceed that of the waves, in the following statement:—"Let him [an observer studying the surface of a sea or large lake, during the successive stages of an increasing wind, from a calm to a storm] begin his observations " in a perfect calm, when the surface of the water is smooth and " reflects like a mirror the images of surrounding objects. This " appearance will not be affected by even a slight motion of the " air, and a velocity of less than half a mile an hour ($8\frac{1}{2}$ in. per " sec.) does not sensibly disturb the smoothness of the reflecting " surface. A gentle zephyr flitting along the surface from point " to point, may be observed to destroy the perfection of the " mirror for a moment, and on departing, the surface remains " polished as before; if the air have a velocity of about a mile an " hour, the surface of the water becomes less capable of distinct " reflexion, and on observing it in such a condition, it is to be " noticed that the diminution of this reflecting power is owing " to the presence of those minute corrugations of the superficial " film which form waves of the *third order*. These corrugations " produce on the surface of the water an effect very similar to the " effect of those panes of glass which we see corrugated for the " purpose of destroying their transparency, and these corrugations " at once prevent the eye from distinguishing forms at a consider- " able depth, and diminish the perfection of forms reflected in " the water. To fly-fishers this appearance is well known as " diminishing the facility with which the fish see their captors. " This first stage of disturbance has this distinguishing circum- " stance, that the phenomena on the surface cease almost " simultaneously with the intermission of the disturbing cause, " so that a spot which is sheltered from the direct action of the " wind remains smooth, the waves of the third order being in- " capable of travelling spontaneously to any considerable distance, " except when under the continued action of the original disturb- " ing force. This condition is the indication of present force, " not of that which is past. While it remains it gives that deep

" blackness to the water which the sailor is accustomed to regard
" as an index of the presence of wind, and often as the forerunner
" of more.

"The second condition of wave motion is to be observed when
" the velocity of the wind acting on the smooth water has in-
" creased to two miles an hour. Small waves then begin to rise
" uniformly over the whole surface of the water; these are waves
" of the second order, and cover the water with considerable
" regularity. Capillary waves disappear from the ridges of these
" waves, but are to be found sheltered in the hollows between
" them, and on the anterior slopes of these waves. The regularity
" of the distribution of these secondary waves over the surface is
" remarkable; they begin with about an inch of amplitude, and
" a couple of inches long; they enlarge as the velocity or duration
" of the wave increases; by and by conterminal waves unite; the
" ridges increase, and if the wind increase the waves become
" cusped, and are regular waves of the *second order*. They con-
" tinue enlarging their dimensions; and the depth to which they
" produce the agitation increasing simultaneously with their
" magnitude, the surface becomes extensively covered with waves
" of nearly uniform magnitude."

The "Capillary waves" or "waves of the third order" referred
to by Russell are what I, in ignorance of his observations on this
branch of his subject, had called "ripples." The velocity of
$8\frac{1}{2}$ inches ($21\frac{1}{2}$ centimetres) per second is precisely the velocity
he had chosen (as indicated by his observations) for the velocity
of propagation of the straight-ridged waves streaming obliquely
from the two sides of the path of a small body moving at speeds
of from 12 to 36 inches per second; and it agrees remarkably
with my theoretical and experimental determination of the
absolute minimum wave-velocity (23 centimetres per second;
see Part V.). Russell has not explicitly pointed out that his
critical velocity of $8\frac{1}{2}$ inches per second was an absolute minimum
velocity of propagation. But the idea of a minimum velocity of
waves can scarcely have been far from his mind when he fixed
upon $8\frac{1}{2}$ inches per second as the minimum of wind that can
sustain ripples. In an article to appear in *Nature* on the
26th of this month, I have given extracts from Russell's Report
(including part of a quotation which he gives from Poncelet
and Lesbros in the memoirs of the French Institute for 1829),

showing how far my observations on ripples had been anticipated. I need say no more here than that these anticipations do not include any indication of the dynamical theory which I have given, and that the subject was new to me when Parts III., IV., and V. of the present communication were written.

PART V.　*Waves under motive power of Gravity and Cohesion jointly, without wind.*

Leaving the question of wind, consider (13), and introduce notation of (16), (17) in it.　It becomes

$$w^2 = \frac{g'}{n} + T'n \quad\ldots\ldots\ldots\ldots\ldots\ldots(19).$$

This has a minimum value,

when
$$\left.\begin{array}{c} w^2 = 2\sqrt{g'T'}, \\[2mm] n = \sqrt{\dfrac{g'}{T'}} \end{array}\right\} \ldots\ldots\ldots\ldots\ldots\ldots(20).$$

In applying these formulæ to the case of air and water, we may neglect the difference between g and g', as the value of σ is about $\frac{1}{820}$; and between T and T', although it is to be remarked that it is T' rather than T that is ordinarily calculated from experiments on capillary attraction.　From experiments of Gay-Lussac's it appears that the value of T' is about ·074 of a gramme weight per centimetre; that is to say, in terms of the kinetic unit of force founded on the gramme as unit of mass,

$$T' = g \times ·073.$$

To make the density of water unity (as that of the lower liquid has been assumed), we must take one centimetre as unit of length. Lastly, with one second as unit of time, we have

$$g = 982;$$

and (18) gives

$$w = \sqrt{982\left(\frac{1}{n} + ·074 \times n\right)}$$

for the wave-velocity in centimetres per second, corresponding to wave-length $\dfrac{2\pi}{n}$.　When $\dfrac{1}{n} = \sqrt{·073} = ·27$ (that is, when the

wave-length is 1·7 centimetre), the velocity has a minimum value
of 23 centimetres per second.

The part of the preceding theory which relates to the effect of
cohesion on waves of liquids occurred to me in consequence of
having recently observed a set of very short waves advancing
steadily, directly in front of a body moving slowly through water,
and another set of waves considerably longer following steadily
in its wake. The two sets of waves advanced each at the same
rate as the moving body; and thus I perceived that there were
two different wave-lengths which gave the same velocity of propagation. When the speed of the body's motion through the
water was increased, the waves preceding it became shorter, and
those in its wake became longer. Close before the cut-water of
a vessel moving at a speed of not more than two or three knots*
through very smooth water, the surface of the water is marked
with an exquisitely fine and regular fringe of ripples, in which
several scores of ridges and hollows may be distinguished (and
probably counted, with a little practice) in a space extending 20
or 30 centimetres in advance of the solid. Right astern of either
a steamer or sailing vessel moving at any speed above four or
five knots, waves may generally be seen following the vessel at
exactly its own speed, and appearing of such lengths as to verify
as nearly as can be judged the ordinary formula

$$l = \frac{2\pi w^2}{g}$$

for the length of waves advancing with velocity w, in deep
water. In the well-known theory of such waves, gravity is
assumed as the sole origin of the motive forces. When cohesion
was thought of at all (as, for instance, by Mr Froude in his
important nautical experiments on models towed through water,
or set to oscillate to test qualities with respect to the rolling of
ships at sea), it was justly judged to be not sensibly influential
in waves exceeding 5 or 10 centimetres in length. Now it
becomes apparent that, for waves of any length less than 5 or
10 centimetres, cohesion contributes sensibly to the motive
system; and that, when the length is a small fraction of a

* The speed 'one knot' is a velocity of one nautical mile per hour, or 51·5 centimetres per second.

centimetre, cohesion is much more influential than gravity as
"motive" for the vibrations.

The following extract from part of a letter to Mr Froude,
forming part of a communication to *Nature* (to appear on the
26th of this month, October 1871), describes observations for an
experimental determination of the minimum velocity of waves in
sea-water :—

"About three weeks later, being becalmed in the Sound of
"Mull, I had an excellent opportunity, with the assistance of
"Professor Helmholtz, and my brother from Belfast, of deter-
"mining by observation the minimum wave-velocity with some
"approach to accuracy. The fishing-line was hung at a distance
"of two or three feet from the vessel's side, so as to cut the water
"at a point not sensibly disturbed by the motion of the vessel.
"The speed was determined by throwing into the sea pieces of
"paper previously wetted, and observing their times of transit
"across parallel planes, at a distance of 912 centimetres asunder,
"fixed relatively to the vessel by marks on the deck and gunwale.
"By watching carefully the pattern of ripples and waves which
"connected the ripples in front with the waves in rear, I had seen
"that it included a set of parallel waves slanting off obliquely on
"each side, and presenting appearances which proved them to be
"waves of the critical length and corresponding minimum speed
"of propagation. Hence the component velocity of the fishing-
"line perpendicular to the fronts of these waves was the true
"minimum velocity. To measure it, therefore, all that was
"necessary was to measure the angle between the two sets of
"parallel lines of ridges and hollows sloping away on the two
"sides of the wake, and at the same time to measure the velocity
"with which the fishing-line was dragged through the water.
"The angle was measured by holding a jointed two-foot rule,
"with its two branches, as nearly as could be judged by the eye,
"parallel to the set of lines of wave ridges. The angle to which
"the ruler had to be opened in this adjustment was the angle
"sought. By laying it down on paper, drawing two straight
"lines by its two edges, and completing a simple geometrical
"construction with a length properly introduced to represent
"the measured velocity of the moving solid, the required
"minimum wave-velocity was readily obtained. Six observa-
"tions of this kind were made, of which two were rejected as

" not satisfactory. The following are the results of the other
" four :—

Velocity of moving solid.		Deduced minimum wave-velocity.		
51 centimetres per second.		23·0	centimetres per second.	
38	„ „	23·8	„	„
26	„ „	23·2	„	„
24	„ „	22·9	„	„
	Mean........	23·22		

" The extreme closeness of this result to the theoretical
" estimate (23 centimetres per second) was, of course, merely a
" coincidence ; but it proved that the cohesive force of sea-water
" at the temperature (not noted) of the observation cannot be
" very different from that which I had estimated from Gay-
" Lussac's observations for pure water."

APPENDIX H.

ON THE MOLECULAR TACTICS OF A CRYSTAL*.

§ 1. MY subject this evening is not the physical properties of crystals, not even their dynamics; it is merely the geometry of the structure—the arrangement of the molecules in the constitution of a crystal. Every crystal is a homogeneous assemblage of small bodies or molecules. The converse proposition is scarcely true, unless in a very extended sense of the term crystal (§ 20 below). I can best explain a homogeneous assemblage of molecules by asking you to think of a homogeneous assemblage of people. To be homogeneous every person of the assemblage must be equal and similar to every other: they must be seated in rows or standing in rows in a perfectly similar manner. Each person, except those on the borders of the assemblage, must have a neighbour on one side and an equi-distant neighbour on the other: a neighbour on the left front and an equi-distant neighbour behind on the right, a neighbour on the right front and an equi-distant neighbour behind on the left. His two neighbours in front and his two neighbours behind are members of two rows equal and similar to the rows consisting of himself and his right-hand and left-hand neighbours, and their neighbours' neighbours indefinitely to right and left. In particular cases the nearest of the front and rear neighbours may be right in front and right in rear; but we must not confine our attention to the rectangularly grouped assemblages thus constituted. Now let there be equal and similar assemblages on floors above and below that which we have been considering, and let there be any indefinitely great number of floors at equal distances from one another above and below. Think of any one person on any intermediate floor and of his nearest neighbours on the floors above and below. These three persons must be exactly

* The Robert Boyle Lecture, delivered before the Oxford University Junior Scientific Club, May 16, 1893.

in one line ; this, in virtue of the homogeneousness of the assemblages on the three floors, will secure that every person on the intermediate floor is exactly in line with his nearest neighbours above and below. The same condition of alignment must be fulfilled by every three consecutive floors, and we thus have a homogeneous assemblage of people in three dimensions of space. In particular cases every person's nearest neighbour in the floor above may be vertically over him, but we must not confine our attention to assemblages thus rectangularly grouped in vertical lines.

§ 2. Consider now any particular person C (fig. 1) on any intermediate floor, D and D' his nearest neighbours, E and E' his next nearest neighbours all on his own floor. His next next nearest neighbours on that floor will be in the positions F and F' in the diagram. Thus we see that each person C is surrounded by six persons, DD', EE', and FF', being his nearest, his next nearest, and his next next nearest neighbours on his own floor. Excluding for simplicity the special cases of rectangular grouping

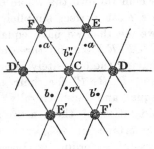

Fig. 1.

we see that the angles of the six equal and similar triangles CDE, CEF, &c., are all acute : and because the six triangles are equal and similar we see that the three pairs of mutually remote sides of the hexagon $DEFD'E'F'$ are equal and parallel.

§ 3. Let now A, A', A'', &c., denote places of persons of the homogeneous assemblage on the floor immediately above, and B, B', B'', &c. on the floor immediately below, the floor of C. In the diagram let a, a', a'' be points in which the floor of CDE is cut by perpendiculars to it through A, A', A'' of the floor above, and b, b', b'' by perpendiculars from B, B', B'' of the floor below. Of all the perpendiculars from the floors immediately above and below, just two, one from each, cut the area of the parallelogram $CDEF$: and they cut it in points similarly situated in respect to the oppositely oriented triangles into which it is divided by either of its diagonals. Hence if a lies in the triangle CDE, the other five triangles of the hexagon must be cut in the corresponding points, as shown in the diagram. Thus, if we think only of the floor of C and of the floor immediately above it, we have points A, A', A'' vertically above a, a', a''. Imagine now a triangular pyramid, or tetrahedron, standing on the base CDE, and having A for vertex: we see that each of its sides ACD, ADE, AEC is an acute-angled triangle because, as we have already seen, CDE is an acute-angled triangle, and because the shortest of the three distances CA, DA, EA is (§ 2) greater than CE (though it may be either greater than or less than DE). Hence the tetrahedron $CDEA$ has all its angles acute; not only the angles of its triangular faces, but the six angles between the planes of its four faces. This important theorem regarding homogeneous assemblages was given by Bravais, to whom we owe the whole doctrine of homogeneous assemblages in its most perfect simplicity and complete generality. Similarly we see that we have equal and similar tetrahedrons on the bases $D'CF$, $E'F'C$; and three other tetrahedrons below the floor of C, having the oppositely oriented triangles $CD'E'$, &c. for their bases and B, B', B'' for their vertices. These three tetrahedrons are equal and heterochirally* similar to the first three. The consideration of these acute-angled tetrahedrons is of fundamental importance in respect to the engineering of an elastic solid, or crystal, according to Boscovich. So also is the consideration of the cluster of thirteen points C, and the six neighbours $DEFD'E'F'$ in the plane of the diagram, and the three neighbours $AA'A''$ on the floor above, and $BB'B''$ on the floor below.

§ 4. The case in which each of the four faces of each of the tetrahedrons of § 3 is an equilateral triangle is particularly interesting. An assemblage fulfilling this condition may conveniently be called an 'equilateral homogeneous assemblage,' or, for brevity, an 'equilateral assemblage.' In an equilateral assemblage C's twelve neighbours are all equi-distant from it. I hold in my hand a cluster of thirteen little black balls, made up by taking one of them and placing the twelve others in contact with it (and therefore packed in the closest possible order), and fixing them all together by fish-glue. You see it looks, in size, colour, and shape, quite like a mulberry. The accompanying diagram shows a stereoscopic view of a similar cluster of balls painted white for the photograph.

* See footnote on § 22 below.

§ 5. By adding ball after ball to such a cluster of thirteen, and always taking care to place each additional ball in some

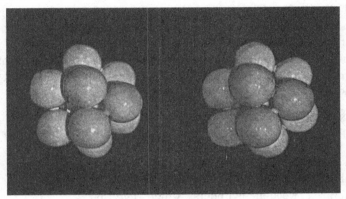

Fig. 2.

position in which it is properly in line with others, so as to make the whole assemblage homogeneous, we can exercise ourselves in a very interesting manner in the building up of any possible form of crystal of the class called 'cubic' by some writers and 'octahedral' by others. You see before you several examples. I advise any of you who wish to study crystallography to contract with a wood-turner, or a maker of beads for furniture tassels or for rosaries, for a thousand wooden balls of about half an inch diameter each. Holes through them will do no harm, and may even be useful; but make sure that the balls are as nearly equal to one another, and each as nearly spherical as possible.

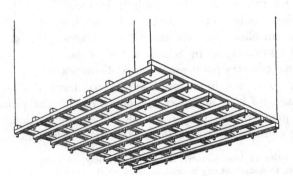

Fig. 3.

§ 6.   You see here before you a large model which I have made to illustrate a homogeneous assemblage of points, on a plan first given, I believe, by Mr William Barlow (*Nature*, December 20 and 27, 1883).   The roof of the model is a lattice-frame (fig. 3) consisting of two sets of eight parallel wooden bars crossing one another, and kept together by pins through the middles of the crossings.   As you see, I can alter it to make parallelograms of all degrees of obliquity till the bars touch, and again you see I can make them all squares.

§ 7.   The joint pivots are (for cheapness of construction) of copper wire, each bent to make a hook below the lattice frame. On these sixty-four hooks are hung sixty-four fine cords, firmly stretched by little lead weights. Each of these cords (fig. 4) bears eight short perforated wooden cylinders, which may be slipped up and down to any desired position*. They are at present actually placed at distances consecutively each equal to the distance from joint to joint of the lattice frame.

§ 8.   The roof of the model is hung by four cords, nearly vertical, of independently variable lengths, passing over hooks from fixed points above, and kept stretched by weights, each equal to one-quarter of the weight of roof and pendants.   You see now by altering the angles of the lattice-work and placing it horizontal or in any inclined plane, as I am allowed to do readily by the manner in which it is hung, I have three independent variables, by varying which I can show you all varieties of homogeneous assemblages, in which three of the neighbours of every point are at equal distances

Fig. 4.

from it.   You see here, for example, we have the equilateral assemblage.   I have adjusted the lattice roof to the proper angle, and its plane to the proper inclination to the vertical, to make a wholly equilateral assemblage of the little cylinders of wood on

---

* The holes in the cylinders are bored obliquely, as shown in fig. 4, which causes them to remain at any desired position on the cord, and allows them to be freed to move up and down by slackening the cord for a moment.

the vertical cords, a case, as we have seen, of special importance. If I vary also the distances between the little pieces of wood on the cords; and the distances between the joints of the lattice work (variations easily understood, though not conveniently producible in one model without more of mechanical construction than would be worth making), I have three other independent variables. By properly varying these six independent variables, three angles and three lengths, we may give any assigned value to each edge of one of the fundamental tetrahedrons of § 3.

§ 9. Our assemblage of people would not be homogeneous unless its members were all equal and similar and in precisely similar attitudes, and were all looking the same way. You understand what a number of people seated or standing on a floor or plain and looking the same way means. But the expression 'looking' is not conveniently applicable to things that have no eyes, and we want a more comprehensive mode of expression. We have it in the words 'orientation,' 'oriented,' and (verb) 'to orient,' suggested by an extension of the idea involved in the word 'orientation,' first used to signify positions relatively to east and west, of ancient Greek and Egyptian temples and Christian churches. But for the orientation of a house or temple we have only one angle, and that angle is called 'azimuth' (the name given to an angle in a horizontal plane). For orientation in three dimensions of space we must extend our ideas and consider position with reference to east and west and up and down. A man lying on his side, with his head to the north and looking east, would not be similarly oriented to a man standing upright and looking east. To provide for the complete specification of how a body is oriented in space we must have in the body a plane of reference, and a line of reference in this plane, belonging to the body and moving with it. We must also have a fixed plane and a fixed line of reference in it, relatively to which the orientation of the moveable body is to be specified; as, for example, a horizontal plane, and the east and west horizontal line in it. The position of a body is completely specified when the angle between the plane of reference belonging to it and the fixed plane is given; and when the angles between the line of intersection of the two planes and the lines of reference in them are also given. Thus we see that three angles are necessary and sufficient to specify the orientation of a moveable body, and we see how the specification is conveniently given in terms of three angles.

§ 10.   To illustrate this take a book lying on the table before you with its side next the title-page up, and its back to the north. I now lift the east edge (the top of the book), keeping the bottom edge north and south on the table till the book is inclined, let us say, 20° to the table.   Next, without altering this angle of 20°, between the side of the book and the table, I turn the book round a vertical axis through 45° till the bottom edge lies north-east and south-west.   Lastly, keeping the book in the plane to which it has been thus brought, I turn it round in this plane through 35°. These three angles of 20°, 45°, and 35° specify, with reference to the horizontal plane of the table and the east and west line in it, the orientation of the book in the position to which you have seen me bring it, and in which I hold it before you.

§ 11.   In figs. 5 and 6 you see two assemblages, each of twelve equal and similar molecules in a plane.   Fig. 5, in which the molecules are all same-ways oriented, is one homogeneous

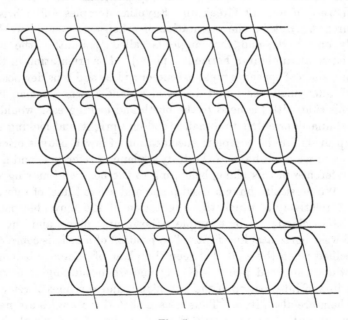

Fig. 5.

assemblage of twenty-four molecules. Fig. 6, in which in one set of rows the molecules are alternately oriented two different ways, may either be regarded as two homogeneous assemblages, each of twelve single molecules; or one homogeneous assemblage of twelve pairs of those single molecules.

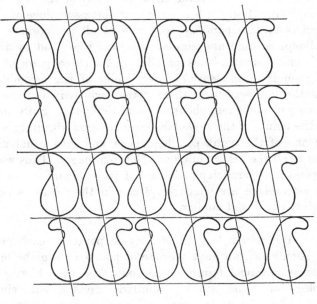

Fig. 6.

§ 12. I must now call your attention to a purely geometrical question* of vital interest with respect to homogeneous assemblages in general, and particularly the homogeneous assemblage of molecules constituting a crystal:—*what can we take as 'the' boundary or 'a' boundary enclosing each molecule with whatever portion of space around it we are at liberty to choose for it, and separating it from neighbours and their portions of space given to them in homogeneous fairness?*

§ 13. If we had only mathematical points to consider we should be at liberty to choose the simple obvious partitioning by three sets of parallel planes. Even this may be done in an infinite

* "On the Homogeneous Division of Space," by Lord Kelvin, *Royal Society Proceedings*, Vol. LV., Jan. 18, 1894.

number of ways, thus:—Beginning with any point P of the assemblage, choose any other three points A, B, C, far or near, provided only that they are not in one plane with P, and that there is no other point of the assemblage in the lines PA, PB, PC, or within the volume of the parallelepiped of which these lines are conterminous edges, or within the areas of any of the faces of this parallelepiped. There will be points of the assemblage at each of the corners of this parallelepiped and at all the corners of the parallelepipeds equal and similar to it which we find by drawing sets of equi-distant planes parallel to its three pairs of faces. (A diagram is unnecessary.) Every point of the assemblage is thus at the intersection of three planes, which is also the point of meeting of eight neighbouring parallelepipeds. Shift now any one of the points of the assemblage to a position within the volume of any one of the eight parallelepipeds, and give equal parallel motions to all the other points of the assemblage. Thus we have every point in a parallelepipedal cell of its own, and all the points of the assemblage are similarly placed in their cells, which are themselves equal and similar.

§ 14. But now if, instead of a single point for each member of the assemblage, we have a group of points, or a globe or cube or other geometrical figure, or an individual of a homogeneous assemblage of equal, similar, similarly dressed, and similarly oriented ladies, sitting in rows, or a homogeneous assemblage of trees closely planted in regular geometrical order on a plane with equal and similar distributions of molecules, and parallel planes above and below, we may find that the best conditioned plane-faced parallelepipedal partitioning which we can choose would cut off portions properly belonging to one molecule of the assemblage and give them to the cells of neighbours. To find a cell enclosing all that belongs to each individual, for example, every part of each lady's dress, however complexly it may be folded among portions of the equal and similar dresses of neighbours; or, every twig, leaf, and rootlet of each one of the homogeneous assemblage of trees; we must alter the boundary by give-and-take across the plane faces of the primitive parallelepipedal cells, so that each cell shall enclose all that belongs to one molecule, and therefore (because of the homogeneousness of the partitioning) nothing belonging to any other molecule. The geometrical problem thus presented, wonderfully complex as it may be in cases such as some of those which I have suggested, is easily performed for any possible case if we begin with any particular parallelepipedal partitioning determined for corresponding points of the assemblage, as explained in § 13, for any homogeneous assemblage of single points. We may prescribe to ourselves that the corners are to remain unchanged, but if so they must to begin with be either in interfaces of contact between the individual molecules, or in vacant space among the molecules. If this condition is fulfilled for one corner it is fulfilled for all, as the corners are essentially corresponding points relatively to the assemblage.

§ 15. Begin now with any one of the twelve straight lines between corners which constitute the twelve edges of the parallelepiped, and alter it arbitrarily to any curved or crooked line between the same pair of corners, subject only to the conditions (1) that it does not penetrate the substance of any member of the assemblage, and (2) that it is not cut by equal and similar parallel curves* between other pairs of corners.

Considering now the three fours of parallel edges of the parallelepiped, let the straight lines of one set of four be altered to equal and similar parallel curves in the manner which I have described; and proceed by the same rule for the other two sets of four edges. We thus have three fours of parallel curved edges instead of the three fours of parallel straight edges of our primitive parallelepiped with corners (each a point of intersection of three edges) unchanged. Take now the quadrilateral of four curves substituted for the four straight edges of one face of the parallelepiped. We may call this quadrilateral a curvilineal parallelogram, because it is a circuit composed of two pairs of equal parallel curves. Draw now a curved surface (an infinitely thin sheet of perfectly extensible india-rubber if you please to think of it so) bordered by the four edges of our curvilineal parallelogram, and so shaped as not to cut any of the substance of any molecule of the assemblage. Do the same thing with an exactly similar and parallel sheet relatively to the opposite face of the parallelepiped; and again the same for each of the two other pairs of parallel faces. We thus have a curved-faced parallelepiped enclosing the

---

* Similar curves are said to be parallel when the tangents to them at corresponding points are parallel.

whole of one molecule and no part of any other; and by similar procedure we find a similar boundary for every other molecule of the assemblage. Each wall of each of these cells is common to two neighbouring molecules, and there is no vacant space anywhere between them or at corners. Fig. 7 illustrates this kind of

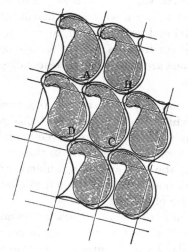

Fig. 7.

partitioning by showing a plane section parallel to one pair of plane faces of the primitive parallelepiped for an ideal case. The plane diagram is in fact a realization of the two-dimensional problem of partitioning the pine pattern of a Persian carpet by parallelograms about as nearly rectilinear as we can make them. In the diagram faint straight lines are drawn to show the primitive parallelogrammatic partitioning. It will be seen that of all the crossings (marked with dots in the diagram) every one is similarly situated to every other in respect to the homogeneously repeated pattern figures: A, B, C, D are four of them at the corners of one cell.

§ 16. Confining our attention for a short time to the homogeneous division of a plane, remark that the division into parallelograms by two sets of crossing parallels is singular in this respect—each cell is contiguous with three neighbours at every corner. Any shifting, large or small, of the parallelograms by relative sliding in one direction or another violates this condition, brings us to a configuration like that of the faces of regularly hewn stones in ordinary bonded masonry, and gives a partitioning which fulfils the condition that at each corner each cell has only two neighbours. Each cell is now virtually a hexagon, as will be seen by the letters A, B, C, D, E, F in the diagram fig. 8. A and D are to be reckoned as corners, each

Fig. 8.

with an interior angle of 180°. In this diagram the continuous heavy lines and the continuous faint lines crossing them show a primitive parallelogrammatic partition by two sets of continuous parallel intersecting lines. The interrupted crossing lines (heavy) show, for the same homogeneous distribution of single points or molecules, the virtually hexagonal partitioning which we get by shifting the boundary from each portion of one of the light lines to the heavy line next it between the same continuous parallels.

Fig. 8 *bis* represents a further modification of the boundary by which the 180° angles A, D become angles of less than 180°. The continuous parallel lines (light) and the short light portions of the crossing lines show the configuration according to fig. 8, from which this diagram is derived.

§ 17. In these diagrams (figures 8 and 8 *bis*) the object enclosed is small enough to be enclosable by a primitive parallelogrammatic partitioning of two sets of continuous crossing parallel straight lines, and by the partitioning of 'bonded' parallelograms both represented in fig. 8, and by the derived hexagonal partitioning represented in fig. 8 *bis*, with faint lines showing the primitive and the secondary parallelograms. In fig. 7 the objects enclosed were too large to be enclosable by any rectilinear parallelogram-

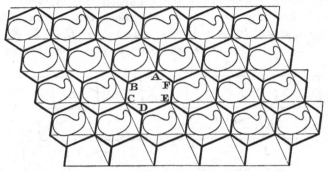

Fig. 8 *bis*.

matic or hexagonal partitioning. The two sets of parallel faint lines in fig. 7 show a primitive parallelogrammatic partitioning, and the corresponding pairs of parallel curves intersecting at the corners of these parallelograms, of which *A, B, C, D* is a specimen, show a corresponding partitioning by curvilineal parallelograms.

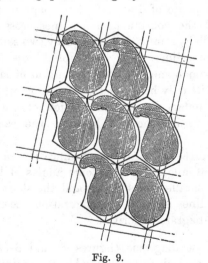

Fig. 9.

Fig. 9 shows for the same homogeneous distribution of objects a better conditioned partitioning by hexagons, in each of which one pair of parallel edges is curved. The sets of intersecting parallel straight lines in fig. 9 show the same primitive parallelo-grammatic partitioning as in fig. 7, and the same slightly shifted to suit points chosen for well-conditionedness of hexagonal partitioning.

§ 18.   For the division of continuous three-dimensional space* into equal, similar, and similarly oriented cells, quite a correspond-ing transformation from partitioning by three sets of continuous mutually intersecting parallel planes to any possible mode of homogeneous partitioning, may be investigated by working out the three-dimensional analogue of §§ 16—17.   Thus we find that the most general possible homogeneous partitioning of space with plane interfaces between the cells gives us fourteen walls to each cell, of which six are three pairs of equal and parallel parallelo-grams, and the other eight are four pairs of equal and parallel hexagons, each hexagon being bounded by three pairs of equal and parallel straight lines.   This figure, being bounded by four-teen plane faces, is called a tetrakaidekahedron.   It has thirty-six edges of intersection between faces; and twenty-four corners, in each of which three faces intersect.   A particular case of it, which I call an orthic tetrakaidekahedron, being that in which the six parallelograms are equal squares, the eight hexagonal faces are equal equilateral and equiangular hexagons, and the lines joining corresponding points in the seven pairs of parallel faces are perpendicular to the planes of the faces, is represented by a stereoscopic picture in fig. 10.   The thirty-six edges and the twenty-four corners, which are easily counted in this diagram, occur in the same relative order in the most general possible partitioning, whether by plane-faced tetrakaidekahedrons or by the generalized tetrakaidekahedron described in § 19.

§ 19.   The most general homogeneous division of space is not limited to plane-faced cells; but it still consists essentially of tetrakaidekahedronal cells, each bounded by three pairs of equal and parallel quadrilateral faces, and four pairs of equal and parallel hexagonal faces, neither the quadrilaterals nor the hexagons being necessarily plane.   Each of the thirty-six edges may be straight or crooked or curved; the pairs of opposite edges, whether of the quadrilaterals or hexagons, need not be equal and

* See footnote to § 12 above.

parallel; neither the four corners of each quadrilateral nor the six corners of each hexagon need be in one plane. But every pair of corresponding edges of every pair of parallel corresponding faces,

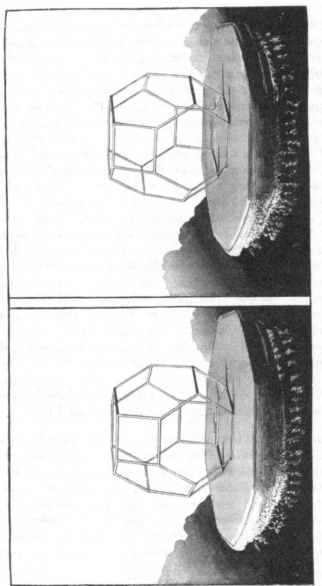

Fig. 10.

whether quadrilateral or hexagonal, must be equal and parallel. I have described an interesting case of partitioning by tetrakai-dekahedrons of curved faces with curved edges in a paper* published about seven years ago. In this case each of the quadrilateral faces is plane. Each hexagonal face is a slightly curved surface having three rectilineal diagonals through its centre in one plane. The six sectors of the face between these diagonals lie alternately on opposite sides of their plane, and are bordered by six arcs of plane curves lying on three pairs of parallel planes. This tetrakaidekahedronal partitioning fulfils the condition that the angles between three faces meeting in an edge are everywhere each 120°; a condition that cannot be fulfilled in any plane-faced tetrakaidekahedron. Each hexagonal wall is an anticlastic surface of equal opposite curvatures at every point, being the surface of minimum area bordered by six curved edges. It is shown easily and beautifully, and with a fair approach to accuracy, by choosing six little circular arcs of wire, and soldering them together by their ends in proper planes for the six edges of the hexagon; and dipping it in soap solution and taking it out.

§ 20. Returning now to the tactics of a homogeneous assemblage remark that the qualities of the assemblage as a whole depend both upon the character and orientation of each molecule, and on the character of the homogeneous assemblage formed by corresponding points of the molecules. After learning the simple mathematics of crystallography, with its indicial system† for defining the faces and edges of a crystal according to the Bravais rows and nets and tetrahedrons of molecules in which we think only of a homogeneous assemblage of points, we are apt to forget that the true crystalline molecule, whatever its nature may be, has sides, and that generally two opposite sides of each molecule may be expected to be very different in quality, and we are almost surprised when mineralogists tell us that two parallel faces on two sides of a crystal have very different qualities in many natural crystals. We might almost as well be surprised to find that an army in battle array, which is a kind of large-grained crystal,

---

* "On the Division of Space with Minimum Partitional Area," *Philosophical Magazine*, Vol. XXIV., 1887, p. 502, and *Acta Mathematica* of the same year.

† A. Levy, *Edinburgh Philosophical Journal*, April, 1822; Whewell, *Phil. Trans. Royal Society*, 1825; Miller, *Treatise on Crystallography*.

presents very different appearance to anyone looking at it from outside, according as every man in the ranks with his rifle and bayonet faces to the front or to the rear or to one flank or to the other.

§ 21. Consider, for example, the ideal case of a crystal consisting of hard, equal and similar tetrahedronal solids all same-ways oriented. A thin plate of crystal cut parallel to any one set of the faces of the constituent tetrahedrons would have very different properties on its two sides; as the constituent molecules would all present points outwards on one side and flat surfaces on the other. We might expect that the two sides of such a plate of crystal would become oppositely electrified when rubbed by one and the same rubber; and, remembering that a piece of glass with part of its surface finely ground but not polished and other parts polished becomes, when rubbed with white silk, positively electrified over the polished parts and negatively electrified over the non-polished parts, we might almost expect that the side of our supposed crystalline plate towards which flat faces of the constituent molecules are turned would become positively electrified, and the opposite side, showing free molecular corners, would become negatively electrified, when both are rubbed by a rubber of intermediate electric quality. We might also from elementary knowledge of the fact of piezo-electricity, that is to say, the development of opposite electricities on the two sides of a crystal by pressure, expect that our supposed crystalline plate, if pressed perpendicularly on its two sides, would become positively electrified on one of them and negatively on the other.

§ 22. Intimately connected with the subject of enclosing cells for molecules of given shape, assembled homogeneously, is the homogeneous packing together of equal and similar molecules of any given shape. In every possible case of any infinitely great number of similar bodies the solution is a homogeneous assemblage. But it may be a homogeneous assemblage of single solids all oriented the same way, or it may be a homogeneous assemblage of clusters of two or more of them placed together in different orientations. For example, let the given bodies be halves (oblique or not oblique) of any parallelepiped on the two sides of a dividing plane through a pair of parallel edges. The two halves are homochirally* similar; and, being equal, we may make a homogeneous assemblage of them by orienting them all the same way and placing them properly in rows. But the closest packing of this assemblage would necessarily leave vacant spaces between the bodies: and we get in reality the closest possible packing of the given bodies by taking them in pairs oppositely oriented and placed together to form parallelepipeds. These clusters may be packed together so as to leave no unoccupied space.

Whatever the number of pieces in a cluster in the closest possible packing of solids may be for any particular shape we may consider each cluster as itself a given single body, and thus reduce the problem to the packing closely together of assemblages of individuals all same-ways oriented; and to this problem therefore it is convenient that we should now confine our attention.

§ 23.   To avoid complexities, such as those which we find in the familiar problem of homogeneous packing of forks or spoons or tea-cups or bowls of any ordinary shape, we shall suppose the given body to be of such shape that no two of them similarly oriented can touch one another in more than one point. Wholly convex bodies essentially fulfil this condition; but it may also be fulfilled by bodies not wholly convex, as is illustrated in fig. 11.

Fig. 11.

§ 24.   To find close and closest packing of any number of our solids S_1, S_2, S_3 ... of shape fulfilling the condition of § 23 proceed thus:—

(1)   Bring S_2 to touch S_1 at any chosen point p of its surface (fig. 12).

(2)   Bring S_3 to touch S_1 and S_2 at r and q respectively.

(3)   Bring S_4 (not shown in the diagram) to touch S_1, S_2, and S_3.

---

* I call any geometrical figure, or group of points, *chiral*, and say that it has chirality if its image in a plane mirror, ideally realized, cannot be brought to coincide with itself. Two equal and similar right hands are homochirally similar. Equal and similar right and left hands are heterochirally similar or ' allochirally ' similar (but heterochirally is better). These are also called ' enantiomorphs,' after a usage introduced, I believe, by German writers. Any chiral object and its image in a plane mirror are heterochirally similar.

(4) Place any number of the bodies together in three rows continuing the lines of S_1S_2, S_1S_3, S_1S_4, and in three sets of equi-distant rows parallel to these. This makes a homogeneous assemblage. In the assemblage so formed the molecules are necessarily found to be in three sets of rows parallel respectively to the three pairs S_2S_3, S_3S_4, S_4S_2. The whole space occupied by an assemblage of n of our solids thus arranged has clearly $6n$ times the volume of a tetrahedron of corresponding points of S_1, S_2, S_3, S_4. Hence the closest of the close packings obtained by the operations (1)...(4) is found if we perform the operations (1), (2), and (3) as to make the volume of this tetrahedron least possible.

§ 25. It is to be remarked that operations (1) and (2) leave for (3) no liberty of choice for the place of S_4, except between two determinate positions on opposite sides of the group S_1, S_2, S_3.

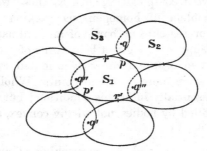

Fig. 12.

The volume of the tetrahedron will generally be different for these two positions of S_4, and, even if the volume chance to be equal in any case, we have differently shaped assemblages according as we choose one or other of the two places for S_4.

This will be understood by looking at fig. 12, showing S_1 and neighbours on each side of it in the rows of S_1S_2, S_1S_3, and in a row parallel to that of S_2S_3. The plane of the diagram is parallel to the planes of corresponding points of these seven bodies, and the diagram is a projection of these bodies by lines parallel to the intersections of the tangent planes through p and r. If the three tangent planes through p, q, and r, intersected in parallel lines, q would be seen like p and r as a point of contact between the outlines of two of the bodies; but this is only a particular case, and in general q must, as indicated in the diagram, be concealed by one or other of the two bodies of which it is the point of contact. Now imagining, to fix our ideas and facilitate brevity of expression, that the planes of corresponding points of the seven bodies are horizontal, we see clearly that S_4 may be brought into proper position to touch S_1, S_2 and S_3 either from above or from below; and that there is one determinate place for it if we bring it into position from above, and another determinate place for it if we bring it from below.

§ 26.   If we look from above at the solids of which fig. 12 shows the outline we see essentially a hollow leading down to a perforation between S_1, S_2, S_3, and if we look from below we see a hollow leading upwards to the same perforation : this for brevity we shall call the perforation pqr. The diagram shows around S_1 six hollows leading down to perforations, of which two are similar to pqr, and the other three, of which $p'q'r'$ indicates one, are similar one to another but are dissimilar to pqr. If we bring S_4 from above into position to touch S_1, S_2 and S_3, its place thus found is in the hollow pqr, and the places of all the solids in the layer above that of the diagram are necessarily in the hollows similar to pqr. In this case the solids in the layer below that of the diagram must lie in the hollows below the perforations dissimilar to pqr, in order to make a single homogeneous assemblage. In the other case S_4 brought up from below finds its place on the under side of the hollow pqr, and all solids of the lower layer find similar places : while solids in the layer above that of the diagram find their places in the hollows similar to $p'q'r'$. In the first case there are no bodies of the upper layer in the hollows above the perforations *similar* to $p'q'r'$, and no bodies of the lower layer in the hollows below the perforations *similar* to pqr. In the second case there are no bodies of the upper layer in the hollows above the perforations *similar* to pqr, and none of the under layer in the hollows below the perforations *similar* to $p'q'r'$.

§ 27.   Going back now to operation (1) of § 24, remark that when the point of contact p is arbitrarily chosen on one of the two bodies S_1, the point of contact on the other will be the point on it corresponding to the point or one of the points of S_1, where its tangent plane is parallel to the tangent plane at p. If S_1 is wholly

convex it has only two points at which the tangent planes are
parallel to a given plane, and therefore the operation (1) is determinate and unambiguous.   But if there is any concavity there
will be four or some greater even number of tangent planes
parallel to any one of some planes, while there will be other
planes to each of which only one pair of tangent planes is parallel.
Hence, operation (1), though still determinate, will have a multiplicity of solutions, or only a single solution, according to the
choice made of the position of p. [Any change of configuration
of the assemblage in stable equilibrium causes expansion! K.;
Holwood, May 23, 1896.]

Henceforth, however, to avoid needless complications of ideas,
we shall suppose our solids to be wholly convex; and of some such
unsymmetrical shape as those indicated in fig. 12 of § 25, and
shown by stereoscopic photograph in fig. 13 of § 36. With or
without this convenient limitation, operation (1) has two freedoms,
as p may be chosen freely on the surface of S_1; and operation (2)
has clearly just one freedom after operation (1) has been performed.
Thus, for a solid of any given shape, we have three disposables, or,
as commonly called in mathematics, three 'independent variables,'
all free for making a homogeneous assemblage according to the
rule of § 22*.

* The following is information regarding ratio of void to whole space in heaps
of gravel, sand, and broken stones, extracted from books of reference for engineers:

Hurst's *Pocket-Book.*

Thames Ballast 20 cub. ft. per ton

Gravel coarse	19	,,	,,
Sand pit	22	,,	,,
Shingle	23	,,	,,

Haswell's *Pocket-Book.*

Voids in a cub. yard of Stone broken to gauge of 2·5″ 10   cub. ft.

,,	,,	,,	,,	,,	2″	10·66	,,
,,	,,	,,	,,	,,	1·5″	11·33	,,
Shingle						9	,,
Thames Ballast (containing sand)						4·5	,,

Law and Burrell's *Civil Engineering.*
(Road Metal.)

"A cub. yard of broken stone metal of an ordinary size—2″ or 2¼″ cube—when
"screened and beaten down in regular layers 6″ thick contains...11 cub. ft. of inter-
"spaces, as tested by filling up the metal with liquid."...Herr E. Bokeberg found
that in a cubic yard of loosely heaped broken stones the void space was 50 °/₀ of
the whole.

§ 28.   In the homogeneous assemblage defined in § 24 each solid, S_1, is touched at twelve points, being the three points of contact with S_2, S_3, S_4, and the three 3's of points on S_1 corresponding to the points on S_2, S_3, S_4, at which these bodies are touched by the others of the quartet.   This statement is somewhat difficult to follow, and we see more clearly the twelve points of contact by not confining our attention to the quartet S_1, S_2, S_3, S_4 (convenient as this is for some purposes), but completing the assemblage and considering six neighbours around S_1 in one plane layer of the solids as shown in fig. 12, with their six points $prq''p'r'q'''$ of contact with S_1; and the three neighbours of the two adjacent parallel layers which touch it above and below.   This cluster of thirteen, S_1 and twelve neighbours, is shown for the case of spherical bodies in the stereoscopic photograph of § 4 above.   We might of course, if we pleased, have begun with the plane layer of which S_1, S_2, S_4 are members, or with that of which S_1, S_3, S_4 are members, or with the plane layer parallel to the fourth side $S_2S_3S_4$ of the tetrahedron : and thus we have four different ways of grouping the twelve points of contact on S_1 into one set of six and two sets of three.

§ 29.   In this assemblage we have what I call 'close order' or 'close packing.'   For closest of close packings the volume of the tetrahedron (§ 24) of corresponding points of S_1, S_2, S_3, and S_4 must be a minimum, and the least of minimums if, as generally will be the case, there are two more different configurations for each of which the volume is a minimum.   There will in general also be configurations of minimax volume and of maximum volume, subject to the condition that each body is touched by twelve similarly oriented neighbours.

§ 30.   Pause for a moment to consider the interesting kinematical and dynamical problems presented by a close homogeneous assemblage of smooth solid bodies of given convex shape, whether perfectly frictionless or exerting resistance against mutual sliding according to the ordinarily stated law of friction between dry hard solid bodies.   First imagine that they are all similarly oriented and each in contact with twelve neighbours, except outlying individuals (which there must be at the boundary if the assemblage is finite, and each of which is touched by some number of neighbours less than twelve).   The coherent assemblage thus defined constitutes a kinematic frame or skeleton for an elastic solid of very peculiar properties. Instead of the six freedoms, or disposables, of strain presented by a natural solid it has only three. Change of shape of the whole can only take place in virtue of rotation of the constituent parts relatively to any one chosen row of them, and the plane through it and another chosen row.

§ 31. Suppose first the solids to be not only perfectly smooth but perfectly frictionless. Let the assemblage be subjected to equal positive or negative pressure inwards all around its boundary. Every position of minimum, minimax, or maximum volume will be a position of equilibrium. If the pressure is positive the equilibrium will be stable if, and unstable unless, the volume is a minimum. If the pressure is negative the equilibrium will be stable if, and unstable unless, the volume is a maximum. Configurations of minimax volume will be essentially unstable.

§ 32. Consider now the assemblage of § 31 in a position of stable equilibrium under the influence of a given constant uniform pressure inwards all round its boundary. It will have rigidity in simple proportion to the amount of this pressure. If now by the superposition of non-uniform pressure at the boundary, for example equal and opposite pressures on two sides of the assemblage, a finite change of shape is produced: the whole assemblage essentially swells in bulk. This is the 'dilatancy' which Osborne Reynolds has described* in an exceedingly interesting manner with reference to a sack of wheat or sand, or an india-rubber bag tightly filled with sand or even small shot. Consider, for example, a sack of wheat filled quite full and standing up open. It is limp and flexible. Now shake it down well, fill it quite full, shake again, so as to get as much into it as possible, and tie the mouth very tightly close. The sack becomes almost as stiff as a log of wood of the same shape. Open the mouth partially and it becomes again limp, especially in the upper parts of the bag. In Reynolds' observations on india-rubber bags of small shot his 'dilatancy' depends, essentially and wholly, on breaches of some of the contacts which exist between the molecules in their configuration of minimum volume: and it is possible that in all his cases

* *Philosophical Magazine*, Vol. xx., 1885, second half-year, p. 469, and *British Association Report*, 1885, Aberdeen, p. 896.

the dilatations which he observed are *chiefly*, if not wholly, due to such breaches of contact.

But it is possible, it almost seems probable, that in bags or boxes of sand or powder, of some kinds of smooth rounded bodies of any shape, not spherical or ellipsoidal, subjected persistently to unequal pressures in different directions, and well shaken, stable positions of equilibrium are found with almost all the particles each touched by twelve others.

Here is a curious subject of Natural History through all ages till 1885, when Reynolds brought it into the province of Natural Philosophy by the following highly interesting statement:—
" A well-marked phenomenon receives its explanation at once
" from the existence of dilatancy in sand. When the falling tide
" leaves the sand firm, as the foot falls on it, the sand whitens and
" appears momentarily to dry round the foot. When this happens
" the sand is full of water, the surface of which is kept up to that
" of the sand by capillary attractions; the pressure of the foot
" causing dilatation of the sand more water is required, which has
" to be obtained either by depressing the level of the surface
" against the capillary attractions, or by drawing water through
" the interstices of the surrounding sand. This latter requires
" time to accomplish, so that for the moment the capillary forces
" are overcome; the surface of the water is lowered below that of
" the sand, leaving the latter white or drier until a sufficient
" supply has been obtained from below, when the surface rises and
" wets the sand again. On raising the foot it is generally seen
" that the sand under the foot and around becomes momentarily
" wet; this is because, on the distorting forces being removed, the
" sand again contracts, and the excess of water finds momentary
" relief at the surface."

This proves that the sand under the foot, as well as the surface around it, must be dry for a short time after the foot is pressed upon it, though we cannot see it whitened, as the foot is not transparent. That it is so has been verified by Mr Alex. Galt, Experimental Instructor in the Physical Laboratory of Glasgow University, by laying a small square of plate-glass on wet sand on the sea-shore of Helensburgh, and suddenly pressing on it by a stout stick with nearly all his weight. He found the sand, both under the glass and around it in contact with the air, all became white at the same moment. Of all the two hundred thousand

40

million men, women, and children who, from the beginning of the world, have ever walked on wet sand, how many, prior to the British Association Meeting at Aberdeen in 1885, if asked, "Is the sand compressed under your foot ? " would have answered otherwise than " Yes ! " ?

(Contrast with this the case of walking over a bed of wet sea-weed !)

§ 33.   In the case of globes packed together in closest order (and therefore also in the case of ellipsoids, if all similarly oriented), our condition of coherent contact between each molecule and twelve neighbours implies absolute rigidity of form and constancy of bulk.   Hence our convex solid must be neither ellipsoidal nor spherical in order that there may be the changes of bulk which we have been considering as dependent on three independent variables specifying the orientation of each solid relatively to rows of the assemblage.   An interesting dynamical problem is presented by supposing any mutual forces, such as might be produced by springs, to act between the solid molecules, and investigating configurations of equilibrium on the supposition of frictionless contacts.   The solution of it of course is that the potential energy of the springs must be a minimum or a minimax or a maximum for equilibrium, and a minimum for stable equilibrium.   The solution will be a configuration of minimum or minimax, or maximum, volume, only in the case of pressure equal in all directions.

§ 34.   A purely geometrical question, of no importance in respect to the molecular tactics of a crystal but of considerable interest in pure mathematics, is forced on our attention by our having seen (§ 27) that a homogeneous assemblage of solids of given shape, each touched by twelve neighbours, has three freedoms which may be conveniently taken as the three angles specifying the orientation of each molecule relatively to rows of the assemblage as explained in § 30.

Consider a solid S_1 and the twelve neighbours which touch it, and try if it is possible to cause it to touch more than twelve of the bodies.   Attach ends of three thick flexible wires to any places on the surface of S_1; carry the wires through interstices of the assemblage, and attach their other ends at any three places of A, B, C, respectively, these being any three of the bodies outside the cluster of S_1 and its twelve neighbours. Cut the wires across at any chosen positions in them; and round off the cut ends, just leaving contact between the rounded ends, which we shall call $f'f$, $g'g$, $h'h$. Do homogeneously for every other solid of the assemblage what we have done for S_1. Now bend the wires slightly so as to separate the pairs of points of contact, taking care to keep them from touching any other bodies which they pass near on their courses between S_1 and A, B, C respectively. After having done this, thoroughly rigidify all the wires thus altered. We may now, having three independent variables at our disposal, so change the orientation of the molecules relatively to rows of the assemblage, as to bring $f'f$, $g'g$, and $h'h$ again into contact. We have thus six fresh points of S_1; of which three are f', g', h'; and the other three are on the three extensions of S_1 corresponding to the single extensions of A, B, C respectively, which we have been making. Thus we have a *real* solution of the interesting geometrical problem :—It is required so to form a homogeneous assemblage of solids of any arbitrarily given shape that each solid shall be touched by eighteen others. This problem is determinate, because the making of the three contacts $f'f$, $g'g$, $h'h$, uses up the three independent variables left at our disposal after we have first formed a homogeneous assemblage with twelve points of contact on each solid. But our manner of finding a shape for each solid which can allow the solution of the problem to be real proves that the solution is essentially imaginary for every wholly convex shape.

§ 35. Pausing for a moment longer to consider afresh the geometrical problem of putting arbitrarily given equal and similar solids together to make a homogeneous assemblage of which each member is touched by eighteen others, we see immediately that it is determinate (whether it has any real solution or not), because when the shape of each body is given we have nine disposables for fixing the assemblage : six for the character of the assemblage of the corresponding points, and three for the orientation of each molecule relatively to rows of the assemblage of corresponding points. These nine disposables are determined by the condition that each body has nine pairs of contacts with others.

Suppose now a homogeneous assemblage of the given bodies, in open order with no contacts, to be arbitrarily made according to any nine arbitrarily chosen values for the six distances between a point of S_1 and the corresponding points of its six pairs of nearest and next nearest neighbours (§ 1 above), and the three angles (§ 9 above) specifying the orientation of each body relatively to rows of the assemblage. We may choose in any nine rows through S_1 any nine pairs of bodies at equal distances on the two sides of S_1 far or near, for the eighteen bodies which are to be in contact with S_1. Hence there is an infinite number of solutions of the problem of which only a finite number can be real. Every solution of the problem of eighteen contacts is imaginary when the shape is wholly convex.

§ 36. Without for a moment imagining the molecules of matter to be hard solids of convex shape, we may derive valuable lessons in the tactics of real crystals by studying the assemblage described in §§ 24 and 25, and represented in figs. 12 and 13. I must for the present forego the very attractive subject of the tactics presented by faces not parallel to one or other of the four

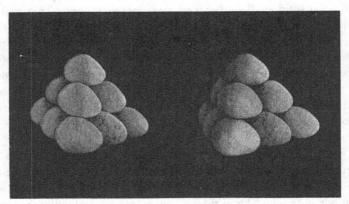

Fig. 13.

faces of the primitive tetrahedrons which we found in § 24, and ask you only to think of the two sides of a plate of crystal parallel to any one of them, that is to say, an assemblage of such layers as those represented geometrically in fig. 12 and shown in stereoscopic view in fig. 13. If, as is the case with the solids *

---

* The solids of the photograph are castings in fine plaster of Paris from a scalene tetrahedron of paraffin wax, with its corners and edges rounded, used as a pattern.

photographed in fig. 13, the under side of each solid is nearly plane but slightly convex, and the top is somewhat sharply curved, we have the kind of difference between the upper and under of the two parallel sides of the crystal which I have already described to you in § 21 above. In this case the assemblage is formed by letting the solids fall down from above and settle in the hollows to which they come most readily, or which give them the stablest position. It would, we may suppose, be the hollows $p'q'r'$, not pqr (fig. 12, § 25), that would be chosen; and thus, of the two formations described in § 25, we should have that in which the hollows above $p'q'r'$ are occupied by the comparatively flat under-sides of the molecules of the layer above, and the hollows below the apertures pqr by the comparatively sharp tops of the molecules of the layer below.

§ 37.   For many cases of natural crystals of the wholly asymmetric character, the true forces between the crystalline molecules will determine precisely the same tactics of crystallization as would be determined by the influence of gravity and fluid viscosity in the settlement from water, of sand composed of uniform molecules of the wholly unsymmetrical convex shape represented in figs. 12 and 13.   Thus we can readily believe that a real crystal which is growing by additions to the face seen in fig. 12 would give layer after layer regularly as I have just described.   But if by some change of circumstances the plate, already grown to a thickness of many layers in this way, should come to have the side facing *from* us in the diagram exposed to the mother-liquor, or mother-gas, and begin to grow from that face, the tactics might probably be that each molecule would find its resting-place with its most nearly plane side in the wider hollows under $p'q'r'$, instead of with its sharpest corner in the narrower and steeper hollows under pqr, as are the molecules in the layer below that shown in the diagram in the first formation.   The result would be a compound crystal consisting of two parts, of differently oriented quality, cohering perfectly together on the two sides of an interfacial plane.   It seems probable that this double structure may be found in nature, presented by crystals of the wholly unsymmetric class, though it may not hitherto have been observed or described in crystallographic treatises.

§ 38.   This asymmetric double crystal becomes simply the

well-known symmetrical 'twin-crystal*' in the particular case in
which each of the constituent molecules is symmetrical on the
two sides of a plane through it parallel to the plane of our
diagrams, and also on the two sides of some plane perpendicular

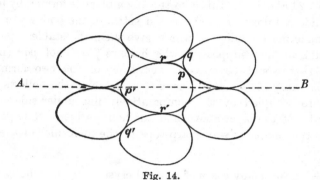

Fig. 14.

to this plane. We see, in fact, that in this case if we cut in two
the double crystal by the plane of fig. 14, and turn one part
ideally through 180° round the intersection of these two planes,
we bring it into perfect coincidence with the other part. This
we readily understand by looking at fig. 14, in which the solid
shown in outline may be either an egg-shaped figure of revolution,
or may be such a figure flattened by compression perpendicular to
the plane of the diagram. The most readily chosen and the most
stable resting-places for the constituents of each successive layer
might be the wider hollows $p'q'r'$: and therefore if, from a single

* "A twin-crystal is composed of two crystals joined together in such a manner
"that one would come into the position of the other by revolving through two right
"angles round an axis which is perpendicular to a plane which either is, or may be,
"a face of either crystal. The axis will be called the twin-axis, and the plane to
"which it is perpendicular the twin-plane." Miller's *Treatise on Crystallography*,
p. 103. In the text the word 'twin-plane,' quoted from the writings of Stokes and
Rayleigh, is used to signify the plane common to the two crystals in each of the
cases referred to: and not the plane perpendicular to this plane, in which one part
of the crystal must be rotated to bring it into coincidence with the other, and which
is the twin-plate as defined by Miller.

Twinning in which each molecule turns (*M. P. P.* Vol. III. xcvII. §§ 60, 61) must
also be considered. *Here* all the molecules are not necessarily of same orientation.
See l. 12 from foot of § 37. K., Oct. 12, 1896.

Turning in the plane of the paper, as contemplated by Miller, could not bring
molecules into coincidence with big and little ends: but turning round the axis AB
does bring them into coincidence. K., Oct. 12, 1896.

layer to begin with, the assemblage were to grow by layer after layer added to it on each side, it might probably grow as a twin-crystal. But it might also be that the presence of a molecule in the wider hollow $p'q'r'$ on one side, might render the occupation of the corresponding hollow on the other side by another molecule less probable, or even impossible. Hence, according to the con-figuration and the molecular forces of the particular crystalline molecule in natural crystallization, there may be necessarily, or almost necessarily, the twin, when growth proceeds simultaneously on the two sides : or the twin growth may be impossible, because the first occupation of the wider hollows on one side may compel the continuity of the crystalline quality throughout, by leaving only the narrower hollows pqr free for occupation by molecules attaching themselves on the other side.

§ 39. Or the character of the crystalline molecule may be such that when the assemblage grows by the addition of layer after layer on one side only, with a not very strongly decided preference to the wider hollows $p'q'r'$, some change of circumstances may cause the molecules of one layer to place themselves in a hollow pqr. The molecules in the next layer after this would find the hollows $p'q'r'$ occupied on the far side, and would thus have a bias in favour of the hollows pqr. Thus layer after layer might be added, constituting a twinned portion of the growth, growing, however, with less strong security for continued homogeneousness than when the crystal was growing, as at first, by occupation of the wider hollows $p'q'r'$. A slight disturbance might again occur, causing the molecules of a fresh layer to settle, not in the narrow hollows pqr, but in the wider hollows $p'q'r'$, notwithstanding the nearness of molecules already occupying the wider hollows on the other side. Disturbances such as these occurring irregularly during the growth of a crystal might produce a large number of successive twinnings at parallel planes with irregular intervals between them, or a large number of twinnings in planes at equal intervals might be produced by some regular periodic disturbance occurring for a certain number of periods, and then ceasing. Whether regular and periodic, or irregular, the tendency would be that the number of twinnings should be even, and that after the disturbances cease the crystal should go on growing in the first manner, because of the permanent bias in favour of the wider hollows $p'q'r'$. These

changes of molecular tactics, which we have been necessarily led
to by the consideration of the fortuitous concourse of molecules,
are no doubt exemplified in a large variety of twinnings and
counter-twinnings found in natural minerals. In the artificial
crystallization of chlorate of potash they are of frequent occurrence,
as is proved, not only by the twinnings and counter-twinnings
readily seen in the crystalline forms, but also by the brilliant
iridescence observed in many of the crystals found among a large
multitude, which was investigated scientifically by Sir George
Stokes ten years ago, and described in a communication to the
Royal Society "On a remarkable phenomenon of crystalline reflection" (*Proc. R. S.*, Vol. XXXVIII., 1885, p. 174).

§ 40. A very interesting phenomenon, presented by what was
originally a clear homogeneous crystal of chlorate of potash, and
was altered by heating to about 245°—248° Cent., which I am
able to show you through the kindness of Lord Rayleigh, and of
its discoverer, Mr Madan, presents another very wonderful case
of changing molecular tactics, most instructive in respect of the
molecular constitution of elastic solids. When I hold this plate
before you with the perpendicular to its plane inclined at 10°
or more to your line of vision you see a tinsel-like appearance,
almost as bright as if it were a plate of polished silver, on this
little area, which is a thin plate of chlorate of potash cemented
for preservation between two pieces of glass; and, when I hold
a light behind, you see that the little plate is almost perfectly
opaque like metal foil. But now when I hold it nearly perpendicular to your line of vision the tinsel-like appearance is lost.
You can see clearly through the plate, and you also see that very
little light is reflected from it. As a result, both of Mr Madan's
own investigations and further observations by himself, Lord
Rayleigh came to the conclusion that the almost total reflection
of white light which you see is due to the reflection of light at
many interfacial planes between successive layers of twinned and
counter-twinned crystal of small irregular thicknesses, and not to
any splits or cavities or any other deviation from homogeneousness
than that presented by homogeneous portions of oppositely twinned-
crystals in thorough molecular contact at the interfaces.

§ 41. When the primitive clear crystal was first heated very
gradually by Madan to near its melting-point (359° according to

Carnelly) it remained clear, and only acquired the tinsel appearance after it had cooled to about 245° or 248° *. Rayleigh found that if a crystal thus altered was again and again heated it always lost the tinsel appearance, and became perfectly clear at some temperature considerably below the melting-point, and regained it at about the same temperature in cooling. It seems therefore certain, that at temperatures above 248° and below the melting-point, the molecules had so much of thermal motions as to keep them hovering about the positions of pqr, $p'q'r'$ of our diagrams, but not enough to do away with the rigidity of the solid; and that when cooled below 248° the molecules were allowed to settle in one or other of the two configurations, but with little of bias for one in preference to the other. It is certainly a very remarkable fact in Natural History, discovered by these observations, that when the molecules come together to form a crystal out of the watery solution, there should be so much more decided a bias in favour of continued homogeneousness of the assemblage than when, by cooling, they are allowed to settle from their agitations in a rigid, but nearly melting, solid.

§ 42. But even in crystallization from watery solution of chlorate of potash the bias in favour of thorough homogeneousness

* " A clear transparent crystal of potassium chlorate, from which the inevitable " twin-plate had been ground away so as to reduce it to a single crystal film about " 1 mm. in thickness, was placed between pieces of mica and laid on a thick iron " plate. About 3 cm. from it was laid a small bit of potassium chlorate, and the " heat of a Bunsen burner was applied below this latter, so as to obtain an indication " when the temperature of the plate was approaching the fusing-point of the substance " (359° C. according to Prof. Carnelly). The crystal plate was carefully watched " during the heating, but no depreciation took place, and no visible alteration was " observed, up to the point at which the small sentinel crystal immediately over the " burner began to fuse. The lamp was now withdrawn, and when the temperature " had sunk a few degrees a remarkable change spread quickly and quietly over the " crystal plate, causing it to reflect light almost as brilliantly as if a film of silver " had been deposited upon it. No further alteration occurred during the cooling; " and the plate, after being ground and polished on both sides, was mounted with " Canada balsam between glass plates for examination. Many crystals have been " similarly treated with precisely similar results; and the temperature at which " the change takes place has been determined to lie between 245° and 248°, by " heating the plates upon a bath of melted tin in which a thermometer was im- " mersed. With single crystal plates no decrepitation has ever been observed, " while with the ordinary twinned-plates it always occurs more or less violently, " each fragment showing the brilliant reflective power above noticed."—*Nature*, May 20, 1886.

is not in every contingency decisive. In the first place, beginning, as the formation seems to begin, from a single molecular plane layer such as that ideally shown in fig. 14, it goes on, not to make a homogeneous crystal on the two sides of this layer, but probably always so as to form a twin-crystal on its two sides, exactly as described in § 38, and if so, certainly for the reason there stated. This is what Madan calls the "inveterate tendency to produce twins (such as would assuredly drive a Malthus to despair)* "; and it is to this that he alludes as "the inevitable twin-plate" in the passage from his paper given in the footnote to § 41 above.

§ 43. In the second place, I must tell you that many of the crystals produced from the watery solution by the ordinary process of slow evaporation and crystallization show twinnings and counter-twinnings at irregular intervals in the otherwise homogeneous crystal on either one or both sides of the main central twin-plane, which henceforth for brevity I shall call (adopting the hypothesis already explained, which seems to me undoubtedly true) the 'initial plane.' Each twinning is followed, I believe, by a counter-twinning at a very short distance from it; at all events Lord Rayleigh's observations† prove that the whole number of twinnings and counter-twinnings in a thin disturbed stratum of the crystal on one side of the main central twin-plane is generally, perhaps always, even; so that, except through some comparatively very small part or parts of the whole thickness, the crystal on either side of the middle or initial plane is homogeneous. This is exactly the generally regular growth which I have described to you (§ 39) as interrupted occasionally or accidentally by some unexplained disturbing cause, but with an essential bias to the homogeneous continuance of the more easy or natural one of the two configurations.

§ 44. I have now great pleasure in showing you a most interesting collection of the iridescent crystals of chlorate of potash, each carefully mounted for preservation between two glass plates, which have been kindly lent to us for this evening by Mr Madan. In March, 1854, Dr W. Bird Herapath sent to

* *Nature*, May 20, 1886.
† *Philosophical Magazine*, 1888, second half-year, p. 260.

Prof. Stokes some crystals of chlorate of potash showing the brilliant and beautiful colours you now see, and, thirty years later, Prof. E. J. Mills recalled his attention to the subject by sending him "a fine collection of splendidly coloured crystals "of chlorate of potash of considerable size, several of the plates "having an area of a square inch or more, and all of them being thick "enough to handle without difficulty." The consequence was that Stokes made a searching examination into the character of the phenomenon, and gave a short but splendidly interesting communication to the Royal Society, of which I have already told you. The existence of these beautifully coloured crystals had been well known to chemical manufacturers for a long time, but it does not appear that any mention of them was to be found in any scientific journal or treatise prior to Stokes's paper of 1885. He found that the colour was due to twinnings and counter-twinnings in a very thin disturbed stratum of the crystal showing itself by a very fine line, dark or glistening, according to the direction of the incident light when a transverse section of the plate of crystal was examined in a microscope. By comparison with a spore of lycopodium he estimated that the breadth of this line, and therefore the thickness of the disturbed stratum of the crystal, ranged somewhere about the one-thousandth of an inch. He found that the stratum was visibly thicker in those crystals which showed red colour than in those which showed blue. He concluded that "the seat of the coloration is certainly a thin twinned stratum" (that is to say, a homogeneous portion of crystal between a twinning and a counter-twinning), and found that "a single twin-plane does not show anything of the kind."

§ 45. A year or two later Lord Rayleigh entered on the subject with an exhaustive mathematical investigation of the reflection of light at a twin-plane of a crystal (*Philosophical Magazine*, September, 1888), by the application of which, in a second paper "On the remarkable Phenomenon of Crystalline Reflection described by Prof. Stokes," published in the same number of the *Philosophical Magazine*, he gave what seems certainly the true explanation of the results of Sir George Stokes's experimental analysis of these beautiful phenomena. He came very decidedly to the conclusion that the selective quality of the iridescent portion of the crystal, in virtue of which it reflects almost totally light nearly of one particular wave-length for one particular direction of incidence (on which the brilliance of the coloration depends), cannot be due to merely a single twin-stratum, but that it essentially is due to a considerable number of parallel twin-strata at nearly equal distances. The light reflected by this complex stratum is, for any particular direction of incident and reflected ray, chiefly that of which the wave-length is equal to twice the length of the period of the twinning and counter-twinning, on a line drawn through the stratum in the direction of either the incident or the reflected ray.

§ 46. It seems to me probable that each twinning is essentially followed closely by a counter-twinning. Probably three or four of these twin-strata might suffice to give colour; but in any of the brilliant specimens as many as twenty or thirty, or more, might probably be necessary to give so nearly monochromatic light as was proved by Stokes's prismatic analysis of the colours observed in many of his specimens. The disturbed stratum of about a one-thousandth of an inch thickness, seen by him in the microscope, amply suffices for the 5, 10, or 100 half wave-lengths required by Rayleigh's theory to account for perceptible or brilliant coloration. But what *can* be the cause of any approach to regular periodicity in the structure sufficiently good to give the colours actually observed? Periodical motion of the mother-liquor relatively to the growing crystal might possibly account for it. But Lord Rayleigh tells us that he tried rocking the pan containing the solution without result. Influence of light has been suggested, and I believe tried, also without result, by several enquirers. We know, by the beautiful discovery of Edmond Becquerel, of the prismatic colours photographed on a prepared silver plate by the solar spectrum, that 'standing waves'. (that is to say, vibrations with stationary nodes and stationary places of maximum vibration), due to co-existence of incident and reflected waves, do produce such a periodic structure as that which Rayleigh's theory shows capable of giving a corresponding tint when illuminated by white light. It is difficult, therefore, not to think that light may be effective in producing the periodic structure in the crystallization of chlorate of potash, to which the iridescence is due. Still, experimental evidence seems against this tempting theory, and we must perforce be content with the question unanswered:—

What can be the cause of 5, or 10, or 100 pairs of twinning and counter-twinning following one another in the crystallization with sufficient regularity to give the colour: and why, if there are twinnings and counter-twinnings, are they not at irregular intervals, as those produced by Madan's process, and giving the observed white tinsel-like appearance with no coloration?

§ 47. And now I have sadly taxed your patience: and I fear I have exhausted it and not exhausted my subject! I feel I have not got half-way through what I hoped I might be able to put before you this evening regarding the molecular structure of crystals. I particularly desired to speak to you of quartz crystal with its ternary symmetry and its chirality*; and to have told you of the etching† by hydrofluoric acid which, as it were, commences to unbuild the crystal by taking away molecule after molecule, but not in the reverse order of the primary up-building; and which thus reveals differences of tactics in the alternate faces of the six-sided pyramid which terminates at either end, sometimes at both ends, the six-sided prism constituting generally the main bulk of the crystal. I must confine myself to giving you a geometrical symbol for the ternary symmetry of the prism and its terminal pyramid.

§ 48. Make an equilateral equiangular hexagonal prism, with its diagonal from edge to edge ninety-five hundredths‡ of its length. Place a number of these close together, so as to make up a hexagonal plane layer with its sides perpendicular to the sides of the constituent hexagonal prisms: see fig. 15, and imagine the semicircles replaced by their diameters. You see in each side of the hexagonal assemblage edges of the constituent prisms, and you see at each corner of the assemblage a face (not an edge) of *one* of the constituent prisms. Build up a hexagonal prismatic assemblage by placing layer after layer over it with the constituent prisms of each layer vertically over those in the layer below; and finish the assemblage with a six-sided pyramid by

---

* See footnote to § 22 above.

† Widmanstätten, 1807. Leydolt (1855), *Wien. Akad. Ber.* 15, 59, T. 9, 10. Baumhauer, *Pogg. Ann.* 138, 563 (1869); 140, 271; 142, 324; 145, 460; 150, 619. For an account of these investigations see Mallard, *Traité de Crystallographie* (Paris, 1884), Tome II. chapitre xvi.

‡ More exactly ·9525, being $\frac{3}{4} \times \cot 38°\ 13'$; see § 48.

building upon the upper end of the prism, layer after layer of diminishing hexagonal groups, each less by one circumferential row than the layer below it. You thus have a crystal of precisely the shape of a symmetrical specimen of rock crystal, with the faces of its terminal pyramid inclined at 38° 13' to the faces of the prism from which they spring. But the assemblage thus constituted has 'senary' (or six-rayed) symmetry. To reduce this to ternary symmetry, cut a groove through the middle of each alternate face of the prismatic molecule, making this groove in the first place parallel to the edges: and add a corresponding projection or fillet to the middles of the other three faces, so that two of the cylinders similarly oriented would fit together, with the projecting fillet on one side of one of them entering the groove in the anti-corresponding side of the other. The prismatic portion of the assemblage thus formed shows (see fig. 15), on its alternate

Fig. 15.

edges, faces of molecules with projections and faces of molecules with grooves; and shows only orientational differences between alternate faces, whether of the pyramid or of the prism. Having gone only so far from 'senary' symmetry we have exactly the triple, or three-pair anti-symmetry required for the piezo-electricity of quartz investigated so admirably by the brothers Curie[*], who found that a thin plate of quartz crystal cut from any position perpendicular to a pair of faces of a symmetrical crystal becomes

[*] J. and P. Curie and C. Friedel, *Comptes Rendus*, 1882, 1883, 1886, 1892.

positively electrified on one side and negatively on the other when pulled in a direction perpendicular to those faces.   But this assemblage has not the *chiral* piezo-electric quality discovered theoretically by Voigt*, and experimentally in quartz and in tourmaline by himself and Riecke†, nor the well-known optic chirality of quartz.

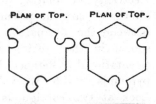

§ 49.   Change now the directions of the grooves and fillets to either of the oblique configurations shown in fig. 16, which I call right-handed, because the directions of the projections are tangential to the threads of a three-thread right-handed screw, and fig. 17 (left-

Fig. 16.   Fig. 17.

handed).   The prisms with their grooves and fillets will still all fit together if they are all right-handed, or all left-handed. fig. 18 shows the upper side of a hexagonal layer of an assemblage

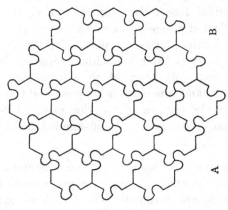

Fig. 18.

thus composed of the right-handed molecule of fig. 16.   Fig. 15 unchanged, still represents a horizontal section through the centres of the molecules.   A prism built up of such layers, and finished

* "Allgemeine Theorie der piëzo- und pyroelectrischen Erscheinungen an Krystallen," W. Voigt, *Königl. Gesellschaft der Wissenschaften zu Göttingen*, August 2, 1890.

† *Wiedemann's Annalen*, 1892, XLV. p. 923.

at each end with a pyramid according to the rule of § 48, has all the qualities of ternary chiral symmetry required for the piezo-electricity of quartz; for the orientational differences of the alternate pairs of prismatic faces; for the absolute difference between the alternate pairs of faces of each pyramid which are shown in the etching by hydrofluoric acid; for the merely orientational difference between the parallel faces of the two pyramids; and for the well-known chiro-optic* property of quartz. Look at two contiguous faces A, B of our geometrical model quartz crystal now before you, with its axis vertical. You will see a difference between them: turn it upside down; B will be undistinguishable from what A was, and A will be undistinguishable from what B was. Look at the two terminal pyramids, and you will find that the face above A and the face below B are identical in quality, and that they differ from the face above B and below A. This model is composed of the right-handed constituent molecules shown in fig. 16. It is so placed before you that the edge of the prismatic part of the assemblage nearest to you shows you filleted faces of the prismatic molecules. You see two pyramidal faces; the one to your right hand, over B, presents complicated projections and hollows at the corners of the constituent molecules; and the pyramidal face next your left hand, over A, presents their unmodified corners. But it will be the face next your left hand which will present the complex bristling corners, and the face next your right hand that will present the simple corners if, for the model before you, you substitute a model composed of left-handed molecules, such as those shown in fig. 17.

§ 50. To give all the qualities of symmetry and anti-symmetry of the pyro-electric and piezo-electric properties of tourmaline, make a hollow in one terminal face of each of our constituent prisms, and a corresponding projection in its other terminal face.

§ 51. Coming back to quartz, we can now understand perfectly the two kinds of macling which are well known to mineralogists as being found in many natural specimens of the crystal, and which I call respectively the orientational macling, and the chiral macling. In the orientational macling all the crystalline molecules

---

* Generally miscalled 'rotational.'

are right-handed, or all left-handed; but through all of some part of the crystal, each of our component hexagonal prisms is turned round its axis through 60° from the position it would have if the structure were homogeneous throughout. In each of the two parts the structure is homogeneous, and possesses all the electric and optic properties which any homogeneous portion of quartz crystal presents, and the facial properties of natural uncut crystal shown in the etching by hydrofluoric acid; but there is a discontinuity at the interface, not generally plane between the two parts, which in our geometrical model would be shown by non-fittings between the molecules on the two sides of the interface, while all the contiguous molecules in one part, and all the contiguous molecules in the other part, fit into one another perfectly. In chiral macling, which is continually found in amethystine quartz, and sometimes in ordinary clear quartz crystals, some parts are composed of right-handed molecules, and others of left-handed molecules. It is not known whether, in this chiral macling, there is or there is not also the orientational macling on the two sides of each interface; but we may say probably *not*; because we know that the orientational macling occurs in nature without any chiral macling, and because there does not seem reason to expect that chiral macling would imply orientational macling on the two sides of the same interface. I would like to have spoken to you more of this most interesting subject; and to have pointed out to you that some of the simplest and most natural suppositions we can make as to the chemical forces (or electrical forces, which probably means the same thing) concerned in a single chemical molecule of quartz SiO_2, and acting between it and similar neighbouring molecules, would lead essentially to these molecules coming together in triplets, each necessarily either right-handed or left-handed, but with as much probability of one configuration as of the other: and to have shown you that these triplets of silica $3(SiO_2)$ can form a crystalline molecule with all the properties of ternary chiral symmetry, typified by our grooved hexagonal prisms, and can build up a quartz crystal by the fortuitous concourse of atoms. I should like also to have suggested and explained the possibility that a right-handed crystalline molecule thus formed may, in natural circumstances of high temperature, or even of great pressure, become changed into a left-handed

Fig. 19.

crystal, or *vice versa*.   My watch, however, warns me that I must not enter on this subject.

§ 52.   Coming back to mere molecular tactics of crystals, remark that our assemblage of rounded, thoroughly scalene tetrahedrons, shown in the stereoscopic picture (§ 36, fig. 13 above), essentially has chirality because each constituent tetrahedron, if wholly scalene, has chirality*.   I should like to have explained to you how a single or double homogeneous assemblage of points has essentially no chirality, and how three assemblages of single points, or a single assemblage of triplets of points, can have chirality, though a single triplet of points cannot have chirality. I should like indeed to have brought somewhat thoroughly before you the geometrical theory of chirality; and in illustration to have explained the conditions under which four points or two lines, or a line and two points, or a combination of point, line and plane, can have chirality: and how a homogeneous assemblage of non-chiral objects can have chirality; but in pity I forbear, and I thank you for the extreme patience with which you have listened to me.

* See footnote to § 22 above.

APPENDIX I.

ON THE ELASTICITY OF A CRYSTAL ACCORDING TO BOSCOVICH*.

§ 1. A CRYSTAL in nature is essentially a homogeneous assemblage of equal and similar molecules, which for brevity I shall call crystalline molecules. The crystalline molecule may be the smallest portion which can be taken from the substance without chemical decomposition, that is to say, it may be the group of atoms kept together by chemical affinity, which constitutes what for brevity I shall call the chemical molecule; or it may be a group of two, three, or more of these chemical molecules kept together by cohesive force. In a crystal of tartaric acid the crystalline molecule may be, and it seems to me probably is, the chemical molecule, because if a crystal of tartaric acid is dissolved and recrystallised it always remains dextro-chiral. In a crystal of chlorate of soda, as has been pointed out to me by Sir George Stokes, the crystalline molecule probably consists of a group of two or more of the chemical molecules constituting chlorate of soda, because, as found by Marbach†, crystals of the substance are some of them dextro-chiral and some of them levo-chiral; and if a crystal of either chirality is dissolved the solution shows no chirality in its action on polarised light; but if it is recrystallised the crystals are found to be some of them dextro-chiral and some of them levo-chiral, as shown both by their crystalline forms and by their action on polarised light. It is possible, however, that even in chlorate of soda the crystalline molecule may be the chemical molecule, because it may be that the chemical molecule in solution has its atoms relatively mobile enough not to remain persistently in any dextro-chiral or levo-

* From *Proc. R. S.* June 8, 1893.

† *Pogg. Ann.* Vol. XCI. pp. 482—487 (1854); or *Ann. de Chimie*, Vol. XLIII. (LV.), pp. 252—255.

chiral grouping, and that each individual chemical molecule settles into either a dextro-chiral or levo-chiral configuration in the act of forming a crystal. See "Molecular Tactics," Oxford Lecture, § 52, reproduced as App. H in the present volume.

§ 2.   Certain it is that the crystalline molecule has a chiral configuration in every crystal which shows chirality in its crystalline form or which produces right- or left-handed rotation of the plane of polarisation of light passing through it. The magnetic rotation has neither right-handed nor left-handed quality (that is to say, no chirality). This was perfectly understood by Faraday and made clear in his writings, yet even to the present day we frequently find the chiral rotation and the magnetic rotation of the plane of polarised light classed together in a manner against which Faraday's original description of his discovery of the magnetic polarisation contains ample warning.

§ 3.   These questions, however, of chirality and magnetic rotation do not belong to my present subject, which is merely the forcive* required to keep a crystal homogeneously strained to any infinitesimal extent from the condition in which it rests when no force acts upon it from without. In the elements of the mathematical theory of elasticity† we find that this forcive constitutes what is called a homogeneous stress, and is specified completely by six generalised force-components, $p_1, p_2, p_3, \ldots, p_6$, which are related to six corresponding generalised components of strain, $s_1, s_2, s_3, \ldots, s_6$, by the following formulas:—

$$w = \tfrac{1}{2}(p_1 s_1 + p_2 s_2 + \ldots + p_6 s_6) \ \ldots\ldots\ldots\ldots(1),$$

where w denotes the work required per unit volume to alter any portion of the crystal from its natural unstressed and unstrained condition to any condition of infinitesimal homogeneous stress or strain :—

$$p_1 = \frac{dw}{ds_1}, \ \ldots, \ p_6 = \frac{dw}{ds_6} \ \ldots\ldots\ldots\ldots\ldots(2),$$

where $\dfrac{d}{ds_1}, \ \ldots, \ \dfrac{d}{ds_6}$ denote differential coefficients on the supposition

---

* This is a word introduced by my brother, the late Professor James Thomson, to designate any system of forces.

† *Phil. Trans.* April 24, 1856, reprinted in Vol. III. *Math. and Phys. Papers* (Sir W. Thomson), pp. 84—112.

that w is expressed as a homogeneous quadratic function of $s_1, \ldots, s_6$:

$$s_1 = \frac{\partial w}{dp_1}, \ldots, s_6 = \frac{\partial w}{dp_6} \ldots\ldots\ldots\ldots\ldots\ldots(3),$$

where $\dfrac{\partial}{dp_1}, \ldots, \dfrac{\partial}{dp_6}$ denote differential coefficients on the supposition that w is expressed as a homogeneous quadratic function of $p_1, \ldots, p_6$.

§ 4. Each crystalline molecule in reality certainly experiences forcive from some of its nearest neighbours on two sides, and probably also from next nearest neighbours and others. Whatever the mutual forcive between two mutually acting crystalline molecules is in reality, and however it is produced, whether by continuous pressure in some medium, or by action at a distance, we may ideally reduce it, according to elementary statical principles, to two forces, or to one single force and a couple in a plane perpendicular to that force. Boscovich's theory, a purely mathematical idealism, makes each crystalline molecule a single point, or a group of points, and assumes that there is a mutual force between each point of one crystalline molecule and each point of neighbouring crystalline molecules, in the line joining the two points. The very simplest Boscovichian idea of a crystal is a homogeneous group of single points. The next simplest idea is a homogeneous group of double points.

§ 5. In the present communication, I demonstrate that, if we take the very simplest Boscovichian idea of a crystal, a homogeneous group of single points, we find essentially six relations between the twenty-one coefficients in the quadratic function expressing w, whether in terms of $s_1, \ldots, s_6$ or of $p_1, \ldots, p_6$. These six relations are such that incompressibility, that is to say infinite resistance to change of bulk, involves infinite rigidity. In the particular case of an equilateral* homogeneous assemblage with such a law of force as to give equal rigidities for all directions of shearing, these six relations give $3k = 5n$, which is the relation

---

* That is to say, an assemblage in which the lines from any point to three neighbours nearest to it and nearest to one another are inclined at 60° to one another; and these neighbours are at equal distances from it. This implies that each point has twelve equidistant nearest neighbours around it, and that any tetrahedron of four nearest neighbours has for its four faces four equilateral triangles.

found by Navier and Poisson in their Boscovichian theory for isotropic elasticity in a solid. This relation was shown by Stokes to be violated by many real homogeneous isotropic substances, such, for example, as jelly and india-rubber, which oppose so great resistance to compression and so small resistance to change of shape, that we may, with but little practical error, consider them as incompressible elastic solids.

§ 6. I next demonstrate that if we take the next simplest Boscovichian idea for a crystal, a homogeneous group of double points, we can assign very simple laws of variation of the forces between the points which shall give any arbitrarily assigned value to each of the twenty-one coefficients in either of the quadratic expressions for w.

§ 7. I consider particularly the problem of assigning such values to the twenty-one coefficients of either of the quadratic formulas as shall render the solid incompressible. This is most easily done by taking w as a quadratic function of $p_1, \ldots, p_6$, and by taking one of these generalised stress components, say p_6, as uniform positive or negative pressure in all directions. This makes s_6 uniform compression or extension in all directions, and makes $s_1, \ldots, s_5$ five distortional components with no change of bulk. The condition that the solid shall be incompressible is then simply that the coefficients of the six terms involving p_6 are each of them zero. Thus, the expression for w becomes merely a quadratic function of the five distortional stress-components, $p_1, \ldots, p_5$, with fifteen independent coefficients : and equations (3) of § 3 above express the five distortional components as linear functions of the five stress-components with these fifteen independent coefficients.

Added July 18, 1893.

§ 8. To demonstrate the propositions of § 5, let OX, OY, OZ be three mutually perpendicular lines through any point O of a homogeneous assemblage, and let x, y, z be the coordinates of any other point P of the assemblage, in its unstrained condition. As it is a homogeneous assemblage of single points that we are now considering, there must be another point P', whose coordinates are $-x, -y, -z$. Let $(x + \delta x, y + \delta y, z + \delta z)$ be the coordinates of the altered position of P in any condition of infinitesimal strain, specified by the six symbols e, f, g, a, b, c, according to the notation of Thomson and Tait's *Natural Philosophy*, Vol. I., Pt. II., § 669. In this notation, e, f, g denote simple infinitesimal elongations parallel to OX, OY, OZ respectively; and a, b, c infinitesimal changes from the right angles between three pairs of planes of the substance, which, in the unstrained condition, are parallel to $(XOY, XOZ), (YOZ, YOX), (ZOX, ZOY)$ respectively (all angles being measured in terms of the radian). The definition of a, b, c may be given, in other words, as follows, with a taken as example: a denotes the difference of component motions parallel to OY of two planes of the substance at unit distance asunder, kept parallel to YOX during the displacement; or, which is the same thing, the difference of component motions parallel to OZ of two planes at unit distance asunder kept parallel to ZOX during the displacement. To avoid the unnecessary consideration of rotational displacement, we shall suppose the displacement corresponding to the strain-component a to consist of elongation perpendicular to OX in the plane through OX bisecting YOZ, and shrinkage perpendicular to OX in the plane through OX perpendicular to that bisecting plane. This displacement gives no contribution to δx, and contributes to δy and δz respectively $\frac{1}{2} az$ and $\frac{1}{2} ay$. Hence, and dealing similarly with b and c, and taking into account the contributions of e, f, g, we find

$$\left. \begin{aligned} \delta x &= ex + \tfrac{1}{2}(bz + cy) \\ \delta y &= fy + \tfrac{1}{2}(cx + az) \\ \delta z &= gz + \tfrac{1}{2}(ay + bx) \end{aligned} \right\} \quad\quad\quad\quad (4).$$

§ 9. In our dynamical treatment below, the following formulas, in which powers higher than squares or products of the infinitesimal ratios $\delta x/r, \delta y/r, \delta z/r$ (r denoting OP) are neglected, will be found useful.

$$\frac{\delta r}{r} = \frac{x\,\delta x + y\,\delta y + z\,\delta z}{r^2} + \tfrac{1}{2}\frac{\delta x^2 + \delta y^2 + \delta z^2}{r^2} - \tfrac{1}{2}\left(\frac{x\,\delta x + y\,\delta y + z\,\delta z}{r^2}\right)^2 \dots (5).$$

Now by (4) we have

$$x\,\delta x + y\,\delta y + z\,\delta z = ex^2 + fy^2 + gz^2 + ayz + bzx + cxy \dots (6),$$

and

$$\delta x^2 + \delta y^2 + \delta z^2 = e^2 x^2 + f^2 y^2 + g^2 z^2$$
$$+ \tfrac{1}{4}\left[a^2(y^2 + z^2) + b^2(z^2 + x^2) + c^2(x^2 + y^2)\right]$$
$$+ [\tfrac{1}{2}bc + (f + g)a]yz + [\tfrac{1}{2}ca + (g + e)b]zx + [\tfrac{1}{2}ab + (e + f)c]xy \dots (7).$$

Using (6) and (7) in (5), we find

$$\frac{\delta r}{r} = r^{-2}(ex^2 + fy^2 + gz^2 + ayz + bzx + cxy) + q\,(e, f, g, a, b, c)\ldots(8),$$

where q denotes a quadratic function of e, f, &c., with coefficients as follows:—

$$\text{Coefficient of } \tfrac{1}{2}e^2 \text{ is } \frac{x^2}{r^2} - \frac{x^4}{r^4}$$

$$\text{,,} \qquad \tfrac{1}{2}a^2 \quad \text{,,} \quad \tfrac{1}{4}\frac{y^2 + z^2}{r^2} - \frac{y^2 z^2}{r^4}$$

$$\text{,,} \qquad fg \quad \text{,,} \quad -\frac{y^2 z^2}{r^4}$$

$$\text{,,} \qquad bc \quad \text{,,} \quad \tfrac{1}{4}\frac{yz}{r^2} - \frac{x^2 yz}{r^4} \qquad\Biggr\} \ldots\ldots(9)$$

$$\text{,,} \qquad ea \quad \text{,,} \quad -\frac{x^2 yz}{r^4}$$

$$\text{,,} \qquad eb \quad \text{,,} \quad \tfrac{1}{2}\frac{zx}{r^2} - \frac{x^3 z}{r^4}$$

and corresponding symmetrical expressions for the other fifteen coefficients.

§ 10. Going back now to § 3, let us find w, the work per unit volume, required to alter our homogeneous assemblage from its unstrained condition' to the infinitesimally strained condition specified by e, f, g, a, b, c. Let $\phi(r)$ be the work required to bring two points of the system from an infinitely great distance asunder to distance r. This is what I shall call the mutual potential energy of two points at distance r. What I shall now call the potential energy of the whole system, and denote by W, is the total work which must be done to bring all the points of it from infinite mutual distances to their actual positions in the system; so that we have

$$W = \tfrac{1}{2}\Sigma\Sigma\phi(r)\ldots\ldots\ldots\ldots\ldots\ldots(10),$$

where $\Sigma\phi(r)$ denotes the sum of the values of $\phi(r)$ for the distances between any one point O, and all the others; and $\Sigma\Sigma\phi(r)$ denotes the sum of these sums with the point O taken successively at every point of the system. In this double summation $\phi(r)$ is taken twice over, whence the factor $\tfrac{1}{2}$ in the formula (10).

§ 11. Suppose now the law of force to be such that $\phi(r)$ vanishes for every value of r greater than $\nu\lambda$, where λ denotes the distance between any one point and its nearest neighbour, and ν any small or large numeric exceeding unity, and limited only by the condition that $\nu\lambda$ is very small in comparison with the linear dimensions of the whole assemblage. This, and the homogeneousness of our assemblage, imply that, except through a very thin surface layer of thickness $\nu\lambda$, exceedingly small in comparison with diameters of the assemblage, every point experiences the same set of balancing forces from neighbours as every other point, whether the system be in what we have called its unstrained condition or in any condition whatever of homogeneous strain. This strain is not of necessity an infinitely small strain, so far as concerns the proposition just stated, although in our mathematical work we limit ourselves to strains which are infinitely small.

§ 12. Remark also that if the whole system be given as a homogeneous assemblage of any specified description, and if all points in the surface-layer be held by externally applied forces in their positions as constituents of a finite homogeneous assemblage, the whole assemblage will be in equilibrium under the influence of mutual forces between the points; because the force exerted on any point O by any point P is balanced by the equal and opposite force exerted by the point P' at equal distance on the opposite side of O.

§ 13. Neglecting now all points in the thin surface layer, let N denote the whole number of points in the homogeneous assemblage within it. We have, in § 10, by reason of the homogeneousness of the assemblage,

$$\Sigma\Sigma\phi(r) = N\Sigma\phi(r)\dots\dots\dots\dots\dots(11),$$

and equation (10) becomes

$$W = \tfrac{1}{2}N\Sigma\phi(r)\dots\dots\dots\dots\dots(12).$$

Hence, by Taylor's theorem,

$$\delta W = \tfrac{1}{2}N\Sigma\left\{\phi'(r)\,\delta r + \tfrac{1}{2}\phi''(r)\,\delta r^2\right\}\dots\dots\dots(13);$$

and using (8) in this, and remarking that if (as in § 14 below) we take the volume of our assemblage as unity, so that N is the number of points per unit volume, δW becomes the w of §3; we find

$$
\begin{aligned}
w = \tfrac{1}{2} N \Sigma \Big\{ & \frac{\phi'(r)}{r} \, (ex^2 + fy^2 + gz^2 + ayz + bzx + cxy) \\
& + r\phi'(r) \, q\,(e, f, g, a, b, c) \\
& + \tfrac{1}{2} \frac{\phi''(r)}{r^2} \, (ex^2 + fy^2 + gz^2 + ayz + bzx + cxy)^2 \Big\} \dots (14).
\end{aligned}
$$

§ 14. Let us now suppose, for simplicity, the whole assemblage, in its unstrained condition, to be a cube of unit edge, and let P be the sum of the normal components of the extraneous forces applied to the points of the surface-layer in one of the faces of the cube.

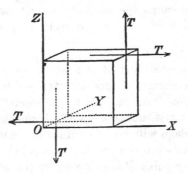

Fig. 1.

The equilibrium of the cube, as a whole, requires an equal and opposite normal component P in the opposite face of the cube. Similarly, let Q and R denote the sums of the normal components of extraneous force on the two other pairs of faces of the cube. Let T be the sum of tangential components, parallel to OZ, of the extraneous forces on either of the YZ faces. The equilibrium of the cube as a whole requires four such forces on the four faces parallel to OY, constituting two balancing couples, as shown in the accompanying diagram. Similarly, we must have four balancing tangential forces S on the four faces parallel to OX, and four tangential forces U on the four faces parallel to OZ.

§ 15. Considering now an infinitely small change of strain in the cube from (e, f, g, a, b, c) to $(e+de, f+df, g+dg, a+da, b+db, c+dc)$; the work required to produce it, as we see by considering the definitions of the displacements e, f, g, a, b, c, explained above in § 8, is as follows,

$$dw = Pde + Qdf + Rdg + Sda + Tdb + Udc \quad \ldots\ldots(15).$$

Hence we have

$$P = dw/de; \qquad Q = dw/df; \qquad R = dw/dg; \left.\right\} \;\ldots(16).$$
$$S = dw/da; \qquad T = dw/db; \qquad U = dw/dc; \left.\right\}$$

Hence, by (14), and taking L, L to denote linear functions, we find

$$P = \tfrac{1}{2}N\Sigma\left\{\frac{\phi'(r)}{r}\,x^2 + L\,(e, f, g, a, b, c)\right\}\left.\right\}$$
$$\qquad\qquad\qquad\qquad\qquad\qquad\qquad\qquad\;\ldots\ldots(17),$$
$$S = \tfrac{1}{2}N\Sigma\left\{\frac{\phi'(r)}{r}\,yz + L\,(e, f, g, a, b, c)\right\}\left.\right\}$$

and symmetrical expressions for Q, R, T, U.

§ 16.   Let now our condition of zero strain be one[*] in which no extraneous force is required to prevent the assemblage from leaving it. We must have $P = 0$, $Q = 0$, $R = 0$, $S = 0$, $T = 0$, $U = 0$, when $e = 0$, $f = 0$, $g = 0$, $a = 0$, $b = 0$, $c = 0$. Hence, by (17), and the other four symmetrical formulæ, we see that

$$\Sigma\frac{\phi'(r)}{r}\,x^2 = 0, \qquad \Sigma\frac{\phi'(r)}{r}\,y^2 = 0, \qquad \Sigma\frac{\phi'(r)}{r}\,z^2 = 0, \left.\right\}$$
$$\qquad\qquad\qquad\qquad\qquad\qquad\qquad\qquad\qquad\;\ldots(18).$$
$$\Sigma\frac{\phi'(r)}{r}\,yz = 0, \qquad \Sigma\frac{\phi'(r)}{r}\,zx = 0, \qquad \Sigma\frac{\phi'(r)}{r}\,xy = 0 \left.\right\}$$

Hence, in the summation for all the points x, y, z, between which and the point O there is force, we see that the first term of the summed coefficients in q, given by (9) above, vanishes in every case, except those of fg and ea, in each of which there is only a single term; and thus from (9) and (14) we find

$$w = \tfrac{1}{2}N\left\{\tfrac{1}{2}e^2\Sigma\varpi\,\frac{x^4}{r^4} + (fg + \tfrac{1}{2}a^2)\,\Sigma\varpi\,\frac{y^2z^2}{r^4}\right.$$
$$\qquad\qquad\qquad\qquad\qquad\qquad\qquad\qquad\;\ldots\ldots(19),$$
$$\left. + (bc + ea)\,\Sigma\varpi\,\frac{x^2yz}{r^4} + eb\,\Sigma\varpi\,\frac{x^3z}{r^4} + \&c.\right\}$$

where $\qquad\qquad -r\phi'(r) + r^2\phi''(r) = \varpi \;\ldots\ldots\ldots\ldots\ldots(20).$

---

[*] The consideration of the equilibrium of the thin surface layer, in these circumstances, under the influence of merely their proper mutual forces, is exceedingly interesting, both in its relation to Laplace's theory of capillary attraction, and to the physical condition of the faces of a crystal and of surfaces of irregular fracture. But it must be deferred. [See App. J, §§ 29, 30.]

The terms given explicitly in (19) suffice to show by symmetry all the remaining terms represented by the " &c."

§ 17. Thus we see that with no limitation whatever to the number of neighbours acting with sensible force on any one point O, and with no simplifying assumption as to the law of force, we have in the quadratic for w equal values for the coefficients of fg and $\frac{1}{2}a^2$; ge and $\frac{1}{2}b^2$; ef and $\frac{1}{2}c^2$; bc and ea; ca and eb; and ab and ec. These equalities constitute the six relations promised for demonstration in § 5.

§ 18. In the particular case of an equilateral assemblage, with axes OX, OY, OZ parallel to the three pairs of opposite edges of a tetrahedron of four nearest neighbours, the coefficients which we have found for all the products except fg, ge, ef clearly vanish; because in the complete sum for a single homogeneous equilateral assemblage we have $\pm x$, $\pm y$, $\pm z$ in the symmetrical terms. Hence, and because for this case

$$\Sigma \varpi \frac{x^4}{r^4} = \Sigma \varpi \frac{y^4}{r^4} = \Sigma \varpi \frac{z^4}{r^4}, \text{ and } \Sigma \varpi \frac{y^2 z^2}{r^4} = \Sigma \varpi \frac{z^2 x^2}{r^4} = \Sigma \varpi \frac{x^2 y^2}{r^4} \dots (21),$$

(19) becomes

$$w = \frac{1}{2}\mathfrak{A}\,(e^2 + f^2 + g^2) + \mathfrak{B}\,(fg + ge + ef) + \frac{1}{2}n\,(a^2 + b^2 + c^2) \dots (22),$$

where $\qquad \mathfrak{A} = \frac{1}{2}N\Sigma \varpi \dfrac{x^4}{r^4},$ and $\mathfrak{B} = n = \frac{1}{2}N\Sigma \varpi \dfrac{y^2 z^2}{r^4} \dots (23).$

§ 19. Looking to Thomson and Tait's *Natural Philosophy*, § 695 (7)*, we see that the $\mathfrak{B}$ of that formula is now proved to be, in our present simplest form of Boscovichian assumption, equal to the n of our present formula (22) which denotes the rigidity-modulus relative to shearings parallel to the planes YOZ, ZOX, XOY: and that if we denote by n_1 the rigidity-modulus relative to shearing parallel to planes through OX, OY, OZ, and cutting (OY, OZ), (OZ, OX), (OX, OY) at angles of 45°, and if k denote the compressibility-modulus, we have

$$\left.\begin{array}{l} \mathfrak{A} = k + \tfrac{4}{3}n_1; \quad \mathfrak{B} = k - \tfrac{2}{3}n_1; \\ n_1 = \tfrac{1}{2}(\mathfrak{A} - \mathfrak{B}); \quad k = \tfrac{1}{3}(\mathfrak{A} + 2\mathfrak{B}) \end{array}\right\} \quad \dots\dots(24);$$

* This formula is given for the case of a body which is wholly isotropic in respect to elasticity moduluses; but from the investigation in §§ 681, 682 we see that our present formula, (22) or (25), expresses the elastic energy for the case of an elastic solid possessing cubic isotropy with unequal rigidities, n_1, n, in respect to these two sets of shearings.

and the previous expression for the elastic energy of the strained solid becomes, for the case of cubic symmetry without any restriction,

$$2w = (k + \tfrac{4}{3}n_1)(e^2 + f^2 + g^2) + 2(k - \tfrac{2}{3}n_1)(fg + ge + ef)$$
$$+ n(a^2 + b^2 + c^2)\ldots\ldots(25).$$

§ 20. Comparing this with (22), for our present restricted case, we find

$$3k = 2n_1 + 3n \ldots\ldots\ldots\ldots(26).$$

This remarkable relation between the two rigidities and the compressibility of an equilateral homogeneous assemblage of Boscovich atoms was announced without proof in § 27 of my paper on the "Molecular Constitution of Matter*." In it n denotes what I called the facial rigidity, being rigidity relative to shearings parallel to the faces of the principal cube† : and n_1 the diagonal rigidity, being rigidity relative to shearings parallel to any of the six diagonal planes through pairs of mutually remotest parallel edges of the same cube. By (24) and (23) we see that if the law of force be such that

$$\Sigma\varpi\frac{x^4}{r^4} = 3\Sigma\varpi\frac{y^2z^2}{r^4} \ldots\ldots\ldots\ldots(27),$$

we have $n = n_1$, and the body constituted by the assemblage is wholly isotropic in its elastic quality. In this case (26) becomes $3k = 5n$, as found by Navier and Poisson; and thus we complete the demonstration of the statements of § 5 above.

§ 21. A case which is not uninteresting in respect to Boscovichian theory, and which is very interesting indeed in respect to mechanical engineering (of which the relationship with Boscovich's theory has been pointed out and beautifully illustrated by M. Brillouin‡), is the case of an equilateral homogeneous assemblage with forces only between each point and its twelve equidistant nearest neighbours. The annexed diagram (fig. 2) represents the point O and three of its twelve nearest neighbours

---

* *R. S. E. Proc.*, July, 1889; Art. xcvii. of my *Math. and Phys. Papers*, Vol. iii.

† That is to say, a cube whose edges are parallel to the three pairs of opposite edges of a tetrahedron of four nearest neighbours.

‡ *Conférences Scientifiques et Allocutions* (Lord Kelvin), traduites et annotées; P. Lugol et M. Brillouin: Paris, 1893, pp. 320—325.

(their distances λ), being in the middles of the near faces of the principal cube shown in the diagram; and three of its six next-nearest neighbours (their distances $\lambda\sqrt{2}$), being at X, Y, Z, the corners of the cube nearest to it; and, at other corners of the cube, three other neighbours K, L, M, which are next-next-next-nearest (their distances 2λ). The points in the middles of the three remote sides of the cube, not seen in the diagram, are next-next-nearest neighbours of O (their distances $\lambda\sqrt{3}$).

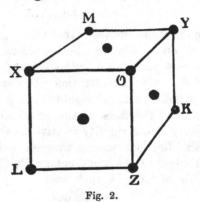

Fig. 2.

§ 22. Confining our attention now to O's nearest neighbours, we see that the nine not shown in the diagram are in the middles of squares obtained by producing the lines YO, ZO, XO to equal distances beyond O and completing the squares on all the pairs of lines so obtained. To see this more clearly, imagine eight equal cubes placed together, with faces in contact and each with one corner at O. The pairs of faces in contact are four squares in each of the three planes cutting one another at right angles through O; and the centres of these twelve squares are the twelve nearest neighbours of O. If we denote by λ the distance of each of them from O, we have for the coordinates x, y, z of these twelve points as follows:—

$$\left. \begin{array}{l} \left(0, \dfrac{\lambda}{\sqrt{2}}, \dfrac{\lambda}{\sqrt{2}}\right), \ \left(0, -\dfrac{\lambda}{\sqrt{2}}, \dfrac{\lambda}{\sqrt{2}}\right), \ \left(0, \dfrac{\lambda}{\sqrt{2}}, -\dfrac{\lambda}{\sqrt{2}}\right), \ \left(0, -\dfrac{\lambda}{\sqrt{2}}, -\dfrac{\lambda}{\sqrt{2}}\right) \\[2mm] \left(\dfrac{\lambda}{\sqrt{2}}, 0, \dfrac{\lambda}{\sqrt{2}}\right), \ \left(-\dfrac{\lambda}{\sqrt{2}}, 0, \dfrac{\lambda}{\sqrt{2}}\right), \ \left(\dfrac{\lambda}{\sqrt{2}}, 0, -\dfrac{\lambda}{\sqrt{2}}\right), \ \left(-\dfrac{\lambda}{\sqrt{2}}, 0, -\dfrac{\lambda}{\sqrt{2}}\right) \\[2mm] \left(\dfrac{\lambda}{\sqrt{2}}, \dfrac{\lambda}{\sqrt{2}}, 0\right), \ \left(-\dfrac{\lambda}{\sqrt{2}}, \dfrac{\lambda}{\sqrt{2}}, 0\right), \ \left(\dfrac{\lambda}{\sqrt{2}}, -\dfrac{\lambda}{\sqrt{2}}, 0\right), \ \left(-\dfrac{\lambda}{\sqrt{2}}, -\dfrac{\lambda}{\sqrt{2}}, 0\right) \end{array} \right\}$$

$$\dots\dots(28).$$

§ 23.   Suppose now O to experience force only from its twelve nearest neighbours: the summations Σ of § 18 (23) will include just these twelve points with equal values of ϖ, which we shall denote by ϖ_0, for all.   These yield eight terms to $\Sigma\,(x^4/r^4)$, and four to $\Sigma\,(y^2z^2/r^4)$; and the value of each term in these sums is $\frac{1}{4}$. Thus we find that

$$\mathfrak{A} = N\varpi_0, \text{ and } \mathfrak{B} = n = \tfrac{1}{2}N\varpi_0\ldots\ldots\ldots\ldots(29).$$

Hence and by (24), we see that

$$n_1 = \tfrac{1}{2}n \ldots\ldots\ldots\ldots\ldots\ldots(30).$$

Thus we have the remarkable result that, relatively to the principal cube, the diagonal rigidity is half the facial rigidity when each point experiences force only from its twelve nearest neighbours. This proposition was announced without proof in § 28 of "Molecular Constitution of Matter*."

§ 24.   Suppose now the points in the middles of the faces of the cubes which in the equilateral assemblage are O's twelve equidistant nearest neighbours to be removed, and the assemblage to consist of points in simplest cubic order; that is to say, of Boscovichian points at the points of intersection of three sets of equidistant parallel planes dividing space into cubes. Fig. 2 shows O; and, at X, Y, Z, three of the six equidistant nearest neighbours which it has in the simple cubic arrangement. Keeping λ with the same signification in respect to fig. 2 as before, we have now for the coordinates of O's six nearest neighbours:

$$(\lambda\sqrt{2}, 0, 0), \ (0, \lambda\sqrt{2}, 0), \ (0, 0, \lambda\sqrt{2}),$$
$$(-\lambda\sqrt{2}, 0, 0), \ (0, -\lambda\sqrt{2}, 0), \ (0, 0, -\lambda\sqrt{2}).$$

Hence, and denoting by ϖ_1 the value of ϖ for this case, we find, by § 18 (23),

$$\mathfrak{A} = N\varpi_1 \text{ and } \mathfrak{B} = n = 0 \ \ldots\ldots\ldots\ldots(31).$$

The explanation of $n = 0$ (facial rigidity zero) is obvious when we consider that a cube having for its edges twelve equal straight bars, with their ends jointed by threes at the eight corners, affords no resistance to change of the right angles of its faces to acute and obtuse angles.

* *Math. and Phys. Papers,* Vol. III. p. 403.

§ 25. Replacing now the Boscovich points in the middles of the faces of the cubes, from which we supposed them temporarily annulled in § 24; putting the results of § 23 and § 24 together; and using (24) of § 19; we find for our equilateral homogeneous assemblage, its elasticity-moduluses as follows:—

$$\left.\begin{aligned}\mathfrak{A} &= N(\varpi_0 + \varpi_1) \\ \mathfrak{B} &= n = \tfrac{1}{2}N\varpi_0 \\ n_1 &= \tfrac{1}{2}N(\tfrac{1}{2}\varpi_0 + \varpi_1)\end{aligned}\right\}\dots\dots\dots\dots\dots(32),$$

where, as we see by § 16 (20) above,

$$\left.\begin{aligned}\varpi_0 &= \lambda F(\lambda) - \lambda^2 F'\lambda \\ \varpi_1 &= \lambda\sqrt{2}F(\lambda\sqrt{2}) - 2\lambda^2 F'(\lambda\sqrt{2})\end{aligned}\right\}\dots\dots\dots(33);$$

$F(r)$ being now taken to denote repulsion between any two of the points at any distance r, which, with $\phi(r)$ defined as in § 10, is the meaning of $-\phi'(r)$. To render the solid, constituted of our homogeneous assemblage, elastically isotropic, we must, by § 19 (24), have $\mathfrak{A} - \mathfrak{B} = 2n$, and therefore, by (32),

$$\varpi_0 = 2\varpi_1 \dots\dots\dots\dots\dots\dots\dots(34).$$

By (33) we see that the distant forces contribute to n_1, and not to n.

§ 26. The last three of the six equilibrium equations, § 16 (18), are fulfilled in virtue of symmetry in the case of an equilateral assemblage of single points whatever be the law of force between them, and whatever be the distance between any point and its nearest neighbours. The first three of them require in the case of § 23 that $F(\lambda) = 0$; and in the case of (24) that $F(\lambda\sqrt{2}) = 0$, results of which the interpretation is obvious and important.

§ 27. The first three of the six equilibrium equations, § 16 (18), applied to the case of § 25, yield the following equation:—

$$\sqrt{\tfrac{1}{2}}F(\lambda\sqrt{2}) = -F(\lambda) \dots\dots\dots\dots(35);$$

that is to say, if there is repulsion or attraction between each point and its twelve nearest neighbours, there is attraction or repulsion of $\sqrt{2}$ of its amount between each point and its six next-nearest neighbours, unless there are also forces between more distant points. This result is easily verified by simple synthetical and geometrical considerations of the equilibrium between a point and its twelve nearest and six next-nearest neighbours in an equilateral homogeneous assemblage. The consideration of it is exceedingly interesting and important in respect to, and in illustration of, the engineering of jointed structures with redundant links or tie-struts.

§ 28.   Leaving, now, the case of an equilateral homogeneous assemblage, let us consider what we may call a scalene assemblage, that is to say, an assemblage in which there are three sets of parallel rows of points, determinately fixed as follows, according to the system first taught by Bravais * :—

I.   Just one set of rows of points at consecutively shortest distances λ_1.

II.   Just one set of rows of points at consecutively next-shortest distances λ_2.

III.   Just one set of rows of points at consecutive distances λ_3 shorter than those of all other rows not in the plane of I. and II.

To the condition $\lambda_3 > \lambda_2 > \lambda_1$ we may add the condition that none of the angles between the three sets of rows is a right angle, in order that our assemblage may be what we may call wholly scalene.

§ 29.   Let $A'OA$, $B'OB$, $C'OC$ be the primary rows thus determinately found having any chosen point, O, in common; we have

$$\left. \begin{aligned} A'O = OA = \lambda_1 \\ B'O = OB = \lambda_2 \\ C'O = OC = \lambda_3 \end{aligned} \right\} \quad \dots\dots\dots\dots\dots(36).$$

Thus A' and A are O's nearest neighbours; and B' and B, O's next-nearest neighbours; and C' and C, O's nearest neighbours not in the plane AOB. (It should be understood that there may be in the plane AOB points which, though at greater distances from O than B and B', are nearer to O than are C and C'.)

§ 30.   Supposing, now, BOC, $B'OC''$, &c., to be the acute angles between the three lines meeting in O; we have two equal and

* *Journal de l'École Polytechnique*, tome xix. cahier xxxiii. pp. 1—128; Paris, 1850.

dichirally similar* tetrahedrons of each of which each of the four faces is a scalene acute-angled triangle. That every angle in and between the faces is acute we readily see, by remembering that OC and OC' are shorter than the distances of O from any other of the points on the two sides of the plane AOB†.

§ 31. As a preliminary to the engineering of an incompressible elastic solid according to Boscovich, it is convenient now to consider a special case of scalene tetrahedron, in which perpendiculars from the four corners to the four opposite faces intersect in one point. I do not know if the species of tetrahedron which fulfils this condition has found a place in geometrical treatises, but I am informed by Dr Forsyth that it has appeared in Cambridge examination papers. For my present purpose it occurred to me thus:—Let QO, QA, QB, QC be four lines of given lengths drawn from one point, Q. It is required to draw them in such relative directions that the volume of the tetrahedron $OABC$ is a maximum. Whatever be the four given lengths, this problem clearly has one real solution and one only; and it is such that the four planes BOC, COA, AOB, ABC are cut perpendicularly by the lines AQ, BQ, CQ, OQ, respectively, each produced through Q. Thus we see that the special tetrahedron is defined by four lengths, and conclude that two equations among the six edges of the tetrahedron in general are required to make it our special tetrahedron.

§ 32. Hence we see the following simple way of drawing a special tetrahedron. Choose as data three sides of one face and the length of the perpendicular to it from the opposite angle. The planes through this perpendicular, and the angles of the triangle, contain the perpendiculars from these angles to the opposite faces of the tetrahedron, and therefore cut the opposite sides of the triangle perpendicularly. (Thus, parenthetically, we have a proof of the known theorem of elementary geometry that the perpendiculars from the three angles of a triangle to the

---

* Either of these may be turned round so as to coincide with the image of the other in any plane mirror. Either may be called a pervert of the other; as, according to the usage of some writers, an object is called a *pervert* of another if one of them can be brought to coincide with the image of the other in a plane mirror (as, for example, a right hand and a left hand).

† See "Molecular Constitution of Matter," § (45), (*h*), (*i*), *Math. and Phys. Papers*, Vol. III. pp. 412—413.

opposite sides intersect in one point.) Let ABC be the chosen triangle and S the point in which it is cut by the perpendicular from O, the opposite corner of the tetrahedron. AS, BS, CS, produced through S, cut the opposite sides perpendicularly, and therefore we find the point S by drawing two of these perpendiculars and taking their point of intersection. The tetrahedron

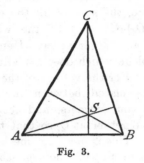

Fig. 3.

is then found by drawing through S a line SO of the given length perpendicular to the plane of ABC. (We have, again parenthetically, an interesting geometrical theorem. The perpendiculars from A, B, C to the planes of OBC, OCA, OAB cut OS in the same point; SO being of any arbitrarily chosen length.)

§ 33. I wish now to show how an incompressible homogeneous solid of wholly oblique crystalline configuration can be constructed without going beyond Boscovich for material. Consider, in any scalene assemblage, the plane of the line $A'OA$ through any point O and its nearest neighbours, and the line $B'OB$ through the same point and its next-nearest neighbours. To fix the ideas, and avoid circumlocutions, we shall suppose this plane to be horizontal. Consider the two parallel planes of points nearest to the plane above it and below it. The corner C of the acute-angled tetrahedron $OABC$, which we have been considering, is one of the points in one of the two nearest parallel planes, that above AOB we shall suppose. And the corner C' of the equal and dichirally similar tetrahedron $OA'B'C'$ is one of the points in the nearest parallel plane below. All the points in the plane through C are corners of equal tetrahedrons chirally similar to $OABC$, and standing on the horizontal triangles oriented as BOA.

All the points C' in the nearest plane below are corners of tetra-hedrons chirally similar to $OA'B'C'$ placed downwards on the triangles oriented as $B'OA'$. The volume of the tetrahedron $OABC$ is $\frac{1}{6}$ of the volume of the parallelepiped, of which OA, OB, OC are conterminous edges. Hence the sum of the volumes of all the upward tetrahedrons having their bases in one plane is $\frac{1}{6}$ of the volume of the space between large areas of these planes: and, therefore, the sum of all the chirally similar tetra-hedrons; such as $OABC$, is $\frac{1}{6}$ of the whole volume of the assemblage through any larger space. Hence any homogeneous strain of the assemblage which does not alter the volume of the tetrahedrons does not alter the volume of the solid. Let tie-struts OQ, AQ, BQ, CQ be placed between any point Q within the tetrahedron and its four corners, and let these tie-struts be mechanically jointed together at Q, so that they may either push or pull at this point. This is merely a mechanical way of stating the Boscovichian idea of a second homogeneous assemblage, equal and similarly oriented to the first assemblage and placed with one of its points at Q, and the others in the other corresponding positions relatively to the primary assemblage. When it is done for all the tetrahedrons chirally similar to $OABC$, we find four tie-strut ends at every point O, or A, or B, or C, for example, of the primary assemblage. Let each set of these four ends be mechanically jointed together, so as to allow either push or pull. A model of the curious structure thus formed was shown at the conversazione of the Royal Society of June 7, 1893. It is for three dimensions of space what ordinary hexagonal netting is in a plane.

§ 34. Having thus constructed our model, alter its shape until we find its volume a maximum. This brings the tetra-hedron, $OABC$, to be of the special kind defined in § 31. Suppose for the present the tie-struts to be absolutely resistant against push and pull, that is to say, to be each of constant length. This secures that the volume of the whole assemblage is unaltered by any infinitesimal change of shape possible to it; so that we have, in fact, the skeleton of an incompressible and inextensible solid*.

---

* This result was given for an equilateral tetrahedronal assemblage in § 67 of "Molecular Constitution of Matter," *Math. and Phys. Papers*, Vol. III. pp. 425—426.

Let now any forces whatever, subject to the law of uniformity in the assemblage, act between the points of our primary assemblage (and, if we please, also between the points of our second assemblage; and between other pairs of points than the nearests of the two assemblages). Let these forces fulfil the conditions of equilibrium; of which the principle is described in § 16 and applied to find the equations of equilibrium for the simple case of a single homogeneous assemblage there considered. Thus we have an incompressible elastic solid; and, as in § 17 above, we see that there are just fifteen independent coefficients in the quadratic function of the strain-components expressing the work required to produce an infinitesimal strain. Thus we realise the result described in § 7 above.

§ 35. Suppose now each of the four tie-struts to be not infinitely resistant against change of length, and to have a given modulus of longitudinal rigidity, which, for brevity, we shall call its stiffness. By assigning proper values to these four stiffnesses, and by supposing the tetrahedron to be freed from the two conditions making it our special tetrahedron, we have six quantities arbitrarily assignable, by which, adding these six to the former fifteen, we may give arbitrary values to each of the twenty-one coefficients in the quadratic function of the six strain-components with which we have to deal when change of bulk is allowed. Thus, in strictest Boscovichian doctrine, we provide for twenty-one independent coefficients in Green's energy-function. The dynamical details of the consideration of the equilibrium of two homogeneous assemblages with mutual attraction between them, and of the extension of §§ 9—17 to the larger problem now before us, are full of purely scientific and engineering interest, but must be reserved for what I hope is a future communication.

APPENDIX J.

MOLECULAR DYNAMICS OF A CRYSTAL*.

§ 1. THE object of this communication is to partially realise the hope expressed at the end of my paper of July 1 and July 15, 1889, on the *Molecular Constitution of Matter*†:—"The mathematical investigation must be deferred for a future communication, when I hope to give it *with some further developments*." The italics are of present date.

Following the ideas and principles suggested in §§ 14—20 of that paper (referred to henceforth for brevity as "M. C. M."), let us first find the work required to separate all the atoms of a homogeneous assemblage of a great number n of molecules to infinite distances from one another. Each molecule may be a single atom, or it may be a group of i atoms (similar to one another or dissimilar, as the case may be) which makes the whole assemblage a group of i assemblages, each of n single atoms.

§ 2. Remove now one molecule from its place in the assemblage to an infinite distance, keeping unchanged the configuration of its constituent atoms, and keeping unmoved every atom remaining in the assemblage. Let W be the work required to do so. This is the same for all the molecules within the assemblage, except the negligible number of those (§ 30 below) which are within influential distance of the surface. Hence $\frac{1}{2}nW$ is the total work required to separate all the n molecules of the assemblage to infinite distances from one another. Add to this n times the work required to separate the i atoms of one of the molecules to infinite distances from one another, and we have the whole work required to separate all the in atoms of the given assemblage.

* *Proc. Roy. Soc. Edin.*, May, 1902.

† *Proc. Roy. Soc. Edin.*, and Vol. III. of *Mathematical and Physical Papers*, Art. xcvii.

Another procedure, sometimes more convenient, is as follows:—
Remove any one atom from the assemblage, keeping all the others
unmoved. Let w be the work required to do so, and let Σw
denote the ,sum of the amounts of work required to do this for
every atom separately of the whole assemblage. The total amount
of work required to separate all the atoms to infinite distances
from one another is $\frac{1}{2}\Sigma w$. This (not subject to any limitation
such as that stated for the former procedure) is rigorously true for
any assemblage whatever of any number of atoms, small or large.
It is, in fact, the well-known theorem of potential energy in the
dynamics of a system of mutually attracting or repelling particles;
and from it we easily demonstrate the item $\frac{1}{2}nW$ in the former
procedure.

§ 3. In the present communication we shall consider only
atoms of identical quality, and only two kinds of assemblage.

I. A homogeneous assemblage of N single atoms, in which the
twelve nearest neighbours of each atom are equidistant from it.
This, for brevity, I call an equilateral assemblage. It is fully
described in "M. C. M.," §§ 46, 50 ... 57.

II. Two simple homogeneous assemblages of $\frac{1}{2}N$ single atoms,
placed together so that one atom of each assemblage is at the
centre of a quartet of nearest neighbours of the others.

For assemblage II., as well as for assemblage I., w is the same
for all the atoms, except the negligible number of those within
influential distance of the boundary. Neglecting these, we therefore have $\Sigma w = Nw$, and therefore the whole work required to
separate all the atoms to infinite distances is—

$$\frac{1}{2}Nw \quad\dots\dots\dots\dots\dots\dots(1).$$

§ 4. Let $\phi(D)$ be the work required to increase the distance
between two atoms from D to ∞; and let $f(D)$ be the attraction
between them at distance D. We have

$$f(D) = -\frac{d}{dD}\phi(D) \quad\dots\dots\dots\dots\dots(2).$$

For either assemblage I. or assemblage II. we have

$$w = \phi(D) + \phi(D') + \phi(D'') + \text{etc.} \quad\dots\dots\dots\dots(3);$$

where D, D', D'', etc., denote the distances from any one atom of
all neighbours, including the farthest in the assemblage, which
exercise any force upon it.

§ 5. To find as many as we desire of these distances for assemblage I. look at figs. 1 and 2. Fig. 1 shows an atom A, and neighbours in one plane in circles of nearest, next-nearest, next-next-nearest, etc. Fig. 2 shows an equilateral triangle of three nearest neighbours, and concentric circles of neighbours in the same plane round it. The circles corresponding to r_4 and r_8 of § 7 below, are not drawn in fig. 2. In all that follows the side of each of the equilateral triangles is denoted by λ.

§ 6. All the neighbours in assemblage I. are found by aid of the diagrams as follows :—

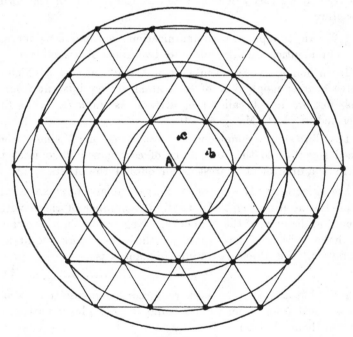

Fig. 1.

(a) The atoms of the net shown in fig. 1. The plane of this net we shall call our "middle plane." Let lines be drawn perpendicular to it through the atom A, and the points marked b, c, to guide the placing of nets of atoms in parallel planes on its two sides.

(b) Two nets of atoms at equal distances $\lambda\sqrt{\tfrac{2}{3}}$ on the two sides of the "middle plane." These nets are so placed that an atom of one of them, say the near one as we look at the diagram, is in the guide line b; and an atom of the far one is in the guide line c.

(c) Two parallel nets of atoms at equal distances, $2\lambda \sqrt{\tfrac{2}{3}}$, on the two sides of the "middle plane," so placed that an atom of the near one is in the guide line c, and an atom of the far one is in the guide line b.

(d) A third pair of parallel planes at equal distances, $3\lambda \sqrt{\tfrac{2}{3}}$, from the "middle plane," and each of them having an atom in guide line A.

(e) Successive triplets of parallel nets with their atoms

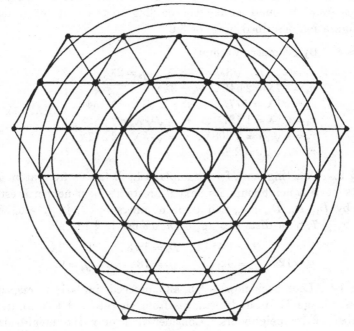

Fig. 2.

cyclically arranged Abc, Abc ... at greater and greater distances from A on the near side of the paper, and Acb, Acb ... at greater and greater distances on the far side.

§ 7. Let q_1, q_2, q_3 ... be the radii of the circles shown in fig. 1, and r_1, r_2, r_3 ... be the radii of the circles shown in fig. 2; and for brevity denote $\lambda \sqrt{\tfrac{2}{3}}$ by κ. The distances from A of all the neighbours around it are :—

In our "middle plane": 6 each equal to q_1; 6, q_2; 6, q_3; 12, q_4; 6, q_5; ....

In the two parallel nets at distances κ from middle : 6 each equal to $\sqrt{(\kappa^2 + r_1^2)}$; 6, $\sqrt{(\kappa^2 + r_2^2)}$; 12, $\sqrt{(\kappa^2 + r_3^2)}$; 12, $\sqrt{(\kappa^2 + r_4^2)}$; 6, $\sqrt{(\kappa^2 + r_5^2)}$; 12, $\sqrt{(\kappa^2 + r_6^2)}$; 6, $\sqrt{(\kappa^2 + r_7^2)}$.

In the two parallel nets at distances 2κ from middle : the same as (B) altered by taking 2κ everywhere in place of κ.

In the two parallel nets at distances 3κ from centre : the same as (A) altered by taking $\sqrt{(9\kappa^2 + q_1^2)}$, $\sqrt{(9\kappa^2 + q_2^2)}$, etc., in place of q_1, q_2, etc.

In nets at distances on each side greater than 3κ : distances of atoms from A, found as above, according to the cycle of atomic configuration described in (e) of § 6.

§ 8.   By geometry we find

$$\left.\begin{aligned}
&q_1 = \lambda; \quad q_2 = \sqrt{3}\lambda = 1\cdot732\lambda; \quad q_3 = 2\lambda; \\
&q_4 = \sqrt{7}\lambda = 2\cdot646\lambda; \quad q_5 = 3\lambda : \\
&r_1 = \sqrt{\tfrac{1}{3}}\lambda = \cdot577\lambda; \qquad r_2 = 2\sqrt{\tfrac{1}{3}}\lambda = 1\cdot154\lambda; \\
&r_3 = \sqrt{\tfrac{7}{3}}\lambda = 1\cdot527\lambda; \qquad r_4 = \sqrt{\tfrac{13}{3}}\lambda = 2\cdot082\lambda; \\
&r_5 = 4\sqrt{\tfrac{1}{3}}\lambda = 2\cdot308\lambda; \quad r_6 = \sqrt{\tfrac{19}{3}}\lambda = 2\cdot517\lambda; \\
&r_7 = 5\sqrt{\tfrac{1}{3}}\lambda = 2\cdot887\lambda.
\end{aligned}\right\} \quad \dots\dots(4).$$

§ 9.   Denoting now, for assemblage I., distances from atom A of its nearest neighbours, its next-nearests, its next-next-nearests, etc., by D_1, D_2, D_3, etc., and their numbers by j_1, j_2, j_3, etc., we find by §§ 7, 8 for distances up to 2λ, for use in § 12 below,

$$D_1 = \lambda, \quad D_2 = 1\cdot414\lambda, \quad D_3 = 1\cdot732\lambda, \quad D_4 = 2\lambda,$$
$$j_1 = 12; \quad j_2 = 6; \qquad j_3 = 18; \qquad j_4 = 6.$$

§ 10.   Look back now to § 5, and proceed similarly in respect to assemblage II., to find distances from any atom A to a limited number of its neighbours. Consider first only the neighbours forming with A a single equilateral assemblage : we have the same set of distances as we had in § 9. Consider next the neighbours which belong to the other equilateral assemblage. Of these, the four nearest (being the corners of a tetrahedron having A at its centre) are each at distance $\tfrac{3}{4}\sqrt{\tfrac{2}{3}}\lambda$, and these are A's nearest neighbours of all the double assemblage II. Three of these four are situated in a net whose plane is at the distance $\tfrac{1}{4}\sqrt{\tfrac{2}{3}}\lambda$ on one side of our "middle plane" through A, and having one of its atoms on either of the guide lines b or c. The distances from A of all the atoms in this net are, according to fig. 2,

$$\sqrt{(\tfrac{1}{16}\kappa^2 + r_1^2)}, \quad \sqrt{(\tfrac{1}{16}\kappa^2 + r_2^2)}, \text{ etc.} \dots\dots\dots\dots(5).$$

The remaining one of the four nearests is on a net at distance $\tfrac{3}{4}\sqrt{\tfrac{2}{3}}\lambda$ from our "middle plane," having one of its atoms on the guide line through A. The distances from A of all the atoms in this net are, according to fig. 1,

$$\tfrac{3}{4}\sqrt{\tfrac{2}{3}}\lambda, \quad \sqrt{(\tfrac{9}{16}\kappa^2 + q_1^2)}, \quad \sqrt{(\tfrac{9}{16}\kappa^2 + q_2^2)}, \text{ etc. } \dots\dots\dots(6).$$

All the other atoms of the equilateral assemblage to which A does not belong lie in nets at successive distances κ, 2κ, 3κ, etc., beyond the two nets we have already considered on the two sides of our "middle plane"; the atoms of each net placed of course according to the cyclical law described in (e) of § 6.

§ 11. Working out for the double assemblage II. for A's nearest neighbours according to § 10, we find four nearest neighbours at equal distances $\tfrac{3}{4}\sqrt{\tfrac{2}{3}}\lambda = \cdot613\lambda$; twelve next-nearests at equal distances λ; and twelve next-next-nearests at equal distances $\sqrt{\tfrac{11}{8}}\lambda = 1\cdot173\lambda$. These suffice for § 12 below. It is easy and tedious, and not at present useful, to work out for D_4, D_5, D_6, etc.

§ 12. Using now §§ 9, 11 in (3) of § 4 we find,—

for assemblage I.,

$$w = 12\phi(\lambda) + 6\phi(1\cdot414\lambda) + 18\phi(1\cdot732\lambda) + 6\phi(2\lambda) + \dots$$

for assemblage II.,

$$w = 4\phi(\cdot613\lambda) + 12\phi(\lambda) + 12\phi(1\cdot173\lambda) + \dots$$

$$\left.\right\} \dots(7).$$

These formulas prepare us for working out in detail the practical dynamics of each assemblage, guided by the following statements taken from §§ 18, 16 of "M. C. M."

§ 13. Every infinite homogeneous assemblage of Boscovich atoms is in equilibrium. So, therefore, is every finite homogeneous assemblage, provided that extraneous forces be applied to all within influential distance of the frontier, equal to the forces which a homogeneous continuation of the assemblage through influential distance beyond the frontier would exert on them. The investigation of these extraneous forces for any given homogeneous assemblage of single atoms—or groups of atoms as explained above (§ 1)—constitutes the Boscovich equilibrium-theory of elastic solids.

It is wonderful how much towards explaining the crystallography and elasticity of solids, and the thermo-elastic properties of solids, liquids, and gases, we find; without assuming, in the Boscovichian law of force, more than one transition from attraction to repulsion. Suppose, for instance, that the mutual force between two atoms is zero for all distances exceeding a certain distance I, which we shall call the diameter of the sphere of influence; is repulsive when the distance between them is $< \zeta$; zero when the distance is $= \zeta$; and attractive when the distance is $> \zeta$ and $< I$.

§ 14. Two different examples are represented on the two curves of fig. 3, drawn arbitrarily to obtain markedly diverse conditions of equilibrium for the monatomic equilateral assemblage I., and also for the diatomic assemblage II. The abscissa (x)

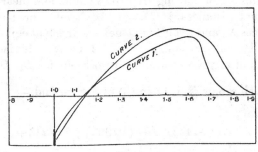

Fig. 3.

of each diagram, reckoned from a zero outside the diagram on the left, represents the distance between centres of two atoms; the ordinates (y) represent the work required to separate them from this distance to ∞. Hence $-\dfrac{dy}{dx}$ represents the mutual attraction at distance x. This we see by each curve is $-\infty$ (infinite repulsion) at distance $1\cdot0$, which means that the atom is an ideal hard ball of diameter $1\cdot0$. For distances increasing from $1\cdot0$ the force is repulsive as far as $1\cdot61$ in curve 1, and $1\cdot55$ in curve 2. At these distances the mutual force is zero; and at greater distances up to $1\cdot8$ in curve 1, and $1\cdot9$ in curve 2, the force is attractive. The force is zero for all greater distances than the last mentioned in the two examples respectively. Thus, according to my old notation, we have $\zeta = 1\cdot61$, $I = 1\cdot8$ in curve 1; and $\zeta = 1\cdot55$, $I = 1\cdot9$ in curve 2. The distances for maximum attractive force (as shown by the points of inflection of the two curves) are $1\cdot68$ for curve 1, and $1\cdot76$ for curve 2.

According to our notation of § 4 we have $y = \phi(D)$, if $x = D$ in each curve.

§ 15. The two formulas (7), § 12, are represented in fig. 4 for curve 1, and in fig. 5 for curve 2; with $x = \lambda$ for Ass. I., and $x = \cdot613\lambda$ for Ass. II. In each diagram the abscissa, x, is distance between nearest atoms of the assemblage. The heavy portions of

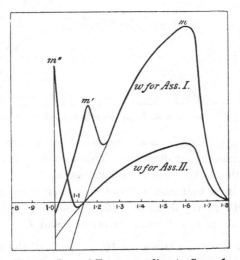

Fig. 4. Law of Force according to Curve 1.

the curves represent the values of w calculated from (7). The light portions of the curves, and their continuations in heavy curves, represent $4\phi(x)$ and $12\phi(x)$ respectively in each diagram. The point where the light curve passes into the heavy curve in each case corresponds to the least distance between neighbours at which next-nearests are beyond range of mutual force. All the diagrams here reproduced were drawn first on a large scale on squared paper for use in the calculations from (7); which included accurate determinations of the maximum and minimum values of w and the corresponding distances between nearest neighbours in

each assemblage. The corresponding densities, given in the last
column of the following table of results, are calculated by the
formula $\sqrt{2}/\lambda^3$ for assemblage I., and $2\sqrt{2}/\lambda^3$ for assemblage II.;

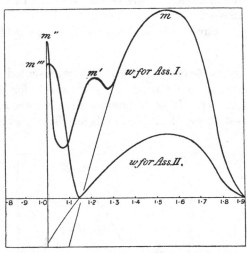

Fig. 5.   Law of Force according to Curve 2.

"density" being in each case number of atoms per cube of the
unit of abscissas of the diagram.   This unit is (§ 14) equal to the
diameter of the atom.   For simplicity we assume the atom to be
an infinitely hard ball exerting (§ 13) on neighbouring atoms, not
in contact with it, repulsion at distance between centres less than
ζ and attraction at any distance between ζ and I.

§ 16.   To interpret these results, suppose all the atoms of the
assemblage to be subjected to guidance constraining them either
to the equilateral homogeneousness of assemblage I., or to the
diatomic homogeneousness of assemblage II., with each atom of
one constituent assemblage at the centre of an equilateral quartet
of the other constituent assemblage.   It is easy to construct
ideally mechanism by which this may be done; and we need not
occupy our minds with it at present.   It is enough to know that
it can be done.   If the system, subject to the prescribed constraining guidance, be left to itself at any given density, the condition
for equilibrium without extraneous force is that w is either a
maximum or a minimum; the equilibrium is stable when w is a
maximum, unstable when a minimum.   It is interesting to see the

two stable equilibriums of assemblage I. according to law of force 1, and the three according to law of force 2; and the two stable equilibriums of assemblage II. with each of these laws of force.

Assemblage I.			Assemblage II.		
Distances between centres of nearest atoms for maximum and minimum values of w.	Maximum and minimum values of w.	Densities.	Distances between centres of nearest atoms for maximum and minimum values of w.	Maximum and minimum values of w.	Densities.
Law of Force according to Curve 1.					
1·16	8·28 (max.)	·904	1·00	11·52 (max.)	·652
1·23	5·22 (min.)	·759	1·10	·76 (min.)	·490
1·61	14·76 (max.)	·338	1·61	4·92 (max.)	·158
Law of Force according to Curve 2.					
1·00	11·58 (max.)	1·414	1·00	12·36 (max.)	·652
1·07	3·78 (min.)	1·146	1·15	0·16 (min.)	·433
1·22	10·44 (max.)	·774	1·53	5·20 (max.)	·184
1·28	9·36 (min.)	·671			
1·53	15·60 (max.)	·393			

§ 17. But we must not forget that it is only with the specified constraining guidance (§ 16) that we are sure of these equilibriums being stable. It is quite certain, however, that without guidance the monatomic assemblage would be stable for the small density corresponding to the point m of each of the diagrams, because for infinitesimal deviations each atom experiences forces only from its twelve nearest neighbours, and these forces are each of them zero for equilibrium. It may conceivably be that each of the maximums of w, whether for the monatomic or the diatomic assemblage, is stable without guidance. But it seems more probable that, for assemblage I. and law of force 2, the intermediate maximum m' (close to a minimum) is unstable. If it is so, the assemblage left to itself in this configuration would fall away, and would (in virtue of energy lost by waves through ether, that is to say, radiation of heat) settle in stable equilibrium corresponding to the maximum m (single assemblage), or either of the maximums m'' (single assemblage), or m''' (double assemblage). It is also possible that for law of force 1 the maximum m' for the single assemblage is unstable. If so, the system left to itself in this configuration would fall away and settle in either of the configurations m (single assemblage) or m'' (double assemblage). Or it is possible that with either of our arbitrarily assumed laws of force there may be stable configurations of equilibrium with the atoms in simple cubic order (§ 21 below): and in double cubic order; that is to say, with each atom in the centre of a cube of which the eight corners are its nearest neighbours.

§ 18.   It is important to remark further, that certainly a law of force fulfilling the conditions of § 13 may be found, according to which even the simple cubic order is a stable configuration; though perhaps not the only stable configuration. The double cubic order, which has hitherto not got as much consideration as it deserves in the molecular theory of crystals, is certainly stable for some laws of force which would render the simple cubic order unstable.   Meantime it is exceedingly probable that there are in nature crystals of elementary substances, such as metals, or frozen oxygen, or nitrogen, or argon, of the simple cubic, and double cubic, and simple equilateral, and double equilateral, classes.   It is also probable that the crystalline molecules in crystals of compound chemical substance are in many cases simply the chemical molecules, and in many cases are composed of groups of the chemical molecules.   The crystalline molecules, however constituted, are, in crystals of the cubic class, probably arranged either in simple cubic, or double cubic, or in simple equilateral, or double equilateral, order.

§ 19.   It will be an interesting further development of the molecular theory to find some illustrative cases of chemical compound molecules (that is to say, groups of atoms presenting different laws of force, whether between two atoms of the same kind or between atoms of different kinds), which are, and others which are not, in stable equilibrium at some density or densities of *equilateral* assemblage.   In this last class of cases the molecules make up crystals not of the cubic class.   This certainly can be arranged for by compound molecules with law of force between any two atoms fulfilling the condition of § 13; and it can be done even for a monatomic homogeneous assemblage very easily, if we leave the simplicity of § 13 in our assumption as to law of force.

§ 20. The mathematical theory wants development in respect to the conditions for stability. If, with the constraining guidance of § 16, w is either a maximum or a minimum, there is equilibrium with or without the guidance. For w a maximum the equilibrium is stable *with the guidance*; but may be stable or unstable without the guidance. A criterion of stability which will answer this last question is much wanted; and it seems to me that though the number of atoms is quasi-infinite the wanted criterion may be finite in every case in which the number of atoms exerting force on any one atom is finite. To find it generally for the equilibrium of any homogeneous assemblage of homogeneous groups, each of a finite number of atoms, is a worthy object for mathematical consideration. Its difficulty and complexity is illustrated in §§ 21, 22 for the particularly simple case of similar atoms arranged in simple cubic order; and in §§ 23–29 for a still simpler case.

§ 21. Consider a group of eight particles at the eight corners of a cube (edge λ) mutually acting on one another with forces all varying according to the same law of distance. Let the magnitudes of the forces be such that there is equilibrium; and in the first place let the law of variation of the forces be such that the equilibrium is stable. Build up now a quasi-infinite number of such cubes with coincident corners to form one large cube or a crystal of any other shape. Join ideally, to make one atom, each set of eight particles in contact which we find in this structure. The whole system is in stable equilibrium. The four forces in each set of four coincident edges of the primitive cubes become one force equal to the force between atom and atom at distance λ. The two forces in either diagonal of the coincident square faces of two cubes in contact make one force equal to the force between atoms at distance $\lambda \sqrt{2}$. The single force in each body-diagonal of any one of the cubes is the force between atom and atom at distance $\lambda \sqrt{3}$. The three moduluses of elasticity (compressibility-modulus, modulus with reference to change of angles of the square faces, and modulus with reference to change of angles between their diagonals) are all easily found by consideration of the dynamics of a single primitive cube, or they may be found by the general method given in "On the Elasticity of a Crystal according to Boscovich*." (In passing, remark that neither in this nor in other

* *Proc. R.S.L.* Vol. LIV. June 8, 1893. App. I.

cases is it to be assumed without proof that stability is ensured by positive values of the elasticity moduluses.)

§ 22. Now while it is obvious that our cubic system is in stable equilibrium if the eight particles constituting a detached primitive cube are in stable equilibrium, it is not obvious without proof that this condition, though sufficient, is necessary for the stability of the combined assemblage. It might be that though each primitive cube by itself is unstable, the combined assemblage is stable in virtue of mutual support given by the joining of eight particles into one at the corners of the cubes which we have put together.

§ 23. The simplest possible illustration of the stability question of § 20 is presented by the exceedingly interesting problem of the equilibrium of an infinite row of similar particles, free to move only in a straight line. The consideration of this linear problem we shall find also useful (§§ 28, 29 below) for investigation of the disturbance from homogeneousness in the neighbourhood of the bounding surface, experienced by a three-dimensional homogeneous assemblage in equilibrium. First let us find a, the distance, or one of the distances, from atom to atom at which the atoms must be placed for equilibrium; and after that try to find whether the equilibrium is stable or unstable.

§ 24. Calling $f(D)$ (as in § 4) the attraction between atom and atom at distance D, we have for the sum, P, of attractions between all the atoms on one side of any point in their line, and all the atoms on the other side, the following finite expression having essentially a finite number of terms, greater the smaller is a:

$$f(a) + 2f(2a) + 3f(3a) + \ldots\ldots = P \ldots\ldots\ldots(8).$$

Hence a, for equilibrium with no extraneous force, is given by the functional equation

$$f(a) + 2f(2a) + 3f(3a) + \ldots\ldots = 0 \ldots\ldots\ldots(9);$$

which, according to the law of force, may give one or two or any number of values for a: or may even give no value (all roots imaginary) if the force at greatest distance for which there is force at all, is repulsive. The solution or all the solutions of this equation are readily found by calculating from the Boscovich curve representative of $f(D)$ a table of values of P, and plotting them on a curve, by formula (8), for values of a from $a = I$ (the limit above which the force is zero for all distances) downwards to the value which makes $P = -\infty$, or to zero if there is no infinite repulsion. The accompanying diagram, fig. 6, copied

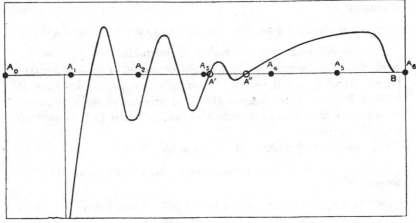

Fig. 6.

from fig. 1 of Boscovich's great book*, with slight modifications (including positive instead of negative ordinates to indicate attraction) to suit our present purpose, shows for this particular curve three of the solutions of equation (8). (There are obviously several other solutions.) In two of the solutions, respectively, A_0, A', and A_0, A'', are consecutive atoms at distances at which the force between them is zero. These are configurations of equilibrium, because A_0B, the extreme distance at which there is mutual action, is less than twice A_0A', and less than twice A_0A''. In the other of the solutions shown, A_0, A_1, A_2, A_3, A_4, A_5, A_6 are seven equidistant consecutive atoms of an infinite row in equilibrium in which A_5 is within range of the force of A_0, and A_6 is beyond it. The algebraic sum of the ordinates with their proper multipliers is zero, and so the diagram represents a solution of equation (9).

---

* Theoria Philosophiæ Naturalis redacta ad unicam legem virium in natura existentium, auctore P. Rogerio Josepho Boscovich, Societatis Jesu, nunc ab ipso perpolita, et aucta, ac a plurimis præcedentium editionum mendis expurgata. Editio Veneta prima ipso auctore præsente, et corrigente. Venetiis, MDCCLXIII. Ex Typographia Remondiniana superiorum permissu, ac privilegio.

§ 25. In the general linear problem to find whether the equilibrium is stable or not for equal consecutive distances, a, let (as in § 4) $\phi(D)$ be the work required to increase the distance between two atoms from D to ∞. Suppose now the atoms to be displaced from equal distances, a, to consecutive unequal distances—

$$\dots a + u_{i-2},\ a + u_{i-1},\ a + u_i,\ a + u_{i+1},\ a + u_{i+2},\ \dots \ \dots(10).$$

The equilibrium will be stable or unstable according as the work required to produce this displacement is, or is not, positive for all infinitely small values of $\dots u_{i-1},\ u_i,\ u_{i+1},\ \dots$. Its amount is $W_0 - W$; where W denotes the total amount of work required to separate all the atoms from the configuration (10) to infinite mutual distances.

According to § 2 above W is given by

$$W = \tfrac{1}{2}(\dots + w_{i-1} + w_i + w_{i+1} + \dots)\dots\dots\dots(11);$$

where

$$w_i = \phi(a + u_i) \quad + \phi(2a + u_{i-1} + u_i) \quad + \phi(3a + u_{i-2} + u_{i-1} + u_i) \quad + \dots$$
$$+ \phi(a + u_{i+1}) + \phi(2a + u_{i+1} + u_{i+2}) + \phi(3a + u_{i+1} + u_{i+2} + u_{i+3}) + \dots$$
$$+ \dots \ \dots\dots\dots\dots\dots\dots\dots\dots\dots\dots\dots\dots\dots\dots\dots(12).$$

Expanding each term by Taylor's theorem as far as terms of the second order, and remarking that the sum of terms of the first order is zero for equilibrium* at equal distances, a, and putting $\phi''(D) = -f'(D)$, we find

$$W_0 - W = \tfrac{1}{4}\Sigma\ \{f'(a)\,(u_i^2 + u_{i+1}^2)$$
$$+ f'(2a)\,[(u_{i-1} + u_i)^2 + (u_{i+1} + u_{i+2})^2]$$
$$+ f'(3a)\,[(u_{i-2} + u_{i-1} + u_i)^2 + (u_{i+1} + u_{i+2} + u_{i+3})^2]$$
$$+ \quad \text{etc.} \qquad \text{etc.} \qquad \text{etc.} \qquad \text{etc.}\}\ \dots(13);$$

where Σ denotes summation for all values of i, except those corresponding to the small numbers of atoms (§§ 28, 29 below) within influential distances of the two ends of the row.

§ 26. Hence the equilibrium is stable if $f'(a), f'(2a), f'(3a)$, etc., are all positive; but it can be stable with some of them

---

* It is interesting and instructive to verify this analytically by selecting all the terms in W which contain u_i, and thus finding $\dfrac{dW}{du_i}$. This equated to zero, for zero values of $\dots u_{i-1},\ u_i,\ u_{i+1},\dots$ gives equation (9) of the text.

negative. Thus, according to the Boscovich diagram, a condition ensuring stability is that the position of each atom be on an up-slope of the curve showing attractions at increasing distances. We see that each of the atoms in each of our three equilibriums for fig. 6 fulfils this condition.

§ 27. Fig. 7 shows a simple Boscovich curve drawn arbitrarily to fulfil the condition of § 13 above, and with the further simpli-fication for our present purpose, of limiting the sphere of influence so as not to extend beyond the next-nearest neighbours in a row of equidistant particles in equilibrium, with repulsions between nearests and attractions between next-nearests. The distance, a, between nearests is determined by

$$f(a) + 2f(2a) = 0 \ldots\ldots\ldots\ldots\ldots\ldots(14),$$

being what (9) of § 24 becomes when there is no mutual force except between nearests and next-nearests. There is obviously one stable solution of this equation in which one atom is at the zero of the scale of abscissas (not shown in the diagram) and its nearest neighbour on the right is at A, the point of zero force with attraction for greater distances and repulsion for less distances. The only other configuration of stable equilibrium is found by solution of (14) according to the plan described in § 24, which gives $a = \cdot680$. It is shown on fig. 7 by A_i, A_{i+1}, as consecutive atoms in the row.

§ 28. Consider now the equilibrium in the neighbourhood of either end of a rectilinear row of a very large number of atoms which, beyond influential distance from either end, are at equal consecutive distances a satisfying § 27 (14). We shall take for simplicity the case of equilibrium in which there is no extraneous force applied to any of the atoms, and no mutual force between any two atoms except the positive or negative attraction $f(D)$. But suppose first that ties or struts are placed between con-secutive atoms near each end of the row so as to keep all their consecutive distances exactly equal to a. For brevity we shall call them ties, though in ordinary language any one of them would be called a strut if its force is push instead of pull on the atoms to which it is applied. Calling A_1, A_2, $A_3 \ldots$ the atoms at one end of the row, suppose the tie between A_1 and A_2 to be removed, and A_1 allowed to take its position of equilibrium. A single equation gives the altered distance A_1A_2, which we shall denote by $a + {}_1x_1$. Let an altered tie be placed between A_1 and A_2 to keep them at this altered distance during the operations which follow. Next remove the tie between A_2 and A_3, and find by a single equation the altered distance $a + {}_1x_2$. After that remove the tie between A_3 and A_4 and find, still by a single equa-

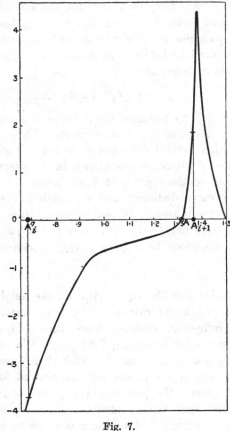

Fig. 7.

tion, the altered distance $a + {}_1x_3$, and so on till we find ${}_1x_7$ or ${}_1x_8$ or ${}_1x_i$, small enough to be negligible. Thus found, ${}_1x_1$, ${}_1x_2$, ${}_1x_3$, ... ${}_1x_i$ give a first approximation to the deviations from equality of distance for complete equilibrium. Repeat the process of removing the ties in order and replacing each one by the altered length as in the first set of approximations, and we find a second set $_2x_1$, $_2x_2$, $_2x_3$ .... Go on similarly to a third, fourth, fifth, sixth ... approximation till we find no change by a repetition of the process. Thus, by a process essentially convergent if the equilibrium with which we started is stable, we find the deviations from equality of consecutive distances required for equilibrium when the system is left free in the neighbourhood of each end, and all through the row (except always the constraint to remain in a straight line). By this proceeding applied to the curve of fig. 7 and the case of equilibrium $a = \cdot680$, the following successive approximations were found :—

	x_1	x_2	x_3	x_4	x_5	x_6	x_7
1st Approximation .	+ ·018	− ·009	+ ·004	− ·002	+ ·001	− ·001	·000
2nd ,, .	+ ·026	− ·014	+ ·007	− ·003	+ ·002		
3rd ,, .	+ ·031	− ·018	+ ·009	− ·005	+ ·003		
4th ,, .	+ ·034	− ·020	+ ·011	− ·006			
5th ,, .	+ ·036	− ·022	+ ·012	− ·007			
6th ,, .	+ ·037	− ·023	+ ·013				
7th ,, .	+ ·038	− ·024					
8th ,, .	+ ·039						

Thus our final solution, with $a = \cdot680$, is

$$x_1 = + \cdot039, \quad x_2 = - \cdot024,$$
$$x_3 = + \cdot013, \quad x_4 = - \cdot007,$$
$$x_5 = + \cdot003, \quad x_6 = - \cdot001,$$
$$x_7 = \cdot000.$$

§ 29. It is exceedingly interesting to remark that the deviations of the successive distances from a are alternately positive and negative, and that they only become less than one-seventh per cent. of a for the distance between A_7 and A_8. Thus, if we agree to neglect anything less than one-seventh per cent. in the distance between atom and atom, the influential distance from either end is $7a$, although the mutual force between atom and atom is null at all distances exceeding $2\cdot2a$.

§ 30. If, instead of $f(D)$ denoting the force between two atoms in a rectilinear row, it denotes the mutual force between two parallel plane nets in a Bravais homogeneous assemblage of single atoms, the work of §§ 27, 28 remains valid ; and thus we arrive at the very important and interesting conclusion that when there is repulsion between nearest nets, attraction between next-nearests, and no force between next-next-nearests or any farther, the disturbance from homogeneousness in the neighbourhood of the bounding plane consists in alternate diminutions and augmentations of density becoming less and less as we travel inwards, but remaining sensible at distances from the boundary amounting to several times the distance from net to net.

APPENDIX K.

ON VARIATIONAL ELECTRIC AND MAGNETIC SCREENING*.

§ 1. A SCREEN of imperfectly conducting material is as thorough in its action, when time enough is allowed it, as is a similar screen of metal. But if it be tried against rapidly varying electrostatic force, its action lags. On account of this lagging, it is easily seen that the screening effect against periodic variations of electrostatic force will be less and less, the greater the frequency of the variation. This is readily illustrated by means of various forms of idiostatic electrometers. Thus, for example, a piece of paper supported on metal in metallic communication with the movable disc of an attracted disc electrometer annuls the attraction (or renders it quite insensible) a few seconds of time after a difference of potential is established and kept constant between the attracted disc and the opposed metal plate, if the paper and the air surrounding it are in the ordinary hygrometric condition of our climates. But if the instrument is applied to measure a rapidly alternating difference of potential, with equal differences on the two sides of zero, it gives very little less than the same average force as that found when the paper is removed and all other circumstances kept the same. Probably, with ordinary clean white paper in ordinary hygrometric conditions, a frequency of alternation of from 50 to 100 per second will more than suffice to render the screening influence of the paper insensible. And a much less frequency will suffice if the atmosphere surrounding the paper is artificially dried. Up to a frequency of millions per second, we may safely say that, the greater the frequency, the more perfect is the annulment of screening by the paper; and this statement holds also if the paper be thoroughly blackened on both sides with ink, although

* *Proc. R.S.*, April, 1891.

possibly in this condition a greater frequency than 50 to 100 per second might be required for practical annulment of the screening.

§ 2. Now, suppose, instead of attractive force between two bodies separated by the screen, as our test of electrification, that we have as test a faint spark, after the manner of Hertz. Let two well insulated metal balls, A, B, be placed very nearly in contact, and two much larger balls, E, F, placed beside them, with the shortest distance between E, F sufficient to prevent sparking, and with the lines joining the centres of the two pairs parallel. Let a rapidly alternating difference of potential be produced between E and F, varying, not abruptly, but according, we may suppose, to the simple harmonic law. Two sparks in every period will be observed between A and B. The interposition of a large paper screen between E, F, on one side, and A, B, on the other, in ordinary hygrometric conditions, will absolutely stop these sparks, if the frequency be less than, perhaps, 4 or 5 per second. With a frequency of 50 or more, a clean white paper screen will make no perceptible difference. If the paper be thoroughly blackened with ink on both sides, a frequency of something more than 50 per second may be necessary; but some moderate frequency of a few hundreds per second will, no doubt, suffice to practically annul the effect of the interposition of the screen. With frequencies up to 1000 million per second, as in some of Hertz's experiments, screens such as our blackened paper are still perfectly transparent, but if we raise the frequency to 500 million million, the influence to be transmitted is light, and the blackened paper becomes an almost perfect screen.

§ 3. Screening against a varying magnetic force follows an opposite law to screening against varying electrostatic force. For the present, I pass over the case of iron and other bodies possessing magnetic susceptibility, and consider only materials devoid of magnetic susceptibility, but possessing more or less of electric conductivity. However perfect the electric conductivity of the screen may be, it has no screening efficiency against a steady magnetic force. But if the magnetic force varies, currents are induced in the material of the screen which tend to diminish the magnetic force in the air on the remote side from the varying magnet. For simplicity, we shall suppose the variations to follow the simple harmonic law. The greater the electric conductivity of the material, the greater is the screening effect for the same frequency of alternation; and, the greater the frequency, the greater is the screening effect for the same material. If the screen be of copper, of specific resistance 1640 sq. cm. per second (or electric diffusivity 130 sq. cm. per second), and with frequency 80 per second, what I have called the "mhoic effective thickness*" is 0·71 of a cm.; and the range of current intensity at depth $n \times 0·71$ cm. from the surface of the screen next the exciting magnet is ϵ^{-n} of its value at the surface.

Thus (as $\epsilon^3 = 20·09$) the range of current-intensity at depth 2·13 cm. is $\frac{1}{20}$ of its surface value. Hence we may expect that a sufficiently large plate of copper $2\frac{1}{4}$ cm. thick will be a little less than perfect in its screening action against an alternating magnetic force of frequency 80 per second.

§ 4. Lord Rayleigh, in his "Acoustical Observations" (*Phil. Mag.*, 1882, first half-year), after referring to Maxwell's statement, that a perfectly conducting sheet acts as a barrier to magnetic force (*Electricity and Magnetism*, § 665), describes an experiment in which the interposition of a large and stout plate of copper between two coils renders inaudible a sound which, without the copper screen, is heard by a telephone in circuit with one of the coils excited by electromagnetic induction from the other coil, in which an intermittent current, with sudden, sharp variations of strength, is produced by a "microphone clock" and a voltaic battery. Larmor, in his paper on "Electromagnetic Induction in Conducting Sheets and Solid Bodies" (*Phil. Mag.*, 1884, first half-year), makes the following very interesting statement :—"If "we have a sheet of conducting matter in the neighbourhood of "a magnetic system, the effect of a disturbance of that system "will be to induce currents in the sheet of such kind as will tend "to prevent any change in the conformation of the tubes [lines] "of force cutting through the sheet. This follows from Lenz's "law, which itself has been shown by Helmholtz and Thomson to "be a direct consequence of the conservation of energy. But if "the arrangement of the tubes [lines of force] *in* the conductor "is unaltered, the field on the other side of the conductor into "which they pass (supposed isolated from the outside spaces by

* *Math. and Phys. Papers*, Vol. III. Art. cii, § 35.

"the conductor) will be unaltered. Hence, if the disturbance is
"of an alternating character, with a period small enough to make
"it go through a cycle of changes before the currents decay
"sensibly, we shall have the conductor acting as a screen.

"Further, we shall also find, on the same principle, that a
"rapidly rotating conducting sheet screens the space inside it from
"all magnetic action which is not symmetrical round the axis of
"rotation."

Mr Willoughby Smith's experiments on "Volta-electric induction," which he described in his inaugural address to the Society
of Telegraph Engineers of November, 1883, afforded good illustrations of this kind of action with copper, zinc, tin, and lead,
screens, and with different degrees of frequency of alternation.
His results with iron are also very interesting : they showed, as
might be expected, comparatively little augmentation of screening
effect with augmentation of frequency. This is just what is to
be expected from the fact that a broad enough and long enough
iron plate exercises a large magnetostatic screening influence ;
which, with a thick enough plate, will be so nearly complete that
comparatively little is left for augmentation of the screening
influence by alternations of greater and greater frequency.

§ 5. A copper shell closed around an alternating magnet
produces a screening effect which on the principle of § 3 we may
reckon to be little short of perfection if the thickness be $2\frac{1}{4}$ cm.
or more, and the frequency of alternation 80 per second.

§ 6. Suppose now the alternation of the magnetic force to be
produced by the rotation of a magnet M about any axis. First,
to find the effect of the rotation, imagine the magnet to be represented by ideal magnetic matter. Let (after the manner of Gauss
in his treatment of the secular perturbations of the solar system)
the ideal magnetic matter be uniformly distributed over the
circles described by its different points. For brevity call I the
ideal magnet symmetrical round the axis, which is thus constituted. The magnetic force throughout the space around the
rotating magnet will be the same as that due to I, compounded
with an alternating force of which the component at any point
in the direction of any fixed line varies from zero in the two
opposite directions in each period of the rotation. If the copper
shell is thick enough, and the angular velocity of the rotation

great enough, the alternating component is almost annulled for external space, and only the steady force due to I is allowed to act in the space outside the copper shell.

§ 7. Consider now, in the space outside the copper shell, a point P rotating with the magnet M. It will experience a force simply equal to that due to M when there is no rotation, and, when M and P rotate together, P will experience a force gradually altering as the speed of rotation increases, until, when the speed becomes sufficiently great, it becomes sensibly the same as the force due to the symmetrical magnet I. Now superimpose upon the whole system of the magnet, and the point P, and the copper shell, a rotation equal and opposite to that of M and P. The statement just made with reference to the magnetic force at P remains unaltered, and we have now a fixed magnet M and a point P at rest, with reference to it, while the copper shell rotates round the axis around which we first supposed M to rotate.

§ 8. A little piece of apparatus, constructed to illustrate the result experimentally, is submitted to the Royal Society and shown in action. In the copper shell is a cylindric drum, 1·25 cm. thick, closed at its two ends with circular discs 1 cm. thick. The magnet is supported on the inner end of a stiff wire passing through the centre of a perforated fixed shaft which passes through a hole in one end of the drum, and serves as one of the bearings; the other bearing is a rotating pivot fixed to the outside of the other end of the drum. The accompanying sections, drawn to a scale of three-fourths full size, explain the arrangement sufficiently. A magnetic needle outside, deflected by the fixed magnet when the drum is at rest, shows a great diminution of the deflection when the drum is set to rotate. If the (triple compound) magnet inside is reversed, by means of the central wire and cross bar outside, shown in the diagram, the magnetometer outside is greatly affected while the copper shell is at rest; but scarcely affected perceptibly while the copper shell is rotating rapidly.

§ 9. When the copper shell is a figure of revolution, the magnetic force at any point of the space outside or inside is steady, whatever be the speed of rotation; but if the shell be not a figure of revolution, the steady force in the external space observable when the shell is at rest becomes the resultant of the force due to a fixed magnet intermediate between *M* and *I* compounded

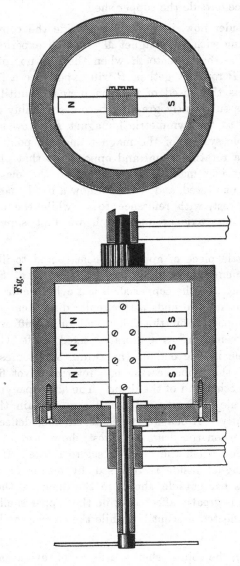

Fig. 1.

with an alternating force with amplitude of alternation increasing to a maximum, and ultimately diminishing to zero, as the angular velocity is increased without limit.

§ 10.   If M be symmetrical, with reference to its northern and southern polarity, on the two sides of a plane through the axis of rotation, I becomes a null magnet, the ideal magnetic matter in every circle of which it is constituted being annulled by equal quantities of positive and negative magnetic matter being laid on it.   Thus, when the rotation is sufficiently rapid, the magnetic force is annulled throughout the space external to the shell.   The transition from the steady force of M to the final annulment of force, when the copper shell is symmetrical round its axis of rotation, is, through a steadily diminishing force, without alternations.   When the shell is not symmetrical round its axis of rotation, the transition to zero is accompanied with alternations as described in § 8.

§ 11.   When M is not symmetrical on the two sides of a plane through the axis of rotation, I is not null; and the condition approximated to through external space with increasing speed of rotation is the force due to I, which is an ideal magnet symmetrical round the axis of rotation.

§ 12.   A very interesting simple experimental illustration of screening against magnetic force may be shown by a rotating disc with a fixed magnet held close to it on one side.   A bar magnet held with its magnetic axis bisected perpendicularly by a plane through the axis of rotation would, by sufficiently rapid rotation, have its magnetic force almost perfectly annulled at points in the air as near as may be to it, on the other side of the disc, if the diameter of the disc exceeds considerably the length of the magnet.   The magnetic force in the air close to the disc, on the side next to the magnet, will be everywhere parallel to the surface of the disc.

APPENDIX L.

ON ELECTRIC WAVES AND VIBRATIONS IN A SUBMARINE TELEGRAPH WIRE*.

§ 1. To simplify our problem, by avoiding the interesting subject of alternating electric currents of electricity in a solid conductor dealt with in §§ 9, 19, 29—35 of Art. CII. of my *Mathematical and Physical Papers*, Vol. III., I suppose for the present the central conductor and the surrounding sheath to be exceedingly thin copper tubes; so thin that the electric current carried by each is uniformly distributed through its substance, with the highest frequency of alternation which we shall have to consider. To simplify farther, by avoiding the exceedingly complex question of electric currents in the water above the cable and wet ground below it, I for the present suppose the outer sheath to be perfectly insulated. This supposition will make exceedingly little difference in respect to the solution of our problem for such frequencies of alternation as are used in submarine signalling; but it makes a vast difference and simplification for the very high frequencies, up to those of the vibrations constituting light, which must be considered. For brevity I shall call the system of two conductors, with air or gutta-percha or other insulating material between them, the cable. For simplicity we shall suppose the cable to be laid straight; and shall specify any place in the cable by x, its distance from any chosen point of reference O in the axis of the inner conductor. But all our calculations will be applicable though the cable be not laid straight; provided its radius of curvature everywhere is very large in comparison with the radius of the sheath in cross-section; and provided x is distance from O measured along the length of the cable.

---

* This is the Appendix promised in the footnote on Lecture IV., page 45, of the present volume. A partial statement of results was given in Nichol's *Cyclopedia* (1860) under title "Electricity, Velocity of," republished *Mathematical and Physical Papers*, Vol. II. Article LXXXI.

§ 2. Let r and r' be the radii of the inner conductor and the sheath ....(1);

R, the sum of the resistances of the two conductors per unit of length of each ....(2);

c, the quantity of electricity on unit length of each conductor when the difference of potentials between them is unity ....(3);

lI, the electromagnetic quasi-inertia of a circuit made ideally by metallically connecting the inner conductor with the sheath at two cross-sections with the length l between them large in comparison with $r' - r$ ....(4);

γ, the current in either of the two conductors at time t and place x ....(5);

φ, the difference of potentials between the two conductors at time t and place x...............(6);

$q\,dx$, the quantity of electricity on a length dx of either conductor at time t and place x......(7).

In all these items of electric reckoning we shall use electromagnetic measure. It is to be borne in mind that in every part of the cable the electric currents in the central conductor, and in the sheath are equal and opposite: also that the total quantities of electricity at every instant on the smaller convex surface of the central conductor, and the larger concave surface of the sheath, are equal and opposite : and that there is no electricity on the inner surface of the central conductor, and no electricity on the outer surface of the sheath. The notation (3) above, in terms of my original definition* of electrostatic capacity, means that c is the electrostatic capacity, reckoned in electromagnetic measure, of unit length of the central conductor; regarded as if it were part of a Leyden phial. By (3) and (7) we have

$$q = c\phi \quad ....(8).$$

§ 3. Changes of electrification of the opposed surfaces of the two conductors can only take place through more electricity

---

* "On the Electrostatical Capacity of a Leyden Phial and of a Telegraph Wire insulated in the Axis of a Cylindrical Conducting Sheath." First published in *Phil. Mag.* 1855, 1st half-year; republished in my volume of *Electrostatics and Magnetism*, §§ 51—56.

flowing into, than flowing out of, any portion. This electro-
kinematic principle gives us the following equation:

$$\frac{dq}{dt} = -\frac{d\gamma}{dx} \quad \dotfill (9).$$

The dynamical equation of our problem is as follows:

$$R\gamma + I\frac{d\gamma}{dt} = -\frac{d\phi}{dx} \quad \dotfill (10).$$

Here the first term of the first member expresses the ohmic
anti-electromotive force; the second term the inertial anti-
electromotive force. Eliminating q and ϕ from these three
equations we find

$$R\frac{d\gamma}{dt} + I\frac{d^2\gamma}{dt^2} = \frac{1}{c}\frac{d^2\gamma}{dx^2} \quad \dotfill (11).$$

§ 4.   For small frequencies, such as those of submarine cable
signalling, the ohmic resistance dominates; and the inertial
resistance is imperceptible: thus we have

$$cR\frac{d\gamma}{dt} \doteqdot \frac{d^2\gamma}{dx^2} \quad \dotfill (12).$$

§ 5.   For very high frequencies the inertial resistance dominates;
and the ohmic is ultimately almost imperceptible: thus we have

$$cI\frac{d^2\gamma}{dt^2} \doteqdot \frac{d^2\gamma}{dx^2} \quad \dotfill (13).$$

§ 6.   According to the principles set forth in Part III., " Electricity in Motion*," of a paper " On the Mechanical Values of
Distributions of Electricity, Magnetism, and Galvanism†," published originally in the *Proceedings of the Glasgow Philosophical
Society* for January 1853, we find lI most readily by imagining
the temporarily closed circuit described in (4) of § 2 to be divided
into a very great number of parts by meridional planes through
the axis of the two cylindrical conductors, and imagining any one
of these parts to be removed to any position outside, while the
currents are kept constant in all the parts: and calculating the
work done in this movement against the attractive force between

---

* *Mathematical and Physical Papers*, Vol. I. pp. 530—533.
† By this ill-chosen word I then meant electric currents in closed circuits.

the portion of the circuit removed and the remainder.   The full synthetic working-out of this plan gives

$$l\gamma^2 \log (r'/r) \dots\dots\dots\dots\dots\dots\dots(14)$$

for the total work spent in taking the whole circuit to pieces separated by infinite distances from one another.   Hence, for the quasi-inertia of the current, per unit length of cable, in the circuit before dissection, we have

$$I = 2 \log (r'/r) \dots\dots\dots\dots\dots\dots(15).$$

§ 7.   By *Electrostatics and Magnetism*, § 55, we have for c in electrostatic measure $\tfrac{1}{2}k/\log (r'/r)$; where k is what Faraday called the specific inductive capacity of the insulating material between the core and the sheath.   Now if we denote by N the number · of electrostatic units in the electromagnetic unit of electric quantity, it is also the number of electrostatic units in the electromagnetic unit of potential.   Hence N^2 is the number of electrostatic units in the electromagnetic unit of capacity : and we have accordingly

$$c = N^{-2}k/2 \log (r'/r) = k/N^2 I \dots\dots\dots\dots(16).$$

Using this in (11) we find

$$\frac{R}{I}\frac{d\gamma}{dt} + \frac{d^2\gamma}{dt^2} = \frac{N^2}{k}\frac{d^2\gamma}{dx^2} \dots\dots\dots\dots(17);$$

and (13) becomes

$$\frac{d^2\gamma}{dt^2} = \frac{N^2}{k}\frac{d^2\gamma}{dx^2} \dots\dots\dots\dots\dots(13');$$

which shows that the propagational velocity of the wave is more and more nearly equal to $N/\sqrt{k}$, the higher the frequency; subject to the restriction of § 8.   Numerous and varied experimental investigations have shown, probably within $\tfrac{1}{8}$ per cent., that N is equal to $3 . 10^{10}$.

§ 8.   Our demonstration [see § 2 (4)] involves essentially the supposition that the wave-length is long in comparison with the thickness $r' - r$ of the dielectric between the outer and inner conductors.

§ 9.   A complete solution of (17) is

$$\gamma = \sin mx \, (A \epsilon^{\rho t} + B \epsilon^{\rho' t}) \dots\dots\dots\dots(18),$$

where m denotes 2π divided by the wave-length, and

$$\rho = \frac{-R}{2I} \pm \sqrt{\left[\left(\frac{R}{2I}\right)^2 - \frac{N^2}{k} m^2\right]} \dots\dots\dots(19);$$

which is real when

$$mN/\sqrt{k} < R/2I \quad \dots\dots\dots\dots\dots(20).$$

But when (20) is not fulfilled, the radical in (19) is zero or imaginary; and, instead of (18), we may take, as a convenient solution for standing waves,

$$\gamma = \sin\ mx \sin t \sqrt{\left[\frac{N^2}{k}\,m^2 - \left(\frac{R}{2I}\right)^2\right]}\ \epsilon^{\frac{-Rt}{2I}} \quad \dots\dots(21).$$

This expresses a subsidential oscillation, with ratio of subsidence per unit of time, $\epsilon^{-\frac{1}{2}R/I}$; which is independent of the wave-length; a very important and interesting result. What we call the period* of the oscillation is

$$2\pi \Big/ \sqrt{\left[\frac{N^2}{k}\,m^2 - \left(\frac{R}{2I}\right)^2\right]} \quad \dots\dots\dots\dots(22);$$

which becomes more and more nearly equal to $(2\pi/m)/(N/\sqrt{k})$, the greater is m, subject to the restriction of § 8.

§ 10. The complicated subsidential law expressed by (18), with two ratios of subsidence, each dependent on the wave-length, one of them with ratio diminishing, the other with ratio increasing, when the wave-length is diminished from infinity, is also very interesting. It is interesting to see how, with very small values of m (very great wave-lengths), the solution blends with the solution of (12) above; and how, with the greatest values of m fulfilling (20), the solution (18) blends into (21).

§ 11. As a single example take the Atlantic Cable of 1865, for which we had

$R \risingdotseq 23000$ c.g.s. per cm. of length ; $r'/r \risingdotseq 3\cdot3$; whence $I \risingdotseq 1\cdot76$

$$\dots\dots\dots(23);$$

$$k \risingdotseq 3 ; \ N/\sqrt{k} \risingdotseq 1\cdot732\,.\,10^{10} ; \ c \risingdotseq 1\cdot894\,.\,10^{-21}\dots\dots(24).$$

By (23) we find $\frac{1}{2}R/I = 6534$. Hence, when (20) is not fulfilled, the ratio of extinction for 1/6534 of a second is ϵ^{-1}, and for a millionth of a second it is $\epsilon^{-\cdot006534}$, or approximately $\cdot99347$. And the limiting condition between fulfilment and non-fulfilment of (20) is, by (24), $m \lesssim 3\cdot772\,.\,10^{-7}$; or, if λ denote the wave-length,

---

* This is according to common usage ; but it is not strictly correct, because subsiding vibratory motion is not periodic.

$\lambda \gtrless 167$ kilometres. Hence, if we take a length of anything less than 167 kilometres of the cable, and somehow manage to give it initially a sinusoidal distribution of current, varying through its length from zero at one end, through maximum and minimum, to zero at the other end, and leave it to itself; the electric distribution will subside vibrationally to zero. A millionth of a second from the beginning the vibrational amplitude will be ·99347 of its initial value; and in 1/6534 of a second it will have subsided to ϵ^{-1} of its initial value. Instead now of the impracticable management to begin with a sinusoidal distribution of current, apply a voltaic battery to maintain, for a very short time, say a hundred-thousandth of a second, a difference of potential between sheath and interior conductor at one end, and leave it to itself. This will give us an initial disturbance represented, according to Fourier, by a sum of sinusoidal currents, each zero at both ends of the cable. Every one of these components will subside vibrationally, all with the same ratio of subsidence in the same time. The periods of the different sinusoidal components will be those expressed by (22); with, for m, the successive values $2\pi/\lambda$, $2 \cdot 2\pi/\lambda$, $3 \cdot 2\pi/\lambda$, ..., where λ denotes the length of the cable. If this length of the cable is 16·7 kilometres, the second term within the brackets of (22) will be only 1/100 of the first, and for this or any shorter length it may be neglected. Thus, for a length of one kilometre, the period of the gravest sinusoidal vibration, as we see by (24) and (22), is $10^5/1·732 \cdot 10^{10}$ or $·577 \cdot 10^{-5}$ of a second. The ratio of subsidence in this time is $\epsilon^{-·0377}$; so that $26\frac{1}{2}$ periods of vibration will be performed before subsidence to ϵ^{-1} of the initial value.

§ 12. Consider, in conclusion, how thin the copper must be to fulfil the first simplifying condition of § 1, when the vibrational period is so short as $·577 \cdot 10^{-5}$ of a second. This is 1/2166 of 1/80, the period for which I found ·714 cm. as the mhoic thickness* of copper of specific resistance 1610 c. g. s. Hence mhoic thickness for period $·577 \cdot 10^{-5}$ is $·714/\sqrt{2166}$, or ·0153 cm. Now without going farther into the theory of the diffusion of alternate currents of electricity through a metal, we may safely guess that, if the thickness of the sheet is anything less than one-third of the

---

* *Mathematical and Physical Papers*, Vol. III, Art. CII, § 35 (May, 1890).

mhoic thickness for any particular period, the current will be nearly uniform through the whole thickness. Hence, to satisfy the first simplifying condition of § 1, the copper need not be thinner than 1/200 of a cm. Taking this as the thickness, let us find what r and r', with the ratio $r'/r = 3\cdot3$, must be to make the sum of the resistances in a cm. of the inner tube and a cm. of the outer tube equal to 23000 c. g. s. If we take the specific resistance of copper as 1610 square centimetres per second, our equation to determine r and r' is

$$\frac{1610 \cdot 200}{2\pi} \left(\frac{1}{r} + \frac{1}{3\cdot3r} \right) = 23000 \quad \ldots\ldots\ldots\ldots(25);$$

by which we have $r = 2\cdot9$ and $r' = 9\cdot6$. It must be noted that it is only for the gravest fundamental vibration that 1/200 cm. thickness of the metal would be small enough to give the law and rate of subsidence determined above.

CAMBRIDGE : PRINTED BY J. AND C. F. CLAY, AT THE UNIVERSITY PRESS.

INDEX.

T. L.   **45**

CAMBRIDGE : PRINTED BY JOHN CLAY, M.A. AT THE UNIVERSITY PRESS.

Printed in the United States
By Bookmasters